T0256790

INTERNATIONAL HARVESTER

SHOP MANUAL IH-201

Models   ■ 100   ■ 130   ■ 140   ■ 200   ■ 230
         ■ 240   ■ 404   ■ 2404

Models   ■ B-275   ■ B-414   ■ 354   ■ 364   ■ 384
         ■ 424   ■ 444   ■ 2424   ■ 2444

Models   ■ 330   ■ 340   ■ 504   ■ 2504

Models   ■ 274   ■ 284

I&T SHOP MANUALS

Information and Instructions

This shop manual contains several sections each covering a specific group of wheel type tractors. The Tab Index on the preceding page can be used to locate the section pertaining to each group of tractors. Each section contains the necessary specifications and the brief but terse procedural data needed by a mechanic when repairing a tractor on which he has had no previous actual experience.

Within each section, the material is arranged in a systematic order beginning with an index which is followed immediately by a Table of Condensed Service Specifications. These specifications include dimensions, fits, clearances and timing instructions. Next in order of arrangement is the procedures paragraphs.

In the procedures paragraphs, the order of presentation starts with the front axle system and steering and proceeding toward the rear axle. The last paragraphs are devoted to the power take-off and power lift systems. Interspersed where needed are additional tabular specifications pertaining to wear limits, torquing, etc.

HOW TO USE THE INDEX

Suppose you want to know the procedure for R&R (remove and reinstall) of the engine camshaft. Your first step is to look in the index under the main heading of ENGINE until you find the entry "Camshaft." Now read to the right where under the column covering the tractor you are repairing, you will find a number which indicates the beginning paragraph pertaining to the camshaft. To locate this wanted paragraph in the manual, turn the pages until the running index appearing on the top outside corner of each page contains the number you are seeking. In this paragraph you will find the information concerning the removal of the camshaft.

More information available at haynes.com
Phone: 805-498-6703

Haynes Group Limited
Sparkford Nr Yeovil
Somerset BA22 7JJ England

Haynes North America, Inc.
2801 Townsgate Road, Suite 340
Thousand Oaks, CA 91361 USA

ISBN-10: 0-87288-789-8
ISBN-13: 978-0-87288-789-3

Disclaimer

There are risks associated with automotive repairs. The ability to make repairs depends on the individual's skill, experience and proper tools. Individuals should act with due care and acknowledge and assume the risk of performing automotive repairs.

The purpose of this manual is to provide comprehensive, useful and accessible automotive repair information, to help you get the best value from your vehicle. However, this manual is not a substitute for a professional certified technician or mechanic.

This repair manual is produced by a third party and is not associated with an individual vehicle manufacturer. If there is any doubt or discrepancy between this manual and the owner's manual or the factory service manual, please refer to the factory service manual or seek assistance from a professional certified technician or mechanic.

Even though we have prepared this manual with extreme care and every attempt is made to ensure that the information in this manual is correct, neither the publisher nor the author can accept responsibility for loss, damage or injury caused by any errors in, or omissions from, the information given.

IH-201, 8S2, 13-280

INTERNATIONAL HARVESTER

Models ■ 100 ■ 130 ■ 140 ■ 200 ■230
　　　　 ■ 240 ■ 404 ■ 2404

Previously contained in I&T Shop Manual No. IH-21

SHOP MANUAL
INTERNATIONAL HARVESTER
SERIES 100-130-140-200-230-240-404-2404

Series 100, 130 and 140 are available with either an adjustable or a non-adjustable front axle. Series 100, 130 and 140 Hi-Clearance are available with an adjustable type front axle only. Series 200, 230, 240 and 404 Farmall are available with single or dual wheel tricycle front system as well as an adjustable type front axle. The 240 International is available only with an adjustable front axle. The 404 and 2404 International is available with an adjustable type front axle as well as a heavy duty fixed tread type front axle.

Engine serial number is stamped on right side of crankcase and the serial number prefix is C-123 for all models except the 404 and 2404 which is C-135.

Tractor serial number is stamped on a plate and on series 100, 130 and 140, the plate is attached on left side of clutch housing. On series 200, 230, 240, 404 and 2404, plate is attached on right side of clutch housing.

Suffix letters of the serial number indicate the following:

A—Distillate burning attachment
B—Kerosene burning attachment
C—L. P. Gas burning attachment (Engine)
D—5000 ft. high altitude attachment
E—8000 ft. high altitude attachment
J—Rockford clutch

T—Cotton picker mounting attachment
U—High altitude attachment
V—Exhaust valve rotator attachment
W—Forward and reverse drive attachment
X—High speed low and reverse attachment
DD—High speed 3rd gear attachment

INDEX (By Starting Paragraph)

CONDENSED SERVICE DATA

GENERAL	Series 100 130	Series 200 230	Series 140 240	Series 404 2404
Engine Make	Own	Own	Own	Own
Engine Model	C-123	C-123	C-123	C-135
Number Cylinders	4	4	4	4
Bore—Inches	$3^1/_8$	$3^1/_8$	$3^1/_8$	$3^1/_4$
Stroke—Inches	4	4	4	$4^1/_{16}$
Displacement—Cu. In.	122.7	122.7	122.7	135
Compression Ratio (Std.)	6.5:1	6.5:1	6.8:1	7.4:1
Pistons Removed From	Above	Above	Above	Above
Main Brgs., Number of	3	3	3	3
Main and Rod Brgs. Adjustable	No	No	No	No
Cylinder Sleeves	Wet	Wet	Wet	Wet
Forward Speeds (standard)	4	4	4	4
Generator & Starter Make	D-R	D-R	D-R	D-R

TUNE-UP

	Series 100 130	Series 200 230	Series 140 240	Series 404 2404
Firing Order	1-3-4-2	1-3-4-2	1-3-4-2	1-3-4-2
Valve Tappet Gap, Intake	0.014H	0.014H	0.014H	0.014H
Valve Tappet Gap, Exhaust	0.014H	0.014H	0.014H	0.020H
Valve Seat Angle	45	45	45	45
Valve Seat Width—Inches	$^1/_8$	$^1/_8$	$^1/_8$	$^5/_{64}$
Ignition Distributor Make	Own	Own	Own	......
Ignition Distributor Symbol	J or X	J or X	X or AB	AB or AH
Ignition Magneto Make	Own	Own	Own	Own
Ignition Magneto Model	H-4	H-4	H-4	......
Distributor Breaker Gap	0.020	0.020	0.020	0.020
Magneto Breaker Gap	0.013	0.013	0.013	......
Distributor Timing, Static	TDC	TDC	TDC	TDC
Distributor Timing, Full Advance		——See Par. 67——		
Magneto Impulse Trip Point	TDC	TDC	TDC	......
Magneto Lag Angle	35°	35°	35°	......
Magneto Running Timing	35° BTC	35° BTC	35° BTC	......
Ignition Retard Mark & Timing Flywheel	"DC1-4"	"DC1-4"	......	......
Fan Belt Pulley	(1)	(1)	DC	DC
Battery Terminal Grounded	Pos.	Pos.	Pos.	Neg.
Spark Plug Make		——Champ., AC or A-L——		
Electrode Gap (Gas)	0.023	0.023	0.023	0.023

TUNE-UP (Continued)	Series 100 130	Series 200 230	Series 140 240	Series 404 2404
Electrode Gap (LP-gas)	......	......	......	0.015
Carburetor Make		Carter, Zenith or Marvel-Schebler		
Float Setting, Carter	17/64	17/64	17/64	......
Float Setting, Zenith	$1^5/_{32}$	$1^5/_{32}$	$1^5/_{32}$	......
Float Setting, M-S	......	......	......	$^5/_4$
Engine Slow Idle—RPM	425	425	425	425
Engine High Idle—RPM	1575	200,1815 230,1980	140,1575 240,2200	2300

SIZES—CAPACITIES—CLEARANCES

(Clearances in thousandths)

	Series 100 130	Series 200 230	Series 140 240	Series 404 2404
Crankshaft Journal Diameter	2.1245	2.1245	2.1245	2.2445(2) 2.6240(3)
Crankpin Diameter	1.7495	1.7495	1.7495	1.8095(4) 2.0595(5)
Camshaft Journal Diameter Front (No. 1)	1.8115	1.8115	1.8115	1.8115
Intermediate (No. 2)	1.5775	1.5775	1.5775	1.5775
Rear (No. 3)	1.4995	1.4995	1.4995	1.4995
Piston Pin Diameter	0.91935	0.91935	0.91935	0.8592
Valve Stem Diameter	0.341	0.341	0.341	0.341
Main Bearings Running Clearance	0.9-3.9	0.9-3.9	0.9-3.9	0.9-3.9
Rod Bearings Running Clearance	0.9-3.4	0.9-3.4	0.9-3.4	0.9-3.9
Rod Side Play	5-14	5-14	5-14	5-14
Piston Skirt Clearance	1.1-1.9	1.1-1.9	1.1-1.9	1.1-1.9
Crankshaft End Play	4-10	4-10	4-10	4-10
Camshaft Bearing Clearance	0.9-5.4	0.9-5.4	0.9-5.4	0.9-5.4
Camshaft End Play	3-12	3-12	3-12	3-12
Cooling System—Gals.	$3^3/_4$	$3^3/_4$	$3^3/_4$	$3^3/_4$
Crankcase Oil—Qts.	5	5	5	5
Transmission & Differential—Qts.	4	19	4(6)	28
Final Drive, each—Quarts	$1^1/_2$	......	$1^1/_2$	......
Add for BP and/or PTO—Pints	2	2	2	......
Touch Control (no Remote Control)—Pints	$8^1/_4$	$8^1/_4$	$8^1/_4$ (7)	......
Hydra-Touch—Pints	......	......	11(8)	......

TIGHTENING TORQUES—Ft.-Lbs.

	Series 100 130	Series 200 230	Series 140 240	Series 404 2404
Cylinder Head		——See Par. 26——		
Rod Nuts (Cotter Pins)	40-45	40-45	40-45	......
Rod Nuts (Self-lock)	43-49	43-49	43-49	......
Rod Bolts (Self-lock)	......	......	......	45
Main Bearing Bolts	75-80	75-80	75-80	75-80

(1) Third notch for series 100 and 200; fourth notch for series 130 and 230. (2) Below Serial No. 100,501. (3) Serial No. 100501 and up. (4) Below Serial No. 100,501. (5) Serial No. 100,501 and up. (6) Applies to 150; Series 240, 19 quarts. (7) Applies to Series 140. (8) Applies to Series 240.

FARMALL 130

FARMALL 230

FRONT SYSTEM-TRICYCLE TYPE

SINGLE WHEEL

Series 200-230-240-404

1. The single front wheel is mounted in a fork which is retained to the steering worm wheel shaft by four stud nuts. Male and female wheel halves are available to accommodate a 7.50x10 tire or, a conventional type wheel is available, accommodating a a 6.00x12 tire. In either case, however, the wheel halves (or hubs) are fitted with bushings. Replacement wheel halves (or hubs) are factory fitted with bushings; or, the bushings are available individually as repair parts. Ream the bushings after installation, if necessary, to provide a free running fit for the axle. Refer to Fig. IH550.

DUAL WHEELS

Series 200-230-240-404

2. The bolster for tricycle type dual front wheels is retained to the steering worm wheel shaft by four stud nuts. The bolster and horizontal axle are available as a pre-riveted assembly; or, the bolster and axle are available as individual repair parts. Refer to Fig. IH551.

When mounting non-adjustable front wheels on 200 and 230 series the short end of the rim clamp should be toward the axle.

Series 240, 404 and 2404 use pressed steel front wheels which bolt to hub assemblies as shown in Fig. IH552.

1. Bolster
2. Axle
3. Dust deflector
4. Felt washer
5. Oil seal
6. Inner bearing
7. Retainer
8. Inner cup
9. Wheel
10. Outer cup
11. Outer bearing
12. Gasket
13. Hub cap

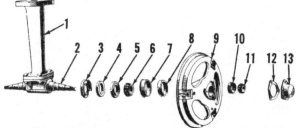

Fig. IH551—Exploded view of Farmall series 200 and 230 dual front wheel bolster, wheel and axle assembly. The bolster and/or axle are available as individual units, or as a pre-riveted assembly. Except for wheels, Farmall series 240 and 404 are similar.

1. Fork
2. Male and female wheel halves (7.50x10)
3. Bushing
7. Axle nut
8. Nut lock
9. Hub shield
10. Felt washer
11. Oil seal
12. Bushing
13. Hub
14. Gasket
15. Wheel (6.00x12)
16. Axle

Fig. IH550—Farmall series 200, 230, 240 and 404 fork mounted single front wheel. The wheel fork is bolted directly to the steering worm wheel shaft.

1. Dust shield
2. Oil seal
3. Bearing cone
4. Wear ring
5. Bearing cup
6. Hub
7. Bearing cup
8. Bearing cone
9. Washer
10. Nut
11. Gasket
12. Hub cap

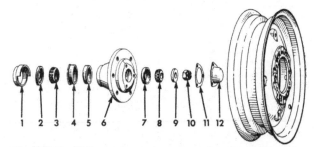

Fig. IH552—Wheel, hub and bearing assembly used on series 240 and 404 Farmall and 404 and 2404 International tractors.

FRONT SYSTEM-AXLE TYPE

AXLE MAIN MEMBER
All Models

3. The 100, 130 and 140 series front axle main member pivots on shaft (17 —Fig. IH553) which is anchored in the steering gear housing base with clamp bolts. The 100HC, 130HC and 140HC series axle pivots on a shaft which passes through the steering gear housing base and extends rearward where the shaft is anchored to the clutch housing. The two pre-sized pivot shaft bushings (18) which are pressed into the axle main member can be renewed after removing member from tractor.

The 200 and 230 series axle member pivots on shaft (17—Fig. IH553) which is retained in the axle mounting bracket by pin (19). Three pre-sized bushings (18) which are pressed into the center member and integral stay rod can be removed after removing center member from tractor.

The 240 and 404 Farmall adjustable front axle is shown in Fig. IH553A. The axle pivots on shaft (17) which

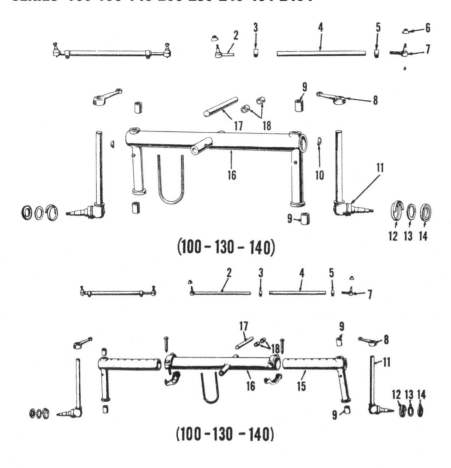

(100 - 130 - 140)

(100 - 130 - 140)

(100HC - 130HC - 140HC)

(200 - 230)

Fig. IH553—Exploded views of the various front axles. Axle pivot shaft bushings (18) are pre-sized. Spindle (11) should have 0.002-0.004 clearance in bushings (9).

1. Center steering arm	8. Knuckle steering arm	12. Dust deflector
2. Tie rod end	9. Bushings	13. Felt washer
3. Tie rod clamp	10. Woodruff key	14. Oil seal
4. Tie rod tube	11. Steering spindle and knuckle	15. Axle extension
5. Tie rod clamp		16. Axle main (or center) member
6. Dust cover		17. Axle pivot shaft
7. Tie rod end		18. Pivot shaft bushings
		19. Pin
		20. Axle mounting bracket

is retained in pivot bracket (20) by pin (19). The pre-sized bushing (18) which is pressed into the center member can be removed after the center member has been removed from tractor.

The 240, 404 and 2404 International tractors have a "V" shaped, tubular, adjustable front axle which pivots on pin (17—Fig. IH553B). The pivot pin is retained in lower bolster (20) by a cap screw. The pre-sized bushing (18) is pressed into the center member and can be removed after the center member has been removed from tractor.

The 404 and 2404 International tractors also have available a heavy duty fixed tread "V" type front axle. This axle is similar to that shown in Fig. IH553B except that tread width is non - adjustable and the steering knuckles and steering arms are splined instead of having Woodruff keys.

TIE-RODS AND TOE-IN

All Models

4. The procedure for renewing the tie-rods and/or tie-rod ends is evident after an examination of the unit and reference to Figs. IH553, IH553A and IH553B. Tie-rod ends are of the non-adjustable type and excessive wear is corrected by renewing the units.

When reassembling, vary the length of each tie-rod, an equal amount, to provide the recommended toe-in of $\frac{1}{8}$-$\frac{1}{4}$-inch for all series except the 240, 404 and 2404 International which should be $\frac{3}{16}$-$\frac{3}{8}$-inch.

STEERING KNUCKLES

All Models

5. The steering spindle and knuckle units are removed from the axle or axle extensions in the conventional manner. Spindle bushings should be installed with a closely fitting mandrel and sized after installation, if necessary, to provide a recommended clearance of 0.002-0.004 for the spindle.

STEERING GEAR

Series 100-130-140

The non-adjustable steering gear is located within the steering gear housing (5—Fig. IH554) which is bolted to the front face of the engine. The front axle support (steering gear housing base) is bolted to the lower portion of the steering gear housing. Steering arm (16) is retained to the sector shaft (10) by a nut.

6. REMOVE AND REINSTALL. To remove the steering gear housing, front axle support (steering gear housing base), axle and wheels as an assembly, proceed as follows: Remove hood, grille and radiator, and jack up front of tractor to remove weight from front wheels. To prevent steering gear housing from tilting when housing is disconnected from engine, place wood wedges between axle and steering gear housing base. Disconnect the head light wires and remove the radiator drain cap and pipe. Remove generator, regulator and mounting bracket from engine. On Hi-Clearance models, unbolt pivot shaft from clutch housing. Remove the steering worm shaft. Remove the bolts retaining the steering gear housing to the engine and move assembly away from tractor.

Reinstall the unit by reversing the removal procedure.

7. OVERHAUL. The steering gear unit can be overhauled without removing the assembly from tractor. Remove steering wheel and bearing retainer (4—Fig. IH554). Rotate worm and shaft (1) forward and out of housing (5). The pre-sized worm bushing (20) and seal (21) can be renewed at this time. Install seal with lip of same facing inward toward

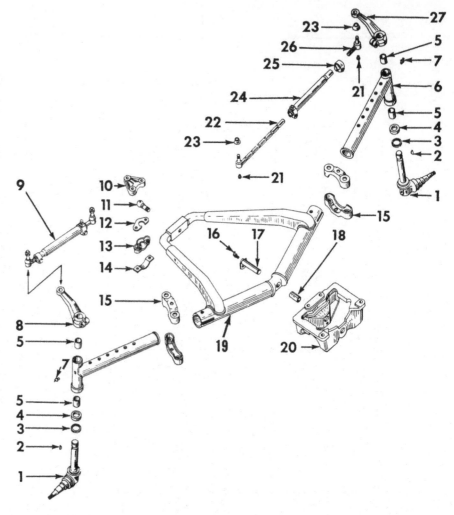

Fig. IH553B—Exploded view of series 240 and 404 International front axle.

1. Steering knuckle	9. Tie-rod	15. Clamp	20. Lower bolster
2. Woodruff key	10. Ball socket	16. Grease fitting	22. Tie-rod end
3. Felt washer	11. Stay rod ball	17. Pivot pin	23. Dust cover
4. Thrust bearing	12. Shim	18. Bushing	24. Tie-rod tube
5. Bushing	13. Socket cap	19. Front axle	25. Clamp
6. Axle extension	14. Lock plate		26. Tie-rod end
8. Right hand steering arm			27. Left hand steering arm

gears. Use tin sleeve or shim stock when installing the worm shaft to avoid damaging the seal. To remove sector (9) and shaft (10), remove axle assembly from tractor and disconnect steering arm (16) from the shaft. Remove the steering gear housing base retaining cap screws and remove the base. Pre-sized bushing (11) and seal (14) can be renewed at this time. Install seal with lip of same facing the gears. Remove the three bearing retainer cap screws and pull sector, shaft and bearing assembly from the housing. To separate the assembly, remove snap ring (6).

Reassemble the gear unit by reversing the disassembly procedure and use tin sleeve or shim stock when assembling the steering gear housing base to the steering gear housing to avoid damaging seal (14).

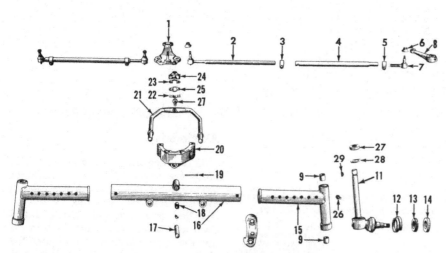

Fig. IH553A—Exploded view of the adjustable wide front axle used on Farmall 240 and 404 series tractors. Refer to Fig. IH553 for legend.

Series 200-230-240-404 Farmall

The steering worm and worm wheel are located in the steering gear housing, which is bolted to the front face of the engine. The bolster or wheel fork is retained directly to the lower portion of the steering worm wheel shaft by four stud nuts. On adjustable axle models, the center steering arm is bolted to the lower portion of the worm wheel shaft, and the axle mounting bracket is retained to the steering gear housing by four cap screws. The unit is non-adjustable; however, excessive backlash between the worm and worm wheel can be partially corrected by changing the position of the worm wheel on the worm wheel shaft splines so as to bring unworn teeth of the worm wheel into mesh with the steering worm.

8. REMOVE AND REINSTALL. To remove the steering gear housing and front axle and support, bolster or wheel fork as an assembly, proceed as follows: Remove hood, grille and radiator. On series 200 and 230, remove

generator, regulator and mounting bracket assembly from engine. Loosen steering shaft support, then disconnect the steering shaft front universal joint, and pry universal joint rearward and off the steering worm shaft. Remove radiator drain cap and pipe. Place jack under torque tube (clutch housing) and remove weight from front wheels. Support steering gear housing and remove

housing to engine retaining bolts. Jack up tractor high enough for crankshaft pulley to clear steering gear housing and move assembly away from tractor.

NOTE: On 404 models equipped with power steering, it will also be necessary to disconnect hydraulic lines from oil cooler and loosen the line clips so lines can be pulled outward.

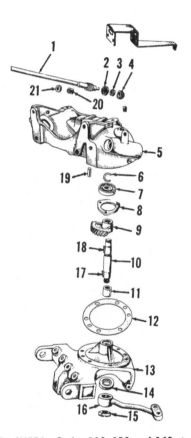

Fig. IH554—Series 100, 130 and 140 steering gear and associated parts. The steering gear used on high clearance models is similarly constructed.

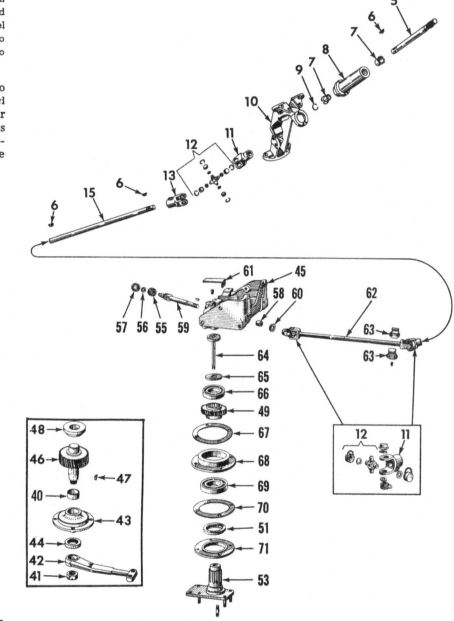

Fig. IH555—Exploded view of Farmall series 200, 230, 240 and 404 steering gear. Inset at lower left shows the parts used in steering gear of 240, 404 and 2404 International series. The unit can be overhauled without removing gear housing from tractor.

1. Worm and shaft	11. Bushing
2. Bearing	12. Gasket
3. Nut	13. Gear housing base
4. Retainer	14. Oil seal
5. Gear housing	15. Oil seal
6. Snap ring	16. Center steering arm
7. Bearing	17. Woodruff key
8. Bearing retainer	18. Woodruff key
9. Sector	19. Dowel pin
10. Sector shaft	20. Bushing
	21. Oil seal

7. Bearing (nylon)	46. Worm gear	57. Bearing retainer	64. Worm wheel shaft
8. Upper support	and shaft	58. Worm bushing	upper bearing stud
10. Lower support	47. Woodruff key	59. Worm	65. Worm shaft seal
40. Bushing	48. Adapter	60. Oil seal	66. Bearing
42. Steering arm	49. Worm wheel	61. Starting crank	67. Gasket
43. Bearing cage	51. Worm wheel	bracket	68. Bearing cage
44. Seal	shaft oil seal	62. Universal joint	69. Bearing
45. Steering gear	53. Worm wheel shaft	shaft and yoke	70. Bearing cage
housing	55. Bearing	63. Steering shaft	cover gasket
	56. Worm nut	center bearing	71. Bearing cage cover

9. OVERHAUL. The steering gear unit can be overhauled without removing the assembly from tractor.

To remove the steering worm, proceed as follows: Remove grille and drain steering gear housing. Loosen steering shaft support, then disconnect the steering shaft front universal joint, and slide universal joint rearward and off the steering worm shaft. Remove Woodruff key from worm shaft. Remove both starting crank bracket front retaining cap screws and block-up between bracket and steering gear housing enough to permit worm to come out. Remove steering worm bearing retainer (57 — Fig. IH555), and turn worm forward and out of housing. Worm bushing (58) and worm shaft oil seal (60) can be renewed at this time. Install worm shaft oil seal with lip of seal facing inward toward steering gears. Use a tin sleeve or shim stock when reinstalling worm shaft to prevent damaging the seal. To remove worm shaft ball bearing, remove cotter key and nut, and bump worm shaft out of bearing.

To remove the steering worm wheel and shaft assembly, proceed as follows: Remove grille and drain steering gear housing. Jack up tractor under torque tube (clutch housing) and remove bolster or wheel fork on tricycle type models; or, on adjustable axle versions, disconnect the center steering arm from worm wheel shaft, remove four axle mounting bracket to steering gear housing retaining cap screws and move axle and wheels assembly away from tractor. Remove the four worm wheel shaft lower bearing cage and cover retaining cap

screws, and bump entire assembly out of steering gear housing. The unit is shown removed in Fig. IH556. To disassemble the unit, remove nut from stud (64) and pull out the stud; place assembly on a suitable press, and press bearings and worm wheel off of worm wheel shaft. At this time, worm wheel shaft lower bearing and oil seal (51—Fig. IH555) can be renewed. Install oil seal with lip of seal facing up toward worm wheel. Use a tin sleeve or shim stock when reinstalling worm wheel shaft to prevent damaging the seal.

Reassemble the unit by reversing the disassembly procedure.

Series 240-404-2404 International

The steering worm and worm wheel are located in the steering gear housing which is bolted to the front face of the engine. The center steering arm is keyed to the lower portion of the worm wheel shaft and the axle support (lower bolster) is retained to the steering gear housing by four cap screws. The unit is non-adjustable.

10. REMOVE AND REINSTALL. The steering gear housing, lower bolster and front axle assembly can be removed with radiator installed; however, some mechanics prefer to remove the radiator.

To remove steering gear housing with radiator attached, proceed as follows: Drain cooling system, remove hood, and disconnect upper and lower radiator hoses. Disconnect the head-

light wires, then remove cap screw retaining radiator brace to thermostat housing. On 240 series, loosen generator adjustment, then unbolt fan blades. Leave blades resting in fan shroud and remove water pump pulley. On 404 and 2404 series, disconnect fan shroud from radiator. Drive roll pin from forward yoke of steering shaft universal and remove cap screws from steering shaft center bearing. Drive front universal from steering worm shaft. Place wood blocks between steering gear housing and axle to prevent tipping, then unbolt stay rod bracket from clutch housing and the steering gear housing from front of engine. Raise engine until crankshaft pulley will clear steering gear housing, then roll complete assembly from tractor. See Fig. IH556A.

NOTE: On 404 and 2404 models equipped with power steering, it will also be necessary to disconnect hydraulic lines from oil cooler and loosen the line clips so lines can be pulled outward.

11. OVERHAUL. The steering gear unit can be overhauled without removing the assembly from tractor.

To remove the steering worm, proceed as follows: Remove grille and side sheet mounting bracket with lower grille pan attached, then drain steering gear housing. Disconnect the steering shaft front universal from worm shaft and the steering shaft center bearing from clutch housing. Drive universal from worm shaft and remove the Woodruff key. Remove

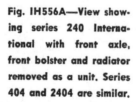

Fig. IH556—Series 200, 230, 240 and 404 (Farmall) steering worm wheel and shaft assembly removed from the steering gear housing. The unit can be disassembled after removing stud (64). See legend for Fig. IH555.

Fig. IH556A—View showing series 240 International with front axle, front bolster and radiator removed as a unit. Series 404 and 2404 are similar.

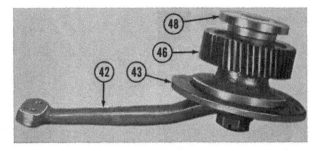

Fig. IH556B—View showing worm wheel assembly of International models removed from steering gear housing. Note position of adapter on top side of worm wheel.

42. Steering arm
43. Cage
46. Worm gear
48. Adapter.

the steering worm retainer (57—Fig. IH555), and turn worm forward and out of housing.

To remove the ball bearing (55), remove cotter key and nut and bump worm shaft out of bearing. Worm shaft bushing (58) and oil seal (60) can be renewed at this time. Install worm shaft oil seal with lip of same facing inward toward steering gears. Use a tin sleeve or shim stock when reinstalling worm shaft to prevent damaging the seal.

To remove the steering worm wheel and shaft assembly, proceed as follows: Remove grille and drain steering gear housing. Place a jack under torque tube (clutch housing), disconnect center steering arm from worm wheel shaft, then unbolt lower bolster from steering gear housing. Disconnect stay rod bracket from clutch housing, then raise tractor and roll front axle assembly forward. Unbolt worm wheel cage from steering gear housing and remove cage, worm wheel and adaptor as shown in Fig. IH556B. Refer also to Fig. IH555. Remove center steering arm (42) and withdraw worm wheel

and shaft (46). Bushing (40) and oil seal (44) can be renewed at this time. Oil seal is installed with lip of same facing toward inside of steering gear housing. When reinstalling center steering arm, torque the retaining nut to 200-250 ft.-lbs. Clamp steering arm in a vise while tightening nut; DO NOT use worm wheel as a stop.

NOTE: A new hardened Woodruff key (47 —Fig. IH555) must be installed in all International 240 tractors prior to chassis serial number 4766. The new key is identified by cross-hatch (knurling) marks on the flat edge.

If, when tightening the steering arm retaining nut, a castellation of the nut does not align with the cotter pin hole at some point between the specified 200-250 ft. lbs. torque, continue to tighten the nut. Do not back-off (loosen) nut to obtain alignment.

When reinstalling the worm wheel assembly, be sure large diameter of adapter (spacer) (48) is on top side as shown in Fig. IH556B.

POWER STEERING

Power steering is available as optional equipment for series 240, 404 and 2404 tractors. Farmall model tractors utilize a torque generating unit (Fig. IH—556C) which applies the power steering assist directly to the steering shaft. International model tractors use a rotary valve, which has the same configuration as the torque generating unit, and a hydraulic power steering cylinder. The power steering assist

on International model tractors is applied to the front axle steering linkage.

Pressurized oil used to operate the power steering system is provided by the tractor hydraulic system. The same lower bolster and steering mechanism is used regardless of whether the tractor does or does not have power steering.

The power unit (Farmall), or rotary valve (International), mounts in the steering shaft lower support and replaces the upper support used on tractors without power steering.

LUBRICATION AND BLEEDING

Series 240-404-2404

12. The tractor hydraulic system is the source of fluid supply to the power steering system. Whenever the power steering oil lines have been disconnected, reconnect the oil lines, fill the reservoir and cycle the power steering system several times to bleed any air from the system.

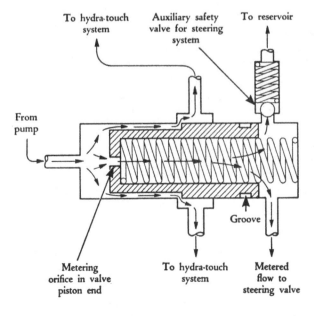

Fig. IH556D—Schematic illustration of the power steering flow control valve used on series 240. The flow control valve satisfies the 2½-3 gpm requirement of the power steering system before any oil flows to the Hydra-Touch system.

Fig. IH556C — View showing location of power steering unit (or rotary valve) on tractors with power steering.

OPERATING PRESSURE, RELIEF VALVE, FLOW CONTROL VALVE

Series 240

13. Working fluid for the hydraulic power steering system is supplied by the same pump which powers the Hydra-Touch system. Interposed between the pump and the Hydra-Touch system is a flow control valve mechanism which is shown schematically in Fig. IH556D. The small metering hole in the end of the flow valve piston passes between 2½ to 3 gallons per minute to the power steering system; but, since the pump supplies considerably more than three gpm, pressure builds up in front of the piston and moves the piston, against spring pressure, until the ports which supply oil to the Hydra-Touch system are uncovered. The steering system, therefore,

Fig. IH556F — Shut-off valve and pressure gage installation diagram for trouble shooting the power steering system on series 240.

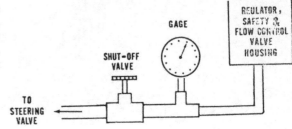

receives priority and the fluid requirements of the steering system are satisfied before any oil flows to the Hydra-Touch system. The auxiliary safety valve for the power steering system maintains a system operating pressure of 1200-1500 psi. The components of the flow control valve are shown exploded from the valve housing in Fig. IH556E.

A pressure test of the power steering circuit will disclose whether the pump, safety valve or some other unit in the system is malfunctioning. To make such a test, proceed as follows: Connect a pressure test gage and shut-off valve in series with the line connecting the flow control valve to the steering valves as shown in Fig. IH-556F. Notice that the pressure gage is connected in the circuit between the shut-off valve and the flow control valve. Open the shut-off valve and run engine at low idle speed until oil is warmed. Advance the engine speed to the specified high idle rpm, close the shut-off valve, observe the pressure gage reading, then open the shut-off valve. If the gage reading is between 1200 and 1500 psi with the shut-off valve closed, the hydraulic pump and auxiliary safety valve are O.K. and any trouble is located elsewhere in the system.

If the gage reading is more than 1500 psi, the auxiliary safety valve may be stuck in the closed position. If the gage reading is less than 1200 psi, renew the auxiliary safety valve spring and recheck the pressure reading. If the gage reading is still less than 1200 psi, a faulty hydraulic pump is indicated.

For information on the regulator, safety and flow control valve assembly, refer to paragraph 131.

Series 404-2404

14. Working fluid for the hydraulic power steering system is supplied by the same pump which powers the hydraulic system. Interposed between the pump and the hydraulic lift housing is a flow control valve assembly which is shown in Fig. IH556G.

Pressurized oil is directed first to the flow control valve and the valve

in turn directs the pressurized oil to the power steering circuit until the needs of the power steering system are satisfied. Thus the power steering system receives priority and its requirements satisfied before any oil flows to the hydraulic system. Hydraulic pumps of 9 gpm capacity are used on tractors with power steering while those tractors with no power steering are equipped with 4½ gpm pumps.

A pressure test of the power steering circuit can be made as follows: Connect a pressure gage in series with the line connecting flow control valve to power unit (Farmall) or rotary valve (International) and be sure gage is between flow control valve and shut-off valve. Open the shut-off valve and run engine at low idle speed until oil is warmed to operating temperature. Advance engine speed to high idle rpm, close shut-off valve and observe gage reading which should be 1350-1550 psi.

Note: To preclude any possibility of overheating oil or damaging hydraulic pump, keep shut-off valve closed only long enough to obtain gage reading.

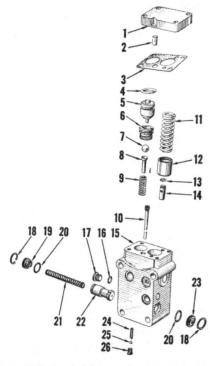

Fig. IH556E—Exploded view of the regulator, safety and flow control valve used on all 240 models with Hydra-Touch system and power steering. The auxiliary safety valve (24, 25 and 26) protects only the power steering system. Valve (22) is not available separately.

1. Cover	14. Safety valve
2. Dowel pin	piston
3. Gasket	15. Valve housing
4. Seal ring	16. Seal ring
5. Regulator valve	17. Plug
piston	18. Snap ring
6. Regulator valve	19. Retainer
seat	20. Seal ring
7. Steel ball	21. Flow control
8. Ball rider	valve spring
9. Ball rider spring	22. Flow control
10. Safety valve	valve
orifice	23. Retainer
11. Safety valve	24. Auxiliary safety
spring	valve spring
12. Spring retainer	25. Steel ball
13. Snap ring	26. Plug

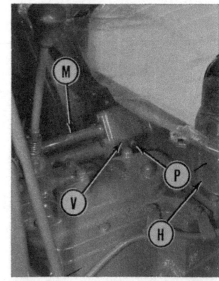

Fig. IH556G — Location of flow control valve on 404 and 2404 series tractors. Note relief valve plug.

H. Hydraulic housing   P. Plug
M. Manifold   V. Flow control valve

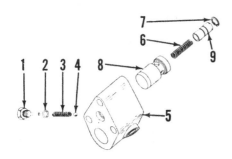

Fig. IH556H—Exploded view of flow control and relief valve used on series 404 and 2404 tractors equipped with power steering.

1. Plug
2. "O" ring
3. Spring
4. Ball (3/16)
5. Valve body
6. Spring
7. Retainer ring
8. Valve spool
9. Retainer

If pressure is lower than specified, remove relief valve plug, add spacer (shim) on stem of plug and recheck. If pressure cannot be brought within limits by the addition of spacer, check hydraulic system relief valve pressure as outlined in paragraph 142A.

If pressure is higher than specified, flow control relief valve is probably stuck and should be removed and cleaned or renewed. Also check hydraulic system pressure relief valve and renew if faulty. Refer to paragraph 142A for information on hydraulic system relief valve.

For information on the flow control valve and relief valve, refer to paragraph 15.

15. Flow control valve and relief valve can be serviced as follows: The relief valve assembly (items 1 through

4—Fig. IH556H) can be removed at any time after removing plug (P—Fig. IH556G). Do not lose shims, if so equipped, from stem of plug. Spring (3—Fig. IH556H) should have a free length of $1\frac{5}{32}$-inches and should test 15.2-17.8 lbs. when compressed to a length of 61/64-inch. Flow control valve spool can be removed as follows: Disconnect manifold line clip from torque tube, then remove cap screws which retain manifold flange. If necessary, shift manifold forward and remove flow control valve assembly. Remove retaining ring (7), then remove spring retainer (9), spring (6) and valve spool (8) from body (5). Valve spool should be a snug fit in valve bore yet slide freely. Small defects can be corrected by polishing valve spool and/or bore with crocus cloth. Renew valve, body and spring retainer if any part shows signs of undue wear, scoring or nicks. Parts are not available separately. Spring (6) should have a free length of $1\frac{1}{4}$ inches and should test 4.9-5.8 lbs. when compressed to a length of $1\frac{1}{16}$ inches. Renew spring if it does not meet specifications or shows signs of fractures or permanent setting. Use new "O" rings when reinstalling assembly.

POWER UNIT OR ROTARY VALVE
Series 240-404-2404

16. **REMOVE AND REINSTALL.** To remove the power steering unit, first remove hood skirts and hood. Un-

bolt instrument panel from instrument panel housing cowl and remove the instrument panel housing. Disconnect steering shaft universal from lower end of power steering unit. Disconnect manifold flange or lines from power steering unit or rotary valve and discard the "O" rings. Remove steering wheel, then unbolt power steering unit from steering shaft lower support and remove the power unit.

Use new "O" ring seals when reinstalling power steering unit.

Series 240-Farmall 404

17. **OVERHAUL POWER UNIT.** With power steering unit removed as outlined in paragraph 16, place a scribe line lengthwise of the unit so it can be reassembled in the same position. Refer to Fig. IH556J and remove cap screws which retain control end cap (20) to spool and sleeve housing (9). Pull cap, control shaft (18) and bearing assembly (15, 16 & 17) from housing (9). Pull sleeve (13) and spool (12) assembly from housing. Remove pin (14) and centering springs (11), then separate sleeve and spool. Pull control shaft (18) from cap (20). Remove "O" ring (19), and if necessary, oil seal (23) and needle bearing (22) from cap.

Remove cap screws, then pull power end housing (2) and power end shaft (5) from spool and sleeve housing. Lift lower plate (7), power end drive (6), gerotor assembly (8) and upper plate (7) from spool and sleeve housing. Remove power end shaft, "O" ring retainer (4) and "O" ring (3) from power end housing.

Wash all parts in a suitable solvent and inspect as follows: Inspect centering springs for fractures or distortion. Springs should return to a minimum height of $\frac{7}{32}$-inch after being depressed to a height of $\frac{1}{16}$-inch. See Fig. IH556K.

Inspect plates (7—Fig. IH556J) for wear and scoring. Polish patterns resulting from rotation of inner rotor are normal and care should be taken not to mistake these patterns for wear. Plates are lapped to within 0.00002 of being flat. Do not attempt to polish out any scratches as sealing of pressure depends upon the plate flatness. NOTE: In cases of emergency plates that are damaged may be turned over so smooth side will be next to gerotor assembly, however, be sure damaged side is lapped until it is flat.

Inspect control valve spool, spool sleeve and valve body for nicks, scoring or undue wear. Damage of any

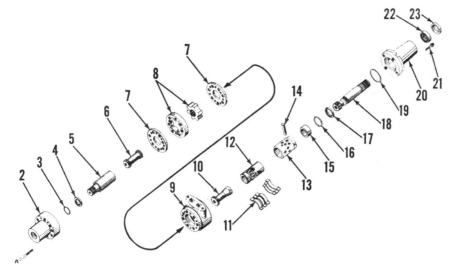

Fig. IH556J—Exploded view of power steering unit used on series 240 and Farmall 404. Items (9, 12 and 13), as well as gerotor set (8) are available as units only.

2. Power end housing
3. "O" ring
4. Retainer washer
5. Power end shaft
6. Power end drive
7. Plate
8. Gerotor set
9. Housing
10. Control end drive
11. Centering springs
12. Valve spool
13. Valve sleeve
14. Pin
15. Bearing race
16. Steel balls (14)
17. Bearing race
18. Control shaft
19. "O" ring
20. Control end cap
22. Needle bearing
23. Oil seal

Fig. IH556K — Centering springs are measured as shown. Refer to text.

ROTARY VALVE

International 404-2404-Farmall 404 (With Adjustable Front Axle)

18. **OVERHAUL ROTARY VALVE.** With unit removed as outlined in paragraph 16, proceed as follows: Remove retaining screws from top end cap and remove cap and steering wheel (input) shaft from valve body as a unit. Pull sleeve, spool and steering (output) shaft from valve body, then remove the cross pin and pull output shaft from sleeve and spool. Start spool out of sleeve and as centering pistons and orifice ball emerge from sleeve, grasp them with fingers to keep parts from flying. Do not attempt to remove check ball assembly from spool unless the assembly is is damaged. If check ball assembly must be renewed, proceed as outlined in paragraph 18A.

18A. Place valve spool in a soft jawed vise with check ball seat upward. Use a 3/64-inch drill with a pin vise and drill through orifice of check valve seat. Use caution not to jam check ball into spring as drill nears inner end of valve seat and DO NOT use any other means of drilling except the pin vise. Select 1/16, 5/64, 3/32 and 7/64-inch drills, and in this sequence, drill through the valve seat with each drill. After drilling with the 7/64-inch drill, remove check ball and spring, then final clean the valve seat bore using a ⅛-inch drill in the pin vise. Use extreme care during this final drilling not to enlarge the valve seat bore.

one part will require renewal of all three as they are not available separately. Spool should fit sleeve and sleeve should fit body snugly yet both should move freely. Cross pin (14) should have a snug fit and if pin is bent or has a diameter of less than 0.2498 at its contacting points, renew pin.

Check gerotor assembly as shown in Fig. IH556L. Clearance between teeth of inner rotor and outer rotor should be 0.001-0.005. Renew rotor assembly if clearance exceeds 0.005.

Inspect output shaft and output shaft housing for wear or damage. Measure outside diameter of output shaft (5—Fig. IH556J) and inside diameter of housing (2) bore. If bore of housing is 0.006 or more larger than diameter of shaft, renew housing and/or shaft.

Inspect control (input) shaft (18), ball bearings (16), race (17), needle bearing (22) and oil seal (23) for wear, scoring or other damage and renew parts as necessary. Use a press to install seal in end cap (20); do not drive on seal. When seal bottoms it will be between flush and 1/64-inch below end of housing. Stake seal in position.

When reassembling, use all new "O" rings and with the exception of needle bearing (22) being coated with Lubriplate, and "O" ring (3) being oiled, all parts are assembled dry. NOTE: "O" ring (3) must be worked into its groove.

Reassemble by reversing disassembly procedure. However, BE SURE the pin slot in control end drive (10) is aligned with the valley of any tooth as indicated in Fig. IH556M. If pin slot is not aligned as shown, unit will attempt to operate in reverse when hydraulic pressure is applied. Tighten cap screws retaining end caps evenly

Fig. IH556L—Measure tooth clearance of gerotor with gerotor set positioned as shown.

Fig. IH556M—When reassembling, pin slot of control end drive must be aligned with any tooth valley of inner rotor.

and to 12 ft.-lbs. torque.

IMPORTANT: When installing the seven screws which retain end housing (2—Fig. IH556J) to housing (9), be sure both cap screws and tapped holes are clean and perfectly dry. Apply one drop of "Loctite" to each cap screw, tighten evenly to 12 ft.-lbs. torque and allow 8 to 12 hours for the "Loctite" to set before attempting to put unit in operation.

18B. Clean all parts in a suitable solvent and dry with compressed air or lint free wipers. Be sure all chips and other foreign material are removed from bores, holes or grooves of valve spool and sleeve. Small nicks or scratches can be removed with a No. 1 Arkansas stone. Sleeve should fit body and spool should fit sleeve snugly yet both should move freely when assembled. Inspect check valve spring and orifice valve spring for fractures, distortion or other damage. Be sure cross pin is straight and not unduly worn. Seals and bearings in end caps can be renewed and the procedure for doing so is obvious.

18C. If check valve assembly was renewed, proceed as follows: Place check valve seat on a flat surface with ball seat (small) end upward. Place check ball on seat and mate ball and seat by striking ball using a soft brass drift and a small ball peen hammer.

Fig. IH556N — Oil cooler for tractors with power steering is mounted as shown.

Fig. IH556P—Power steering cylinder used on International 404 and 2404 tractors is mounted as shown.

Place valve spool in a soft jawed vise with check valve bore upward. Install check valve spring in bore then install ball on top of spring. Place check valve ball seat on installing tool (IHC No. FES-64-5) with the large flat end toward shoulder of installing tool, and holding installing tool at a 45-degree angle, drive check ball seat into its bore until outer end is flush. Rotate installing tool when removing. Remove any burrs which may be present from seat, slot and/or bore.

NOTE: Do not use any tool other than IHC installing tool No. FES-64-5 and do not use air, or probe through hole, to unseat check ball as check ball could be forced into spring and the assembly rendered inoperative.

18D. When reassembling, use all new "O" rings and dip all parts in IHC Hy-Tran fluid, or its equivalent. Install "O" rings and back-up washers on centering pistons with "O" rings toward flat end of piston. Place centering piston spring in bore, then install centering pistons in ends of bore with rounded ends toward outside of spool using care not to cut "O" rings as pistons are installed.

Align orifice hole of valve spool with cross pin hole of valve sleeve and while depressing centering pistons, start valve spool into valve sleeve. Push valve spool into sleeve far enough to retain centering pistons in their slots and yet leave orifice hole exposed. Insert orifice wire through its spring, then insert wire and spring in orifice bore and be sure orifice hole wire is in the orifice hole. Place orifice ball in orifice hole, depress same and complete insertion of valve spool in valve sleeve. Install output shaft in lower end of valve spool, align pin holes of output shaft, valve spool and valve sleeve and install valve cross pin. Install lower end cap and seal on valve body, if previously removed, then install valve and output shaft assembly in valve body. Install steering wheel (input) shaft in top end cap and install assembly on valve body. Torque both end cap retaining cap screws to 150 in.-lbs.

Complete assembly by reversing the disassembly procedure and bleed steering system as outlined in paragraph 12.

PUMP
Series 240-404-2404

19. The regular hydraulic pump supplies fluid to both the power steering and the hydraulic system. Refer to paragraphs 122, 123, 129, 148A and 148B for R&R and/or resealing of the pump.

OIL COOLER AND RELIEF VALVE
Series 404-2404

20. Models equipped with power steering are fitted with an oil cooler mounted in front of radiator as shown in Fig. IH556N. This unit is used to cool the oil used by the power steering system. Also incorporated into the oil cooler system is a relief (by-pass) valve (V—Fig. IH556C) which opens and by-passes the oil to reservoir, should the oil cooler become plugged or if oil is too cold to circulate.

Oil cooler can be unbolted from radiator after removing hood and grille and disconnecting hoses. Relief valve can be removed at any time and procedure for doing so is obvious.

Relief valve can be bench tested and should open at 85 psi. Correct faulty units by renewing same.

POWER STEERING CYLINDER
Series 404-2404 International

21. R&R AND OVERHAUL. To remove power steering cylinder (Fig. IH556P), first disconnect hoses and immediately plug same to prevent oil drainage. Disconnect cylinder from center steering arm and axle and remove from tractor.

With cylinder removed, extend and retract piston rod to clear cylinder of oil. Remove locking wire and cylinder head retainer, then remove cylinder head, piston rod and piston from cylinder tube. Remove piston retaining nut and remove piston and cylinder head from piston rod.

Renewal of piston ring, "O" rings, back-up washers and wiper seal will be obvious after an examination. Renew piston rod if same is bent, heavily scored or otherwise damaged. Cylinder tube bore should be smooth and free of scoring. Renew cylinder tube if scored or unduly worn.

ENGINE AND COMPONENTS

R&R ENGINE WITH CLUTCH

25. To remove the engine and clutch as an assembly, proceed as follows: Support tractor under clutch housing and remove radiator, steering gear unit and front axle assembly as outlined in paragraphs 6, 8 or 10. On models equipped with hydraulic "Touch-Control", drain hydraulic cylinder and remove the hydraulic lines. On models equipped with "Hydra-Touch" remove hydraulic lines between hydraulic pump and control valve support. On models with draft and position control, disconnect hydraulic pressure line at pump and return hose at clutch housing. Disconnect fuel lines,

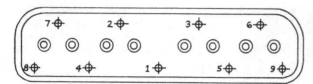

Fig. IH557—On C123 engines with 9-stud cylinder heads, tighten retaining nuts in sequence shown.

wiring harness and controls from engine and engine accessories. Remove oil cup from air cleaner on all except 404 and 2404 series tractors. Remove the starting motor, remove the fuel tank front support bolts, loosen the fuel tank rear support bolts, and block-up fuel tank. On models with underslung exhaust, disconnect exhaust pipe at manifold, clutch housing and left brake housing and remove exhaust pipe and muffler as a unit. Support engine in a hoist, remove clutch housing cover, bolts retaining engine to clutch housing, and pull engine forward and away from tractor.

CYLINDER HEAD

26. To remove the cylinder head, first drain cooling system and remove hood, valve cover, rocker arms assembly and push rods. Loosen upper radiator hose and on all series except 140, 240, 404 and 2404, disconnect the combination fan bracket and water outlet casting from head. Disconnect fuel lines and remove carburetor and manifold assembly. Remove the cylinder head retaining stud nuts and lift cylinder head from tractor.

When reinstalling cylinder head, refer to Figs. IH557 and IH558 and tighten the stud nuts to a torque of 65 ft.-lbs. on C123 engines prior to engine serial number 65001, and to 80-90 ft.-lbs. on C123 engines after 65000 and all C135 engines.

VALVES AND SEATS

27. Intake and exhaust valves are not interchangeable and on all except series 404 and 2404, seat directly in the cylinder head. Series 404 and 2404 exhaust valves seat on renewable in-

serts which are available in standard size as well as oversizes of 0.015 and 0.030. All valves have a seat angle of 45 degrees. Seat width is 5/64-inch for all models except those having a C123 engine with a serial number prior to 65001 which should be 1/8-inch. Valves of the earlier models are equipped with stem safety retainers to prevent valve from dropping into combustion chamber should a valve spring break. Valves have stem diameter of 0.3405-0.3415 and a clearance of 0.0015-0.0035 in the guides with a maximum allowable clearance of 0.006.

Valve tappet gap should be set to 0.014 hot for all tractors except series 404 and 2404 which should be 0.014 hot for the intake and 0.020 hot for exhaust.

When removing valve seat inserts, use the proper puller or pry them out with the edge of a large chisel. Do not attempt to drive chisel under seat insert as counterbore will be damaged. Chill new seat insert with dry ice or liquid Freon and when insert is properly bottomed, it should be 0.008-0.030 below edge of counterbore. After installation, peen the cylinder head material around the complete outer circumference of the valve seat insert.

VALVE GUIDES AND SPRINGS

28. Intake and exhaust valve guides are interchangeable. Valve guides are pre-sized and if not distorted during installation, will require no final sizing. Installed height of valve guides is $1\frac{3}{16}$-inch above surface of cylinder head. The 0.3405-0.3415 diameter valve stems should have a clearance of 0.0015-0.0035 in the installed guides.

Renew any spring which is rusted,

discolored or does not meet the following specifications:

Free Length (inches)
  100-200 (Gaso., Kero., Dist.)
    Int. & Exh. (no rotocap)..2 17/32
    Exhaust (with rotocap)....2 9/64
  130-140-230-240-404-2404 (Gaso.)
    Int. & Exh. (no rotocap)..2 47/64
    Exhaust (with rotocap).....2 1/4
  404-2404 (LPG)
    Int. & Exh. (no rotocap)..2 47/64
  130-230 (Kero., Dist.)
    Int. & Exh. (no rotocap)..2 17/32
    Exhaust (with rotocap).....2 1/4
  140-240 (Kero., Dist.)
    Int. & Exh. (no rotocap)..2 47/64
    Exhaust (with rotocap).....2 1/4

Test Load (lbs.) @ Length (in.)
  100-200 (Gaso., Kero., Dist.)
    Int. & Exh.
      (no rotocap) ......43@1 43/64
    Exh. (with rotocap)....36@1 3/4
  130-140-230-240-404-2404 (Gaso.)
    Int. & Exh. (no rotocap)....36@2
    Exh. (with rotocap)..36@1 27/32
  404-2404 (LPG)
    Int. & Exh. (no rotocap)....36@2
  130-230 (Kero., Dist.)
    Int. & Exh.
      (no rotocap) ......43@1 43/64
    Exh. (with rotocap)..36@1 27/32
  140-240 (Kero., Dist.)
    Int. & Exh. (no rotocap)....36@2
    Exh. (with rotocap)..36@1 27/32

VALVE TAPPETS
(CAM FOLLOWERS)

29. The 0.560-0.561 diameter mushroom type tappets operate directly in the unbushed crankcase bores and can be removed after removing the camshaft as outlined in paragraph 35. Clearance of tappets in the crankcase bores should be 0.0005-0.003. Oversize tappets are not available.

VALVE LEVERS
(ROCKER ARMS)

30. The valve levers and hollow shaft assembly is pressure lubricated from the center camshaft bearing via an oiler stud in top surface of cylinder head.

Replacement valve levers are available with installed bushings; or, bushings are available for field installation in cast iron levers which are in otherwise good condition. When installing the bushings, make certain that the oil hole in bushing is in register with the oil spurt hole in lever and ream the bushing to an inside

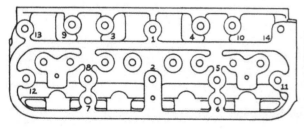

Fig. IH558—On C123 and C135 engines with 14-stud cylinder heads, tighten retaining nuts in sequence shown.

diameter of 0.751-0.752 to provide 0.002-0.004 clearance for the 0.748-0.749 diameter lever shaft. Maximum allowable clearance of lever on shaft is 0.006. Some models may be equipped with levers of welded construction, with a non-serviceable bushing. These levers, which are interchangeable with the equivalent cast iron lever, must be renewed when the lever to shaft clearance is excessive.

VALVE ROTATORS

31. Positive type exhaust valve rotators ("Rotocaps") are factory installed on some models.

Normal servicing of the valve rotators consists of renewing the units. It is important, however, to observe the valve action after the engine is started. Rotator action can be considered satisfactory if the valve rotates a slight amount each time the valve opens. A cut-away view of a typical 'Rotocap' installation is shown in Fig. IH559.

VALVE TIMING

32. Valves are properly timed when single punch marked tooth on camshaft gear is meshed with the single punch marked tooth space on crankshaft gear as shown in Fig. IH560.

TIMING GEAR COVER

33. To remove the crankcase front cover, first drain cooling system and remove the complete front end assembly as outlined in paragraph 6, 8 or 10.

On series 100, 130, 200 and 230, remove the fan assembly, governor housing assembly and water pump. Attach a suitable puller as shown in Fig. IH561 and remove the crankshaft pulley. Unbolt and remove cover from engine.

On series 140, 240, 404 and 2404, remove water pump and governor hous-

Fig. IH560 — Timing gear train of IH C-123 engine. Gears must be installed with timing marks in register. C135 engines are similar.

ing. Remove crank nut (or crankshaft nut), attach suitable puller and remove crankshaft pulley and Woodruff key. Unbolt and remove cover from engine.

Extra care must be taken when installing the oil seal in the crankcase front cover, so as not to distort or bend the cover. Install seal with lip of same facing inward toward timing gears.

When reassembling, leave the cover retaining cap screws loose until crankshaft pulley has been installed—this will facilitate centering the seal with respect to the pulley.

TIMING GEARS

34. To renew the camshaft gear and/or crankshaft gear, it is necessary to use a suitable press after the respective shaft has been removed from engine.

When reassembling, mesh the single punch marked tooth on camshaft gear with the single punch marked tooth space on crankshaft gear and the

double punch marked tooth on camshaft gear with the similarly marked tooth space on the governor and ignition unit drive gear.

CAMSHAFT

35. To remove the camshaft, first drain cooling system and remove the crankcase front cover as outlined in paragraph 33. Remove the valve cover, valve levers and shaft assembly, push rods, oil pan and oil pump. Push tappets up into their bores. Working through openings in camshaft gear, remove the cap screws retaining the shaft thrust plate to crankcase and carefully withdraw the camshaft and gear unit from engine. Gear can be removed from camshaft by using a suitable press.

Normal camshaft end play of 0.003-0.012 is controlled by the thrust plate located between the cam gear and crankcase. Excessive end play is corrected by renewing the plate.

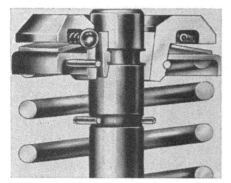

Fig. IH559—Cut-away view showing typical installation of a valve rotator.

Fig. IH561—Puller installation for removing crankshaft pulley from early C-123 engines. The pulley has a very tight fit on the shaft. Pulleys on engines after serial number 65,-000 are retained by a crank nut. C135 engines are similar.

The three camshaft bearing journals ride directly in the crankcase bores with a normal clearance of 0.0009-0.0054 with a maximum allowable clearance of 0.006.

NOTE: On series 240 engines after serial number C123-65,000 and 404, 2404 engines, a bushing has been added at the front camshaft journal. The intermediate and rear camshaft journals ride directly in the crankcase bores. Camshaft journal diameters and operating clearances have not changed.

Shaft journal sizes are as follows:

No. 1 (front)..............1.811-1.812
No. 2 ....1.577-1.578
No. 3 ....1.499-1.500

Oil leakage at the shaft rear bearing journal is prevented by an expansion plug. Renewal of this plug requires splitting engine from clutch housing as outlined in paragraph 78 and removing the flywheel.

ROD AND PISTON UNITS

36. Connecting rod and piston units are removed from above in the conventional manner after removing the the cylinder head, oil pan and oil pump. Cylinder numbers are stamped on connecting rod and cap. When reassembling, make certain that the numbers are in register and face toward camshaft side of engine. Tighten the rod bolt nuts to a torque of 40-45 Ft.-Lbs. for nuts locked with cotter pins, 43-49 Ft.-Lbs. for self-locking nuts or 45 Ft.-Lbs. for self-locking bolts.

PISTONS, SLEEVES AND RINGS

37. Pistons are not available as individual replacement parts, but only as matched units with the wet type sleeves for standard compression ratio engines, special compression ratio engines and engines for operation at 5000 and 8000 foot altitudes. The matched units are available individually or in sets of four.

Recommended clearance of new pistons in new sleeves is 0.0011-0.0019 when measured between piston skirt and cylinder sleeve at 90 degrees to piston pin.

38. The wet type cylinder sleeves should be discarded when the out-of-round exceeds 0.006, or taper exceeds 0.005.

Special pullers are available to remove the wet type sleeves from above after the pistons have been removed. Before installing sleeves, check to make certain that the counterbore at top and sealing ring groove at the bottom are clean and free from foreign material. All sleeves should enter crankcase bores full depth and should be free to rotate by hand when tried in bores without sealing rings. After making trial installation without sealing rings, remove the sleeves and install new sealing rings dry into the grooves in crankcase. Wet the end of the sleeve with a thick soap solution or equivalent and install sleeves. If sealing ring is in place and not pinched, very little hand pressure is required to press the sleeve completely into place. Normally, the top of the sleeves will extend 0.003-0.007 above the machined top surface of the cylinder block. If sleeve stand out is excessive, check for foreign material under the sleeve flange.

Fig. IH562—Rear view of C-123 engine showing the installation of the crankshaft rear oil seal retainer plates.

Note: The cylinder head gasket forms the upper cylinder sleeve seal, and excessive sleeve stand out will result in coolant leakage. To test lower sealing rings for proper installation, fill crankcase (cylinder block) water jacket with cold water and check for leaks near bottom of sleeves.

39. On engines prior to engine serial number 501-36001, pistons were fitted with three compression rings and one oil ring. On engines serial number 501-36001 and up, the gas & LPG models are fitted with two compression rings and one oil ring, while distillate and kerosene models have three compression rings and one oil ring. All rings should have an end gap of 0.010-0.020. Side clearance of rings in the piston grooves is 0.003-0.0045 for the top compression ring; 0.0015-0.003 for the other compression rings and the oil control ring.

Rings stamped top should be installed with word "TOP" facing toward top of engine.

Replacement ring sets are available to correct high oil consumption without renewing the pistons and sleeves. These rings, however, are not to be used with new pistons and sleeves nor if the wear exceeds the values which follow:

Sleeve out-of-round...........0.006
Sleeve taper 0.005
Top ring clearance in groove...0.009

PISTON PINS

40. The full floating type piston pins are retained in the piston bosses by snap rings and are available in standard size as well as one oversize of 0.005. The 0.9192-0.9195 (C123) or 0.8591-0.8593 (C135) diameter pins should have a clearance of 0.0002 in piston bosses and a clearance of 0.0004 in the connecting rod bushing.

CONNECTING RODS AND BEARINGS

41. Connecting rod bearings are of the slip-in, precision type, renewable from below after removing the oil pan and connecting rod bearing caps. When installing new inserts, make certain that the projections on same engage slots in connecting rod and cap and that the cylinder identifying numbers on rod and cap are in register and face toward camshaft side of engine. Connecting rod bearings are available in standard size as well as undersizes of 0.002, 0.003, 0.030 and 0.032 for all except series 404 and 2404 which have bearings available in standard size as well as undersizes of 0.002, 0.010, 0.020 and 0.030.

Bearing inserts should have a running clearance of 0.0009-0.0034 for all models except 404 and 2404 which should be 0.0009-0.0039, and all connecting rods should have a side play of 0.005-0.014 on the 1.749-1.750 diameter crankshaft crankpins. Tighten the rod bolt nuts to a torque of 40-45 Ft.-Lbs. for nuts locked with cotter pins, 43-49 Ft.-Lbs. for self-locking nuts or 45 Ft.-Lbs. for self-locking bolts.

CRANKSHAFT AND MAIN BEARINGS

42. The crankshaft is supported in three slip-in, precision type main bearings, renewable from below after removing the oil pan, rear oil seal lower retainer plate and main bearing caps. Normal crankshaft end play of 0.004-0.010 is controlled by the flanged rear main bearing inserts. Renewal of the crankshaft requires R&R of engine. Main bearings are available in standard size as well as undersizes of 0.002, 0.003, 0.030 and 0.032 for all models except series 404 and 2404 which have bearings available in undersizes of 0.002, 0.010, 0.020 and 0.030.

Bearing inserts should have a running clearance of 0.0009-0.0039 on the 2.124-2.125 diameter crankshaft main journals. Tighten the main bearing bolts to a torque of 75-80 Ft.-Lbs.

CRANKSHAFT REAR OIL SEAL

43. A two piece felt rear oil seal contained in a two piece retainer as shown in Fig. IH562 is used on early engines. Later engines have a one piece retainer and lip type seal as shown in Fig. IH562A. The procedure for renewing the seal is evident after splitting engine from clutch housing and removing the clutch and flywheel.

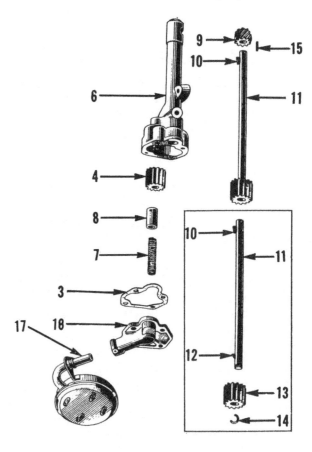

Fig. IH563 — Exploded view of oil pump. Inset shows separate drive shaft and body drive gear used prior to 140 and 240 series.

    3. Gasket
    4. Idler gear
    6. Pump body
    7. Relief valve spring
    8. Relief valve
    9. Drive pinion
    10. & 12. Key
    11. Drive shaft
    13. Body drive gear
    14. Retainer
    15. Pin
    16. Cover

When installing the two piece retainer, be sure to renew all gaskets; and tighten the two vertical screws (S—Fig. IH562) before tightening those securing the retainer halves to crankcase.

FLYWHEEL

44. The flywheel can be removed after splitting engine from clutch housing and removing the clutch. To install the flywheel ring gear, heat same to approximately 500 degrees F.

OIL PUMP AND RELIEF VALVE

45. The gear type oil pump is gear driven from a pinion on the camshaft and can be removed after removing the oil pan. The procedure for overhauling the pump is evident after an examination of the unit and reference to Fig. IH563. Gaskets (3), between cover and pump body can be varied to obtain the recommended body gear end play. Check the pump parts against the values which follow:

NOTE: The drive shaft and gear are now an integral part and the separate shaft and gear are no longer available. The later shaft and gear assembly is interchangeable with the earlier shaft and gear.

Gear to body
    clearance ....0.007 -0.013
Body gears end play.....0.0035-0.006
Body gear backlash......0.003 -0.006
Relief valve spring
    free length 2.398
Lbs. test @ 1.674 inches........24.2
Oil pressure at
    governed speed ....45-55 psi

Fig. IH562A—Rear view of C135 engine showing installation of one piece seal retainer and lip type oil seal.

CARBURETOR (GASOLINE)

46. Tractors are equipped with either Carter, Zenith or Marvel Schebler carburetors. Their applications and specifications are as follows:

Carter

Series 100 gasoline..........UT2222S
   Float setting ....17/64 (in.)
   Gasket set 207
   Inlet needle & seat.........25-33S
   Main metering jet.........224-24S
   Nozzle bleed screw plug....11B-320
   Low speed jet...........11B-2085
   Idle needle 30A-37
   Main metering screw
     & needle 224-30S
   Nozzle and gasket.........12-348S

Series 200 gasoline.........UT771SB
   Float setting ....17/64 (in.)
   Gasket set 207
   Inlet needle & seat.........25-33S
   Main metering jet.........224-25S
   Nozzle bleed screw plug....11B-243
   Low speed jet............11-208S
   Idle needle 30A-37
   Nozzle and gasket.........12-336S

Series 100-130-140-200-230-240
   Kero. & Dist...........UT925SA
   Float setting ....17/64 (in.)
   Gasket set 207
   Inlet needle & seat.........25-33S
   Nozzle bleed screw plug....11B-235
   Low speed jet............11-208S
   Idle needle 30A-37

Main metering screw &
   needle 159-144S
Nozzle and gasket.........12-366S

Zenith

Series 100 gasoline.......67X7 11704
   *Float setting ....1 5/32 (in.)
   Basic repair kit...........K11704
   Gasket set C181-314
   Inlet needle & seat.....C81-17-35
   Idle jet C55-22-18
   Discharge nozzle C66-90-55-19
   Well vent jet...........C77-18-26

Series 130-140 gasoline..68X7 12122A
   *Float setting ....1 5/32 (in.)
   Basic repair kit..........K12122A
   Gasket set C181-329
   Inlet needle & seat.....C81-17-35
   Idle jet C55-22-11
   Main jet or needle......C52-6-1-23
   Discharge nozzle C66-114-38
   Well vent jet...........C77-18-20

Series 200 gasoline.......67X7 11115
   *Float setting ....1 5/32 (in.)
   Basic repair kit...............K790
   Gasket set C181-314
   Inlet needle and seat.....C81-17-35
   Idle jet C55-22-18
   Main jet or needle......C66-90-20
   Discharge nozzle C66-90-20-55
   Well vent jet...........C77-18-29

Series 230 gasoline........68X7 12115
   *Float setting ....1 5/32 (in.)
   Basic repair kit...........K12115
   Gasket set C181-329
   Inlet needle and seat.....C81-17-35
   Idle jet C55-22-11
   Main jet or needle.......C52-6-21
   Discharge nozzle C66-114-50
   Well vent jet...........C77-18-23

Series 240 gasoline........68X7 12285
   *Float setting ....1 5/32 (in.)
   Basic repair kit............K12285
   Gasket set C181-329
   Inlet needle and seat.....C81-17-35
   Idle jet C55-22-11
   Main jet or needle.......C52-7-21
   Discharge nozzle C66-114-50
   Well vent jet...........C77-18-23

Series 100-130-140-200-230-240
   kerosene and distillate..67X7 11340
   *Float setting ....1 5/32 (in.)
   Basic repair kit...............K790
   Gasket set C181-314
   Inlet needle and seat.....C81-17-35
   Idle jet C55-22-18
   Discharge nozzle C66-90-50-18
   Well vent jet...........C77-18-18

Series 404-2404 gasoline.68X7-12225C
   *Float setting ....$1\frac{5}{32}$ (in.)
   Basic repair kit...........K12225
   Gasket set C181-329
   Inlet needle and seat....C81-66-35
   Idle jet C55-22-11
   Discharge nozzle C66-114-50
   Well vent jet...........C77-18-25
   *Float setting is measured from furthest face of float to gasket surface of bowl cover.

Marvel-Schebler

Series 404-2404 gasoline......TSX748
   Float setting ....1/4 (in.)
   Basic repair kit...........286-1245
   Gasket set 16-613
   Inlet needle and seat......233-536
   Idle jet 49-101-L
   Discharge nozzle 47-490

LP-GAS SYSTEM

Series 404 and 2404 tractors are available with LP-Gas equipment manufactured by Ensign. Carburetor is a model Mg 1 and regulator is a model W. As with other LP-Gas systems, these systems are designed to operate with the fuel tank no more than 80% filled. It is important when starting these tractors, to open the vapor valve on the supply tank slowly; if opened too fast, the fuel supply to the regulator will be automatically shut off.

CAUTION: Before disconnecting fuel lines or removing any of the system components, close both fuel tank withdrawal valves and run engine until all fuel is exhausted from the fuel lines and components and engine stops.

ADJUST SYSTEM
Series 404-2404

47. The LP-Gas system has four adjustments; three are located on the carburetor and one on the regulator. Adjustments located on the carburetor are main fuel (load) adjustment, starting fuel adjustment and the throttle idle stop screw. The idle mixture adjusting screw is located on top side of regulator.

The adjustments are pre-set at the factory and normally do not need readjustment; however, if adjustments have been disturbed, or any of the components serviced, readjust system as follows: Loosen lock nuts on starting fuel adjustment and main fuel adjustment on carburetor. Turn both of these adjustment screws and the idle mixture adjusting screw on the regulator all the way in; then make the following initial adjustments. Starting adjustment screw 1¼ turns open. Main fuel adjustment screw 1½ turns open. Idle mixture adjustment screw (on regulator) 1½ turns open. Start engine and bring to operating temperature, then place throttle control lever in low idle position. Adjust throttle stop screw to obtain an engine low idle speed of approximately 425 rpm, then turn the idle mixture needle on the regulator in or out as required to obtain the highest and smoothest engine operation. Readjust throttle stop screw, if necessary. Place hand throttle in the high idle position and turn the main fuel (load) adjustment screw in or out as required until the smoothest engine operation is obtained.

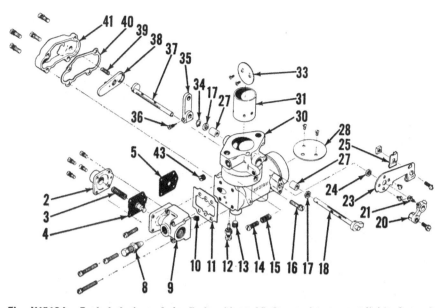

Fig. IH563A—Exploded view of the Ensign Mg 1 LP-Gas carburetor available for series 404 and 2404 tractors.

2. Cover
3. Spring
4. Economizer diaphragm assembly
5. Economizer diaphragm
8. Fuel adjusting screw
9. Inlet housing
10. Bleed screw
11. Gasket

12. Starting adjustment screw
13. Pipe plug
14. Throttle stop screw
15. Spring
16. Venturi set screw
17. Dust seal
18. Throttle shaft
20. Choke lever

21. Set screw
23. Choke cable support
24. Dust seal
25. Clamp
27. Bushing
28. Choke disc
30. Body
31. Venturi
33. Throttle disc

34. Snap ring
35. Throttle lever
36. Set screw
37. Choke shaft and cam
38. Valve lever
40. Gasket
41. Valve cover

NOTE: The load screw adjustment may have to be altered slightly after engine is loaded. If a dynamometer is available, use it when making the load screw adjustment.

Starting adjustment should be approximately correct; however, it may be varied as needed if cold starts are not satisfactory. Counter-clockwise rotation of the needle richens the mixture.

CARBURETOR
Series 404-2404

48. **R&R AND OVERHAUL.** Removal and reinstallation procedure for the LP-Gas carburetor is obvious after an examination of the unit.

With unit removed, refer to Fig. IH563A and completely disassemble the carburetor. Wash all parts in a suitable solvent and blow out all passages with compressed air. Be sure economizer diaphragm is in satisfactory condition with no ruptures or tears. Use new gaskets when reassembling.

After carburetor is assembled and installed, adjust same as outlined in paragraph 47.

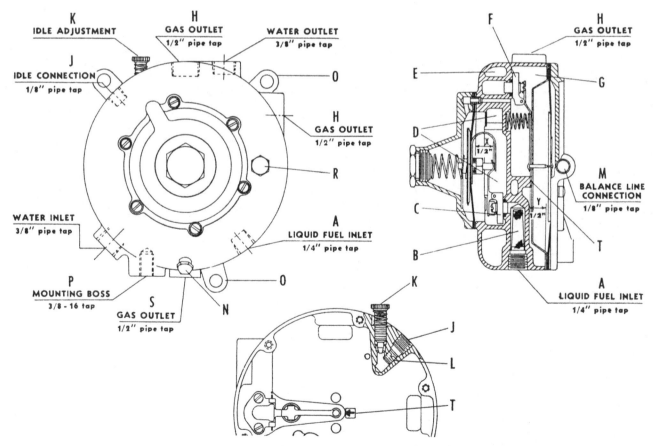

Fig. IH563B—Ensign model W regulator used on series 404 and 2404 tractors.

B. Strainer
C. High pressure valve
D. Vaporizing chamber

E. Water jacket
F. Low pressure valve

G. Low pressure chamber
L. Idle orifice

N. Drain
O. Support lugs
T. Boss

REGULATOR
Series 404-2404

49. **HOW IT OPERATES.** Fuel from the supply tank enters the regulating unit inlet (A—Fig. IH563B) at tank pressure and is reduced from tank pressure to about 4 psi at the high pressure reducing valve (C) after passing through the strainer (B). Flow through high pressure reducing valve is controlled by the adjacent spring and diaphragm. When the liquid fuel enters the vaporizing chamber (D) via the valve (C) it expands rapidly and is converted from a liquid to a gas by heat from the water jacket (E) which is connected to the cooling system of the engine. The vaporized gas then passes (at a pressure slightly below atmospheric pressure) via the low-pressure reducing valve (F) into the low-pressure chamber (G) where it is drawn off to the carburetor via outlet (H). The low pressure reducing valve is controlled by the larger diaphragm (T) and small spring.

Fuel for the idling range of the engine is supplied from a separate outlet (J) which is connected by tubing to a separate idle fuel connection on the carburetor. Adjustment of the carburetor idle mixture is controlled by the idle fuel screw (K) and the calibrated orifice (L) in the regulator. The balance line (M) is connected to the air inlet horn of the carburetor so as to reduce the flow of fuel and thus prevent over-richening of the mixture which would otherwise result when the air cleaner or air inlet system becomes restricted.

TROUBLE SHOOTING

50. SYMPTOM—Engine will not idle with Idle Mixture Adjustment Screw in any position.

CAUSE AND CORRECTION — A leaking valve or gasket is the cause of the trouble. Look for leaking low pressure valve caused by deposits on valve or seat. To correct the trouble wash the valve and seat in gasoline or other petroleum solvent.

If the foregoing remedy does not correct the trouble, check for leak at high pressure valve by connecting a low reading (0 to 20 psi) pressure gage at point (R) on the regulator. If the pressure increases **after** a warm engine is stopped, it proves a leak in the high pressure valve. Normal pressure is 3½-5 psi.

52. SYMPTOM—Cold regulator shows moisture and frost after standing.

CAUSE AND CORRECTION—Trouble is due either to leaking valves as per paragraph 50 or the valve levers are not properly set. For information on setting of valve lever refer to paragraph 54.

REGULATOR OVERHAUL

If an approved station is not available, the model W regulator can be overhauled as outlined in paragraphs 53 and 54.

53. Remove the unit from the engine and completely disassemble using Fig. IH563B as reference. Thoroughly wash all parts and blow out all passages with compressed air. Inspect each part carefully and discard any that are worn.

54. Before reassembling the unit, note dimension (X) which is measured from the face on the high pressure side of the casting to the inside of the groove in the valve lever when valve is held firmly shut as shown in Fig. IH563C. If dimension (X) which can be measured with Ensign gage No. 8276, or with a depth rule, is more or less than ½-inch, bend the lever until this setting is obtained. A boss or post (T—Fig. IH563B) is machined and marked with an arrow to assist in setting the lever. Be sure to center the lever on the arrow before tightening the screws holding the valve block. The top of the lever should be flush with the top of the boss or post (T).

LP-GAS FILTER
Series 404-2404

55. Filters (Fig. IH563D) used on these systems are designed for a working pressure of 375 psi and should be able to stand this pressure without leakage. Filter element should be renewed when indications are that element is plugged sufficiently to prevent proper fuel flow. A clogged element

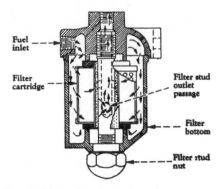

Fig. IH563C—Using Ensign gage No. 8276 to measure dimension (X) on model W regulator. Dimension (X) is shown in Fig. IH563B.

Fuel inlet

Filter cartridge

Filter stud outlet passage

Filter bottom

Filter stud nut

Fig. IH563D—Sectional view showing construction of LP-Gas filter.

causes a pressure drop and results in vaporization of the fuel which may cause regulator freezing and subsequent fuel starvation of the engine. When major engine work is being performed it is advisable to remove the lower part of the filter, thoroughly clean the interior and renew the treated paper cartridge if same is not in good condition.

GOVERNOR

The centrifugal flyweight type governor is mounted on front face of engine and is driven by the engine timing gear train. Before attempting any governor adjustments, check the operating linkage and remove any binding or lost motion.

All Models

56. **ADJUSTMENT.** To adjust the governor proceed as follows: With engine stopped, place the speed change lever in wide open position and re-

move clevis pin (12—Fig. IH564) from the governor rockshaft arm. Hold the rockshaft arm (13) and the carburetor throttle rod (10) as far toward carburetor as they will go; at which time, pin (12) should slide freely into place. If pin holes are not in alignment, adjust the length of rod (10) until proper register is obtained.

With engine running and speed change lever in wide open position, turn screw (8) on top of governor housing either way as required to obtain the recommended speeds which follow:

Engine High Idle (No Load) rpm
  Series 100-130-140 1575
  Series 200 1815
  Series 230 1980
  Series 240 2200
  Series 404-2404 2300

Engine Rated (Loaded) rpm
  Series 100-130-140 1400
  Series 200 1650
  Series 230 1800
  Series 240-404-2404 2000

Engine Slow Idle rpm
  All Series 425

Belt Pulley High Idle (No Load) rpm
  Series 100-130-140 1301
  Series 200 1499
  Series 230 1636
  Series 240 1817
  Series 404-2404 1043

Belt Pulley Rated (Loaded) rpm
  Series 100-130-140 1157
  Series 200 1363
  Series 230 1487
  Series 240 1652
  Series 404-2404 948

Belt Pulley Slow Idle rpm
  All Series (except 404-2404)... 351
  Series 404-2404 201

PTO High Idle (No Load) rpm
  Series 100-130-140 609
  Series 200 593
  Series 230 @ 1815 engine rpm.. 593
  Series 240 @ 1952 engine rpm.. 637
  Series 404-2404 @ 2300
    engine rpm 623

PTO Rated (Loaded) rpm
  Series 100-130-140 541
  Series 200 539
  Series 230 @ 1650 engine rpm.. 539
  Series 240 @ 1654 engine rpm.. 540
  Series 404-2404 @ 2000
    engine rpm 542

PTO Slow Idle rpm
  Series 100-130-140 164
  Series 200-230-240 139
  Series 404-2404 115

Hunting or unsteady running can be eliminated by removing cap (22) and turning the bumper spring screw (21) in. Do not turn the bumper spring screw in too far as it will interfere with the low idle speed adjustment. CAUTION: Stop the engine before making the bumper spring adjustment.

57. R&R AND OVERHAUL. Remove the magneto or distributor and mark location of slots in the ignition unit drive coupling on governor gear (1—Fig. IH564) in relation to crankcase. This procedure will facilitate reinstallation of governor gear in proper

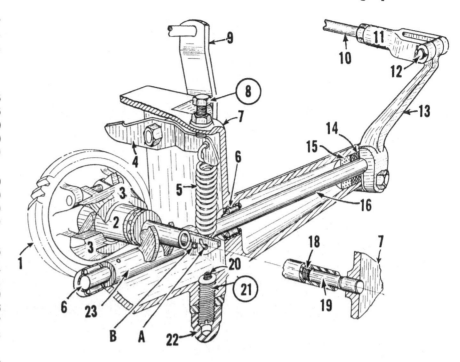

Fig. IH564—Cut-away view of a typical governor assembly. The unit is driven from the engine timing gear train.

1. Ignition governor gear
2. Sleeve and thrust bearing
3. Governor weights
4. Spring lever
5. Governor spring
6. Needle bearings
7. Governor housing
8. Speed adjusting screw
9. Speed change lever
10. Throttle rod
11. Throttle rod clevis (yoke)
12. Clevis pin
13. Rockshaft arm
14. Felt seal
15. Oilite bushing
16. Rockshaft
18. Thrust spring
19. Thrust pin
20. Surge (bumper) spring
21. Surge spring adjusting screw
22. Spring body cap
23. Rockshaft lever
A-B. Spring position holes

timing mesh with camshaft gear if position of crankshaft is not changed while governor is removed. Remove governor speed control rod then on all tractors except series 404 and 2404, unbolt and remove the governor assembly from timing gear cover. On series 404 and 2404, remove radiator assembly before removing governor.

When disassembling, do not bend hooks of governor spring (5). The spring lever (4) should be removed from the speed change lever shaft to remove and install the spring. Upper hook of spring should be inserted in lever shaft so that open end of hook will face the side of governor housing.

Place the lower hook of spring in hole (A) of rockshaft lever.

Two needle bearings (6) and one "Oilite" bushing (15) which support the rockshaft (16) can be renewed when worn. The felt seal (14) which is assembled outside the bushing should not place any drag on the rockshaft.

The bushing in the crankcase which supports the governor and ignition unit drive gear hub may be renewed when governor and ignition unit are off. The I&T recommended clearance of gear hub in bushing is 0.0015 to 0.002.

COOLING SYSTEM

RADIATOR

58. To remove the radiator, first drain cooling system and remove the hood and grille. Disconnect the radiator hoses and unbolt the radiator brace. Remove nuts from the lower support bolts and on **series 404 and** 2404 loosen fan shroud from radiator, then lift radiator from tractor.

On series 404 and 2404 equipped with power steering, also disconnect lines from the oil cooler.

Radiator is of one piece construction with non-detachable top and bottom tanks.

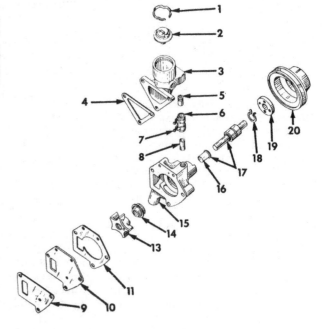

1. Gasket
2. Bracket
3. Thermostat
4. Retainer ring
5. Washer
6. Spindle
7. Fan hub
8. Oil retaining screw
9. Gasket
10. Hub gasket
11. Fan bearing
12. Impeller
13. Nut
14. Gasket
15. Bearing cap
16. Fan
17. Cap screw

Fig. IH565—Exploded view of fan shaft and associated parts used on all models except 140, 240, 404 and 2404.

THERMOSTAT

59. Thermostat is located in the engine water outlet casting and the removal procedure is evident. Standard thermostat for the 100, 130, 200 and 230 series begins to open at 165 deg. F. and is fully open at 195 deg. F. Standard thermostat for 140 and 240 series begins to open at 155 deg. F. and is fully open at 185 deg. F. Standard thermostat for 404 and 2404 series begins to open at 170 deg. F. and is fully open at 192 degrees F. A special (130-150 degree) thermostat is available for use with low boiling point anti-freeze for 100, 130 and 140 series.

FAN
Series 100-130-200-230

60. An exploded view of the fan and hub is shown in Fig. IH565. The procedure for overhauling the unit is evident. The fan bearing should have an I&T suggested clearance of 0.0045-0.005 on its spindle.

After unit is installed on tractor, remove the oil retaining screw (8), turn the unit until the screw hole in hub is in the 9 o'clock position and add oil until it runs out the screw hole. Turn hub until screw hole is down and install the retainer screw when oil stops dripping from the screw hole.

WATER PUMP
Series 100-130-200-230

61. R&R AND OVERHAUL. To remove the water pump, first drain

Fig. IH566—Sectional view of water pump used on all models except 140, 240, 404 and 2404.

A. 3 21/32 inches
B. 0.031-0.041
1. Pulley
2. Snap ring
3. Seal assembly
4. Impeller
5. Gasket
6. Rear body
7. Front body
8. Shaft and bearing assembly
9. Flinger

cooling system and remove the water drain pipe. Remove drive belts from generator and water pump. Loosen the lower radiator hose, remove the pump retaining cap screws and lift water pump from tractor.

To disassemble the pump, remove the cap screws which retain rear body (6—Fig. IH566) to body (7) and separate the two bodies. Remove the bearing retainer wire (2) and using a suitable puller, remove pulley (1). Press the shaft, seal and impeller out of pump body, then press the shaft out of the impeller, and remove seal (3).

Note: The shaft and bearing assembly are available as a preassembled unit only and the unit is factory fitted with a flinger (9).

To assemble the water pump, install the seal (3), insert the shaft and bearing unit and install retaining wire (2). Support the pump drive shaft and press the impeller on the shaft until the clearance between the impeller vanes and rear face of front body is 0.031-0.041. The clearance is shown at (B). Using a new gasket, install rear body to front body. Press the drive pulley on the drive shaft until the distance between rear face of rear body and front face of pulley is $3\frac{21}{32}$ inches. This dimension is shown at (A).

Series 140-240

62. R&R AND OVERHAUL. To remove the water pump, first drain cooling system. Unbolt and remove generator. Disconnect inlet hose and by-pass hose. Note: While by-pass hose can be removed without remov-

Fig. IH566A — Exploded view of the water pump used on series 140, 240, 404 and 1404. The pump shaft and bearing are available as an assembled unit only.

1. Retainer
2. Thermostat
3. Outlet elbow
4. Gasket
5. Nipple
6. By-pass hose
8. Nipple
9. Gasket
10. Plate
11. Gasket
13. Impeller
14. Seal
15. Pump body
16. Flinger
17. Shaft and bearing assembly
18. Snap ring
19. Hub
20. Pulley

ing any other parts, some mechanics prefer to remove hood skirts. Unbolt fan blades and pulley from water pump hub. Let fan blades rest in shroud. Unbolt water pump from engine and withdraw same from left side of tractor.

To disassemble pump, use a suitable puller and remove hub (19—Fig. IH-566A). Remove plate (10) and snap ring (18), then, support pump body and press shaft and bearing assembly (17) from impeller (13) and pump body. Remove seal (14).

The shaft and bearing are available only as a pre-assembled unit. Flinger (16) is available separately. When renewing seal (14), press only on outer diameter.

Reassembly is evident, keeping in mind that vanes of impeller (13) and flange of hub (19) are toward front of tractor. Press impeller and hub on shaft until bottom of chamfer at each end shows. Clearance from face of body to face of impeller hub should be 0.031.

Series 404-2404

63. **R&R AND OVERHAUL.** To remove the water pump, first remove hood and grille and drain the cooling system. Disconnect fan shroud from radiator. Disconnect upper and lower radiator hoses from radiator, then unbolt and remove radiator. Loosen and remove generator belt and the fan blades. Disconnect by-pass hose, then unbolt and remove water pump from engine.

To disassemble pump, use a suitable puller and remove hub (19—Fig. IH-566A). Remove plate (10) and snap ring (18); then, support pump body and press shaft and bearing assembly (17) from impeller (13) and pump body. Remove seal (14).

The shaft and bearing are available only as a pre-assembled unit. Flinger (16) is available separately. When renewing seal (14), press only on outer diameter.

Reassembly is evident, keeping in mind that vanes on impeller (13) and flange of hub (19) are toward front of tractor. Press impeller and hub on shaft until bottom of chamfer at each end shows. Clearance from face of body to face of impeller hub should be 0.031.

IGNITION AND ELECTRICAL SYSTEM

DISTRIBUTOR

65. **INSTALLATION AND TIMING.** Some 100 and 200 models have timing pointer located on the right side of the crankcase front cover and the TDC mark is the third notch on the back flange of the crankshaft fan pulley as shown in Fig. IH567. On other 100 and 200 models, the timing pointer and "DC 1-4" mark are visible through hand hole in bottom of clutch housing as shown in Fig. IH568.

On series 130 and 230 the fan pulley has four notches and the TDC mark is the fourth notch on the back flange of the crankshaft fan pulley as shown in Fig. IH569.

On series 140, 240, 404 and 2404, degree marks are stamped on the crankshaft pulley and the TDC position is indicated by a "DC" mark. See Fig. IH569A.

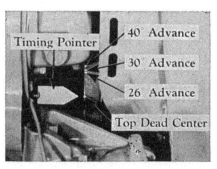

Fig. IH569—Series 130 and 230 tractors have the timing pointer located on the right side of the crankcase front cover. Note that pulley has four notches.

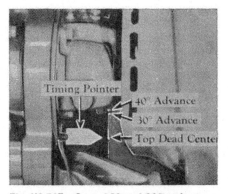

Fig. IH 567—Some 100 and 200 series tractors have the timing pointer located on the right side of the crankcase front cover. Refer also to Fig. IH568.

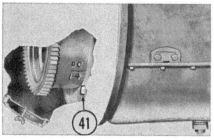

Fig. IH568—Some series 100 and 200 tractors have ignition timing marks which can be viewed through hand hole in bottom of clutch housing. Refer to Fig. IH567 for later construction. Item (41) is pointer on clutch housing cover.

Fig. IH569A—Series 140 and 240 tractors have the timing pointer located on the right side of the crankcase front cover. Crankshaft pulley has degree marks stamped on outer diameter. Series 404 and 2404 are similar.

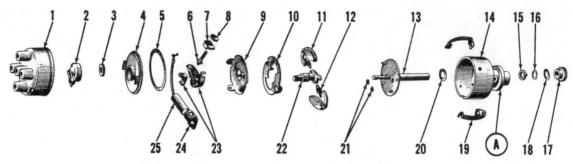

Fig. IH570—Exploded view of typical International Harvester battery ignition distributor. The distributor identification symbol is stamped on the outside diameter of mounting flange (A). Shaft (13) rides directly in unbushed housing (14)—wear is corrected by renewal of one or both parts.

1. Distributor cap	6. Primary terminal screw	11. Governor weight	17. Distributor gear	21. Weight arm spacer
2. Rotor	7. Insulator	12. Governor spring	18. Thrust washer	22. Cam
3. Breaker cover felt seal	8. Insulating washer	13. Distributor shaft	19. Cap retaining spring	23. Breaker contact set
4. Breaker cover	9. Breaker plate	15. Oil seal retainer	and support	24. Condenser clamp
5. Breaker cover gasket	10. Governor weight guard	16. Oil seal	20. Thrust washer	25. Condenser

66. To install the battery ignition unit, first crank engine until No. 1 piston is coming up on compression stroke, then continue cranking until the TDC timing marks previously outlined are in register.

Turn the distributor drive shaft until rotor arm is in the No. 1 firing position and mount the ignition unit on the engine, making certain that lugs on ignition unit engage slots in the drive coupling.

Note: If the driving lugs on the battery ignition unit will not engage the coupling drive slots, when rotor is in No. 1 firing position, it will be necessary to remesh the drive gears as follows: Grasp the distributor drive shaft and pull same outward to disengage the gears. Turn drive shaft until lugs will engage the drive slots, then push drive shaft inward to engage the gears.

Adjust the breaker contact gap to 0.020.

67. To time the battery ignition distributor after same is installed as outlined above, proceed as follows: Loosen the distributor mounting cap screws and retard distributor about 30 degrees by turning distributor assembly in same direction as the cam rotates. Disconnect coil secondary cable from distributor cap and hold free end of cable $\frac{1}{16}$-$\frac{1}{8}$ inch from distributor primary terminal. Advance the distributor by turning the distributor body in opposite direction from cam rotation until a spark occurs at the gap. Tighten the distributor mounting cap screws at this point.

To check distributor timing, position the coil secondary cable as outlined above and crank engine slowly. A spark should occur at the gap when TDC marks are in register. Assemble the spark plug cables to the distributor cap in the proper firing order of 1-3-4-2.

With engine running at high idle, the running timing can be checked with a neon timing light. Refer to the following table when checking running timing.

Series	BTDC
100-200-230-240	30°
130	26°
140 (prior engine No. 73050)	26°
140 (after engine No. 73049)	22°
404-2404	20°

68. **OVERHAUL.** Defects in the battery ignition system may be approximately located by simple tests which can be performed in the field; however, complete ignition system analysis and component unit tests require the use of special testing equipment. All of the distributors have automatic spark advance which is obtained by a centrifugal governor built into the unit.

Identification and advance curve data are as follows:

Series 100-200

Distributor part No. 357 935 R91
Distributor symbol ....J
Number of cylinders................4

*Advanced data (degrees @ rpm)
　Start .... 0 @ 200
　Intermediate ....4.5 @ 400
　Intermediate .... 12 @ 600
　Maximum .... 15 @ 800
Breaker contact gap........ 0.020
Breaker arm
　spring pressure .... 21-25 oz.

Series 130-140 (prior 73050)-230-240

Distributor part No. 366 928 R91
Distributor symbol X
No. Cyls. 4

*Advance data (degrees @ rpm)
　Start .... 5 @ 400
　Intermediate .... 9 @ 600
　Intermediate ....13 @ 800
　Maximum ....15 @ 900
Breaker contact gap........ 0.020
Breaker arm
　spring pressure ....21-25 oz.

Series 140 (after 73049)

Distributor part No......372 712 R91
Distributor symbol ....AB
Number of cylinders................4

*Advance data (degrees @ rpm)
　Start .... 5 @ 400
　Intermediate ....10 @ 600
　Maximum ....11 @ 800
Breaker contact gap............0.020
Breaker arm spring pressure.21-25 oz.

Series 404-2404

Distributor part No. 380 622 R91
Distributor symbol ....AH
No. cylinders 4

*Advance data (degrees @ rpm)
　Start .... 0 @ 200
　Intermediate .... 3 @ 300
　Intermediate .... 6 @ 600
　Maximum ....10 @ 1000
Breaker contact gap...............0.020
Breaker arm spring pressure.21-25 oz.

Distributor part No. 375 423 R91
Distributor symbol ....AE
　No. cylinders 4

*Advance data (degrees @ rpm)
　Start .... 0- 1 @ 200
　Intermediate .... 7- 9 @ 600
　Intermediate ....10-12 @ 800
　Maximum ....12-13 @ 1000
Breaker contact gap............0.020
Breaker arm spring pressure.21-25 oz.

*Advance data are given in distributor degrees and rpm. For crankshaft degrees and rpm, double listed values.

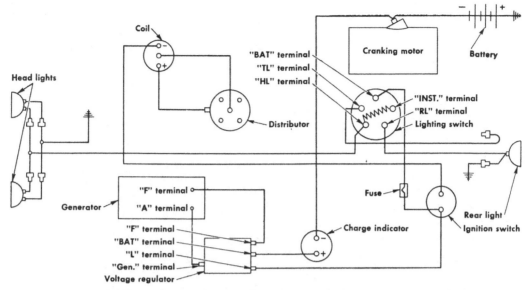

Fig. IH570A—View showing the wiring diagram of the series 140 tractors.

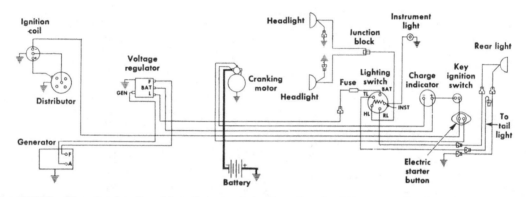

Fig. IH570B—View showing the wiring diagram of the Farmall series 240 tractors. The International series 240 tractors are similar.

MAGNETO
All Models So Equipped

69. INSTALLATION AND TIMING. To install the magneto, first crank engine until No. 1 piston is coming up on compression stroke and the TDC timing marks described in paragraph 65 are in register.

Turn the magneto drive lugs until rotor arm is in the No. 1 firing position and mount ignition unit on engine, making certain that lugs on ignition unit engage slots in drive coupling.

Adjust the breaker contact gap to 0.013.

70. To time the magneto after same is installed as outlined above, proceed as follows: Loosen the magneto mounting bolts and retard the magneto by turning top of magneto in the normal direction of rotation. Crank engine one complete revolution and align timing marks as before. Advance mag-

neto by turning top of magneto opposite to normal rotation until impulse coupling snaps; then, tighten the magneto mounting bolts. To check magneto timing crank engine until No. 1 piston is again coming up on compression stroke and impulse coupling snaps. At this time, the aforementioned timing marks should be in register. Assemble the spark plug cables to the distributor cap in the proper firing order of 1-3-4-2.

Running timing can be checked with a neon light, with engine running faster than the impulse coupling cut-out speed. Impulse coupling lag angle is 35 degrees.

Refer to the following specifications when servicing magnetos:

Model H4
Breaker point gap...............0.013
Breaker arm spring pressure.19-23 oz.
Rotation C
Impulse trips ....TDC
Lag angle ....35°

GENERATOR, REGULATOR AND STARTING MOTOR

71. Delco-Remy electrical units are used. Their applications and specifications are as follows:

Generator 1100501
(Series 100-130-200-230)
Brush spring tension (ounces)....27
Field draw
  Volts 6.0
  Amperes 2.5-2.72
Output
  Hot or Cold...................Hot
  Amperes 16.0-19.0
  Volts 6.9-7.1
  Generator rpm 2500

Generator 1100531
(Series 100-130-200-230)
Brush spring tension (ounces)....16
Field draw
  Volts 6.0
  Amperes 2.5-2.72
Output
  Hot or Cold.................Hot
  Amperes 16.0-19.0

Volts 6.9-7.1
Generator rpm 2500

Generator 1100042
(Series 140-240)
Brush spring tension (ounces)....28
Field draw
Volts 6.0
Amperes 1.85-2.03
Output
Hot or Cold................Cold
Amperes 35.0
Volts 8.0
RPM 2650

Generator 1100409
(Series 404-2404)
Field draw
Volts 12
Amperes 1.58-1.62
Output
Hot or Cold..............Cold
Amperes 20
Volts 14
Generator rpm 2300

Regulator 1118780
(All Series except 404-2404)
Cut-Out Relay
Air gap 0.020
Point gap 0.020
Closing voltage range......5.9-7.0
Adjust to 6.4
Voltage Regulator
Air gap 0.075
Setting (volts) range.....6.6-7.2
Adjust to 6.9
Ground polarity ....Positive

Regulator 1119241E
(Series 404-2404)
Cut-Out Relay
Air gap 0.020
Point gap 0.020
Closing voltage range....11.8-13.5
Adjust to 12.6
Voltage Regulator
Air gap 0.060
Setting (volts) range @
85° F. 14.2-15.2
Adjust to 14.7
Current Regulator
Air gap 0.075
Setting (amperes) range.18.5-21.5
Adjust to 20
Ground polarity ....Negative

Starting Motor 1107093
(All Series Except 240-404-2404)
Volts 6.0
Brush spring tension (ounces) 24-28
No Load Test
Volts 5.67
Amperes 65.0
RPM 5000
Lock Test
Volts 3.37
Amperes 525.0
Torque (ft.-lbs.) 12

Starting Motor 1107174
(Series 240)
Volts 6.0

Brush spring tension (ounces)....24
No Load Test
Volts 5.6
Amperes 80.0
RPM 5500
Lock Test
Volts 3.25
Amperes 550.0
Torque (ft.-lbs.) 11

Starting Motor 1107228
(Series 404-2404)
Volts 12

Brush spring tension (ounces).....35

No Load Test
Volts 10.6
Amps., Min. (includes solenoid).49
Amps., Max. (includes solenoid).76
RPM, Min. 6200
RPM, Max. 9400

Resistance Test
Volts 4.3
Amps., Min. (includes solenoid).270
Amps., Max. (includes solenoid).310

CLUTCH

75. Tractors may be equipped with either an Auburn or Rockford 9-inch clutch.

76. **ADJUST.** Adjustment to compensate for lining wear is accomplished by adjusting the clutch pedal linkage not by adjusting the position of the clutch release levers.

On series 100, 130 and 140, vary length of the clutch actuating rod (18 —Fig. IH572) with nuts (19) until the clutch pedal free travel is 1-1¼ inches for series 100 and 130; 1⅜ inches for series 140.

On series 200, 230, 240, 404 and 2404, vary length of the clutch pedal rod by means of clevis (C—Fig. IH571) until the clutch pedal free travel is 1$\frac{7}{16}$ inches for series 200 and 230; 1$\frac{5}{16}$ inches for series 240, 404 and 2404.

77. **REMOVE AND REINSTALL** The procedure for removing and reinstalling the clutch is evident after splitting the engine from the clutch housing as outlined in the following paragraph:

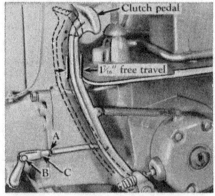

Fig. IH571—Series 200 and 230 clutch pedal free travel is adjusted with clevis (C). Series 240, 404 and 2404 are adjusted in a similar manner.

78. **TRACTOR SPLIT.** To split the clutch housing from the engine, remove hood and on series 200, 230, 240, 404 and 2404, disconnect the steering shaft. On series 100 and 130, remove steering wheel. On models equipped with hydraulic "Touch Control," drain the hydraulic reservoir and remove the hydraulic lines unit. On models equipped with "Hydra-Touch," disconnect hydraulic lines between pump and control valve support. On models equipped with "Draft and Position Control," disconnect hydraulic pressure and return lines at filter end of lines and if tractor has power steering, disconnect oil cooler lines at rear connections located at lower right of steering shaft lower support. On International models of the 404 and 2404 series, also disconnect power steering cylinder lines and the radius rod support bracket from bottom of clutch housing, then remove the exhaust pipe and muffler as a unit. On all tractors except series 404 and 2404, remove oil cup from air cleaner and starting motor. On all models, disconnect fuel lines, wiring, wiring harness and controls from engine and engine accessories. Remove fuel tank front support bolts and block up fuel tank and support. Suport both halves of tractor and remove the clutch housing lower dust cover. On axle type tractors, wedge both sides of front axle between axle and axle support, then remove engine to clutch housing cap screws and separate tractor.

79. **OVERHAUL.** With tractor split as outlined in paragraph 78, refer to the following data when overhauling clutch.

Rockford

Series 100 (prior ser. no. 501-2955)

Series 200 (prior ser. no. 501-2289)

  IH part number.......351 447 R91

  Distance of pressure plate to
    base of clutch cover (inch)..27/32

  Distance from release lever
    face to friction face of
    pressure plate (inches) ....$2\frac{3}{8}$

  Pressure springs

    Number used 6

    Pounds test @ height
      (inches)............175 @ $1\frac{7}{16}$

Series 100 (ser. no 501-2955 and up)—
130-140

Series 200 (ser. no. 501-2289 and up)—
230-240

  IH part number.......353 242 R91

  Distance of pressure plate to
    base of clutch cover (inch)..27/32

  Distance from release lever
    face to friction face of
    pressure plate (inches)...2 23/64

  Pressure springs

    Number used 6

    Pounds test @ height
      (inches)........185-194 @ $1\frac{13}{16}$

Series 404-2404

  IH part number 375 493 R91

  Distance of pressure plate to
    base of clutch cover (inches)..$\frac{27}{32}$

  Distance from release lever
    face to friction face of
    pressure plate (inches)....2 23/64

  Pressure springs

    Number used 6

    Pounds test @ height
      (inches) ....185-194 @ $1\frac{13}{16}$

Auburn

Series 100 (prior ser. no. 501-2955)

Series 200 (prior ser. no. 501-2289)

  IH part number.......351 296 R92

  Distance from release lever
    face to friction face of new
    plate (inches)...........2 23/64

  Thickness of key stock in lieu
    of the new plate (inch).....27/32

  Pressure springs

    Number used 3

    Color ....Brown

    Pounds test @ height
      (inches)............185 @ $1\frac{11}{16}$

Series 100 (ser. no 501-2955 and up)—
130-140

Series 200 (ser. no. 501-2289 and up)—
230-240

  IH part number.......362 138 R91

  Distance from release lever
    face to friction face of new
    plate (inches)...........2 23/64

  Thickness of key stock in lieu
    of the new plate (inch).....27/32

Pressure springs

  Number used 3

  Color ....Gold

  Pounds test @ height
    (inches)..............196 @ $1\frac{25}{32}$

RELEASE BEARING

80. The procedure for removing the clutch release bearing is evident after splitting the engine from the clutch housing as outlined in paragraph 78.

Note: A ball thrust type clutch release bearing must be used on all models equipped with a "Hydra-Creeper" attachment because, when the "Hydra-Creeper" is in use, the clutch release bearing is in continuous operation. The regular "Graphitar" clutch release bearing used on models without "Hydra-Creeper", was not designed for continuous operation and the heat caused by continuous operation would cause the bearing to fail quickly.

CLUTCH SHAFT

81. The procedure for removing the clutch shaft is evident after splitting the clutch housing from the transmission as follows:

Series 100-130-140-200-230

82. Disconnect the "Touch-Control" or "Hydra-Touch" rods, starter rod and governor control rod at their rear attaching points. Disconnect the choke rod, tail light wire and on series 200 and 230, disconnect the steering shaft universal joint. On series 230 disconnect the hydraulic lines from right hand control valve and seat support. Remove the battery and battery box. Remove the instrument panel and steering shaft support attaching cap screws and slide the instrument panel and steering shaft support assembly forward enough to clear the attaching cap screws. On series 100 and 130, disconnect the brake rods and unbolt platform from rear half of tractor. On series 200 and 230, disconnect the clutch rod. On all models, support engine half of tractor and install a roll-

ing floor jack under transmission. Unbolt the transmission from the clutch housing and separate the units.

Series 240

83. Disconnect hydraulic lines from control valve or Tel-A-Depth valve. If so equipped, disconnect follow-up cable and control rod from Tel-A-Depth valve. Disconnect tail light wire and remove loom from clips. Disconnect battery cable and brake pawl rod. Unbolt and remove platforms, then, if so equipped, disconnect rear muffler clamp. Disconnect clutch rod. Support clutch housing and place a rolling floor jack under transmission. Unbolt the transmission from clutch housing and separate the units.

Series 404-2404

84. Drain rear center frame. Place rolling floor jack under rear center frame and take weight. Disconnect Dual Range, or Forward and Reverse, shifter rod, then remove left step plate and shifter lever and pull shifter arm from shifter shaft. Unhook return spring and disconnect clutch control rod. Remove battery housing covers, disconnect battery cables and remove battery. Remove battery tray. Disconnect tail light wires under left side of instrument panel. On right side of tractor, remove the shield over hydraulic lines and the brake lock rod bracket. Unhook brake return springs and remove right step plate. Disconnect hydraulic lines manifold from hydraulic lift reservoir (or power steering flow control valve), disconnect clip from steering shaft support, then disconnect lines at their forward ends and remove lines. If tractor is equipped with underslung exhaust system, remove exhaust pipe and muffler as a unit. Wedge front axle, if axle type tractor. Support front section of tractor, unbolt torque tube from rear center frame and separate tractor.

TRANSMISSION

Series 100-130-140

85. **MAJOR OVERHAUL.** Data for removing and overhauling the various transmission components are as follows:

86. **SHIFTER RAILS AND FORKS.** To remove transmission case cover and shifter rails and forks, remove cap screws retaining cover to transmission case and lift off the cover. Methods of checking and overhauling shifter rails

and forks are conventional. Shifter rails may be removed by spreading the four cap screw locks and removing the cap screws which retain the two shift rail guide brackets to the cover.

87. SPLINE SHAFT. To remove main spline shaft (13—Fig. IH572) proceed as follows: Disconnect transmission case from clutch housing and remove rear cover or power take-off drive housing. Remove transmission top cover. Disconnect transmission drive flange (16) and using a suitable puller, remove flange from spline shaft. Remove front bearing retainer (15) and riveted pin which positions reverse gear (7) on spline shaft. Buck up reverse gear and drive main spline shaft (13) rearward and out of case.

When reinstalling, be sure to insert and rivet the pin in reverse gear (7). If special tool for this work is not available, a bar can be bent to hook over side of case to buck up while peening rivet .

88. COUNTERSHAFT. The integral countershaft and bevel pinion (30 —Fig. IH572) can be removed after removing the main spline shaft, as in paragraph 87, and differential unit as outlined in paragraph 102.

Remove the front bearing retainer (20) and bearing retaining cap screw (21) and washer. Bump shaft rearward and out of transmission case. Remove bearing cage (22) and shims (A) from front of case and save shims for reinstallation.

When reinstalling countershaft, use original number of shims (A) under bearing cage (22). Shims are available in thickness of 0.004, 0.007, 0.015 and 0.030. The fore and aft position of the bevel pinion is controlled by these shims. Check and adjust mesh of bevel pinion and ring gears as outlined in paragraph 101.

89. REVERSE IDLER. To remove the reverse idler gear and shaft, remove main spline shaft as outlined in paragraph 87. Remove the lock screw that retains shaft in position and bump shaft forward.

Replacement bushings are pre-sized to provide 0.002-0.004 clearance between bushing and reverse idler shaft.

Series 200-230-240-404-2404

90. MAJOR OVERHAUL. Data on overhauling the various transmission components are given in the following paragraphs. Before any overhaul work can be accomplished, however, the transmission (rear frame cover on series 200, 230 and 240) cover must

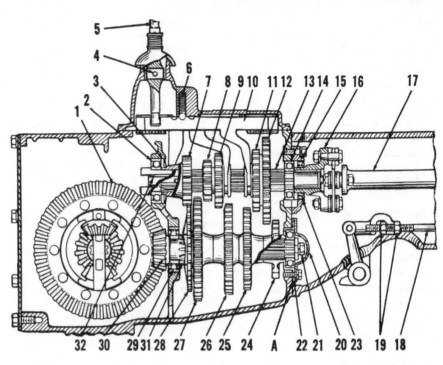

Fig. IH572—Series 100, 130 and 140 transmission and differential side sectional view. The countershaft is integral with the main drive bevel pinion.

A. Shims	8. First speed gear
1. Main drive bevel ring gear	9. Second speed gear
2. Bearing	10. Shifter rails
3. Oil retainer	11. Third speed gear
4. Gear shift swivel pin	12. Fourth speed gear
5. Gear shift lever	13. Spline shaft
6. Detent	14. Bearing
7. Reverse gear	15. Bearing retainer
	16. Transmission drive flange
17. Clutch shaft	24. Fourth speed gear
18. Clutch actuating rod	25. Third speed gear
19. Clutch adjustment	26. Second speed gear
20. Countershaft bearing retainer	27. First speed gear
21. Retaining cap screw	28. Oiler gear
22. Bearing cage	29. Roller bearing
23. Oil seal	30. Countershaft and bevel pinion
	31. Spacer
	32. Bushing

be removed as outlined in the following paragraph.

91. Remove the rear frame top cover on series 200 and 230 as follows: Disconnect the "Touch Control" or "Hydra-Touch", starter, choke and governor control rods at their rear attaching points. Disconnect the tail light wires and the steering shaft universal joint. Remove battery and battery box. Remove the instrument panel attaching cap screws and slide the instrument panel and steering shaft support forward enough to clear the rear frame top cover; then, unbolt and remove the cover.

92. Remove the top cover on series 240 as follows: Disconnect tail light wires and remove loom from its retaining clip. Remove seat and seat support. If so equipped, disconnect Tel-A-Depth control rod and follow-up cable. Disconnect hydraulic lines, then, unbolt and remove top cover.

93. Transmission top cover on series 404 and 2404 is a separate unit and can be removed at any time after

moving seat rearward.

94. SHIFTER RAILS AND FORKS. To remove the shifter rails and forks, first remove the transmission cover as in paragraph 91, 92 or 93. Shifter rails and forks can be removed from the transmission (rear frame) cover by spreading the locks and removing the cap screws which retain the shifter rail guide brackets to the cover. Methods of checking and overhauling the shifter rails and forks are conventional.

95. SPLINE SHAFT. To remove the transmission spline shaft (8 — Fig. IH573), remove transmission cover as in paragraph 91, 92 or 93, detach transmission from clutch housing (torque tube) and remove differential. On series 404 and 2404 without Dual Range transmission or Forward and Reverse Drive, remove the cap screw which retains the coupling (drive) flange to the spline shaft and using a suitable puller, remove the flange. Remove the Woodruff key from front of spline shaft and remove the spline shaft front bearing retainer (11). On series 404 and 2404 with Dual Range

Transmission or Forward and Reverse Drive, unbolt and remove unit then remove shaft and bearing retaining snap rings. Remove rivet which positions the reverse gear (6) on the spline shaft, buck up behind the reverse gear, drive the spline shaft out rear of case and remove the gears from above. The bushing in the rear of the spline shaft can be renewed at this time.

Reinstall the spline shaft and gears by reversing the removal procedure and when installing the reverse spline gear rivet, buck up same while peening.

96. COUNTERSHAFT. The countershaft and integral main drive bevel pinion (22—Fig. IH573) can be removed after removing the spline shaft as per preceding paragraph. Remove the countershaft front bearing cage retainer and nut (13). Drive the shaft out rear of transmission case and remove gears from above. Remove bearing cage (15) and shims (A) from front of case and save the shims for reinstallation. When reinstalling the countershaft, use the original number of shims (A) under the bearing cage. Shims are catalogued as light, medium, heavy, and extra heavy. The fore and aft position of the bevel pinion is controlled by these shims. Check and adjust the mesh of the bevel pinion and ring gears as outlined in paragraph 101.

97. REVERSE IDLER. The reverse idler gear and shaft can be removed after removing the transmission spline shaft as in paragraph 95. Remove the reverse idler shaft lock pin and drive the shaft forward. Replacement bushings in the reverse idler gear are pre-sized.

Fig. IH573—Series 200, 230 and 240 transmission shafts and differential assembly. Shims (A) control fore and aft position of the main drive bevel pinion. Series 404 and 2404 are basically similar but do not include oiler gear (20).

1. Bull pinion and brake shaft	6. Reverse spline gear	10. Gasket
2. Bearing	7. First and second sliding gear	11. Bearing retainer
3. Differential carrier bearing	8. Spline shaft	12. Drive flange
4. Bevel ring gear	9. Third and fourth sliding gear	14. Bearing
5. Bearing		15. Bearing cage
		16. Fourth speed gear

17. Third speed gear
18. Second speed gear
19. First speed gear
20. Oiler gear
21. Roller bearing
22. Bevel pinion and countershaft

Fig. IH573A—View showing mounting of Dual Range transmission. Forward and Reverse unit will be similar as same housing is used.

DUAL RANGE OR FORWARD AND REVERSE UNIT

The Dual Range transmission, or Forward and Reverse unit, is bolted to the front face of the tractor transmission case as shown in Fig. IH573A. Both units are similar in that the same housing, output shaft and countershaft are used. Forward and reverse operation is obtained by the addition of a reverse idler gear assembly on the shifter fork shaft. Tractors which are equipped with pto, have the pto unit drive shaft pinned in a counterbore of the unit input shaft as shown in Fig. IH573B. Service on the Dual Range transmission, or Forward and Reverse unit requires splitting the tractor between torque tube and rear frame (transmission).

98. REMOVE AND REINSTALL. To remove the dual range transmission, or forward and reverse unit, first split rear frame from torque tube as follows: Drain rear frame. On models equipped with underslung exhaust, disconnect exhaust pipe from elbow at exhaust manifold, remove clamps at clutch housing and left brake housing and remove exhaust pipe, muffler and tail pipe assembly. Disconnect dual range shifter rod clevis, then unbolt and remove left step plate and dual range shifter lever. Remove shifter arm. Disconnect clutch control rod. Remove battery housing top cover, left side plate and rear plate, then disconnect battery cables and remove battery. Disconnect tail light wires under left side of instrument panel

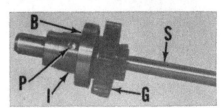

Fig. IH573B—View showing pto drive shaft which is pinned to unit input shaft.

B. Bearing	P. Pin
G. Input gear	S. Pto shaft
I. Input shaft	

housing and remove wires from clips. Remove cap screws from left side of battery tray. Remove hydraulic pump manifold heat shield from right side of torque tube. Disconnect hydraulic pump manifold at front and rear and from center bracket on torque tube and remove lines. Unbolt right step

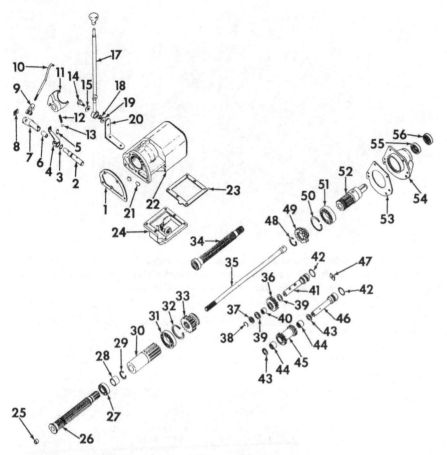

Fig. IH573C—Exploded view of Dual Range transmission and/or Forward and Reverse unit. Items (36), (37), (38), (39) and (40) are used in Forward and Reverse units.

1. Gasket	17. Control lever	31. Ball bearing
2. Shaft	18. Bushing	32. Snap ring
3. "O" ring	20. Bracket	33. Output shaft gear
4. Shifter fork arm	21. Plug	34. Transmission spline
5. Set screw	22. Housing	shaft (no/pto)
6. Spacer	23. Gasket	35. Input shaft (w/pto)
7. Actuating lever	24. Cover	36. Reverse idler gear
8. Locking plate	25. Bushing	37. Spacer
9. Clevis	26. Quill shaft	38. Snap ring
10. Actuating rod	(w/pto)	39. Thrust washer
11. Shifter fork	27. Ball bearing	40. Bearing
12. Poppet spring	28. Spacer (w/pto)	41. Reverse idler shaft
13. Poppet ball	29. Snap ring	42. "O" ring
14. Pivot pin	30. Output shaft	43. Thrust washer

44. Bearing	
45. Spool gear	
46. Countershaft	
47. Lock	
48. Snap ring	
49. Input shaft gear	
50. Snap ring	
51. Ball bearing	
52. Input shaft	
53. Gasket	
54. Bearing cage	
55. Bearing	
56. Oil seal	

99. OVERHAUL. With unit removed as outlined in paragraph 98, disassemble as follows: Disconnect clutch shaft joint from input shaft drive flange, then remove cap screw, washer, gasket, drive flange and Woodruff key from input shaft. Remove bottom cover and shifter fork arm assembly. If necessary, the shifter fork arm and shaft can be removed from bottom cover after set screw in shifter fork arm has been removed. Remove countershaft and shifter fork shaft retaining lock and pull countershaft from case. Remove spool gear and the two nylon thrust washers from case. See Figs. IH573C and IH573D. The two needle bearings can now be removed from spool gear, if necessary.

On models with dual range transmission, pull shifter fork shaft from case and shifter fork and follow the shaft with a socket of proper diameter to fit shifter fork bore to keep detent ball and spring from flying.

On models with forward and reverse unit, lift snap ring which positions the reverse idler gear from its groove and slide it rearward on shifter fork shaft, then pull shaft from reverse idler, spacer, the nylon thrust washers and shifter fork. Use a socket of the proper diameter to fit shifter fork bore and follow shifter shaft as it comes out of shifter fork to prevent detent ball and spring from flying.

To remove input shaft, lift snap ring from groove at inner end of input shaft and slide it rearward. Remove front cage (retainer) retaining cap screws. On models with pto, bump aft end of pto input shaft with a soft hammer and drive input shaft and bearing assembly from case. On models with no pto, use a brass drift inserted through the hollow output shaft. Remove snap ring from bearing bore of cage, bump forward end of input shaft and remove input shaft and bearing assembly from cage. See Fig. IH573B. Bearing and pto drive

Fig. IH573D — View of removed countershaft and spool gear. Note nylon thrust washers (T).

plate and brake lock rod bracket and remove step plate and brake lock assembly. Remove battery tray. Place jack under rear frame and take rear frame weight. Use wooden blocks to

wedge front axle of axle type tractors, or make provision to prevent tipping on tricycle type tractors, then attach hoist to front section of tractor. Unbolt torque tube from rear frame and separate tractor.

Dual range transmission, or forward and reverse unit can now be unbolted and removed from front face of rear frame.

Fig. IH573E — View of Dual Range unit assembled except for bottom cover. Note position of shifter fork. Forward and Reverse units will be similar except that an additional gear will be added to shifter fork shaft.

shaft (if so equipped) can now be removed from input shaft, if necessary. Oil seal and needle bearing can also be removed from cage and procedure for doing so is obvious.

Remove snap ring which retains output shaft bearing in case and bump

output shaft and bearing out toward inside of case. Bearing can now be pressed from output shaft.

Reassembly is the reverse of disassembly; however, keep the following points in mind: Install oil seal in retainer with lip toward inside of case.

When installing needle bearing, press on end which carries identification markings. Shifter fork groove of output gear faces input shaft. Shoulder of input gear is toward front of case. Larger gear of spool gear is toward rear of case. Shoulder of shifter fork is toward rear of case. See Fig. IH-573E.

MAIN DRIVE BEVEL GEARS AND DIFFERENTIAL

RENEW BEVEL GEARS

100. To renew the bevel pinion, follow the procedure for overhaul of the transmission countershaft, paragraph 88 for series 100, 130 and 140 and paragraph 96 for series 200, 230, 240, 404 and 2404. To renew the main drive bevel ring gear, follow the procedure for overhaul of differential, paragraph 102 for series 100, 130 and 140 and paragraph 103 for series 200, 230, 240, 404 and 2404.

After renewing either of the bevel gears, check and adjust, if necessary, the mesh position and backlash as follows:

101. **MESH AND BACKLASH.** The first step in adjusting a new set of bevel gears is to arrange shims (B), as shown in Figs. IH577 and IH578, to provide the desired backlash between the main drive bevel pinion and ring gear. Desired bevel gear backlash is 0.005-0.007 for series 100, 130 and 140, 0.006-0.008 for series 200, 230, 240, 404 and 2404.

The next step is to arrange shims (A), as shown in Figs. IH572 and 573, to provide the proper tooth contact (mesh) position of the bevel gears.

Paint the bevel pinion teeth with Prussian blue or red lead and rotate pinion in normal direction of rotation under zero (light) load and then observe the contact pattern on the tooth surfaces.

The area of **heaviest contact** will be indicated by the coating being **removed** at such points. On the actual pinion the tooth contact areas shown in black on the illustrations will be bright; that is, there will be no blue or red coating on them.

The desired condition is indicated in Fig. IH574, which shows that the paint has been removed from the toe end of the teeth over the distance A to B as shown.

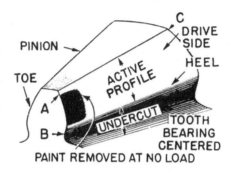

Fig. IH574—Tooth contact should be centered between "A" and "B" (no load). When a heavy load is placed on the gears, the tooth contact will extend from the toe almost to the heel of the tooth.

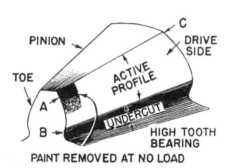

Fig. IH575—High tooth contact at "A" (no load) indicates that the pinion has been set too far in.

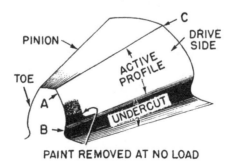

Fig. IH576—Low tooth contact at "B" (no load) indicates that the pinion has been set too far out.

When the heavy contact is concentrated high on the toe at A as shown in Fig. IH575, the pinion should be moved toward front of tractor by adding a shim behind the pinion bearing cage.

When the heavy contact is concentrated low on the toe of the pinion tooth as in Fig. IH576, remove a shim from the pinion bearing cage.

After obtaining desired tooth contact, recheck the backlash and if not within the desired limits as listed, adjust by shifting a shim or shims from behind one differential bearing cage to the other bearing cage until desired backlash is obtained.

Do not expect the contact pattern to extend much farther toward the heel of the pinion than shown in Fig. IH574. The teeth are purposely ground to produce a toe bearing or contact under zero or light load so that when the gear supports deflect and the teeth deform under heavy load conditions, the contact pattern will increase in area along the teeth toward the heel and thus automatically increase the load carrying capacity of the gear.

DIFFERENTIAL AND CARRIER BEARINGS
Series 100-130-140

Differential is of the two pinion type mounted back of a dividing wall in the transmission case. The bevel ring gear is held to the one piece case by rivets. Differential carrier bearings are of the non-adjustable, ball type.

102. **R&R AND OVERHAUL.** To remove the differential assembly, first detach the final drive and differential shaft housing assemblies and the rear cover or power take-off and/or belt pulley attachment from the transmission case. Remove the differential bearing retainers (10—Fig. IH577) and lift the main drive bevel ring gear and differential unit from rear of transmission case. Note: The differential

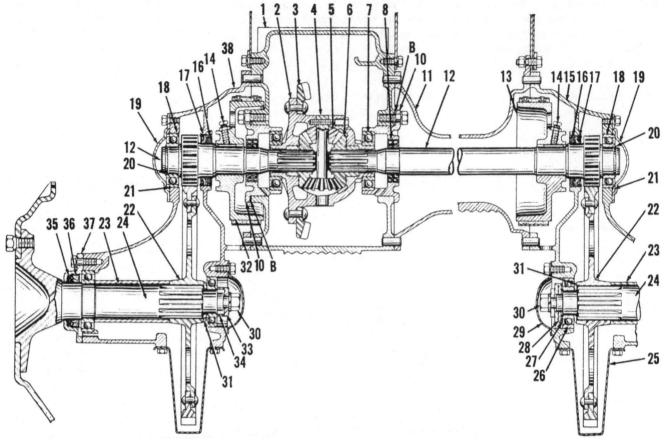

Fig. IH 577—Sectional view of main drive bevel ring gear, differential and final drive assembly which is typical of that used on series 100, 130 and 140 tractors.

B. Shims	6. Differential gear	13. Brake drum	20. Snap ring	26. Bearing	32. Brake drum
1. Transmission case	7. Bearing	14. Set screw	21. Snap ring	27. Spacer	33. Washer
2. Differential case	8. Oil seal	15. Rear axle housing	22. Bull gear	28. Spacer	34. Spacer
3. Main drive bevel	10. Bearing retainer	16. Felt seal	23. Spacer	29. Bearing	35. Felt seal
ring gear	11. Differential shaft	17. Oil seal	24. Rear axle	cap	36. Oil seal
4. Pinion shaft	housing	18. Bearing	25. Rear axle housing	30. Cap screw	37. Bearing retainer
5. Differential pinion	12. Differential shaft	19. Bearing cap	pan	31. Spacer	38. Rear axle housing

carrier bearings (7) can be renewed at this time.

To disassemble the differential, remove the pinion shaft lock pin, then, remove pinion shaft, pinions, side gears and thrust washers.

To reassemble, reverse the disassembly procedure and reinstall unit in transmission case. The differential carrier bearings are non-adjustable, but shims (B) should be arranged to provide a bevel gear backlash of 0.005-0.007.

Refer to paragraph 101 for method of adjusting tooth contact (mesh pattern) of the main drive bevel gears.

Series 200-230-240-404-2404

Differential unit is of the four pinion type enclosed in a two piece case which is mounted behind a dividing wall in the transmission housing as shown in Fig. IH579. The bolts holding the two halves of the differential case together also secure the main drive bevel ring gear to the case. The carrier bearings are adjustable.

103. **R&R AND OVERHAUL.** To remove the differential assembly from tractor, drain rear frame (transmission housing), and proceed as follows:

On series 404 and 2404, remove the pto unit. On all models remove the final drive bull gears as outlined in paragraph 107, both bull pinions and bull pinion shaft bearing cages and lift the differential and bevel ring gear assembly from the housing. Note: The differential carrier bearing cups can be removed from the bull pinion shaft bearing cages and the bearing cones can be removed from the differential case halves at this time.

To disassemble the unit, remove the assembly bolts, and separate the two halves. Differential assembly case bolt nuts are of the self-locking type. The shims (B—Fig. IH578), which are located between the bull pinion shaft bearing cages and the transmission case, control the differential carrier bearing adjustment and the backlash

between the main drive bevel gears. Shims are catalogued as light, medium, heavy, and extra heavy. Adjust the carrier bearings to a slight pre-load (10-18 In.-Lbs. rolling torque); then, transfer shims from one retainer to the other to obtain a bevel gear backlash of 0.006-0.008.

Refer to paragraph 101 for method of adjusting the tooth contact (mesh pattern) of the main drive bevel gears.

FINAL DRIVE

Series 100-130-140

As treated in this section, the final drive is assumed to be the differential shafts (bull pinion shafts) and housings and the rear axle housings containing drive (bull) gears and axles and brake assemblies. Refer to Fig. IH577.

104. **DIFFERENTIAL (BULL PINION) SHAFTS.** To remove the left differential (bull pinion) shaft, sup-

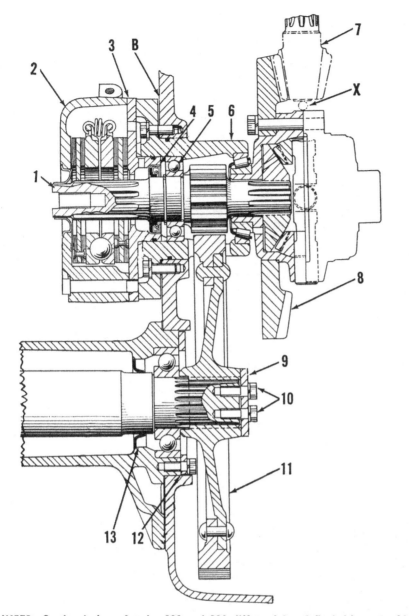

Fig. IH578—Sectional view of series 200 and 230 differential and final drive assembly. Series 240, 404 and 2404 are similar except that axle is retained in bull gear by a snap ring.

B. Shims	4. Oil seal	9. Washer
1. Bull pinion and brake shaft	5. Bearing	10. Cap screws
2. Brake housing	6. Bearing quill	11. Bull gear
3. Bearing retainer	7. Bevel pinion	12. Retainer plate
	8. Bevel ring gear	13. Oil seal

Fig. IH579—Top view of series 200 and 230 transmission case, showing installation of the transmission shafts, differential and bull gears. Series 240, 404 and 2404 are similar.

port rear of tractor, remove the left rear wheel and tire and disconnect the brake rod. Remove the nuts retaining left rear wheel axle housing (38—Fig. IH577) to transmission case and remove rear axle housing assembly. Remove brake drum.

Remove the cap screws retaining differential shaft bearing cap (19) to the housing and bump shaft on inner end and out of housing.

To remove the right differential (bull pinion) shaft, support rear of tractor and remove the right rear wheel and tire. Disconnect the brake

rod and unbolt platform and fender from housing. Remove the cap screws retaining differential shaft housing (11) to transmission case and remove as an assembly with rear wheel axle housing. Separate differential shaft housing from rear wheel axle housing and remove the brake drum. Remove the cap screws retaining differential shaft bearing cover (19) to rear axle housing and bump shaft on inner end and out of housing.

105. R&R WHEEL AXLE SHAFT AND/OR BULL GEAR. To remove either rear wheel axle (24—Fig. IH-577) or drive (bull) gear (22), pro-

ceed as follows: Support rear of tractor and remove the respective rear wheel. Remove outer bearing retainer (37) cap screws, inner bearing cover (29) (or seed plate drive shaft and retainer), cap screw and washer (30) and housing cover pan (25). Bump shaft on inner end and out of inner bearing, spacer, drive gear and rear axle housing.

Rear axle shaft bearings are non-adjustable. Install rear axle outer seal with lip toward the bearing.

Series 200-230-240-404-2404

The final drive consists of two bull pinions and integral brake shafts and two bull gears which are splined to the inner ends of the rear axle shafts. Each axle shaft is carried in a sleeve which is bolted to the side of the transmission case.

106. R&R BULL PINION (DIFFERENTIAL) SHAFTS. To remove either bull pinion and integral shaft, first remove the respective brake unit. Remove the pinion shaft bearing retainer plate (inner brake drum) (3—Fig. IH 578) and withdraw the bull pinion shaft and bearing from the bearing cage.

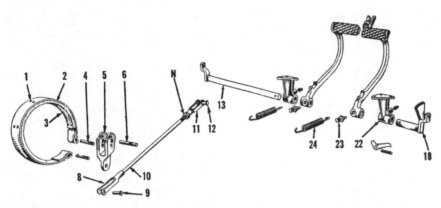

Fig. IH580—Exploded view of series 100, 130 and 140 brake bands, pedal mechanism and associated linkage. Brake adjustment is accomplished by varying the length of rod (10).

1. Rivet	8. Clevis	18. Right brake shaft
2. Band	9. Clevis pin	22. Brake shaft
3. Lining	10. Brake rod	bracket
4. Anchor pin	11. Adjusting clevis	23. Pedal locking
5. Toggle	12. Clevis pin	screw
6. Pivot pin	13. Left brake shaft	24. Return spring

Fig. IH581—Cut-away view of series 200 and 230 disc type brake. Brake for series 240, 404 and 2404 is basically similar.

1. Adjusting rod
2. Spring
3. Nut
4. Nut
5. Nut
6. Housing
7. Lined disc
8. Actuating disc
9. Spring
10. Bearing retainer (inner brake drum)
11. Bearing quill
12. Ball

108. R&R WHEEL AXLE SHAFT. To remove either wheel axle shaft on series 200, 230 and 240, first remove the respective bull gear as outlined in the preceding paragraph and proceed as follows: Remove cap from outer end of the axle sleeve and bump shaft on inner end and out of the sleeve. On series 404 and 2404, remove pto unit, then remove snap ring from inner end of axle. Remove cap from outer end of axle sleeve and bump axle on inner end and out of the sleeve. The axle inner bearing and seal can be renewed after removing the bearing retainer ring halves from the inner end of the sleeve. The procedure for renewing the outer seals and bearing is evident. Install seals with lips of same facing the bull gear.

BRAKES

Series 100-130-140

109. Brakes are of the external band type contracting on drums mounted on the differential (bull pinion) shafts and enclosed by the rear wheel axle housings. Relining of the brakes requires removal of the final drive assemblies which gives free access to the bands.

110. **ADJUST.** Adjustment is accomplished by changing the length of the brake rods (10—Fig. IH580) by means of an adjustable clevis at pedal end of rods. To make the adjustment, jack up rear of tractor, remove pin (12), loosen jam nut (N) and turn clevis (11) either way as required until a slight drag is felt when wheels are turned by hand.

To equalize the brakes, first block-up rear of tractor securely, start engine and shift the transmission into third or fourth gear. With the pedals latched together, apply the brakes. If both wheels stop at the same time the adjustment is O.K. If not, loosen the adjustment slightly on the tight brake until equalization is obtained.

Series 200-230-240-404-2404

111. Brakes are of the double disc, self energizing type, which are splined to the outer ends of the bull pinion and integral brake shaft. The linings are of the moulded type, riveted to the discs on series 200, 230 and 240. Series 404 and 2404 brake discs have bonded linings. Procedure for removing the lined discs is evident after an examination of the unit.

112. **ADJUSTMENT.** To adjust the brakes on series 200 and 230, loosen jam nut (5—Fig. IH581) and turn rod (1) either way as required to obtain

107. **R&R BULL GEAR.** To remove the left bull gear, drain the transmission case and remove the transmission case top cover on series 200, 230 and 240. On series 404 and 2404, remove the hydraulic lift unit and pto unit. On all models, remove the rear wheel and cap screws, or snap ring, retaining bull gear to inner end of the wheel axle shaft. Remove the axle carrier retaining cap screws and withdraw axle and carrier assembly from the transmission case. Lift the bull gear from the case. To remove the right bull gear, follow the same procedure as for the left, except it is usually necessary to remove the belt pulley and power take-off unit and remove the belt pulley oiler tube on series 200, 230 and 240.

the correct pedal free travel of 1½ inches.

On series 240, 404 and 2404 loosen jam nut and turn brake operating ball nut (B—Fig. IH581A) either way as required to obtain the correct pedal free travel of 1½ inches. Secure ball nut by tightening the jam nut (A).

To equalize the brakes, first block-up rear of tractor securely, start engine and shift the transmission into third or fourth gear. With the pedals latched together, apply the brakes. If both wheels stop at the same time, the adjustment is O.K. If not, loosen the adjustment slightly on the tight brake until equalization is obtained. Tighten

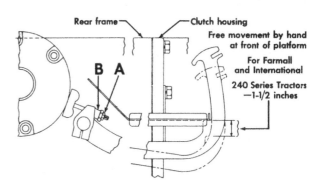

Fig. IH581A — Brakes on series 240, 404 and 2404 tractors are correctly adjusted when pedal has 1½ inches free travel as indicated.

jam nut on series 200 and 230 when adjustment is completed.

If for any reason the brake operating rod has been disassembled on se-

ries 200 and 230, spring (2—Fig. IH-581) should be adjusted to a pre-load dimension of $1\frac{1}{16}$ inches. Adjustment is made by loosening jam nut (4) and turning nut (3).

BELT PULLEY AND POWER TAKE-OFF

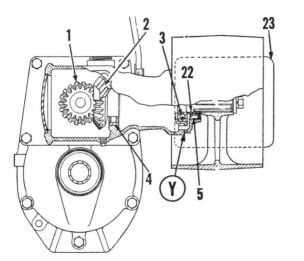

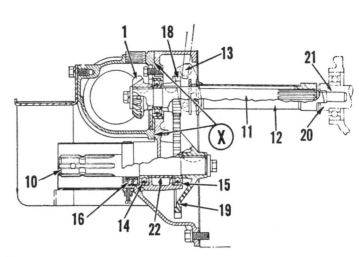

Fig. IH582—Sectional views showing the combination belt pulley and power take-off unit used on all series except 404 and 2404.

X. Shims	5. Pulley shaft oil seal	14. PTO shaft outer bearing	18. Drive shaft pinion	21. Drive shaft bushing
Y. Shims	10. PTO shaft	15. PTO shaft inner bearing	19. PTO shaft drive gear	22. Bearing spacer
1. Bevel drive gear	11. Drive shaft	16. PTO shaft oil seal	20. Transmission spline shaft	23. Belt pulley guard
2. Pulley shaft driven gear	12. Shifter coupling			
3. Pulley shaft outer bearing				
4. Inner bearing				

Series 100-130-140-200-230-240

113. The belt pulley and power take-off assembly shown in Fig. IH582 is driven by shaft (11) and coupling (12) which engage the transmission spline shaft (20). The forward end of drive shaft (11) is supported in pilot bushing (21) in the end of the transmission spline shaft. Disassembly and reassembly procedure is evident after an examination of the unit and reference to Fig. IH582. Oil seals should be installed with lips of same facing inward. Vary the number of shims (x and y) so that heels of gears are in register and backlash is 0.004-0.006 (I&T recommended).

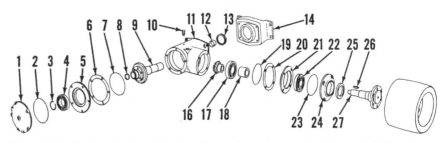

Fig. IH583—Exploded view of belt pulley unit used on series 404 and 2404 when so equipped. Unit is driven from pto shaft.

1. Retainer	7. "O" ring	14. Adapter	21. Bearing cage
2. "O" ring	8. Cup plug	16. Pulley shaft gear	22. Ball bearing
3. Snap ring	9. Drive gear	17. Ball bearing	23. "O" ring
4. Ball bearing	10. Vent plug	18. Spacer	24. Retainer
5. Bearing cage	11. Housing	19. "O" ring	25. Oil seal
6. Shims (.007, .012, .030)	12. Bushing	20. Shims (.007, .012, .030)	26. Woodruff key
	13. Oil seal		27. Pulley shaft

Shims (x) are available in thickness of extra light, 0.004; light, 0.007; medium, 0.015, and heavy, 0.030. Shims (y) are available in thicknesses of light, medium and heavy.

BELT PULLEY

Series 404-2404

114. The belt pulley is mounted on rear of tractor rear frame by means of an adapter (14—Fig. IH583) and receives its drive from the tractor pto shaft.

115. **R&R AND OVERHAUL.** Procedure for removal, disassembly and reassembly is obvious after an examination of the unit and reference to Fig. IH583. Mesh position and backlash of bevel gears (9 and 16) are controlled by shims (6 and 20). Recommended bevel gear backlash is 0.008-0.010. Shims are available in thicknesses of 0.007, 0.012 and 0.030.

POWER TAKE-OFF

116. Pto used on series 404 and 2404 with standard transmission, is the same as the unit on those models with Dual Range or Fast Reverser except for the coupling (29 or 30—Fig. IH-584) on forward end of pto input shaft (25).

Series 404-2404

117. **REMOVE AND REINSTALL.** On models so equipped, hydraulically raise the three point hitch to its upper position and drain the transmission and differential housing. Remove pto safety shield and right fender. Remove the cap screw which retains torsion bar to bracket and pull torsion bar from bracket and draft control bellcrank. Remove retainer from bellcrank pin, then remove pin and bellcrank from draft control rod nut. Unbolt and remove torsion bar bracket from hydraulic lift housing. Disconnect return spring and control rod from pto shifter arm. Unbolt and remove pto unit from rear center frame.

Reinstall by reversing the removal procedure, however, BE SURE splines of coupling (29 or 30) are engaged with its driving shaft before mounting cap screws are tightened. Failure to mate these splines will result in shearing of the coupling pin when mounting cap screws are tightened. The torsion bar is master splined to the draft control bellcrank and when correctly aligned during installation, the hole in anchor plate of torsion bar will mate with dowel of bracket and tor-

Fig. IH584—Exploded view of pto unit used on series 404 and 2404 when so equipped. Coupling (29) used on tractors equipped with Dual Range or Forward and Reverse unit.

2. Shield	11. Spring clip	19. Set screw	27. Snap ring
3. Oil seal	12. Control rod	20. Shifter fork	28. Snap ring
4. Needle bearing	13. Clevis	21. Shifter fork block	29. Coupling
5. Snap ring	14. Control handle	22. Bushing	30. Coupling
6. Cover	15. Bracket	23. Driven gear	31. Housing
7. Dowel pin	16. Gasket	24. Output shaft	32. Needle bearing
8. "O" ring	17. Spring pin	25. Input shaft	33. Ball bearing
9. Cross shaft	18. Brake	26. Ball bearing	34. Snap ring
10. Return spring			35. Lubricating tube

sion bar will slide into position without forcing.

118. **OVERHAUL.** With unit removed as outlined in paragraph 117, remove lubricating tube (35—Fig. IH-584) from housing (31), then unbolt and separate housing (31) and cover (6). Remove bearing snap ring (27), then using a suitable puller, push input shaft (25) and bearing (26) from housing. Use caution not to damage needle bearing (32). If necessary to remove bearing (26) from input shaft, remove roll pin, coupling (29 or 30) and snap ring (28). Bearing can now be removed from shaft. Remove snap ring (5), brake plate (18) and lock pin (17). Remove snap ring (34), then engage the internal splines of gear (23) with splines of output shaft (24) and press output shaft from bearing (33) and gear (23). If necessary, bushing (22) can now be pressed from gear (23). Remove lock screw (19) from shifter fork (20) and pull shifter shaft (9) from fork (20) and cover (6). Oil seal (3) and needle bearing (4) can be removed from cover and the procedure for doing so is obvious.

Clean all parts in a suitable solvent

and inspect all splines and gear teeth for pitting, burrs, fractures or excessive wear. Inspect all bearings and the driven gear bushing for excessive wear or other damage. Inspect output shaft brake for excessive scoring, wear or other damage. Refer to the following table for pertinent specifications. Renew parts as necessary.

Input Shaft

O. D. at front bearing . . 1.1810-1.1815
O. D. at rear bearing . . 0.9995-1.0000

Output Shaft

O. D. at front bearing . . 1.3780-1.3785
O. D. at rear bearing . . 1.4995-1.5000
O. D. at driven gear
    bushing 1.7490-1.7500
O. D. at oil seal 1.3750-1.3780

Driven Gear

Bushing bore
    (installed) . . . . 1.751 -1.755
Bushing operating
    clearance . . . . 0.001 -0.006
Shifter fork slot width 0.500 -0.508

Shifter Fork Blocks

Height . . . . 0.495 -0.498
Width . . . . 0.495 -0.498
Length 0.500

Backlash

Input shaft to
    driven gear 0.0014-0.0043
Driven gear to
    output shaft 0.0024-0.0064
Reassemble by reversing the disassembly procedure and keep the following points in mind. Bearing (4) is installed in cover with forward end flush with forward end of bearing bore. Oil seal (3) is installed with lip facing front of tractor and is pressed into bore until it bottoms. Bushing (22) is installed with aft end flush with aft end of driven gear.

Note: A new driven gear is fitted with a bushing, however, a new bushing is available separately.

Use new "O" ring when installing the shifter fork shaft.

HYDRAULIC LIFT (Touch-Control)

Series 100-130-140-200

The hydraulic power lift ("Touch Control") system is composed of three basic units: The gear type pump, which is driven from the engine timing gear train; a double work cylinder and valves unit which is mounted on the clutch housing (torque tube); and the rockshaft assembly which is bolted to the forward end of the cylinder and valves unit.

Tractors are also available with a remote control system which consists of a control valve unit, break-away coupling and hydraulic cylinder. Hydraulic pressure for the remote control system is supplied by the same pump used with the "Touch Control" system.

NOTE: The maintenance of absolute cleanliness of all parts is of utmost importance in the operation and servicing of the hydraulic system. Of equal importance is the avoidance of nicks or burrs on any of the working parts.

LUBRICATION AND BLEEDING

119. To refill the reservoir and bleed the system, after the system has been drained, proceed as follows: Refill the reservoir with IH "Touch Control" or "Hydra-Touch and Touch-Control" fluid. Start the engine and run at approximately 650 rpm. With the filler plug removed, move the "Touch Control" levers and the remote control lever (when so equipped) back and forth 10-12 times; then, place the "Touch Control" levers in the rear position. Then, with the remote control lever in neutral position and the cylinder retracted, add sufficient fluid to bring the reservoir fluid level to within ½-inch of the filler opening.

CAUTION: If the system is to be flushed, do not use kerosene. It is recommended that only IH "Touch Control" or Hydra-Touch and Touch-Control" fluid be used.

TROUBLE-SHOOTING

120. If the "Touch Control" system does not operate properly, it is advisable to apply a few quick checks to determine which unit is at fault. To check the system, install a Schrader (IH No. SE-1338-A) gage or equivalent in the position shown in Fig. IH585. Load the system with two rear wheel weights attached to a rear mounted rockshaft. Start the engine, operate the lift system and refer to trouble diagnosis chart which follows:

Note: Keep in mind that the gage should show high pressure (1100-1500 psi) only during the movement of rockshaft arms. When the rockshaft arms have completed their travel, the system will return to low pressure (15-40 psi). The fact that this low range of pressure will not be indicated on the SE-1338-A gage is of no importance, since the low range of pressure is not a factor in the trouble shooting procedure.

A. System Will Not Lift Load, High Gage Pressure.
  1. Binding or scored rockshaft bearings.

2. Damaged implement.
3. Defective cylinder head gasket.

B. System Will Not Lift Load, Low Gage Pressure.
  1. No oil in system — check for leaks.
  2. Faulty pump.
  3. Regulator and/or safety valves stuck.
  4. Weak or broken safety valve spring.
  5. Excessive safety valve clearance in its bore.

C. Lift Cycle Slow (More than 2 seconds required to lift load), Low Gage Pressure.
  1. Same as 2, 3, 4 and 5 under condition B.
  2. Orifice plug opening too large.
  3. Internal pipe plugs (clean-out plugs) loose or missing.

D. Gage Shows High Pressure When Control Levers Are Stopped At Either Of The Extreme Positions, Low Pressure When Levers Are Stopped At Intermediate Positions.
  1. Faulty implement. Implement preventing full rockshaft travel.
  2. Stop clips out of adjustment.

E. Gage Shows High Pressure With Control Levers And Rockshaft In Any Position.
  1. Orifice plug stopped up.
  2. Stuck regulator valve.
  3. Internal pipe plugs (clean-out plugs) loose or missing.
  4. Cracked or faulty work cylinder block.

F. Load Oscillates When Engine Is Running, Drops Slowly When Engine Is Stopped.
  1. Oil leaking past work cylinder piston due to:
    a. Check valves not seating in bushings.
    b. Defective seal rings on check valve bushings.
    c. Defective work cylinder piston seal rings.
    d. Defective cylinder head gasket.
    e. Cracked or faulty work cylinder block or head.

Fig. IH585—Schrader (IH No. SE-1338-A) gage installation for checking the "Touch Control" system on series 100, 130, 140 and 200.

G. Same As Condition F, Except Load Stays In Raised Position.
1. Refer to condition F.
2. Work cylinder piston inner seal ring leaking.
3. Leak at weld between piston head and sleeve.
4. Thermal relief valves leaking.

H. Loss Of Oil from System, No External Leaks.
1. Pump drive shaft seal leaking oil into crankcase.

I. Control Levers Creep When Rockshaft Is In Motion.
1. Insufficient friction at control levers.
2. Sprung or bent control rods.
3. Binding control spool—free up walking beam.

J. Gage Pressure Too High (More than 1500 psi).
1. Stuck or binding safety valve.
2. Faulty safety valve spring.

ADJUSTMENT

121. STOP CLIPS. To adjust the "Touch Control" stop clips, proceed as follows: Install a Schrader (IH No. SE-1338A) gage or equivalent in place of the ¼-inch Allen head pipe plug in the pump output side of the manifold rear flange as shown in Fig. IH-585. Place the left hand control lever and its front stop about mid-position of the quadrant, tighten the stop thumb screw and wire the control lever to the stop. Start engine and run at half speed. Slip the right front quadrant stop past the right hand control lever and move the control lever fully forward. At this time, when the rockshaft has completed its stroke, the system should remain on high pressure (1100 - 1500 psi) as shown on the gage. If the system does not remain on high pressure, adjust the length of the right hand control rod until proper condition is obtained. Now, watching the gage, slowly move the right hand control lever toward rear until system returns to low pressure but not far enough to move the rockshaft from its extreme position. Then, scribe a line across the top edge of both rockshaft arm shields. Now again move the right hand control lever toward rear until the outer rockshaft arm has moved back approximately ½-inch as shown by the distance between the scribed lines on the rockshaft arm shields.

Without moving the rockshaft arms from the previously mentioned position, move the right hand stop clip (T — Fig. IH586) forward until the clip firmly contacts the control valve

Fig. IH586 — Adjust the "Touch Control" stop clip (T) with screws (S).

operating lever pin and lock the clip in this position with the two Allen head cap screws (S).

Operate the right hand control lever back and forth several times and check the extreme forward position of the rockshaft to make certain that the scribe lines on the arm shields stop ½-inch apart. Note: It may be necessary to readjust the clip slightly to maintain the ½-inch dimension. Paint over the scribe lines so as not to confuse with new lines made when adjusting the left stop.

Procedure for adjustment of the left hand stop clip is the same as for the right, except the right hand control lever must be wired in its mid-position on the quadrant.

PUMP UNIT

122. REMOVE AND REINSTALL. To remove the gear type hydraulic pump, which is gear driven from the camshaft gear, drain the hydraulic system and remove the hydraulic lines (manifold). Remove the pump attaching cap screws and lift pump from tractor.

Install the hydraulic pump by reversing the removal procedure and bleed the "Touch Control" system as outlined in paragraph 119.

123. OVERHAUL. See Fig. IH587. Overhaul of the hydraulic pump is limited to disassembling, cleaning and installing a gasket and seals package. If parts, other than gaskets and seals are excessively worn or damaged, it will be necessary to renew the complete pump unit.

To renew the pump gaskets and seals, proceed as follows: Mark the pump body and cover so they can be reassembled in the same relative position and remove the drive gear and cover. Mark the exposed end of the driven (idler) gear so it can be installed in the same position and remove the drive and driven gears. Identify the bearings with respect to the pump body and cover so they can be reinstalled in the same position and remove the bearings. The procedure for further disassembly is evident after an examination of the unit.

Check the pump parts for damage or wear. If any of the parts are excessively worn, the pump should be renewed.

Note: Small nicks and/or scratches can be removed from the body cover, shafts and gears, and bearings by using crocus cloth or an oil stone. When dressing the bearing flanges, however, make certain that the flange thickness of both bearings in either pair are identical. Be sure that the small drilled passages in pump body and cover are open and clean.

Lubricate all pump parts with IH "Touch Control" or "Hydra-Touch and Touch-Control" fluid prior to reassembly. Use care when installing the drive shaft to avoid damaging the shaft seal lip. A seal jumper, shown in Fig. IH588 can be made to facilitate installation of the shaft.

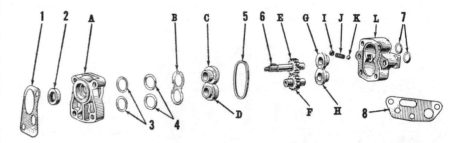

Fig. IH587—Exploded view of the Pesco hydraulic pump. Some models are equipped with a Thompson pump. The two pumps are interchangeable. Numbered items are contents of the gasket and seal package.

A. Cover	F. Idler gear	K. Ball
B. Bearing spring	G. Body bearing	L. Body
C. Cover bearing	H. Body bearing	1. Mounting gasket
D. Cover bearing	I. Plug	2. Seal
E. Driver gear	J. Spring	3. Back-up washers

4. Bearing seal rings
5. Body seal ring
6. Drive gear key
7. Flange seal rings
8. Manifold gasket

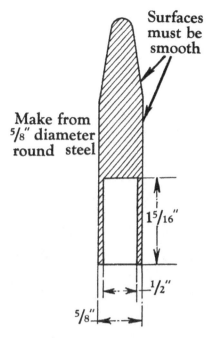

Surfaces must be smooth

Make from 5/8" diameter round steel

1 5/16"

1/2"

5/8"

Fig. IH588—A seal jumper, which can be home-made, facilitates installation of the hydraulic pump drive shaft.

CYLINDER AND VALVES UNIT

124. REMOVE AND REINSTALL. To remove the hydraulic cylinder and valves unit, drain the system and remove the hydraulic lines (manifold). Remove the hood and fuel tank and disconnect wires and control rods from the cylinder. On models so equipped, remove the heat indicator sending unit from the cylinder block. Remove the cap screws retaining the unit to the torque tube and lift the unit from the tractor.

Caution: If for any reason the tractor must be started after the hydraulic cylinder block is removed, the hydraulic pump must be removed. The pump will fail quickly when it is disconnected from the oil supply.

125. OVERHAUL. Overhaul of the "Touch Control" cylinder and valves block is limited to completely disassembling the unit, cleaning and renewing any damaged parts. Refer to Fig. IH589. To disassemble the unit,

proceed as follows: Remove the cylinder head, bottom and top inspection plate and oil strainer. Note the position of all parts so they can be reinstalled in their same position and remove check valves, screw plug, pressure regulator piston, etc., from the block. Turn the block around and remove the rockshaft assembly, dust boots, control valves and pistons. The balance of the disassembly procedure is evident after an examination of the unit.

When reassembling the unit, dip all parts in clean "Touch Control" or "Hydra-Touch and Touch-Control" fluid and lubricate all "O" ring seals with Vaseline or equivalent. When assembling the piston and connecting rod, tighten the rod yoke retaining nut to a torque of 110 ft.-lbs. Tighten the cylinder head retaining cap screws to a torque of 45 ft.-lbs.

Reinstall the unit on the tractor, bleed the system as outlined in paragraph 119 and adjust the unit as in paragraph 121.

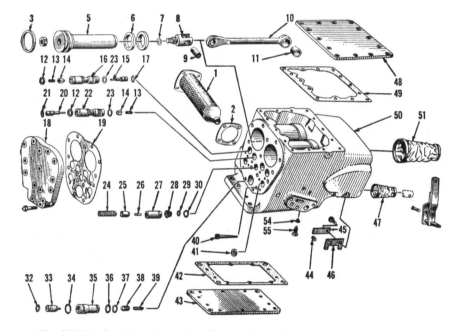

Fig. IH589—Exploded view of the "Touch Control" valves and cylinder assembly.

1. Strainer	16. Rear check valve bushing
2. Strainer gasket	17. Actuator seal
3. Piston head seal	18. Cylinder head
5. Piston	19. Head gasket
6. Piston sleeve seal	20. Check valve actuator
7. Rod yoke seal	21. Actuator seal
8. Rod yoke	22. Front check valve bushing
9. Pin	23. Bushing seal
10. Rod	24. Safety valve spring
11. Rod bushing	25. Safety valve sleeve
12. Check valve bushing seal	26. Safety valve piston
13. Check valve spring	27. Safety valve outer bushing
14. Check valve	28. Safety valve inner bushing
15. Check valve actuator	

29. Inner bushing seal	40. Orifice plug and screen
30. Bushing tension spring	41. Control valve seal
32. Regulator valve piston seal ring	42. Gasket
33. Pressure regulator valve piston	43. Bottom cover
34. Bushing seal ring	44. Plug
35. Regulator valve bushing	45. Stop block
36. Bushing seal ring	46. Operating lever stop
37. Regulator valve tension spring	47. Control valve boot
38. Pressure regulator check valve	48. Top cover
39. Regulator check valve spring	49. Gasket
	50. Cylinder block
	51. Piston sleeve boot
	54. Relief valve screen
	55. Relief valve

Fig. IH591—Remote control valve parts.

6. Control valve lever	14. Control valve spring
7. Lever guide	15. Snap ring
8. Lever post	16. Bottom cap gasket
9. Valve boot	17. Bottom cap
10. Oil seal	22. Manifold flange seals
12. Valve body	
13. Washer	

HYDRAULIC LIFT (Hydra-Touch)

NOTE: The maintenance of absolute cleanliness of all parts is of utmost importance in the operation and servicing of the hydraulic system. Of equal importance is the avoidance of nicks or burrs on any of the working parts.

LUBRICATION
Series 230-240

126. It is recommended that only IH "Touch-Control" or "Hydra-Touch and Touch-Control" fluid be used in the hydraulic system and the reservoir fluid level should be maintained at ½-inch below level of filler plug hole. Whenever the hydraulic lines have been disconnected, reconnect the lines. fill the reservoir and with the reservoir filler plug removed, cycle the system several times to bleed air from the system; then, refill reservoir to ½-inch below level of filler plug hole and install the filler plug.

TESTING
Series 230-240

127. The unit construction of the Hydra-Touch system permits removing and overhauling any component of the system without disassembling the others. However, before removing a suspected faulty unit, it is advisable to make a systematic check of the complete system to make certain which unit (or units) are at fault. To make such a check, use IH test fixture No. SP-121-A (Fig. IH592) or equivalent and proceed as follows:

NOTE: High pressure (up to 1200-1500 psi) in the system is normal only when one or more of the control valves is in either lift or drop position. When the control valve (or valves) are returned to neutral, the pressure regulator valve is automatically opened and the system operating pressure is returned to a low, by-pass pressure of approximately 30-60 psi. The fact that the SE-1338-A gage, which is part of the SP-121-A test fixture, will not register this low pressure is of no importance since this low, by-pass pressure is not a factor in the test procedure.

Should improper adjustment, overload or a malfunctioning pressure regulator valve prevent the system from returning to low pressure, continued operation will cause a rapid temperature rise in the hydraulic fluid, damaging "O" rings and seals. Excessively high temperatures will cause discoloration of the paint on the hydraulic pump and manifold.

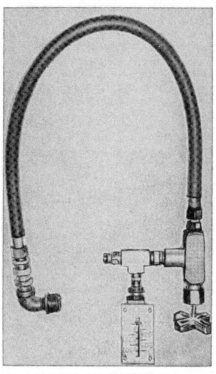

Fig. IH592—IH text fixture No. SP-121-A used for testing and trouble-shooting the Hydra-Touch system.

Before proceeding with the test, first make certain that the reservoir is properly filled with fluid. If not, add sufficient fluid to bring the fluid level to ½-inch below level of filler plug hole. If necessary to add more than one quart, the pump drive shaft seal may be damaged and fluid may be leaking into the engine crankcase.

Remove the pipe plug from the regulator-safety valve block and install the SP-121-A test fixture. Hose from test fixture is connected to the reservoir filler opening. Be sure the test fixture shut-off valve is open; then, with all hydraulic cylinders disconnected from the system and with all of the hydraulic control valves in neutral position, start engine and run until the hydraulic fluid is at normal operating temperature.

Step 1. Advance the engine speed to ¾ throttle, move one of the Hydra-Touch control valve levers to lift position and while observing the pressure gage, rapidly close the shut-off valve until the control valve unlatches and returns to neutral position. The control valve should unlatch and return to neutral when the pressure gage reads 900-1200 psi.

NOTE: The shut-off valve must be closed with sufficient speed to give a rapid pressure rise, since a slow rise in pressure may not actuate the unlatching mechanism.

When the control valve unlatches and returns to neutral, the pressure gage should drop to its minimum reading, indicating the pressure regulator valve is functioning properly and returning the system to low pressure.

Step 2. Re-open the test fixture shut-off valve and move the same control valve lever (as used in step 1) to the drop position. Make certain, however, that the valve is set on "D" for a double acting cylinder. Rapidly close the shut-off valve until the control valve unlatches and returns to neutral. Here again, the control valve should unlatch at 900-1200 psi, from which the gage should drop to its minimum reading, indicating that the system is returned to low pressure.

Step 3. Using the same control valve as in steps 1 and 2, hold the control valve lever in the lift position, close the shut-off valve and observe the maximum pressure gage reading which should be 1200-1500 psi. Release the control valve lever and open the shut-off valve just as soon as the maximum pressure reading is noted. Continued operation at high pressure will cause rapid rise in the fluid temperature and damage to the system "O" rings and seals.

Step 4. If the tractor is equipped with dual or triple (240) control valves, repeat steps 1, 2 and 3 for the other control valve.

Step 5. Stop the tractor engine and check all connections for evidence of leakage. Also check the reservoir fluid level and engine crankcase oil for evidence of internal fluid leakage around the pump shaft seal.

Step 6. Attach all hydraulic cylinders to the system and load the cylinders with their attached implements. Check each cylinder in turn as follows: With engine running at ¾ speed, control valve levers in neutral and the test fixture shut-off valve closed, operate each cylinder in turn and observe the pressure gage reading. Pressures exceeding the control valve unlatching range of 900-1200 psi indicates there is trouble in the cylinder, attaching linkage, couplings and/or attached implements.

Step 7. If the hydraulic system does not function as outlined in steps 1, 2, 3, 4, 5 or 6, refer to the trouble-shooting paragraph 128 and locate points requiring further checking.

TROUBLE-SHOOTING

Series 230-240

128. The following trouble-shooting chart lists troubles which may be encountered in the operation and servicing of the hydraulic power lift system. The procedure for correcting many of the causes of trouble is obvious. For those remedies which are not so obvious, refer to the appropriate subsequent paragraphs.

A. System unable to lift load, gage shows high pressure (900 psi or higher). Could be caused by:
1. System is overloaded
2. Damaged hydraulic cylinder
3. Implement damaged in a manner to restrict free movement
4. Interference restricting movement of cylinder or implement
5. Hose couplings not completely coupled

B. System unable to lift load, gage shows little or no pressure. Could be caused by:
1. Loss of fluid
2. Pump failure
3. Faulty pressure regulator valve
4. Failure of safety valve to close
5. Leakage past cylinder piston seals
6. Plugged pump intake screen

C. System lifts load slowly, gage shows low pressure. Could be caused by:
1. Pump failure
2. Plugged pump intake screen
3. Faulty pressure regulator valve
4. Failure of safety valve to close
5. Pressure regulator orifice enlarged or loose in block

D. With all control valves in neutral, gage shows high pressure. Could be caused by:
1. Pressure regulator piston stuck in its bore
2. Regulator orifice plugged

E. Loss of hydraulic fluid, no external leakage. Could be caused by:
1. Failure of the pump drive shaft seal

F. Operating pressure exceeds 1500 psi. Could be caused by:
1. Safety valve piston stuck in its bore
2. Failure of safety valve spring

G. Control valve will not latch in either lift or drop position. Could be caused by:
1. Broken garter spring
2. Orifice in upper unlatching piston plugged
3. Unlatching valve leakage

H. Control valve cannot be readily moved from neutral. Could be caused by:
1. Control valve retaining bolts too tight, causing valves to bind in valve bodies.
2. Control valve linkage binding
3. Orifice in upper unlatching piston plugged
4. Scored control valve and body

I. Control valve unlatches before cylinder movement is completed, gage shows high pressure. Could be caused by:
1. System is overloaded
2. Damaged hydraulic cylinder
3. Implement damaged in a manner to restrict free movement
4. Interference restricting movement of cylinder or implement
5. Hose couplings not completely coupled

J. Control valve unlatches before cylinder movement is completed, gage shows low pressure. Could be caused by:
1. Weak unlatching valve spring
2. Unlatching valve leakage

K. Control valve unlatches from lift position but not from drop position. Could be caused by:
1. Valve set for single acting cylinder, wherein unlatching valve is inoperative in drop position
2. Channel between upper and lower unlatching valve is plugged

L. Control valve will not center itself in neutral position. Could be caused by:
1. Control valve gang retaining bolts and nuts too tight, causing valves to bind in valve bodies
2. Control valve linkage binding
3. Scored control valve and body
4. Centering spring weak or broken
5. Unlatching pistons restricted in movement

M. Control valve will not automatically unlatch from either lift or drop position. Could be caused by:
1. Loss of fluid
2. Pump failure
3. Faulty pressure regulator valve
4. Plugged channels in control valve which lead to unlatching valve
5. Leakage past the unlatching pistons
6. Loose upper unlatching piston retainer

N. High noise level in operation of pump. Could be caused by:
1. Insufficient fluid in reservoir
2. Pump manifold tubes contacting some foreign part of tractor
3. Plugged intake screen

O. Cylinder will not support load. Could be caused by:
1. External leakage from cylinder, hoses or connections
2. Leakage past cylinder piston rings
3. Internal leaks in control valve

PUMP

Series 230-240

129. Hydraulic pump is same as pump used on series 100, 130, 140 and 200. For information pertaining to same, refer to paragraphs 122 and 123.

CONTROL VALVES

Series 230-240

130. Tractors may be equipped with either a single, dual or triple (240) valve system. All valves are similar except that the control valve lever shaft and yoke assembly may be reversed depending on the position valve is mounted. Therefore, this section will cover the overhaul of only one valve.

The procedure for removing the control valves is evident. Be sure to mark the relative position of the small lever with respect to the shaft (29—Fig. IH593) to insure correct assembly.

Thoroughly clean the removed control valve unit in a suitable solvent, refer to Fig. IH593 and proceed as follows:

Remove body cover (23) and gasket (26). Drift out roll pin (22) and re-

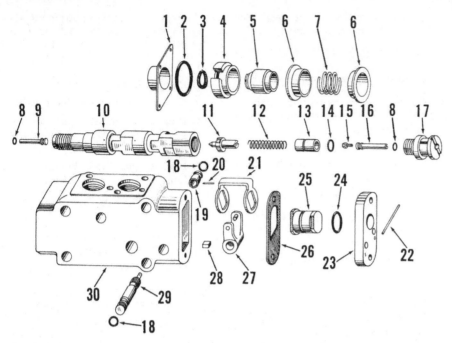

Fig. IH593—Exploded view of a Hydra-Touch control valve. Tractors may be equipped with a single, dual or triple valve system.

1. Valve cap	8. Seal ring	15. Control valve orifice plug and screen
2. Seal ring	9. Lower unlatching piston	16. Upper unlatching piston
3. Garter spring	10. Control valve spool	17. Upper unlatching piston retainer
4. Garter spring sleeve	11. Unlatching valve	18. Seal ring
5. Lower unlatching piston retainer	12. Unlatching valve spring	19. Bushing
6. Control valve centering spring retainer	13. Unlatching piston guide	20. Roll pin
7. Control valve centering spring	14. Seal ring	21. Guide
		22. Roll pin
		23. Body cover
		24. Seal ring
		25. Indexing bushing
		26. Gasket
		27. Yoke
		28. Body plug
		29. Control valve lever shaft
		30. Control valve body

move indexing bushing (25). Remove roll pin (20) and withdraw lever shaft (29), bushing (19) and yoke (27). Lift out the valve guide (21) and body plug (28). Remove cap (1), garter spring sleeve (4) and garter spring (3). Push the control valve spool assembly from the valve body. Clamp the spool in a soft jawed vise and remove the upper retainer (17); then withdraw the unlatching valve (11) and its spring (12). Remove the unlatching piston (16) and guide (13) from the upper retainer. Unscrew the orifice plug (15) from piston (16). Turn the spool over in the vise and remove the lower retainer (5), retainer (6), centering spring (7) and lower unlatching piston (9).

The unlatching valve spring (12) should have a free length of 1⅝ inches and should test 12 lbs. at ⅞-inch. The garter spring (3) should be renewed if outer diameter shows excessive wear. Original clearance of valve spool (10) in body (30) is 0.0004-0.0007. Renew the matched spool and body units if clearance is excessive or if either part shows evidence of scoring or galling. Thoroughly clean channels and the small bores in the valve spool (10). Inspect the unlatching valve (11) and

its seat in the control valve spool for damage. If seat in spool (10) is destroyed, renew the spool and valve. If the seat is not damaged beyond repair, recondition the seat as follows: Place a new valve (11) on the seat and using a piece of ⅜-inch copper tubing slipped over the valve stem, rotate the valve on its seat and lightly tap the tubing with a hammer four or five times. The new valve used for reseating purposes must be discarded and another new valve used for assembly. Width of valve seat should be as narrow as possible to insure a good seal.

Clean the orifice plug and screen (15) with compressed air and make certain that bore in upper piston (16) as well as the passages in body (30) are open and clean.

When reassembling, dip all parts in clean hydraulic fluid, then use new "O" ring seals and gaskets and reverse the disassembly procedure. Special bullet tool No. ED-3396 will facilitate installation of the garter spring (3) and spring sleeve (4). Note: Garter springs in later control valves are of two-piece construction but are interchangeable with early springs. When

installing guide (21), make certain that strap of guide is toward same side of body as shown in Fig. IH593. Install yoke (27) with serrated end of yoke bore toward same side of body as bushing (19). Insert shaft (29) with pin hole up; then install bushing (19) and roll pin (20). Install body cover (23) so that indexing bushing (25) engages guide (21). Install the small lever on outer serrations of shaft (29) in its original position. Note: On multiple valve systems, the relative position of the lever with respect to shaft (29) must be the same on all valves.

When installing the control valve, tighten the retaining bolts to a torque of 25 ft.-lbs. Over-tightening will result in distorted body and binding valve spool.

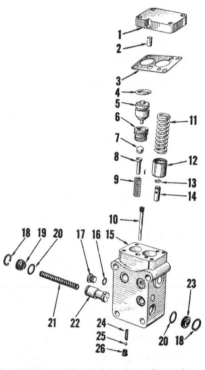

Fig. IH594 — Exploded view of regulator and safety valve unit used on models equipped with power steering and Hydra-Touch. On those models without power steering the unit is same except that flow control valve (numbers 16 through 26) is not used.

1. Cover	14. Safety valve
2. Dowel	15. Housing
3. Gasket	16. "O" ring
4. "O" ring	17. Plug
5. Piston, regulator valve	18. Snap ring
6. Seat	19. Spring retainer
7. Steel ball 9/16"	20. "O" ring
8. Ball rider	21. Spring
9. Spring	22. Flow control valve
10. Plug and screen	23. Valve retainer
11. Spring	24. Spring
12. Spring retainer	25. Steel ball 3/16"
13. Snap ring	26. Seat

REGULATOR AND SAFETY VALVE BLOCK

Series 230-240

131. On models with power steering the regulator and safety valve block is fitted with a flow control valve composed of items (16 through 26—Fig. IH594). On models without power steering, the unit is similar except the flow control valve is not used.

The procedure for removing the valve block is evident.

Thoroughly clean the removed control valve unit in a suitable solvent, refer to Fig. IH594 and proceed as follows:

Remove two diagonally opposite cap screws retaining the cover (1) to the body and insert in their place 2-inch long cap screws and tighten them finger tight. Remove the two remaining short cap screws; then relieve the pressure of the safety valve spring gradually by alternately unscrewing the 2-inch long screws. Remove cover (1) and gasket (3). Withdraw the safety valve spring (11), spring retainer (12) and safety valve (14). Remove the regulator valve piston (5), unscrew the valve seat (6) and remove the ball (7), rider (8) and spring (9). Unscrew and remove the orifice plug and screen (10). On models with power steering, remove snap rings (18) and withdraw the flow control valve spring and valve. Remove the auxiliary safety valve (24, 25 and 26).

Inspect ball valve (7) and valve seat (6) for damaged seating surfaces. Piston (5) and safety valve (14) must be free of nicks or burrs and must not bind in the block bores. Clean the orifice plug and screen (10) with compressed air and make certain that bores in block (15) are open and clean.

When reassembling, dip all parts in clean hydraulic fluid, use new "O" ring seals and gaskets and reverse the disassembly procedure. Be sure that all openings in block, gasket (3) and cover (1) are aligned.

TEL-A-DEPTH SYSTEM

Series 240

A Tel-A-Depth hydraulic system is available on series 240 tractors equipped with an implement hitch. The system provides a means whereby the implement returns to the previously determined working depth which has been selected on the control quadrant. An adjustable stop on the quadrant allows the operator to select any desired working depth and to return to the same depth by moving control lever against stop. The stop can also be by-passed should the operator desire.

When a tractor is equipped with the Tel-A-Depth system, the Tel-A-Depth valve replaces the Hydra-Touch control valve used for rear implements.

Refer to Fig. IH595 for schematic view of the Tel-A-Depth system.

SYSTEM ADJUSTMENT

132. Check hydraulic fluid reservoir and fill to proper level, if necessary. Start engine and cycle system at least ten times; then, place hand control lever in the forward position. The system should go off pressure and the distance between center of cylinder rod pin and face of cylinder should be 1¾-1⅞ inches as shown in Fig. IH-595A. Now move the hand control lever to the rearward position. The system should again go off pressure and the distance between center of cylinder rod pin and face of cylinder should be 9⅜-9½ inches as shown in Fig. IH595B.

If above conditions are not met, adjust the system as follows: Start tractor engine and place hand control lever in its forward position on the quadrant. Disconnect the ball joint from the valve follow-up lever and fully retract (collapse) hitch cylinder by moving the follow-up lever rearward. Shut off engine and be sure cylinder remains retracted. Loosen lock nut

Fig. IH595A—With hand control lever in the forward position, system should be off pressure and distance between face of cylinder and centerline of cylinder rod pin should be 1¾-1⅞ inches.

1. Compensating lever	4. Scale
2. Ball joint	5. Piston rod
3. Compensating rod	6. Gear box

Fig. IH595B—With hand control lever in the rearward position, system should be off pressure and distance between face of cylinder and center line of cylinder rod pin should be 9⅜-9½ inches. Refer to Fig. IH595A for legend.

under ball joint on compensating rod (3—Fig. IH595A); then, disconnect ball joint (2) from compensating lever (1). Move compensating lever downward to be sure that rockshaft drive unit slip clutch is engaged. If

Fig. IH595 — Schematic view of Tel-A-Depth system. Lever (1) opens valve (2) which actuates cylinder (3) to raise or lower implement. When movement of lever (1) is stopped, the follow-up cable attached to rockshaft closes valve (2).

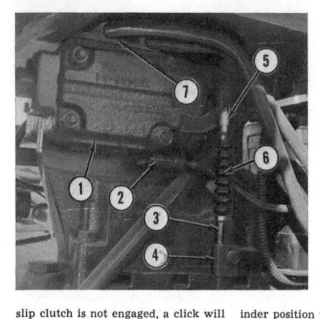

Fig. IH595D — Tel-A-Depth valve mounts to control valve support as shown. On multi-valve systems, a Hydra-Touch control valve may mount on outer side of Tel-A-Depth valve.

1. Tel-A-Depth valve
2. Ball joint
3. Cable assembly
4. Pivot clamp
5. Ball joint
6. Rubber boot
7. Manifold

Fig. IH597—View showing shafts and gears removed from Tel-A-Depth valve.

slip clutch is not engaged, a click will be heard as compensating lever is moved downward and the clutch engages. NOTE: Slip clutch should engage every 90 degrees. If clutch does not engage, on the downward movement of the compensating lever, move lever in the opposite (upward) direction until it does engage. Pull upward on compensating lever and pull follow-up cable into conduit until approximately ⅜-inch of cable is left exposed between end of conduit and back of lock nut at valve end of cable. Hold the compensating lever in this position to maintain the ⅜-inch dimension, then, adjust length of compensating rod so that ball joint can be engaged on ball of compensating lever without binding. In some cases, it may be necessary to reposition the compensating lever on the splined shaft which extends from the rockshaft drive unit (6—Fig. IH595A). When this occurs, mark the position of lever on shaft prior to removal of lever and re-establish the previously mentioned ⅜-inch dimension after lever is reinstalled. Tighten ball joint lock nut on compensating rod and, if necessary, the cap screw on compensating lever; then, reconnect ball joint to follow-up lever on valve.

Start tractor engine, cycle system several times and place hand control lever in the forward position. If system remains on pressure with hand control forward and cylinder retracted (collapsed), loosen conduit pivot clip and move the conduit back and forth until an off pressure position is found and the distance between center of cylinder pin and cylinder face is 1¾-1⅞ inches. Tighten pivot clip. Cycle system several times and recheck cyl-

inder position with hand lever in forward position. If additional adjustment is needed to maintain the 1¾-1⅞ inches cylinder measurement, the quadrant limit stop can be adjusted until proper cylinder measurement is obtained.

NOTE: The ⅜-inch exposed cable measurement may not be maintained after making above adjustments.

With engine running, place hand control in rearward position, check to see that system is off pressure and that the distance between center of cylinder pin and face of cylinder is 9⅜-9½ inches. Adjust the quadrant stop to obtain the proper cylinder measurement.

TEL-A-DEPTH CONTROL VALVE
Series 240

133. **R&R AND OVERHAUL.** Refer to Fig. IH595D and disconnect the ball joints from control valve input and follow-up levers. Remove the nuts from the through bolts which hold ganged valves to valve support and if necessary, remove the outer "Hydra-Touch" control valve. Withdraw through bolts from right side of trac-

tor until Tel-A-Depth valve is free; then, pull valve downward and remove manifold from top side of valve. Use caution during this operation not to deform manifold.

With valve removed, disassembly is as follows: Remove valve cover (18—Fig. IH596). Valve spool can be removed for cleaning and/or inspection by removing pin which retains spool to yoke. If spool and/or valve body are damaged, renew the valve assembly as spool and body are mated parts. Remove gears and shafts by driving roll pins from hubs of control lever and follow-up shaft gears. Unscrew yoke from link and separate yoke, idler gear and link. Refer to Fig. IH597.

NOTE: If valve is the early type, without markings, pay close attention to the position of gear yoke, input lever and follow-up lever. Place correlation marks on shafts and levers prior to removing the levers.

On later valves, the idler gear has two index marks 180 degrees apart on back face of gear and the input and follow-up gears have one index mark on the outside diameters. In addition, the input and follow-up levers and their shafts have index marks to insure correct reassembly. Refer to Fig. IH598.

Centering spring and spring cup can be removed after removing cap (2—Fig. IH596)

Inspect all parts for damage and/or wear. If input shaft, input gear, follow-up shaft or follow-up gear are to be renewed, it will be necessary to renew complete valve assembly as the above parts are not catalogued separately.

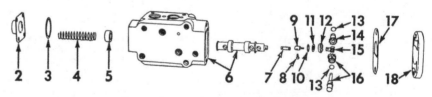

Fig. IH596—Exploded view of Tel-A-Depth valve.

2. Cap
3. "O" ring
4. Centering spring
5. Spring cup
6. Valve body and spool
7. Spool link
8. Pin
9. Link yoke
10. Washer
11. Shim (light and heavy)
12. Idler gear
13. "O" ring
14. Control shaft and gear
15. Eyebolt shaft
16. Follow-up shaft and gear
17. Gasket
18. Cover

Reassembly is the reverse of disassembly, however, it is recommended that new "O" rings be used. Adjust yoke until idler gear is snug yet will turn freely. On the early unmarked valves be sure parts are reassembled in the same position as they were originally. Axis of yoke and gears must be parallel to cover surface. On later marked valves, mate all index marks. Refer to Fig. IH599 for an assembled view.

If necessary, readjust system as outlined in paragraph 132.

ROCKSHAFT DRIVE UNIT

Series 240

134. R&R AND OVERHAUL. Remove conduit shield. Disconnect ball joint from valve follow-up lever and remove pivot clip retaining cap screw. Remove conduit from retaining clips located along tractor rear frame. Disconnect compensating rod from compensating lever and remove lever. Unbolt rockshaft drive unit from rockshaft drive side plate and pull unit rearward. Remove ball joint and rubber boot from valve end of cable assembly. Loosen conduit retaining nut from drive unit and remove cable and conduit by turning cable counterclockwise. Pull cable from conduit. Remove cover from drive unit, which in some cases will require removal of staking. Remove parts from drive unit as shown in Fig. IH600.

Clean all parts and inspect same for excessive wear and/or damage.

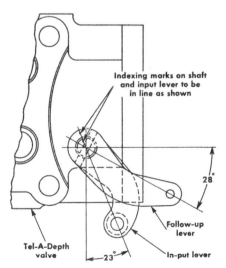

Fig. IH598 — When reassembling later model valves, align scribe marks as shown. Use the above illustration as a guide when reassembling the earlier, unmarked valves.

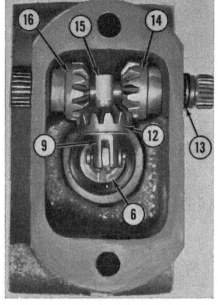

Fig. IH599 — View of Tel-A-Depth valve with end cover removed. Note that shaft on right hand side is partially removed to show "O" ring.

6. Spool	15. Eyebolt (idler)
9. Yoke	shaft
12. Idler gear	16. Follow-up shaft
13. "O" ring	gear
14. Control lever gear	

NOTE: Early units had gears made of powdered metal while later gears are steel. When renewing planet gear or input gear, order IH service package number 373 623 R91. Other parts are available as service items.

With cable and conduit cleaned, lubricate cable with Lubriplate (105-V) or equivalent, and install cable in conduit. Temporarily install ball joint on threaded end of cable and attach a low reading (1-10 lbs.) spring scale to same. Measure effort required to move cable through conduit and if effort exceeds four pounds, renew cable and conduit assembly.

Lubricate all drive unit parts with Lubriplate (105-V) or equivalent, and proceed as follows: Place drive wheel in housing and position same so cable

anchor hole aligns with cable lead-in hole in housing. Hold wheel in position, install cable and turn cable clockwise until it bottoms. Rotate wheel and push cable into housing until conduit is positioned; then, tighten conduit retaining nut. Balance of reassembly is the reverse of disassembly. The planet gear carrier may be installed in any position with respect to the drive wheel.

After installing compensating lever and prior to installing the compensating rod, check operation of rockshaft slip clutch as follows: Pull up on compensating lever until ball joint butts against conduit. Attach a spring scale to lever and while keeping scale at 90 degrees to compensating lever, measure the force required to cause clutch to slip. This force should be 38 to 50 pounds. If not within these limits, refer to paragraph 135.

Connect compensating rod and lever and adjust system as outlined in paragraph 132.

ROCKSHAFT SLIP CLUTCH

Series 240

135. R&R AND OVERHAUL. To remove the rockshaft slip clutch, disconnect the compensating rod from compensating lever, loosen the conduit retaining nut, then, unbolt the drive unit from the rockshaft side plate and swing unit away from rockshaft. Use a tool such as that shown in Fig. IH601 and compress clutch spring and ring retainer into rockshaft. Remove the retaining snap ring, slowly release tool and remove ring retainer, spring and coupling. See Fig. IH604A.

Clean all parts including the coupling insert which is pressed into rockshaft. Inspect all parts for excessive wear and/or damage. Pay particular attention to mating surfaces of clutch coupling and insert and remove any burrs which may be present.

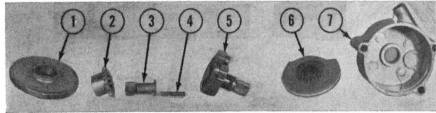

Fig. IH600—View showing the component parts of the rockshaft drive unit (gear box) and their relative position. Later units have steel gears instead of powdered metal gears.

1. Cover	4. Spring	6. Cable drive wheel
2. Input gear	5. Planet gear assy.	7. Housing
3. Coupling		

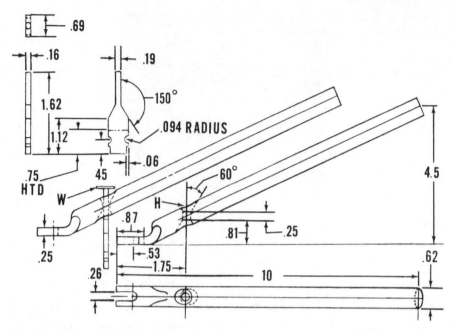

Fig. IH601—View of special tool used to facilitate removal of rockshaft slip clutch. Tool can be made locally. "H" is hole of 0.250 diameter. "HTD" means harden this distance. "W" is washer.

Apply a light coat of light oil to clutch faces and inside bore of rockshaft; then, reassemble unit by reversing the disassembly procedure. Check operation of clutch after assembly as follows: Use a torque wrench fitted with a screw driver adaptor and check the torque required to cause clutch to slip. Clutch should slip in either direction of rotation between 80-100 inch-pounds. Turn clutch coupling 180 degrees and repeat operation. If clutch does not meet above specifications, renew assembly.

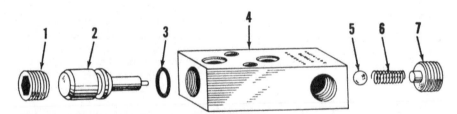

Fig. IH603—Exploded view of the Hydra-Touch system check valve block used on early models. Later models use a double check valve which is similar in construction.

1. Plug	4. Block	6. Check valve spring
2. Piston	5. Ball	7. Plug
3. Seal ring		

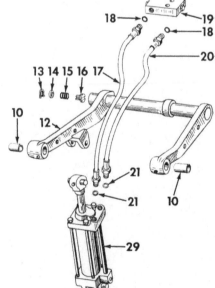

Fig. IH604A—View showing slip clutch exploded from rockshaft.

Fig. IH604 — Exploded view of the hydra-creeper motor used on series 100, 130, 140 and 200 tractors. Items 3, 5, 6, 7 and 8 are not available separately.

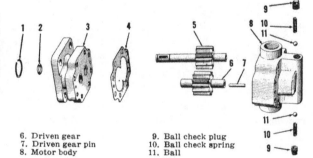

1. Snap ring
2. Drive gear seal ring
3. Cover
4. Gasket (0.001, 0.0015 and 0.003)
5. Drive gear
6. Driven gear
7. Driven gear pin
8. Motor body
9. Ball check plug
10. Ball check spring
11. Ball

10. Bushing	16. Coupling
12. Rockshaft	18. Ring seal
13. Snap ring	19. Check valve
14. Retainer	21. Ring seal
15. Spring	29. Cylinder

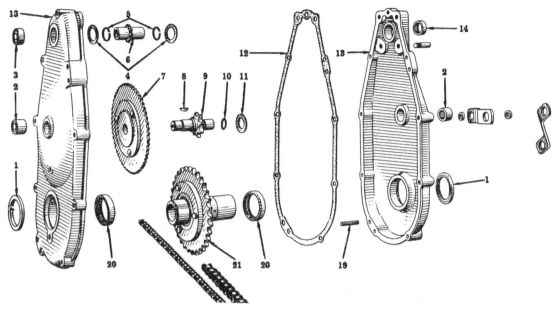

Fig. IH605—Exploded view of the hydra-creeper transmission.

1. Driven sprocket oil
   seal
2. Idler pinion needle
   bearing
3. Drive sprocket needle
   bearing
4. Washer

5. Retainer rings
6. Drive sprocket
7. Idler sprocket
8. Woodruff key
9. Idler pinion
10. Retainer ring
11. Washer

12. Gasket
13. Housing
14. Needle bearing
19. Dowel pin
20. Driven sprocket needle
    bearing
21. Driven sprocket

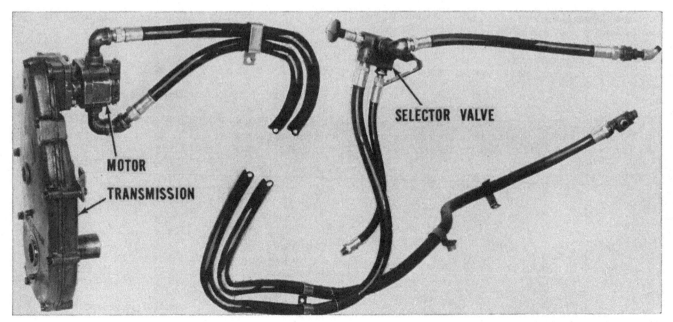

Fig. IH606—Layout of hydra-creeper attachment, showing the relative location of the main components. Hydraulic power is supplied by the hydraulic lift pump.

HYDRAULIC LIFT

The series 404 and 2404 hydraulic lift unit, which provides both draft and position control, also serves as a top cover for the differential and final drive portions of the tractor rear center frame. Contained within the hydraulic lift housing are the rockshaft, work cylinder, control valves and the linkage for operation of the components. See Fig. IH610.

Pressurized oil for the operation of the hydraulic lift unit is supplied by an externally mounted pump which is driven from the engine gear train. Pumps on tractors without power steering have a 4½ gpm capacity while pumps on tractors with power steering have a capacity of 9 gpm. Both pumps operate at a pressure of 1550-1600 psi, controlled by the system pressure relief valve. A hydraulic fluid filter is mounted at aft end of hydraulic pump.

The hydraulic lift system draws its oil supply from the transmission and differential housing and thus, the tractor rear center frame and hydraulic lift system share a common reservoir.

LUBRICATION AND BLEEDING

140. To drain and refill the reservoir, and bleed the hydraulic system, proceed as follows: Remove drain plug from bottom of rear center frame and right rear bottom of hydraulic lift housing. Pull coil wire from center of distributor cap and crank engine briefly to clear pump and connecting lines.

Remove filler plug from transmission top cover and level plug from right rear (back of axle housing) of

rear center frame. Fill rear center frame to level plug opening with IH Hy-Tran Fluid, or a mixture of IH Torque Amplifier Additive and SAE 10W engine oil in the ratio of one quart of additive to each four gallons of engine oil. Start engine and with filler plug out, cycle lift system, and remote cylinders if so equipped, about ten or twelve times, then place position control lever and remote control levers in the forward position. If tractor is equipped with power steering, cycle steering system from one extreme position to the other, then place wheels in straight ahead position. Recheck fluid level and add as necessary to bring to level plug opening. Install and tighten level plug and filler plug.

FILTER

141. The hydraulic system filter is bolted to the aft end of the hydraulic pump. Filter element should be renewed after the first 50 hours of operation and then every 250 hours thereafter, under normal operating conditions. Any abnormal operations will, of course, require more frequent changes.

Renew element as follows: Place catch pan under filter, remove case retaining bolt and remove case with contained parts. Remove filter screen and element and discard element. Clean filter screen and all other parts. Inspect cover seal, screen gasket, case gasket and valve gasket. Renew parts as necessary.

Assemble filter as follows: Place case bolt, with gasket, in case. Install remaining parts over case bolt in the following order. Element cover spring

(large), bolt seal, element cover, element cover spring (small), bolt seal, by-pass valve assembly, valve gasket, element and filter screen. Place case gasket and screen gasket in filter base then install case with combined parts and tighten case bolt securely.

Start engine, bring oil to operating temperature and check filter for leaks.

TROUBLE SHOOTING

142. The following trouble-shooting chart lists troubles which may be encountered in the operation and servicing of the hydraulic lift system. The procedure for correcting many of the causes of trouble is obivous. For those remedies which are not so obvious, refer to the appropriate subsequent paragraphs.

A. Hitch will not raise. Could be caused by:
1. Unloading valve orifice plugged.
2. Unloading valve piston sticking.
3. Flow control valve spring broken, piston sticking or check ball stuck in orifice.
4. Unloading valve ball not seating.
5. Main relief valve spring broken or valve is leaking.
6. Cylinder safety (cushion) valve faulty.
7. Auxiliary valve cover "O" ring damaged.
8. Linkage disconnected from control lever or valve.

B. Hitch lifts load too slow. Could be caused by:
1. Flow control valve piston sticking or faulty valve spring.
2. Flow control valve piston stop broken.
3. Unloading valve ball seat leaking.
4. Faulty main relief valve.
5. Cylinder safety (cushion) valve leaking.
6. Scored lift cylinder or piston "O" ring damaged.
7. Excessive load.

C. Hitch will not lower. Could be caused by:
1. Control valve spool sticking or "O" ring damaged.
2. Drop poppet valve sticking or "O" ring damaged.

D. Hitch lowers too slow. Could be caused by:
1. Drop control valve spool sticking.

Fig. IH610 — View of hydraulic draft and position control unit with top cover removed.

C. Cylinder
N. Adjusting collar
R. Relief valve
S. Safety (cushion) valve
T. Timing switch
V. Control valve

2. Damaged "O" ring on drop poppet valve.
3. Linkage out of adjustment.

E. Hitch lowers too fast. Could be
   caused by:
   1. Faulty drop control valve piston.

F. Hitch lowers too fast in slow action
   position. Could be caused by:
   1. Drop control linkage out of ad-
      justment.

G. Hitch raises and lowers but will
   not maintain position (hiccups).
   Could be caused by:
   1. Check valve in main control
      valve leaking.
   2. Cylinder safety (cushion) valve
      leaking.
   3. Cylinder scored or piston "O"
      ring damaged.
   4. Drop poppet valve sticking.
   5. Check valve actuating rod ad-
      justing screw improperly ad-
      justed.

H. System stays on high pressure.
   Could be caused by:
   1. Broken, disconnected or im-
      properly adjusted linkage.
   2. Control valve spools faulty.
   3. Auxiliary valve not in neutral.
   4. Restraining chains adjusted too
      short.
   5. Core plug in control valve body
      missing or leaking badly.

I. Malfunction of position control.
   Could be caused by:
   1. Control valve spool or linkage
      binding.
   2. Control lever quadrant not
      mounted correctly.
   3. Incorrect valve link - to - spool
      adjustment.

J. Response time too slow. Could be
   caused by:
   1. Unloading valve piston sticking.
   2. Unloading valve orifice plugged
      (partially).

K. Hitch has too much depth variation
   (over-travels). Could be caused
   by:
   1. Torsion bar bearings not lubri-
      cated.
   2. Flow control valve check ball
      missing.
   3. Unloading valve orifice parti-
      ally plugged.

L. Inadequate depth control during
   deep operation. Could be caused
   by:
   1. Foreign material between tor-
      sion bar bellcrank and its stops.
   3. Incorrect adjustment of hitch
      lift link.
   2. Interference between top link
      and rear frame.
   3. Control lever quadrant mounted
      in wrong position.
   4. Top link pin excessively worn.

M. Insufficient transport clearance of
   mounted implement. Could be
   caused by:
   1. Incorrect valve link-to-control
      valve spool adjustment.
   2. Safety shut-off improperly ad-
      justed.
   3. Incorrect adjustment of hitch
      lift link.

N. Auxiliary circuit will not lift load
   or lifts load slowly. Could be
   caused by:
   1. Faulty main relief valve.
   2. Excessive leakage past valve
      spool.
   3. Faulty auxiliary valve check
      ball.
   4. Faulty cylinder relief valve
      (industrial valve).

O. Auxiliary valve does not automatically unlach. Could be caused by:
   1. Incorrect unlatching adjust-
      ment.
   2. Faulty unlatching piston.
   3. Faulty detent sleeve.
   4. Spool sticking due to improp-
      erly tightened mounting bolts

P. Auxiliary valve unlatches prematurely. Could be caused by:
   1. Incorrect unlatching adjust-
      ment.
   2. Broken unlatching spring.
   3. Faulty detent sleeve.

Q. Load drops slightly when auxiliary
   valve is put in lift position. Could
   be caused by:
   1. Valve check ball leaking or not
      seating.

R. Auxiliary valve will not hold load
   in position. Could be caused by:
   1. Excessive leakage past valve
      spool.
   2. Faulty cylinder or piston.
   3. Faulty cylinder relief valve (in-
      dustrial valve).

S. Fluid leaking from detent breather.
   Could be caused by:
   1. Faulty unlatching piston "O"
      ring.
   2. "O" ring at detent end of valve
      spool leaking.

SYSTEM OPERATING PRESSURE AND RELIEF VALVE

Series 404-2404

142A. Tractors which are equipped
with ONLY draft and position control
(no auxiliary control valves) must be
tested as outlined under SYSTEM
ADJUSTMENTS (paragraphs 143
through 148) as there is no opening
available to attach a pressure checking gage.

On tractors which are equipped with
draft and position control and auxiliary control valves, system operating

pressure can be checked as follows:
Use a gage capable of registering at
least 2000 psi and install gage in series
with a shut-off valve in a line to
which has been attached the male half
of a quick coupler. Gage must be in
the line between shut-off valve and
quick coupler half. Install test assembly male end in quick coupler of tractor and place open end in reservoir
filler hole. Be sure to fasten open end
securely in filler hole. Start engine
and operate until hydraulic fluid is
warmed to operating temperature;
then, with engine operating at high
idle speed, move auxiliary control
valve lever to the position that will
direct fluid to the test gage. Manually
hold control valve lever in this position and close the shut-off valve only
long enough to observe the system
relief valve pressure. System pressure
should be 1550-1600 psi.

NOTE: Before removing test fixture,
test auxiliary control valve unlatching
pressure as follows: Operate engine at
low idle, move auxiliary control valve
lever to position that will direct fluid
to test gage, then slowly close the
shut-off valve and observe the pressure at which the valve control lever
returns to neutral. This pressure
should be not less than 1000 psi nor
more than 1250 psi.

If system pressure is not as specified, remove seat, seat brackets and
hydraulic lift unit top cover; then, remove and renew relief valve (R—Fig.
IH610). System relief valve is preset
and heavily staked at factory and cannot be adjusted. For information concerning unlatching mechanism of
auxiliary control valve, refer to paragraph 157.

NOTE: Both system relief valve and cylinder cushion (safety) valve can be tested
after removal by using an injector tester
and the proper adapters. However, bear in
mind that the pressures obtained will be
toward the low side of the specified pressure ranges due to the low volume of fluid
being pumped.

SYSTEM ADJUSTMENTS

143. Whenever hydraulic lift unit
has been serviced and adjustment is
required, the system should be cycled,
by using the position control lever, at
least ten or twelve times to insure
purging any air which might be present in system. All checking and adjusting of the system should be done
with a load on the hitch and the
hydraulic fluid at operating temperature.

IMPORTANT: Adjustments are made with the top cover of unit removed. The system pressure relief valve located in the right front corner of the unit housing discharges upward when it relieves, therefore, it is necessary to install a shield over the relief valve when operating the system with the top cover removed.

While some early model tractors may not be so equipped, the later model tractors have a drain plug located in the left front corner of unit housing to permit lowering of the fluid level so adjustments can be made without working below the surface of fluid.

144. DROP POPPET ADJUSTMENT. Place the draft control lever in the full forward position, raise the hitch to its maximum position with the position control lever, then lower hitch to mid-position. System should go off high pressure with no cycling (hiccups). If unit cycles (hiccups), adjust the drop poppet actuating rod adjusting screw (located between legs of main control valve link) as follows: Again fully raise hitch and lower to mid-position. Turn adjusting screw in until unit begins to cycle (hiccup), then back-out screw until the point is reached where the unit stops hiccupping. See Figs. IH611 and IH612. Now turn the screw out an additional ¾-turn to obtain the proper clearance (0.020) between push rod and check ball.

NOTE: Move the position control handle a small amount after each turn of the adjusting screw to make sure control valve spool is in the normal centered position.

145. POSITION CONTROL LINK-AGE ADJUSTMENT. Place draft control lever in full forward position. Loosen cap screw (B—Fig. IH613) and slide hydraulic shut-off switch assembly (C) rearward temporarily. Place the position control lever at the offset on forward end of quadrant at which time the hitch should be fully lowered. Check hitch by by-passing the quadrant offset with the position control handle and moving it to the bottom of the slot after loosening lock nut (L) and backing out drop control screw (M). If hitch drops any further, the control valve actuator tube must be turned into the control valve link. See Fig. IH611. Be sure to tighten lock nut after each adjustment. This adjustment can be checked by moving the position control lever to the rearward position at which time the rock-shaft lift arms should be 45-46 degrees above vertical position. Angle can be determined by measuring from center

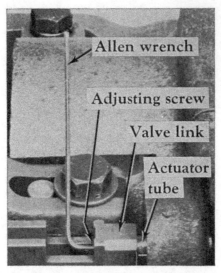

Fig. IH611 — View showing method of adjusting drop poppet actuating rod. Refer to text.

of lower link pivot pin to the center of the lift link pivot pin. This distance should be 32⅛ inches with a plus or minus tolerance of ⅛-inch.

Note: On some early model tractors, it is permissible to set the position control lever not more than ⅝-inch beyond offset in quadrant to put hitch in the fully lowered position. On these early models, use shims between valve link and actuator tube when adjusting, or install a new actuator tube which has the adjusting nut.

Without moving the position control lever after obtaining the 45-46 degree rockshaft arm position, move the hydraulic switch assembly (C—Fig. IH613) forward until lever (F) is in contact with pin (G) and lever (H) is in contact with pin (E) and tighten cap screw (B). Distance between end of slot and dowel should be approximately 11/32-inch as shown

Fig. IH612—Drop poppet actuating rod adjusting screw is located as shown. Valve link has been removed for illustrative purposes.

in Fig. IH613. System should go off high pressure when the rockshaft lift arms reach the 45-46 degree above vertical position when hitch is raised with draft control lever.

With position control lever position established as outlined, set bumper stop (Fig. IH614) on the quadrant so lever contacts it 0.020-0.030 before lever reaches end of its travel and tighten stop securely.

Cycle system several times and recheck the 45-46 degree rockshaft arm position. Rockshaft arms should have approximately 1-inch additional free travel, when lifted manually, after hitch has reached its highest point. If hitch doesn't have the additional 1-inch free travel, recheck the rock-shaft lift arm angle and the actuator tube adjustment.

146. DROP CONTROL ADJUST-MENT. Loosen lock nut (L—Fig. IH613) and back-out adjusting screw (M). Move position control lever to offset of quadrant, then turn adjusting screw until it just contacts the cam which operates the drop control valve spool. Tighten lock nut.

This adjustment can be checked by moving the draft control lever completely forward. The drop control valve spool should bottom when the lever is about ⅛-inch from full forward.

147. DRAFT CONTROL LINKAGE ADJUSTMENT. Remove cap screw which retains torsion bar flange to bracket and pull torsion bar out until flange clears dowel and bellcrank is free to rotate. Set the position control lever at approximately the offset of the quadrant. Set the draft control lever at the full forward position. Rotate top of bellcrank to the rear until it contacts its stop and hold in this position with a small bar. Rock-shaft should not move; however, if it does move, remove pin from draft control rod adjusting nut and turn nut out (counter-clockwise) until bellcrank, with pin installed, will bottom without causing rockshaft to move.

Continue to hold bellcrank against stop and move draft control lever to the rear edge of word "OFF" on the quadrant. The rockshaft arms should raise to the 45-46 degree angular position and the system should go off high pressure. If system does not react as stated, turn draft control rod nut in (clockwise) until operation is correct. Reinstall pin, spring washers and retaining pin. Reinstall torsion bar.

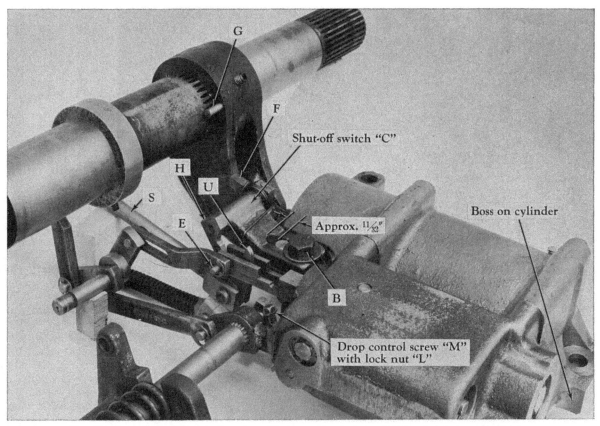

Fig. IH613 — View of the draft and position control unit internal linkage and component parts.

148. **CONTROL LEVERS ADJUST-MENT.** Position control lever is adjusted by the two locking rings (collars) at inside of control levers and draft control lever is adjusted by the two nuts on the outside of the levers. Procedure for adjusting is obvious. Levers should be adjusted until 4 to 6 lbs. for used friction discs; or 6 to 8 lbs. for new friction discs, is required to move levers. Measurement is taken at control lever knob.

PUMP UNIT

The hydraulic system pump may be either Cessna or Thompson. Tractors without power steering are fitted with 4½ gpm pumps while those with power steering are fitted with 9 gpm pumps. Overhaul of the hydraulic pump is limited to disassembly and cleaning and installing a gasket and seals package. If parts, other than gaskets and seals, are excessively worn or damaged, it will be necessary to renew the complete pump unit. See Fig. IH614A.

148A. **REMOVE AND REINSTALL.** To remove the hydraulic pump, first disconnect the lines from the hydraulic system filter and make provision for catching oil, or plug lines. Remove filter assembly, as a unit, from hydraulic pump. Remove pump retaining cap screws and pull pump from engine.

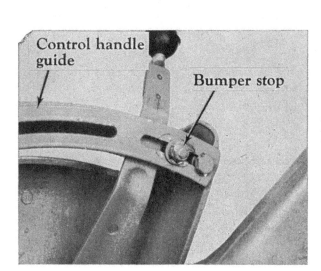

Fig. IH614 — View showing bumper stop. Refer to text for adjustment.

Fig. IH614A—View showing hydraulic pump and filter assembly installation on series 404 and 2404.

Fig. IH615—Seal tube (T) and seal (S) assembly can be removed after torsion bar bracket is off.

Install hydraulic pump by reversing the removal procedure and bleed hydraulic system as outlined in paragraph 140.

148B. OVERHAUL. To renew pump seals and gaskets remove pump as outlined in paragraph 148A and proceed as follows: Place a scribe line across pump cover and body so they can be reassembled in the same relative position. Remove pump drive gear. Remove pump cover and do not lose check ball and spring (Cessna pump) or the pressure plate spring (Thompson pump) as the cover is removed. Any further disassembly will be dictated by the condition of the

pump and procedure for doing so is obvious.

Small nicks and/or scratches can be removed from the body, cover, shafts, gears and bearings by using crocus cloth or a fine oil stone. Be sure the small drilled passages in pump body and cover are open and clean.

Lubricate all parts with hydraulic fluid prior to reassembly. Use care not to damage shaft seal when installing drive shaft.

HYDRAULIC LIFT UNIT

149. REMOVE AND REINSTALL. To remove the hydraulic lift unit from tractor, first remove plug from lower rear right hand corner of lift housing and drain housing. Remove seat, upper lift link and right fender. Remove retainer from right end of draft control bellcrank pin and slide pin to left until it clears draft control rod nut. Unbolt torsion bar flange from torsion bar bracket and remove torsion bar and bellcrank. Unbolt and remove torsion bar bracket from lift housing.

Note: At this time, bushings in torsion bar bracket, and seal and bearings in draft control rod tube (Fig. IH615) can be renewed. Control rod tube can be removed by unscrewing after draft control rod nut is off.

Disconnect hydraulic pump manifold flange at lift housing, and if necessary, remove remote control valves. Disconnect lift links from rockshaft arms. Remove cap screws which retain lift housing to rear center frame, attach chain and hoist and lift unit from tractor.

DRAFT CONTROL CYLINDER AND VALVE ASSEMBLY

150. REMOVE AND REINSTALL. To remove the cylinder and valve assembly, remove plug from rear lower right hand corner of lift housing and drain housing. Remove seat, seat brackets and top cover. Remove "C" ring and pin from control valve link and disconnect control linkage from control valve link, move control levers rearward, then unbolt and remove the complete cylinder and valve assembly from housing.

Note: Catch safety switch assembly as mounting cap screws are loosened. Tip forward end of cylinder and valve assembly upward when removing from housing.

NOTE: When reinstalling, it is necessary that the cylinder be seated against the boss in lift housing. Use a small bar to hold cylinder against boss while tightening retaining cap screws. In some cases it may be necessary to reposition the draft control valve.

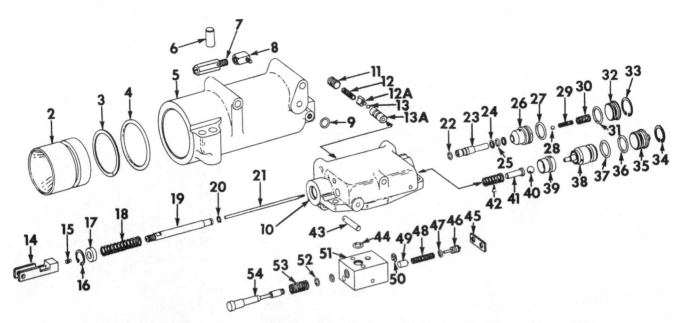

Fig. IH616—Draft control cylinder and valve assembly showing component parts and their relative positons.

2. Piston	11. Retaining screw	17. Spring retainer	26. Drop poppet valve	36. "O" ring	45. Plug retainer
3. Back-up washer	12. Spring	18. Spring	27. "O" ring	37. "O" ring	46. Plug
4. "O" ring	12A. Ball retainer	19. Actuator tube	28. Ball	38. Unloading valve	47. "O" ring
5. Cylinder	13. Check ball	20. Retaining ring	29. Spring	piston	48. Spring
6. Dowel	13A. Flow control	21. Drop valve	30. Spring	39. Valve seat	49. Drop valve piston
7. Relief (safety)	valve	actuating rod	31. "O" ring	40. Ball	50. Retaining ring
valve	14. Actuator	22. "O" ring	32. Plug	41. Ball rider	51. Drop valve body
8. Elbow	15. Drop valve	23. Pilot valve seat	33. Snap ring	42. Spring	52. "O" ring
9. "O" ring	adjusting screw	24. Back-up washer	34. Snap ring	43. Pivot pin	53. Spring
10. Draft control	16. Snap ring	25. "O" ring	35. Plug	44. "O" ring	54. Variable orifice
valve body					spool

Fig. IH617—Cylinder and control valve assembly shown removed. Note that valve link is removed.

C. Cylinder
D. Drop control valve
P. Piston
V. Control valve

Fig. IH618 — View of drop control valve with internal parts removed.

After unit is installed, refer to paragraph 143 for adjustment. See Figs. IH616 and IH617.

151. OVERHAUL. Because of the inter-relation of the cylinder and piston, draft control valve and drop control valve, they will be treated as sub-assemblies and each will be removed from the complete assembly, serviced and reinstalled.

152. DROP CONTROL VALVE. To remove and overhaul the drop control valve assembly, remove the complete cylinder and valve assembly as outlined in paragraph 150.

Remove the Allen screws which retain drop control valve, lift same from draft control valve and remove "O" ring. Remove "C" retainer from end of variable orifice spool, pull spool from bore and remove the "O" rings from each end of body bore. Remove end plate (retainer) and remove the spacer, plug and guide (esna pin), spring and piston. Remove "O" ring from plug assembly. See Fig. IH618.

Inspect spool, piston and bores for nicks, burrs, scoring and undue wear. Variable orifice spool and valve should fit bores snugly, yet slide freely. Inspect springs for fractures, distortion or signs of permanent set-

ting. Refer to following specifications and renew parts as necessary.

Use all new "O" rings, reassemble valve by reversing disassembly procedure and install unit on draft control valve.

Variable Orifice Spool Spring
　Free length—In. ....$1\frac{1}{8}$
　Test load lbs. at
　　length—In.　.......13.5-16.5 @ $\frac{7}{8}$

Piston Spring
　Free length—In. ....$1\frac{3}{32}$
　Test load lbs. at
　　length—In. ....10.5-12.3 @ 61/64

153. DRAFT CONTROL VALVE. To remove and overhaul the draft control valve, remove the complete cylinder and valve assembly as outlined in paragraph 150. Remove drop control valve from draft control valve, then remove draft control valve from work cylinder and piston assembly.

With valve removed, refer to Fig. IH619 and remove internal snap ring, then remove plug and unloading valve assembly. Unscrew valve seat and remove ball, ball rider and ball rider spring.

Refer to Fig. IH620 and remove internal snap ring, then remove plug, drop poppet valve spring (large), pilot valve spring (small), ball and the drop poppet valve and pilot valve seat assembly. Pull pilot valve seat from poppet valve.

Fig. IH619—Unload valve assembly removed from control valve body.

Fig. IH620—Drop poppet valve assembly removed from control valve body.

Fig. IH621 — Main control valve assembly removed from control valve body. Note drop control valve cam.

ring is toward pressure (closed) end of piston. Coat piston assembly with oil prior to installation in cylinder. Use new "O" ring between cylinder and control valve assembly.

Note: The safety relief (cushion) valve (S—Fig. IH610) attached to left side of work cylinder can be removed after lift housing top cover has been removed and procedure for doing so is obvious. Relief valve can be bench tested by using an injector tester; however, unit is preset at the factory and is non-adjustable. Renew valve if found to be faulty. Valve is set to relieve at 1650-1900 psi.

Refer to Fig. IH621 and remove link from control valve, then remove internal snap ring, and remove spring retainer, main valve spring, main valve actuator tube and valve spool. Remove the small retaining ring and pull drop valve actuating rod from actuator tube. Do not remove drop valve adjusting screw (Fig. IH612) from outer end of actuator tube unless necessary.

Refer to Fig. IH622 and remove plug, spring, the nylon check valve ball retainer and ball and the flow control valve.

Inspect all parts for nicks, burrs, scoring and undue wear. Refer to the following specifications and renew parts as necessary. Valve body and spool are not available separately. Use caution when renewing "O" rings to insure that proper size is installed and be sure the back-up rings used in bore of drop poppet valve are positioned on each side of the "O" ring. Be sure all check balls and their respective seats are in good condition.

Coat all parts with lubricating oil and reassemble by reversing disassembly procedure. Use new "O" ring and install valve to cylinder assembly.

bly. Install drop control valve assembly to draft control valve.

Unloading Valve
Spring free length—In..........$1\frac{7}{16}$
Test load lbs. at
length—In. ....16.0-17.4 @ $1\frac{1}{32}$

Drop Poppet Valve
Spring free length—In.........41/64
Test load lbs. at
length—In. ....9-11 @ $\frac{17}{32}$
Check ball spring free
length—In. 59/64
Check ball spring test load
lbs. at length—In.....3.5-4.1 @ ¾

Control Valve
Spring (return) free
length—In. 2 61/64
Test load lbs. at
length—In. ....18.4-21.6 @ $1\frac{23}{32}$

Flow Control Valve
Spring free length—In..........1¼
Test load lbs. at
length—In. ....4.9-5.8 @ $1\frac{1}{16}$

154. WORK CYLINDER AND PISTON. To remove and overhaul the work cylinder and piston, first remove the complete cylinder and valve assembly as outlined in paragraph 150, then remove the draft control valve and drop control valve assembly from work cylinder.

Piston can be removed from cylinder by bumping open end of cylinder against a wood block, or by carefully applying compressed air to the cylinder oil inlet port. With piston removed, the piston "O" ring and back-up washer can be removed.

Inspect piston and cylinder bore for nicks, burrs, scoring or undue wear. Small defects can be corrected by using crocus cloth. Renew parts which are unduly scored or worn. Piston outside diameter is 3.497-3.499 inches. Cylinder inside diameter is 3.500-3.502 inches.

When installing piston "O" ring and back-up washer, be sure "O"

155. CONTROL LEVERS AND SHAFTS AND CONTROL LINKAGE. Control levers, quadrant and control lever shafts can be removed as an assembly as follows: Drain hydraulic lift housing and remove seat and lift housing top cover. Remove cylinder and valve assembly as outlined in paragraph 150. Remove control handle knobs, then remove cap screw from one end of control handle guide, loosen other at opposite end and allow guide to hang. Remove quadrant and note position the quadrant was mounted (i.e., which two holes used) so it can be reinstalled the same way. Pull draft control handle rearward, then drive out roll pin at inner end of shaft and remove control lever. Remove snap ring from inner end of position control lever shaft. Unbolt control lever and quadrant support from hydraulic housing and pull support and control levers from housing.

Draft control linkage can now be removed after removing adjusting nut from aft end of draft control rod. Any further disassembly of linkage will be evident after an examination of same. Removal of rockshaft bellcrank and actuating hub will require removal of rockshaft as outlined in paragraph 156.

Oil seal and bearing for control lever shaft can now be removed from lift housing, if necessary.

Any disassembly and/or overhaul required on the control levers and quadrant assembly will be obvious after an examination of the unit. Note that the inner control lever shaft is sealed to the outer control lever shaft by an "O" ring.

NOTE: The eccentric shaft can also be removed at this time. Disconnect internal linkage, if not already done, then remove outer retainer and remove eccentric shaft from inside of housing. Eccentric shaft is fitted with

Fig. IH622—Flow control valve and check ball shown removed. Plug has a nylon locking insert.

an "O" ring seal. Shaft bushing can also be pulled from lift housing.

When reassembling control levers and drop valve actuating lever to shafts, be sure to align the register marks. Belleville washers on outer end of inner (draft control) shaft are installed as follows: Outer washer, dish toward inside; center washer, dish toward outside; inner washer, dish toward inside.

156. ROCKSHAFT. If rockshaft seal renewal is all that is required, seals can be renewed as follows: Remove control lever quadrant and both lift arms. Use a screwdriver, or similar tool, and pry out old seals. Use a suitable driver and drive new seals in until they bottom. Note that left seal has a smaller inside diameter than the right seal.

To remove the rockshaft, remove the right fender, in addition to the quadrant and rockshaft lift arms. Remove the cylinder and valve assembly as outlined in paragraph 150. Remove Allen screws from actuating hub and bellcrank and slide rockshaft from left to right out of housing, bellcrank and actuating hub. Remove actuating hub key as soon as it is exposed. If actuating hub sticks on rockshaft, either pry against is with a heavy screwdriver, or use a spacer between hub and housing. DO NOT drive on rockshaft without supporting actuating hub as damage to linkage will occur.

Always renew oil seals whenever rockshaft is removed; however, do not install the seals until after the rockshaft is installed. Rockshaft bushings can be removed and reinstalled using a proper sized bushing driver. Outside edge of bushings should be flush with bottom of oil seal counterbore. Inside diameter of left bushing is 2.090-2.095; inside diameter of right bushing is 2.315-2.320. Outside diameter of rockshaft at bearing surfaces is 2.085-2.087 for the left and 2.310-2.312 for the right.

Prior to reassembly, it is recommended that the following identification marks be made even though the rockshaft and rockshaft bellcrank are master splined. These marks will provide visibility and aid in reassembly. Use yellow paint and paint the "V" notch in the actuating hub key, the rockshaft master spline and the allen screw seat in the rockshaft. Also paint a line straight up from the master spline in the rockshaft bellcrank.

Start rockshaft into housing from right side of housing and start ac-

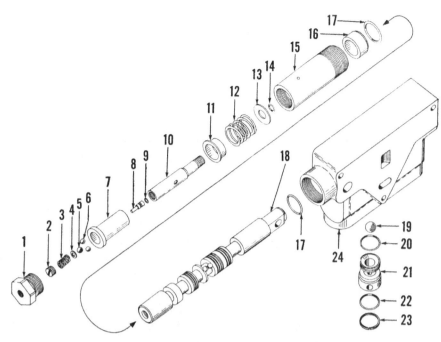

Fig. IH 623—Exploded view of hydraulic auxiliary control valve used for remote cylinders. Valves can be stacked for dual valve operation. Items (18 & 24) are serviced as a unit.

1. Cap	6. Detent ball	12. Centering spring
2. Plug	7. Position control	13. Washer
3. Detent spring	sleeve	14. "O" ring
4. Detent actuating	8. Unlatching piston	15. Sleeve
ball washer	9. "O" ring	16. Retainer
5. Detent actuating	10. Actuator	17. "O" ring
ball	11. Spring retainer	18. Spool

19. Check ball
20. "O" ring
21. Retainer
22. "O" ring
23. Back-up washer
24. Body

tuating hub and rockshaft bellcrank over rockshaft. Align the affixed markings (master splines) of rockshaft and bellcrank and position bellcrank until set screw seat in rockshaft is aligned with set screw hole in bellcrank, then install the allen screw. Install the actuating hub key and slide the actuating hub over key until the "V" notch is visible through set screw hole, then install the set screw. NOTE: Use a mirror during these operations. Install new oil seals. Install rockshaft lift arms and torque the retaining bolts to 170-190 ft.-lbs.

Complete reassembly by reversing the disassembly procedure.

AUXILIARY CONTROL VALVE

157. R&R AND OVERHAUL. To remove the auxiliary control valve, or valves, first remove the banjo bolts which retain the manifold tubes to the valve. Remove mounting bolts and pull cover and valve, or valves, from lift housing.

When reinstalling, torque mounting bolts to 20-25 ft.-lbs. Do not overtighten mounting bolts as valve body may be distorted and valve sticking could result.

To disassemble, use Fig. IH623 as a guide. Remove control handle and

bracket. Remove end cap (1), then unscrew the actuator (10) and remove the actuator and detent assembly. Remove sleeve (15) and pull balance of parts from body. Check ball and retainer (21) can be removed at any time. Note that retainer (21) is held in position by one of the mounting bolts. Align holes in retainer and body during reassembly.

NOTE: Some valves do not include the detent assembly. When disassembling these valves, sleeve (15) must be removed before removing actuator (10).

In addition, industrial valves have a circuit relief valve located directly below sleeve (15) and valve can be removed at any time.

Detent (3, 4, 5 and 6) can be disassembled after removing plug (2). Push unlatch piston (8) out of actuator (10) by using a long thin punch.

Inspect all parts for nicks, burrs, scoring and undue wear and renew parts as necessary. Spool (18) and body (24) are not available separately. Check detent spring (3) and centering spring (12) against the following specifications.

Detent spring

    Free length—In. . . . . $1\frac{7}{16}$

Test load lbs. at

    length—In. . . . . 23.5-28.5 @ 45/64

Centering spring
  Free length—In. .... $2\frac{5}{16}$
  Test load lbs. at
    length—In. ....26.5-33.5 @ 1 7/64

Use all new "O" rings and reassem-

ble by reversing the disassembly procedure. Detent unlatching pressure is adjusted by plug (2). Unit must unlatch at not less than 1000 nor more than 1250 psi. The circuit relief valve on industrial valves is a cartridge

type with the pressure setting stamped on end of body. Faulty relief valves are corrected by renewing the complete unit. Be sure filter in end cap (1) is clean (no paint) and in satisfactory condition.

NOTES

||||||||||||||||||||||||||||||||||||||||||||||||||||||||||||||||||||||||||||

INTERNATIONAL
HARVESTER

Models   ■ B-275   ■ B-414   ■ 354   ■ 364   ■ 384
              ■ 424   ■ 444   ■ 2424   ■ 2444

Previously contained in I&T Shop Manual No. IH-45

|||||||||||||||||||||||||||||||||||||||||||||||||||||||||||||||||||||||||||||

SHOP MANUAL
INTERNATIONAL HARVESTER

SERIES B-275-B-414-354-364-384-424-444-2424-2444

All tractors are powered by four cylinder, four cycle engines. The Series B-275 tractors are equipped with a 144 cubic inch displacement diesel engine while the Series B-414 is available with either a 154 cubic inch displacement diesel engine or a 144 cubic inch displacement gasoline engine. Series 364 and 384 have a diesel 154 cubic inch displacement engine. Series 354 has a 144 cubic inch displacement engine in both diesel and gasoline models. Series 424 and 2424 are available with either a 154 cubic inch displacement diesel engine or a 146 cubic inch displacement gasoline engine. Series 444 and 2444 are available with either a 154 cubic inch displacement diesel engine or a 153 cubic inch displacement gasoline engine. A dual range transmission is standard equipment on all tractors and provides eight forward and two reverse speeds. In addition, the Series B-414, 364, 384, 424, 444, 2424 and 2444 tractors have available a forward and reverse transmission designed to work in conjunction with the main tractor transmission.

Engine serial number is stamped on a pad on right side of engine crankcase. Tractor serial number is stamped on a name plate on the right side of the clutch housing.

INDEX (By Starting Paragraph)

INDEX CONT.

CONDENSED SERVICE DATA

	Series B-275, 354 Diesel	Series 424, 2424, 444, 2444, B-414, 364, 384 Diesel	Series B-414, 354 Non-Diesel	Series 424, 2424 Non-Diesel	Series 444, 2444 Non-Diesel
GENERAL					
Engine Make	Own	Own	Own	Own	Own
Number of Cylinders	4	4	4	4	4
Bore—Inches	3-3/8	3½	3-3/8	3-3/8	3-3/8
Stroke—Inches	4	4	4	4-1/16	4¼
Displacement—Cubic Inches	144	154	144	146	153
Compression Ratio	19.3:1 (B-275) 20.1:1 (354)	23:1	6.3:1	7.6:1	7.7:1
Compression Pressure at 200 rpm Cranking Speed	330-335	445-470	80-105	180	180
Cylinder Sleeves Wet or Dry?	Wet	Wet	Wet	None	None
Forward Speeds—Number of	8	8	8	8	8
Main Bearings—Number of	5	5	5	3	3
Alternator, Generator and Regulator Make	Lucas	Delco-Remy (1)	Lucas	Delco-Remy	Delco-Remy
Starter Make	Lucas	Delco-Remy (1)	Lucas	Delco-Remy	Delco-Remy
Distributor Make	....	....	Lucas (2)	IH	IH

(1) B-414, 364 and 384 Lucas.   (2) 354 Delco-Remy.

CONDENSED SERVICE DATA CONT.

	Series B-275, 354 Diesel	Series 2424, 444 2444, B414, 364, 384 Diesel	Series B-414, 354 Non-Diesel	Series 424, 2424 Non-Diesel	Series 444, 2444 Non-Diesel
TUNE-UP					
Firing Order	1-3-4-2	1-3-4-2	1-3-4-2	1-3-4-2	1-3-4-2
Valve Tappet Gap (Hot)	0.020	0.020	0.020	——In. 0.014-Ex. 0.020——	
Valve Face and Seat Angle	45°	45°	45°	45°	45°
Breaker Contact Gap	......	......	0.014	0.020	0.020
Distributor Timing—Retard	......	......	5° BTDC	TDC	TDC
Distributor Timing—Advanced	......	......	39° BTDC	17° BTDC	17° BTDC
Timing Mark Location	Crankshaft Pulley	Flywheel	————————Crankshaft Pulley————————		
Injection Pump Make	C.A.V.	C.A.V.	......	......	......
Injection Pump Model	BPE or DPA	DPA	......	......	......
Injection Pump Timing—Static	20° BTDC	16° BTDC	......	......	......
Injection Nozzle Make	C.A.V.	C.A.V.	......	......	......
Injection Nozzle Model	BDN8S1 (3)	BDN8S1 (3)	......	......	......
Injection Nozzle Pop Pressure	See Par. 95	See Par. 95	......	......	......
Spark Plug Electrode Gap	......	......	0.024	0.023	0.023
Carburetor Make	......	......	Zenith	Marvel Schebler	Marvel Schebler
Carburetor Model	......	......	VNN or VNP	TXS896	TSX896
Engine High Idle Rpm	See	See	2200	2200	2200
Engine Rated Rpm	Paragraph	Paragraph	2000	2000	2000
Engine Low Idle Rpm	113	113	500-525	425	425

(3) 354, 364 and 384BDN-4SD

SIZES—CAPACITIES—CLEARANCES					
Crankshaft Main Journal Diameter	2.124-2.125	2.124-2.125	2.124-2.125	2.6235-2.6245	2.6235-2.6245
Crankshaft Rod Journal Diameter	1.7495-1.750	1.7495-1.750	1.7495-1.750	2.059-2.060	2.059-2.060
Piston Pin Diameter	1.1021-1.1024	1.1021-1.1024	1.1021-1.1024	0.8591-0.8593	0.8591-0.8593
Valve Stem Diameter	0.341-0.342	0.341-0.342	0.341-0.342	0.3405-0.3415	0.3405-0.3415
Rocker Arm Shaft Diameter	0.748-0.749	0.748-0.749	0.748-0.749	0.748-0.749	0.748-0.749
Camshaft Journal Diameter, No. 1	1.811-1.812	1.811-1.812	1.811-1.812	1.811-1.812	1.811-1.812
Camshaft Journal Diameter, No. 2	1.577-1.578	1.577-1.578	1.577-1.578	1.577-1.578	1.577-1.578
Camshaft Journal Diameter, No. 3	1.499-1.500	1.499-1.500	1.499-1.500	1.499-1.500	1.499-1.500
Main Bearings—Diametral Clearance	0.002-0.004	0.002-0.004	0.002-0.004	0.0009-0.0039	0.009-0.0039
Rod Bearings—Diametral Clearance	0.001-0.0029	0.001-0.0029	0.001-0.0029	0.0009-0.0039	0.0009-0.0039
Camshaft Bearings—Diametral Clearance	0.0015-0.0035	0.0015-0.0035	0.0015-0.0035	0.0009-0.0054	0.0009-0.0054
Piston Skirt Clearance	0.0015-0.0035(4)	0.0031-0.0039	0.0031-0.0039	0.001-0.002	0.001-0.002
Crankshaft End Play	0.004-0.008	0.004-0.008	0.004-0.008	0.004-0.01	0.004-0.01
Camshaft End Play	0.008-0.017	0.008-0.017	0.008-0.017	0.003-0.012	0.003-0.012
Cooling System—Quarts	10.8	10.8(5)	10.8(5)	13	13
Crankcase Oil—Quarts	8(6)	5(7)	5(8)	5.5	5.5
Transmission and Differential—Quarts	20	20	20	20	20
Hydraulic Reservoir—Gallons	3	3	3	3	3

(4) B-276-0.0048-0.0056. (5) B-414—9.5. (6) B-275—5.4. (7) B-414—7.5 (Engine Serial No. 30945 and up).
(8) B-414—7.5 (Engine Serial No. 5595 and up).

TIGHTENING TORQUES—FT.-LBS.					
Cylinder Head	75-80(9)	75-80	75-80	80-90	80-90
Main Bearing Bolts			See Paragraphs 70 and 80		
Connecting Rod Bolts	40-45(10)	40-45	40-45	43-49	43-49

(9) B-275—70-75. (10) B-275—30-35.

FRONT SYSTEM

AXLE MAIN MEMBER

Series B-275-B-414-354-364-384

1. The axle main member pivots on pin (5—Fig. 1 or 2) which is retained in front axle support (7) by a groove pin or bolt (9). Diameter of pivot pin is 1.115-1.116. Press new bushing (4) into place and ream to 1.118-1.120. Normal operating clearance of pivot pin in the bushing is 0.002-0.005.

To renew the axle pivot pin and bushings on Series B-275 and B-414, raise hood and remove the cotter pins from rear of stay rod slides, then disconnect headlight wires from connectors located on forward side of front support. Unbolt and remove hood, being careful not to lose the two pivot spacers.

On Series 354, 364 and 384, remove hood, grille, grille support and front end weight support.

On models equipped with stay rod attachment, unbolt and remove stay rod ball socket cap (3—Fig. 3). Do not

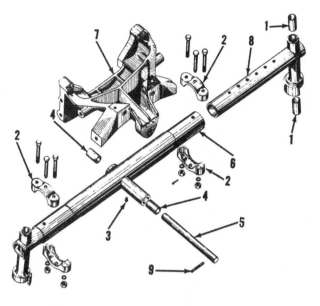

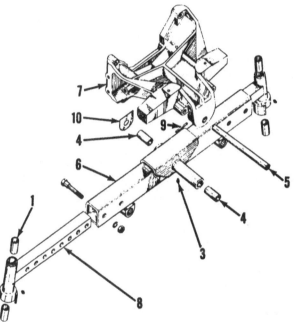

Fig. 1—Exploded view of Series B-275 adjustable front axle and component parts. Bushings (1 and 4) are pre-sized. Shims (10—Fig. 2) are also used with this axle.

1. Knuckle bushings
2. Axle clamps
3. Grease fitting
4. Pivot bushings
5. Pivot pin
6. Axle main member
7. Front support
8. Axle extension
9. Groove pin

Fig. 2—Exploded view of Series B-414, 354, 364 and 384 adjustable front axle and component parts.

1. Knuckle bushings
3. Grease fitting
4. Pivot bushings
5. Pivot pin
6. Axle main member
7. Front support
8. Axle extension
9. Groove pin
10. Shims

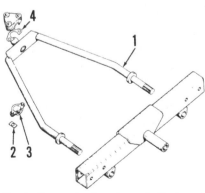

Fig. 3—Power steering stay rod attachment.

1. Stay rod
2. Lock plate
3. Socket cap
4. Shim

Fig. 4—Exploded view of Series 424, 444, 2424 and 2444 standard adjustable front axle and component parts.

1. Steering knuckle
2. Woodruff key
3. Felt washer
4. Thrust bearing
5. Bushing
6. Axle extension
8. Steering arm R.H.
9. Tie rod assy. (2 used)
10. Ball socket
11. Ball
12. Shim
13. Cap
14. Lock plate
15. Clamp
17. Pivot pin
18. Axle main member
19. Pivot bushing
20. Lower bolster
22. Tie rod extension
24. Tube
25. Clamp
26. Tie rod end
27. Steering arm L.H.
28. Spacer

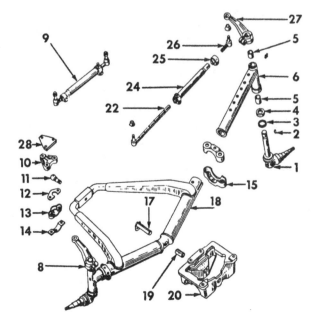

damage shim (or shims) (4) between ball socket and socket cap.

On all models, disconnect drag link or power steering cylinder from left steering arm, then drive out the groove pin or bolt (9—Fig. 1 or 2) which retains axle pivot pin (5) in front support (7). Jack up tractor enough to take the weight off the front axle, drive pivot pin forward out of front support and remove the axle and wheels assembly. Do not damage the shim (or shims) which are between axle main member and front support. Shims are available in thicknesses of 0.002, 0.0032, 0.0048, 0.010, 0.028 and 0.036. Any further disassembly is evident.

Series 424-444-2424-2444

2. The axle main member (18—Fig. 4, 5 and 6) pivots on pin (17) which is retained in lower bolster (20) by a cap screw. The pre-sized pivot bushing (19) is pressed into position and can be removed after the axle main member has been removed from tractor. Normal clearance between pivot pin and bushing is 0.008-0.014.

To remove the axle main member assembly, support front of tractor using a jack under the clutch housing. Disconnect tie-rods from steering arm at center of tractor. On models equipped with standard axle (Fig. 4) or heavy duty axle (Fig. 6) and with power steering, disconnect hydraulic cylinders from axle main member. On models equipped with narrow tread axle (Fig. 5) and with power steering, unpin anchor end of cylinder and unbolt and remove lower cylinder arm (33). Remove cylinder or secure with wire on left side of tractor. Then on all models, adjust the height of the jack to remove weight from front tires. Remove pivot pin retaining cap screw and drive out pivot pin. Unbolt and remove stay rod ball socket cap (13). Raise front of tractor and roll front axle assembly forward from under tractor.

STEERING KNUCKLES

Series B-275-B-414-354-364-384

3. The steering knuckles can be removed from the axle extensions after steering arms and wheel assemblies have been removed.

After being pressed into place, ream inside diameter of new knuckle bushings (1—Fig. 1 or 2) to 1.359-1.360. Outside shaft diameter of new steering knuckle (8 or 11—Fig. 7) is 1.357-1.358. Operating clearance between knuckle shafts and bushings is 0.001-0.003. Use a piloted drift when installing bushings and install same with outer ends flush with bore.

Series 424-444-2424-2444

4. To remove the steering knuckles (1—Fig. 4 or 5) from tractors equipped with adjustable front axle, first support front of tractor and remove front wheel assemblies. Disconnect tie rods (9) from steering arms (8 and 27) and remove the arms from steering knuckles.

Remove Woodruff keys and lower steering knuckles out of axle.

The procedure for removing steering knuckles (1—Fig. 6) from the non-adjustable heavy duty axle is the same as the adjustable axle with the following exceptions: The steering arms and steering knuckles are splined instead of having Woodruff keys. Snap rings (7) must also be removed before steering knuckles can be removed from axle.

Steering knuckle bushings (5—Fig. 4, 5 and 6) should be installed with a closely fitting mandrel and sized after installation, if necessary, to provide a recommended clearance of 0.002-0.004

for the steering knuckle.

When installing thrust bearings (4) on steering knuckles, make certain that open end of outer race is facing downward.

TIE-ROD, DRAG LINK AND TOE-IN

All Models

5. On early B-275 models the tie-rod and drag link ends are of the non-adjustable type and excessive wear is corrected by renewal of the complete tie-rod and/or drag link. On later B-275

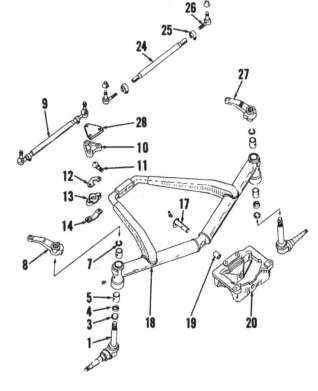

Fig. 6—Exploded view of Series 424, 444, 2424 and 2444 heavy duty non-adjustable front axle and component parts.

1. Steering knuckle
3. Felt washer
4. Thrust bearing
5. Bushing
7. Snap ring
8. Steering arm R.H.
9. Tie rod assy.
10. Ball socket
11. Ball
12. Shim
13. Cap
14. Lock plate
17. Pivot pin
18. Axle main member
19. Pivot bushing
20. Lower bolster
24. Tube
25. Clamp
26. Tie rod end
27. Steering arm L.H.
28. Spacer

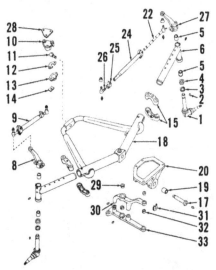

Fig. 5—Exploded view of Series 424, 444, 2424 and 2444 narrow tread adjustable front axle and component parts.

1. Steering knuckle
2. Woodruff key
3. Felt washer
4. Thrust bearing
5. Bushing
6. Axle extension
8. Steering arm R.H.
9. Tie rod assy.
10. Ball socket
11. Ball
12. Shim
13. Cap
14. Lock plate
15. Axle clamp
17. Pivot pin
18. Axle main member
19. Pivot bushing
20. Lower bolster
22. Tie rod extension
24. Tube
25. Clamp
26. Tie rod end
27. Steering arm L.H.
28. Spacer
29. Rear bushing
30. Cylinder arm (upper)
31. Pivot pin lock
32. Front bushing
33. Cylinder arm (lower)

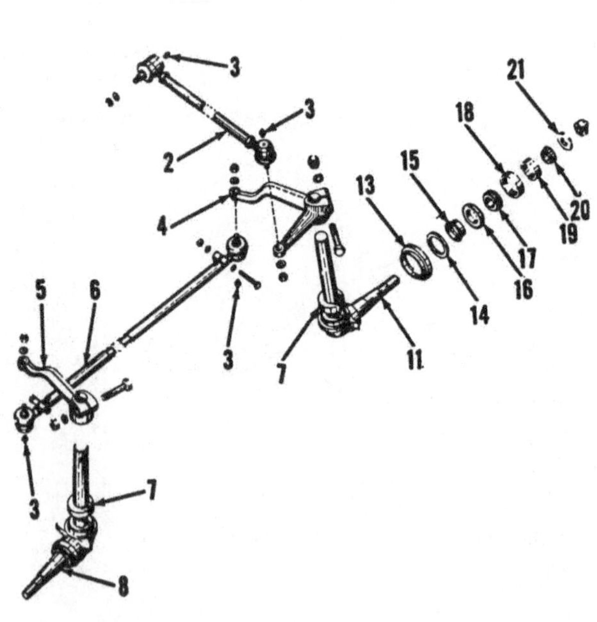

Fig. 7—Exploded view of Series B-275, B-414, 354, 364 and 384 steering knuckles, steering arms, tie rod and drag link used on models with no power steering. Models with power steering are similar.

2. Drag link
3. Grease fitting
4. L.H. steering arm
5. R.H. steering arm
6. Tie rod
7. Thrust bearing
8. R.H. steering knuckle
11. L.H. steering knuckle
13. Dust shield
14. Felt washer
15. Bearing spacer
16. Oil seal
17. Bearing cone
18. Bearing cup
19. Bearing cup
20. Bearing cone
21. Retainer washer

and all other models, component parts are available and can be purchased separately.

On Models B-275 and B-414, correct toe-in of ¼ to 3/8 inch is obtained by loosening the tie-rod clamps and rotating tie-rod either way as required. If necessary, adjust the drag link in a similar manner until there is equal contact at the stops when front wheels are turned full left and full right.

The correct toe-in for Models 354, 364, 414, 2424 and 2444 is 3/16 to 5/16 inch and 0 to 1/16 inch for Model 384. Both tie-rods must be adjusted equally so that full left and full right turning radius of the tractor will be the same.

STEERING GEAR

Series B-275-B414

The cam and lever steering gear assembly is provided with shim adjustments to compensate for wear of the steering worm (cam) shaft and the rocker shaft. The complete assembly is doweled and bolted to the clutch housing.

6. ADJUST WORM SHAFT. To adjust the steering worm shaft it is recommended that the unit be removed from tractor as follows:

Unlatch and raise hood, then remove the battery cover plate. Remove battery hold-down, disconnect battery cables and lift out batteries.

NOTE: Removal of left battery will be simplified if the hold-down studs are unscrewed from battery tray. It will also be easier to handle the fuel tank if the fuel is drained at this time.

Remove knob from fuel shut-off rod, then unbolt and remove the lower instrument panel. Remove filler panel from center of upper instrument panel then unbolt upper instrument panel and lay same on right hand foot plate.

NOTE: It is not necessary to disconnect the oil pressure line from gage providing care is exercised.

Remove steering wheel and if force is required, use a puller. Do not use a hammer. Disconnect link from lower end of hand throttle lever, then unbolt bracket and friction disc assembly from upper end of steering shaft tube and pull hand throttle lever from fuel tank tunnel. Straighten tabs of lock plates, then unbolt tank and move same up and over steering column. Disconnect fuel shut-off rod at forward end and remove. Remove the cross shaft to bellcrank control rod. Disconnect drag link

from drop (Pitman) arm, then unbolt and remove steering gear assembly from tractor.

With assembly removed as outlined, vary shims under worm shaft end cap (15—Fig. 8) to provide zero end play of the worm shaft. Shims are available in thicknesses of 0.0024, 0.005 and 0.010.

7. ADJUST ROCKER SHAFT. Steering gear assembly need not be removed to adjust rocker end play. Remove side cover (1—Fig. 8) and vary shims (2) until a slight drag can be felt when steering gear passes through its mid (straight ahead) position. Disconnect drag link from drop (Pitman) arm when making this adjustment. Shims are available in thicknesses of 0.0024, 0.005 and 0.010.

8. OVERHAUL. To overhaul the steering gear assembly, first remove same as outlined in paragraph 6.

With unit removed, proceed as follows: Remove drop (Pitman) arm (6—Fig. 8) by using a suitable puller. Remove side cover (1) and be careful not to lose or damage shims (2). Rocker shaft (3) can now be removed.

NOTE: Do not drive on drop (Pitman) arm to remove it from rocker shaft as damage to internal parts could result. If a suitable puller is not available, remove side cover first, then drive rocker shaft from drop arm using the retaining nut, or some other means, to prevent damage to the rocker shaft threads.

Remove worm shaft end cover (15) and be careful not to damage or lose shims (14). Withdraw worm shaft (13) and bearing assemblies (10 and 11). Balance of disassembly is evident and will be dictated by the repair needed.

If new bushings (4) are installed, align ream same to an inside diameter of 1.2495-1.2510. If felt bushing (7) is renewed, it is recommended that the new felt bushings be soaked in warm graphite for 12 hours before installation. Oil seal (5) is installed with lip facing toward inside.

Reassemble by reversing the disassembly procedure. Adjust worm shaft to zero end play by varying shims (14) located under end cap (15). Adjust rocker shaft (3) by varying shims (2) under side cover (1) until a slight drag

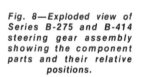

Fig. 8—Exploded view of Series B-275 and B-414 steering gear assembly showing the component parts and their relative positions.

1. Side cover
2. Shim (0.0024, 0.005, 0.010)
3. Rocker shaft
4. Bushings
5. Oil seal
6. Drop (pitman) arm
7. Felt bushing
8. Housing & tube
9. Filler plug
10. Bearing cup
11. Bearing
13. Cam (worm) shaft
14. Shim (0.0024, 0.005, 0.010)
15. End cover

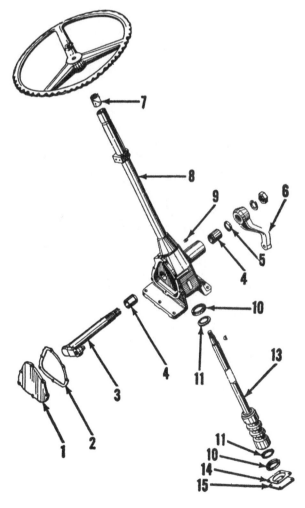

is felt on steering wheel when gear passes through the straight ahead position. All shims are available in thicknesses of 0.0024, 0.005 and 0.010. Use paper gaskets on end cover (15) and side cover (1) when assembling. Torque drop arm retaining nut to 150 ft.-lbs.

Series 354-364-384

9. **R&R AND OVERHAUL.** To remove steering gear assembly, remove hood, battery and battery carrier. On diesel tractors remove leak-off pipe and shut-off lever. Close shut-off tap and disconnect fuel pipe and sender unit cable, then unbolt and remove fuel tank. Remove steering wheel, unbolt instrument panel and support from steering gear assembly, disconnect foot accelerator and necessary wires, lift unit up and set to one side.

NOTE: It is not necessary to disconnect oil pressure line from gage providing care is exercised.

Disconnect drag link from drop (Pitman) arm, then unbolt and remove steering gear assembly from tractor. With unit removed proceed as follows: Punch mating marks on the drop arm and rockershaft to aid assembly. Remove drop arm (13—Fig. 9) by using suitable puller, then remove side plate (16). Rockershaft (14) can now be removed. Remove the tube assembly (2) and being careful not to damage or lose shims (5), remove wormshaft (7) and bearing assembly.

After being pressed into place the inside diameter of new rocker shaft bushings (10—Fig. 9) is 1.3748-1.3758. Oil seal (11) is installed with lip facing toward inside.

Reassemble by reversing the disassembly procedure. Adjust worm shaft to zero end play by varying shims (5) located under tube assembly (2).

NOTE: Bearing must not be preloaded.

Adjust rockershaft (14) by turning thrust screw (18) in side cover (16) until a slight drag is felt on steering wheel when gear passes through the straight ahead position. Torque drop arm retaining nut to 150 ft.-lbs.

Series 424-444-2424-2444

The manual steering worm and worm wheel are located in the steering gear housing which is bolted to the front face of the engine. The center steering arm is keyed to the lower portion of the worm wheel shaft and the axle support (lower bolster) is retained to the steering gear housing by four cap screws. The manual steering unit is non-adjustable.

10. **REMOVE AND REINSTALL.** The steering gear housing, lower bolster and front axle assembly can be removed with radiator installed; however, some mechanics prefer to remove the radiator.

To remove steering gear housing with radiator attached, proceed as follows: Drain cooling system, remove hood and disconnect upper and lower radiator hoses. Disconnect the headlight wires and unbolt the radiator brace and fan shroud from radiator. Drive roll pin from forward yoke of steering shaft universal and remove cap screws from front steering shaft support. Drive front universal from steering worm shaft. Place wood blocks between steering gear housing and axle to prevent tipping. Using a suitable jack under the clutch housing raise front of tractor to remove most of the weight from the front tires. Unbolt the stay rod bracket from clutch housing the steering gear housing from front of engine. Raise engine until crankshaft pulley will clear steering gear housing, then roll complete front end assembly from tractor.

11. **OVERHAUL.** The steering gear can be overhauled without removing the assembly from tractor.

To remove the steering worm, proceed as follows: Remove hood, grille, side panels and grille housing and then, drain steering gear housing. Disconnect

the steering shaft front universal from worm shaft and the steering shaft front support from the fuel tank bracket. Drive universal from worm shaft and remove the Woodruff key. Remove the steering worm retainer (5—Fig. 10) and turn worm shaft forward out of housing.

To remove ball bearing (7), remove cotter pin and nut and bump worm shaft out of bearing. When reinstalling the bearing on worm shaft, tighten nut to a torque of 75-120 ft.-lbs. Do not back off nut to below 75 ft.-lbs. to install cotter pin. Worm shaft bushing (11) and oil seal (12) can be renewed at this time. If needle bearing (10) is being renewed, it must be installed to the dimension shown in Fig. 11.

To remove the steering worm wheel and shaft assembly, proceed as follows: Disconnect the tie-rods from center steering arm and unbolt stay rod bracket from clutch housing. Place a jack under clutch housing to support front of tractor. Remove axle pivot pin, raise front of tractor and roll the front axle assembly forward. Unbolt and remove lower bolster from steering gear housing. Remove the cap screws retaining the worm wheel cage (21—Fig. 10) to steering gear housing and remove the worm wheel and shaft assembly. Remove center steering arm (23) and withdraw worm wheel and shaft (18). Bushing and oil seal can be renewed at this time. Oil seal is

Fig. 9—Exploded view of Series 354, 364 and 384 steering gear assembly showing the component parts and their relative positions.

1. Bushing assy.
2. Tube assy.
3. Bushing
4. Oil level plug
5. Shims (0.0024, 0.005 & 0.010)
6. Bearing assy.
7. Shaft assy.
8. Bearing assy.
9. Housing
10. Bushings
11. Seal
12. Ring
13. Drop (pitman) arm
14. Rocker shaft
15. Gasket
16. Side cover
17. Locknut
18. Thrust screw

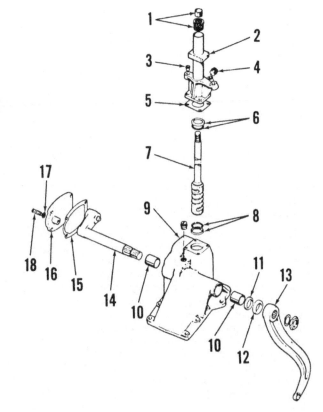

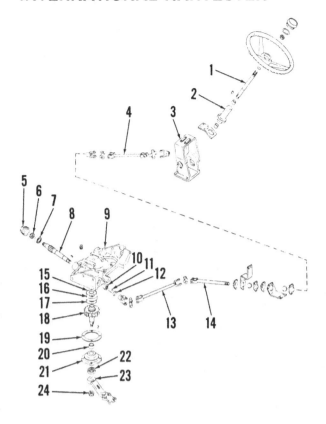

1. Rear steering shaft
2. Rear shaft support
3. Pedestal
4. Rear steering joint
5. Steering worm retainer
6. Nut
7. Ball bearing
8. Worm shaft
9. Steering gear housing
10. Needle bearing
11. Bushing
12. Oil seal
13. Front steering joint
14. Center steering joint
15. Shim (0.008)
16. Adapter
17. Shim (0.021 & 0.033)
18. Worm wheel
19. Gasket
20. Bushing
21. Cage
22. Oil seal
23. Center steering arm
24. Nut

installed with lip of same facing toward inside of steering housing. When reinstalling center steering arm, clamp the arm in a vise and tighten the retaining nut (24) to a torque of 200-250 ft.-lbs. If, when tightening the steering arm retaining nut, a castellation of the nut does not align with the cotter pin hold at some point between the specified

200-250 ft.-lbs. torque, continue to tighten nut. Do not back-off (loosen) nut to obtain alignment.

When reinstalling the worm wheel assembly, be sure large diameter of adater (16) is on top side. Install shims (15 and 17) as required to remove excessive end play of worm wheel and shaft assembly (18).

POWER STEERING

Series B-414-354-364-384

12. The power steering system used on B-414 prior to serial number 21196 is comprised of the steering gear

assembly, a flow control valve and a power steering cylinder and control valve assembly. See Fig. 12 for a schematic view showing the general arrangement of the component parts. Pressurized oil for the operation of the power steering system is furnished by the same engine driven pump which supplies the hydraulic lift system.

Late production B-414 tractors (serial number 21196 and up) and all 354, 364 and 384 tractors are equipped with a dual stage pump and a relief valve block. Refer to Fig. 13. Flow control (divider) valve is not used on this system.

For information pertaining to the steering gear assembly, which is the same as that used on models with no power steering, refer to paragraphs 6 through 9.

Series 424-444-2424-2444

13. A full time (Hydrostatic) power steering system is used on this series. Refer to Figs. 14 and 15 for a schematic view showing the general layout of components and tubing for the system.

NOTE: The maintenance of absolute cleanliness of all parts is of utmost importance in the operation and servicing of the hydraulic power steering system. Of equal importance is the avoidance of nicks or burrs on any of the working parts.

The power steering system is composed of the following components: An engine driven gear type pump, a flow divider valve, a spool type directional control valve, a gerotor type pump on the steering wheel shaft and two ram type single acting steering cylinders on models equipped with the standard axle (Fig. 4) or the heavy duty axle (Fig. 6) and one double acting cylinder on models equipped with narrow tread front axle (Fig. 5).

4.200

Install needle bearing to dimension shown

Fig. 11—When renewing needle bearing in Series 424 or 444 mechanical steering gear housing, install the bearing to dimension shown.

Fig. 12—View showing general arrangement of early Series B-414 power steering components.

1. Reaction bracket
2. Cylinder & valve assy.
3. Drag link
4. Connecting link
5. Banjo
6. Banjo bolt
7. Hose
8. Hose
9. Flow control valve
10. Suction line
11. Hydraulic pump
12. Pressure line
13. Adapter
14. Shield
15. Rubber insert

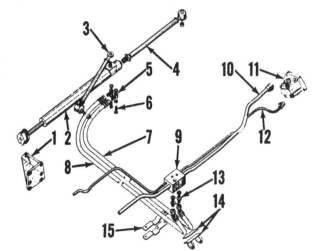

LUBRICATION AND BLEEDING

Series B-414-424-444-2424-2444-354-364-384

14. The hydraulic lift housing serves as the fluid reservoir for both the power steering system and the hydraulic lift system. Capacity of the reservoir is 3.0 gallons. Recommended fluid is IH "Hy-Tran" fluid.

To bleed the power steering system on B-414, 354, 364 and 384 tractors, fill reservoir to bottom of filler hole, start engine and cycle the steering system from lock to lock position until all air is bled from system. Recheck fluid level in reservoir and add fluid as required.

On Models 424, 444, 2424 and 2444, fill reservoir to proper level and then, start engine and run at low idle speed. Rotate steering wheel (hand pump) as rapidly as possible in order to activate the control valve and continue to rotate the steering wheel until the front wheels reach the stop in the direction in which the steering wheel is being turned. Now quickly reverse the direction of steering wheel and follow same procedure until front wheels reach stop in opposite direction. Continue to turn front wheels from lock to lock until steering wheel has no wheel spin (free wheeling) and has a solid feel with no skips or sponginess. Check and add fluid to reservoir, if necessary.

TROUBLESHOOTING

Series B-414-354-364-384

15. Some of the troubles that may arise in the operation of the power steering system and their possible causes are given as follows:
1. Loss of Power Assist.
   a. Fluid level low.
   b. Connections loose or damaged.
   c. Relief valve sticking.
   d. Power cylinder internal parts worn.
   e. Control valve sticking.
   f. Low pump output.
2. Binding.
   a. Power cylinder internal parts worn.
   b. Ball pin binding.
   c. Control valve sticking.
   d. Worn ball pin cups.
   e. Loose locating sleeve.
3. Heavy Steering.
   a. Relief valve sticking.
   b. Power cylinder internal parts worn.
   c. Control valve sticking.
   d. Worn ball pin cups.
   e. Loose locating sleeve.
   f. Low pump output.
4. Noisy Operation.
   a. Fluid level low.

b. Relief valve sticking.
c. Control valve sticking.
5. Steering Chatter.
   a. Worn piston rod ball cups.
   b. Loose reaction bracket.

Series 424-444-2424-2444

16. The following list shows some of the troubles, and their possible causes, which may occur in the operation of the Hydrostatic steering system.
1. No Power Steering or Steers Slowly.
   a. Excessive load on front wheels and/or air pressure low in front tires.

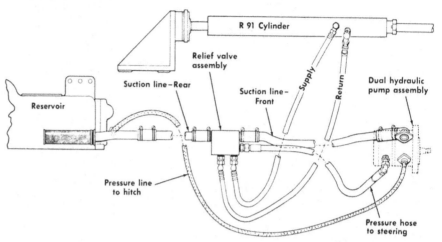

Fig. 13—View showing general arrangement of power steering components used on all Series 354, 364, 384 and Series B-414 S/N 21196 and later. On tractors equipped with R92 cylinder, supply and return hoses are connected to cylinder in reverse of connections shown.

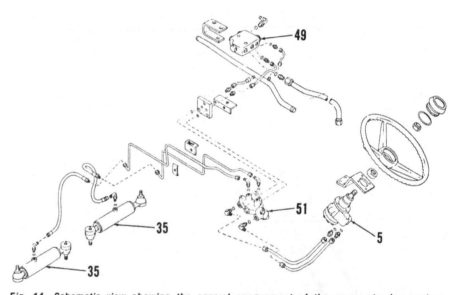

Fig. 14—Schematic view showing the general arrangement of the power steering system components of Series 424, 444, 2424 and 2444 used with front axles shown in Figs. 4 and 6.

5. Hand pump	49. Flow divider valve
35. Steering cylinder	51. Control (pilot) valve

Fig. 15—Schematic view showing the general arrangement of the power steering system components used on Series 424, 444, 2424 and 2444 equipped with the narrow tread front axle shown in Fig. 5.

5. Hand pump
35. Steering cylinder
49. Flow divider valve
51. Control (pilot) valve

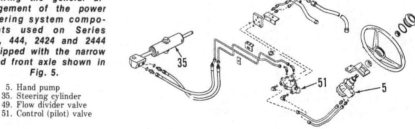

b. Steering cylinder faulty.

c. Faulty commutator in hand pump.

d. Flow divider valve spool sticking or leaking excessively.

e. Control (pilot) valve spool sticking.

2. Will Not Steer Manually.

a. Excessive load on front wheels and/or air pressure low in front tires.

b. Pumping element in hand pump faulty.

c. Steering cylinder faulty or damaged.

d. Pressure check valve leaking.

e. Control (pilot) valve spool binding or centering spring broken.

3. Hard Steering Through Complete Cycle.

a. Low pressure from supply pump.

b. Internal or external leakage.

c. Line between hand pump and control (pilot) valve obstructed.

d. Faulty steering cylinder.

e. Excessive load on front wheels and/or air pressure low in front tires.

4. Momentary Hard or Lumpy Steering.

a. Air in power steering circuit.

b. Control (pilot) valve sticking.

5. Shimmy.

a. Control (pilot) valve spring weak or broken.

b. Control (pilot) valve spring washers bent, worn or broken.

SYSTEM OPERATING PRESSURE

Series B-414-354-364-384

17. A pressure test of the power steering circuit can be made as follows: Connect a pressure gage, capable of registering at least 3000 psi, in series with the pressure line running to the power steering cylinder. Run engine and cycle steering system until fluid is at operating temperature, advance the engine speed to 2000 rpm, turn front wheels against either stop, continue to apply turning effort to steering wheel and observe the pressure gage reading which should be 900-1100 psi.

If pressure is as stated, hydraulic pump and relief valve can be considered satisfactory and any trouble is located in the power steering cylinder, control valve or connections.

If pressure is higher than stated, relief valve is probably stuck in the closed position and should be removed and cleaned or renewed. If pressure is lower than specified, remove and inspect relief valve assembly (items 6 through 10—Fig. 16 or items 13 through 19—Fig. 17). Spring (8—Fig. 16) should have a free length of 1.250 inches and test 21 lbs. when compressed to a length of 1.087 inches. Plug (6), spring (8), ball (9) and sleeve (10) are catalogued separately. Spring (15—Fig. 17) should have a free length of 2.250 inches and should test 47.5 lbs. when compressed to a length of 2.012 inches. If servicing the relief valve does not correct the pressure reading, overhaul hydraulic pump as outlined in paragraph 198 or 199.

Series 424-444-2424-2444

18. To check power steering operating pressure, refer to paragraph 34 which gives the method of checking the flow divider relief valve pressure.

POWER STEERING OPERATIONAL TESTS

Series 424-444-2424-2444

19. The following tests are valid only when the power steering system is completely void of any air. If necessary, bleed system as outlined in paragraph 14 before performing any operational tests.

20. MANUAL PUMP. With engine driven pump inoperative (engine not running), attempt to steer manually in both directions.

NOTE: Manual steering with engine driven pump not running will require high steering effort. If manual steering can be accomplished with engine driven pump inoperative, it can be assumed that the manual pump will operate satisfactorily with the engine driven pump operating.

Refer also to paragraph 22 for information regarding steering wheel (manual pump) slip.

21. CONTROL (PILOT) VALVE. Attempt to steer manually (engine not running). Manual steering will require high steering effort but if steering can be accomplished, control (pilot) valve is working.

No steering can be accomplished if control valve is stuck on center. A control valve stuck off center will allow steering in one direction only.

22. STEERING WHEEL SLIP TEST. Steering wheel slip is the term used to describe the inability of the steering wheel to hold a given position without further steering movement. Wheel slip is generally due to leakage, either internal or external, or a faulty hand pump, steering cylinder or control (pilot) valve. Some steering wheel slip, with hydraulic fluid at operating temperature, is normal and permissible. A maximum of three revolutions per min-

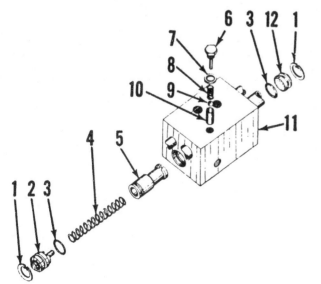

Fig. 16—Exploded view of early Series B-414 flow control and relief valve assembly used with single stage pump.

1. Snap ring
2. Plug
3. "O" ring
4. Spring
5. Flow control valve
6. Plug & pin
7. Seal washer
8. Spring
9. Relief valve ball
10. Valve guide
11. Valve body & tubes assy.
12. Plug

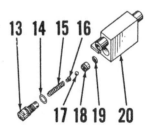

Fig. 17—Exploded view of Series B-414, 354, 364 and 384 power steering relief valve assembly used with dual stage pump.

13. Spring housing	17. Relief valve ball
14. "O" ring	18. Valve seat
15. Spring	19. Washer
16. Ball rider	20. Relief valve block

ute is acceptable. By using the steering wheel slip test and a process of elimination, a faulty unit in the power steering system can be located.

However, before making a steering wheel slip test to locate faulty components, it is imperative that the complete power steering system be free of air before any testing is attempted. Hydraulic fluid must also be at the correct level in reservoir.

To check for steering wheel slip, cycle steering system until all components and fluid are at operating temperature (150°F). Remove steering wheel cap (monogram), then turn front wheels until they are against stop. Attach a torque wrench to steering wheel nut.

NOTE: Either an inch-pound, or a foot-pound wrench may be used; however, an inch-pound wrench is recommended as it is easier to read.

Advance throttle until engine reaches 2000 rpm, then apply 72 inch-pounds (6 foot-pounds) to torque wrench in same direction as the front wheels are positioned against stop. Keep this pressure (torque) applied for a period of one minute and count the revolutions of the steering wheel. Use same procedure and check the steering wheel slip in the opposite direction. A maximum of three revolutions per minute in either direction is acceptable and system can be considered as operating satisfactorily. If the steering wheel revolutions per minute exceed three, record the total rpm for use in checking the steering cylinder on models equipped with double acting steering cylinder (Fig. 19).

NOTE: While three revolutions per minute of steering wheel slip is acceptable, it is generally considerably less in normal operation.

23. STEERING CYLINDER TEST. If steering wheel slip is more than three revolutions per minute on models equipped with double acting steering cylinder, check the steering cylinder for internal leakage as follows: Be sure operating temperature is being maintained, then disconnect and plug the steering cylinder lines. Repeat the steering wheel slip test, in both directions, as described in paragraph 22. If steering wheel slip is ½ rpm or more, **below** that recorded in paragraph 22, overhaul or renew the steering cylinder.

POWER STEERING CYLINDER AND CONTROL VALVE

Series B-414-354-364-384

24. **REMOVE AND REINSTALL.** To remove the power steering cylinder and control valve assembly, first disconnect hoses from cylinder and either plug hoses or suspend the disconnected ends high enough to prevent fluid drainage. Turn front wheels full left and full right several times to clear oil from cylinder and control valve. Disconnect drag link from ball pin, then disconnect cylinder from steering arm and reaction bracket and remove assembly from tractor.

CAUTION: Do not use a wedge or pinch bar to remove drag link from ball pin as damage to control valve could result. Use a puller.

When reinstalling, be sure anchor ball (50—Fig. 18) moves freely but without slack in its cups. Vary shims (52) to adjust anchor ball.

After cylinder and valve are installed, fill and bleed system as outlined in paragraph 14.

25. **OVERHAUL.** The overhaul procedure given is for the R91 steering cylinder. Overhaul procedure for the R92 cylinder is similar. The pressure and return ports are reversed on the

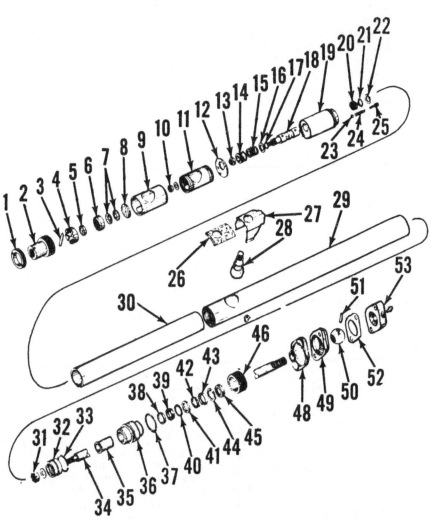

Fig. 18—Exploded view of Series B-414, 354, 364 and 384 power steering cylinder and valve assembly. Refer to Fig. 12 for view of connecting link which fits into take-off extension (2).

1. Lock ring	14. Reaction ring	27. Spring cover	40. Vellumoid washer
2. Take-off extension	15. Reaction spring	28. Ball pin	41. Washer
3. Spring clip	16. Washer	29. Outer tube	42. Scraper
4. Ball cup holder	17. "O" ring	30. Inner tube	43. Scraper housing
5. Ball cup	18. Valve spool	31. Nut	44. Snap ring
6. Ball cup	19. Valve body	32. Piston	45. Scraper
7. Belleville washers	20. End cover	33. Piston ring	46. Lock ring
8. Backing washer	21. "O" ring	34. Piston rod	48. Rubber boot
9. Locating sleeve	22. Snap ring	35. Bushing	49. Inner ball cup
10. Nut	23. Relief valve ball	36. Bearing housing	50. Anchor ball
11. Operating sleeve	24. Spring	37. "O" ring	51. Pin
12. Spacer	25. Plug & pin	38. Seal spreader	52. Shim
13. Collar	26. Felt pad	39. Gland seal	53. Outer ball cup

R92 cylinder and the by-pass valve is installed in the opposite end of valve body. The reaction ring was also eliminated on the R92 cylinder. With power steering cylinder and control valve assembly removed as outlined in paragraph 24, refer to Fig. 18 and proceed as follows: Drive pin (51) from anchor ball (50) and remove anchor ball, inner ball cup (49) and rubber cover (48) from piston rod. Remove cotter pins from cylinder ends. Remove bearing lock ring (46), then pull piston and bearing assembly (36) from cylinder. Remove retaining nut and washer and remove piston (32) from piston rod, then remove bearing assembly (36) from piston end of piston rod. Snap ring (44), scraper (45) and seal assembly (items 38 through 43) can now be removed from bearing assembly (36).

Loosen jam nut and remove connecting link from forward end of cylinder. Remove spring cover (27) and felt pad (26). Remove lock ring (1) and take-off extension (2), then remove spring clip (3), ball cup holder (4) and ball cup (5). Remove ball pin (28) from end of cylinder, then remove ball cup (6), Belleville washers (7) and backing washer (8). Remove control valve and inner tube (30) assembly from outer tube (29) and remove inner tube from control valve body. Remove the cotter pin, nut (10) and washer from forward end of spool (18). The operating sleeve (11), spacer (12) and collar (13) can now be removed. Remove reaction ring (14), spring (15), washer (16) and "O" ring (17) from spool. Remove snap ring (22), end cover (20) and "O" ring. Remove relief valve plug and pin (25), spring (24) and ball (23).

26. Thoroughly clean all parts and inspect for wear and/or damage.

Inspect spool and valve body for burrs and scoring. Burrs can be removed using crocus cloth or very fine emery cloth. Fit of spool in body is considered satisfactory when spool will fall freely by its own weight into the valve body when coated with a light film of oil.

NOTE: Valve spool and body are not catalogued separately and must be purchased as a mated pair.

Use caution when depressing spool not to round off any of the sharp edges. To do so may affect the operation of the valve.

Inspect the mating surfaces of the operating and locating sleeves and remove any burrs or scoring with crocus cloth or very fine emery cloth. The operating sleeve should slide freely in the locating sleeve when coated with a light film of oil.

Inspect the inner tube, piston, piston ring, piston rod and bearings for signs of wear, scoring and distortion. Renew as necessary.

NOTE: Piston rod (34), bushing (35), bearing housing (36), anchor ball (50) and anchor ball pin (51) are not available separately.

Inspect anchor ball, anchor ball pin and anchor ball cups for signs of wear or hammering and renew as necessary.

Use all new "O" rings when reassembling.

27. To reassemble the power steering cylinder and control valve assembly, the following procedure is recommended: Install relief valve ball and spring in valve body and tighten the plug and pin assembly securely. Any renewal of dowel pins can be accomplished by driving the new pins into place. Install new "O" ring on end cover (20) of valve body end cover, then install end cover in valve body and secure with snap ring (22). Install inner tube (30) on valve body and be sure slot in inner tube mates with dowel of valve body. Place new "O" rings on forward end of valve spool and outside diameter of reaction ring, then install the washer, spring and reaction ring on spool and be sure the chamfers of reaction washer and reaction ring face valve spool. Install the collar on the threaded portion of spool, then install spacer over collar. Position spacer so that the holes in same align with corresponding holes and dowel pin in valve body. Assemble the operating sleeve on the collar, then install the steel washer and nut. Tighten nut to a torque of 110 in.-lbs. and install cotter pin. Lubricate the operating sleeve, then slide locating sleeve over same.

Install seal spreader into the piston rod bearing, flat side first; then, using the proper size socket, install gland seal. Start from piston end of piston rod and slide snap ring, scraper housing, square section scraper, flat washer and vellumoid washer on piston rod. Install bearing assembly on piston rod, install seals and washers in counterbore of bearing assembly and secure in place by installing snap ring. Install scraper. Install new "O" ring on outside diameter of bearing assembly.

Install piston ring on piston, then install piston to piston rod, flat side first and install washer and castellated nut. Tighten nut to a torque of 34-35 ft.-lbs. and install cotter pin.

NOTE: Do not overtighten nut or piston may distort and bind in tube.

Compress the piston ring and slide the piston rod assembly into the inner tube assembly as far as it will go, making sure shoulder of bearing assem-

bly enters inner tube. Lubricate spool assembly and carefully slide same into valve body making sure small hole in spacer is located on the dowel of the valve body.

Slide the inner tube and valve assembly into the outer tube from the anchorage (aft) end. Align the hose ports radially. Screw piston rod bearing lock ring into outer tube until hose ports are also aligned longitudinally and a slot in the lock ring is aligned with cotter pin hole, then install cotter pin.

Install backing plate, chamfer side first, into the operating sleeve and be sure it is correctly seated. Place heavy grease on the Belleville washers and place them in recess of the rear ball cup with the convex sides together; then install ball cup and washers in the operating sleeve. Install ball pin, front ball cup, ball cup holder, spring clip and take-off extension. Tighten take-off extension until snug, then back-off ¼-turn. Install take-off extension lock ring, tighten and secure with cotter pin.

Install rubber cover and inner anchor ball cup on piston rod, then install anchor ball and retaining pin.

Mount power steering cylinder on tractor as outlined in paragraph 24. Fill and bleed power steering system as outlined in paragraph 14.

STEERING CYLINDER

Series 424-444-2424-2444

28. **DOUBLE ACTING CYLINDER.** To remove the double acting steering cylinder (Fig. 19), disconnect and immediately plug the hydraulic lines. Remove cap screws from lower steering arm (33—Fig. 5) and disengage cylinder from steering arms. Remove pin retaining anchor assembly (15—Fig. 19) to axle main member and remove cylinder assembly from tractor.

Move piston rod back and forth several times to clear oil from cylinder. Place end of piston rod which has the flats in a vise, then unscrew and remove anchor assembly (15). Remove cylinder head retaining ring (7) as follows: Lift end of retainer ring out of slot, then using a pin type spanner, rotate cylinder head (2) and work retainer ring out of its groove. Cylinder head and the piston and rod assembly (10) can now be removed from cylinder tube (12). Remove remaining cylinder head in the same manner. All seals, "O" rings and back-up washers are now available for inspection and/or renewal. Clean all parts in a suitable solvent and inspect. Check cylinder tube for scoring, grooving and out-of-roundness.

Light scoring can be polished out by using a fine emery cloth and oil, providing a rotary motion is used during the polishing operation. A cylinder tube that is heavily scored or grooved, or that is out-of-round, should be renewed. Check piston rod and piston for scoring, grooving and straightness. Polish out very light scoring with fine emery cloth and oil, using a rotary motion. Renew rod and piston assembly if heavily scored or grooved, or if piston rod is bent. Inspect piston ring (9) for frayed edges, wear and imbedded dirt or foreign particles. Renew piston ring if any of the above conditions are found.

NOTE: Do not remove "O" ring (8) located under the piston ring unless renewal is indicated as it is not necessary to renew this "O" ring unless it is damaged.

Inspect balance of "O" rings, back-up washers and seals and renew as necessary. Inspect bores of cylinder heads and renew same if excessively worn or out-of-round.

Reassemble steering cylinder as follows: Place "O" ring (14), with back-up washer (13) on each side, in groove at inner end of anchor assembly oil tube. Install piston rod "O" ring (11) in groove at threaded end of piston rod.

Install wiper seal (1), back-up washer (3), "O" ring (4), back-up washer (5) and cylinder head "O" ring (6) to cylinder head, then install cylinder head assembly over threaded end of piston rod. Lubricate "O" ring and back-up washers on inner end of anchor assembly oil tube and carefully insert into threaded end of piston rod. Lubricate piston rod "O" ring (11) and push anchor assembly toward piston rod. As "O" ring on inner end of oil tube approaches the drilled hole (port) in piston rod (located near piston), use IHC tool FES 65, or equivalent, to depress "O" ring and washers so they will pass the port without being damaged. Screw anchor assembly onto piston rod and tighten to a torque of 150 ft.-lbs. Lubricate piston ring (9) and cylinder head "O" ring (6), then using a ring compressor, or a suitable hose clamp, install piston and rod assembly into cylinder tube. Install cylinder head in cylinder tube so hole in groove will accept nib of retaining ring. Position retaining ring and pull same into its groove by rotating cylinder head. Complete balance of assembly by reversing disassembly procedure.

Reinstall unit on tractor, then fill and bleed the power steering system as outlined in paragraph 14.

29. SINGLE ACTING CYLINDERS. To remove the single acting steering cylinders, disconnect the lines from steering cylinders and cap the lines to prevent dirt from entering system. Disconnect the cylinders from the center steering arm and axle main member, then remove the cylinders from tractor.

To disassemble the removed cylinders, pull the rod (5—Fig. 20) from the cylinder barrel (1). Normal service on the single acting cylinders consists of renewing the "O" ring (2), back-up washer (3) and wiper seal (4).

After reinstalling the cylinders, fill and bleed the system as outlined in paragraph 14.

HAND PUMP

Series 424-444-2424-2444

30. REMOVE AND REINSTALL. To remove the hand pump, remove steering wheel and panel below instrument panel.

NOTE: Use a puller to remove steer-

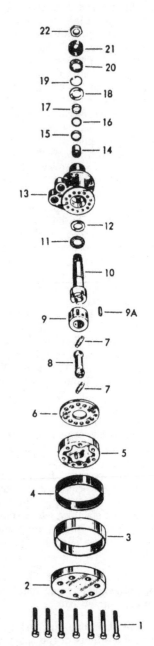

Fig. 21—Exploded view of power steering hand pump assembly.

1. Cap screws	11. Thrust bearing
2. End plate	12. Bearing race
3. Seal retainer	13. Body
4. Seal	14. Needle bearing
5. Rotor set	15. Seal
6. Spacer	16. Back-up washer
7. Link pin	17. Spacer
8. Drive link	18. Washer
9. Commutator	19. Snap ring
9A. Commutator pin	20. Felt seal
10. Coupling (input)	21. Water seal
shaft	22. Wheel nut

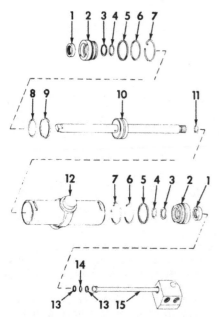

Fig. 19—Exploded view of double acting steering cylinder used on Series 424, 444, 2424 and 2444 equipped with narrow tread front axle.

1. Wiper seal	8. Piston "O" ring
2. Cylinder head	9. Piston ring
3. Back-up washer	10. Piston & rod
4. "O" ring	11. "O" ring
5. Back-up washer	12. Cylinder tube
6. "O" ring	13. Back-up washer
7. Cylinder head	14. "O" ring
retaining ring	15. Anchor assy.

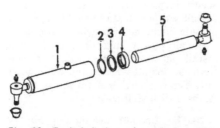

Fig. 20—Exploded view of a single acting cylinder used on Series 424, 444, 2424 and 2444 equipped with standard adjustable front axle or heavy duty non-adjustable front axle.

1. Barrel	
2. "O" ring	4. Wiper seal
3. Back-up washer	5. Rod

ing wheel, do not bump on upper end of steering wheel shaft.

Disconnect lines from hand pump, then unbolt and remove hand pump from under instrument panel.

Reinstall by reversing the removal procedure and bleed power steering system as outlined in paragraph 14.

31. **OVERHAUL MANUAL (HAND) PUMP.** Remove the manual pump as outlined in paragraph 30. Clear fluid from unit by rotating steering wheel (input) shaft back and forth several times. Place unit in a soft jawed vise with end plate on top side, then remove end plate retaining cap screws and lift off end plate (2—Fig. 21).

NOTE: Lapped surfaces of end plate (2), pumping element (5), spacer (6), commutator (9) and pump body (13) must be protected from scratching, burring or any other damage as sealing of these parts depends only on their finish and flatness.

Remove seal retainer (3), seal (4), pumping element (5), link pin (7) and spacer (6) from body (13). Remove commutator (9) and drive link (8), with link pin (7) and commutator pin (9A), from body. Smooth any burrs or nicks which may be present on input shaft (10), wrap spline with masking tape, then remove input shaft from body. Remove bearing race (12) and thrust bearing (11) from input shaft. Remove snap ring (19), washer (18), spacer (17), back-up washer (16) and seal (15). Do not remove needle bearing (14) unless renewal is required. If it should be necessary to renew bearing, press same out pumping element end of body.

Clean all parts in a suitable solvent and if necessary, remove paint from outer edges of body, spacer and end plate by passing these parts lightly over crocus cloth placed on a perfectly flat surface. Do not attempt to dress out any scratches or other defects since these sealing surfaces are lapped to within 0.0002 of being flat. However, in

cases of emergency, a spacer that is damaged on one side only may be used if the smooth side is positioned next to the pumping element and the damaged side is lapped flat.

Inspect commutator and housing for scoring and undue wear. Bear in mind that burnish marks may show, or discolorations from oil residue may be present, on commutator after unit has been in service for some time. These can be ignored providing they do not interfere with free rotation of commutator in body.

Check fit of commutator pin in the commutator. Pin should be a snug fit and if bent, or worn until diameter at contacting points is less than 0.2485, renew pin.

Measure inside diameter of input shaft bore in body and outside diameter of input shaft bearing surface. If body bore is 0.006, or more, larger than shaft diameter, renew shaft and/or body and commutator.

NOTE: Body and commutator are not available separately.

Check thrust bearing and race for excessive grooving, flat spots or any other damage and renew bearing assembly if necessary.

Place pumping element on a flat surface and in the position shown in Fig. 22. Use a feeler gage and check clearance between ends of rotor teeth and high points of stator. If clearance exceeds 0.003, renew pumping element. Use a micrometer and measure width (thickness) of rotor and stator. If stator is 0.002 or more wider (thicker) than the rotor, renew the pumping element. Pumping element rotor and stator are available only as a matched set.

Check end plate for wear, scoring and flatness. Do not confuse the polish pattern on end plate with wear. This pattern, which results from rotor rotation, is normal. Renew end plate if worn or scored and is not within 0.0002 of being flat.

When reassembling, use all new seals and back-up washers. All parts, except those noted below, are installed dry. Reassemble as follows: If needle bearing (14—Fig. 21) was removed, lubricate with IH Hy-Tran fluid, install from pumping element end of body and press bearing into bore until inside end measures 3-13/16 to 3-7/8 inches from pumping element end of body as shown in Fig. 23. Lubricate thrust bearing assembly with IH Hy-Tran fluid and install assembly on input shaft with race on top side. Install input shaft and bearing assembly in body and check for free rotation. Install a link pin in one end of the drive link, then install drive link in input shaft by engaging the flats

on link pin with slots in input shaft. Use a small amount of grease to hold commutator pin in commutator, then install commutator and pin in body while engaging pin in one of the long slots of the input shaft. Commutator is correctly installed when edge of commutator is slightly below sealing surface of body. Clamp body in a soft jawed vise with input shaft pointing downward. Again make sure surfaces of spacer, pumping element, body and end plate are perfectly clean, dry and undamaged. Place spacer on body and align screw holes with those of body. Put link pin in exposed end of drive link, then install pumping element rotor while engaging flats of link pin with slots in rotor. Position pumping element stator over rotor and align screw holes of stator with those of spacer and body. Lubricate pumping element seal lightly with IH Hy-Tran fluid and install seal in seal retainer, then install seal and retainer over pumping element stator. Install end cap, align screw holes of end cap with those in pumping element, spacer and body, then install cap screws. Tighten cap screws evenly to a torque of 18-22 ft.-lbs.

NOTE: If input shaft does not turn evenly after cap screws are tightened, loosen and retighten them again. However, bear in mind that the unit was assembled dry and some drag is normal.

If stickiness or binding cannot be eliminated, disassemble unit and check for foreign material, nicks or burrs which could be causing interference.

Lubricate input shaft seal with IH Hy-Tran fluid and with input shaft splines taped to protect seal, install seal, back-up washer, spacer, washer and snap ring. The felt washer and

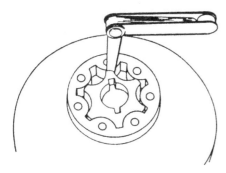

Fig. 22—Position pumping element as shown to check tooth clearance. Refer to text.

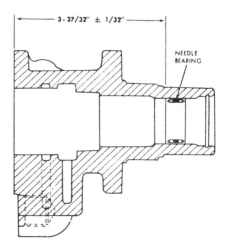

Fig. 23—When renewing needle bearing in pump body, install same to dimension shown.

water seal may be installed at this time but there will be less chance of loss or damage if installation is postponed until the time the steering wheel is installed.

After unit is assembled, turn unit on side with hose ports upward. Pour unit full of oil and work pump slowly until interior (pumping element) is thoroughly coated. Drain excess oil.

Reinstall unit by reversing the removal procedure and bleed power steering system as outlined in paragraph 14.

CONTROL (PILOT) VALVE

Series 424-444-2424-2444

32. **R&R AND OVERHAUL.** To remove the control (pilot) valve, drain the hydraulic reservoir and disconnect the hydraulic lines. Unbolt and remove the control (pilot) valve.

NOTE: Plug hydraulic lines and openings immediately to prevent dirt from entering system.

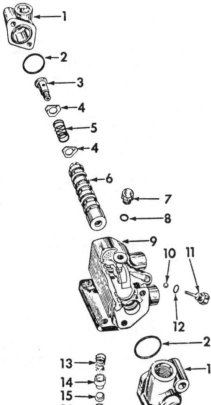

Fig. 24—Exploded view of the control (pilot) valve.

1. End cap	8. "O" ring
2. "O" ring	9. Valve body
3. Centering spring screw	10. Check ball
4. Centering spring washer	11. Plug
5. Centering spring	12. "O" ring
6. Valve spool	13. Spring
7. Plug	14. Check valve
	15. Seal
	16. Retainer

With valve removed, disassemble as follows: Refer to Fig. 24 and remove end caps (1) with "O" rings (2). Pull spool and centering spring assembly from valve body. Place a punch or small rod in hole of centering spring screw (3) and remove screw, centering spring (5) and centering spring washers (4) from spool. Remove plug (11), "O" ring (12) and circulating check ball (10). Remove retainer (16), seat (15), and pressure check valve (14) and spring (13).

Wash all parts in a suitable solvent and inspect. Valve spool and spool bore in body should be free of scratches, scoring or excessive wear. Spool should fit its bore with a snug fit and yet move freely with no visible side play. If spool or spool bore is defective, renew complete valve assembly as spool and valve body are not available separately.

Inspect pressure check valve and seat. Renew parts if grooved or scored.

Reassembly is the reverse of disassembly and the following points should be observed. Coat all parts with IH Hy-Tran fluid, or its equivalent, prior to installation. Install spool assembly in valve body so that centering spring is at end opposite the recirculating valve. Measure distance between gasket surface of circulating check ball plug and inner end of roll pin. This distance should be 15/16-inch, and if necessary, obtain this measurement by adjusting roll pin in or out. Tighten end cap retaining cap screws to a torque of 186 in.-lbs.

Reinstall valve by reversing removal procedure and bleed power steering system as outlined in paragraph 14.

Fig. 25—Exploded view of Series B-414 flow control and relief valve used with single stage pump.

1. Snap ring
2. Plug
3. "O" ring
4. Spring
5. Flow control valve
6. Plug & pin
7. Seal washer
8. Spring
9. Relief valve ball
10. Valve guide
11. Valve body & tubes assy.
12. Plug

FLOW DIVIDER VALVE

Series B-414

33. **R&R AND OVERHAUL.** Procedure for removal of the flow control valve and tubing assembly will be obvious after an examination of the unit.

Removal of relief valve (items 6 through 10—Fig. 25) can be accomplished by removing plug (6). Removal of flow control valve (5) and spring (4) can be accomplished after removal of snap rings (1) and plugs (2 and 12).

Free length of spring (4) is 3¾ inches and spring should test 18.0 lbs. when compressed to a length of 2-3/8 inches.

Spring (8) should have a free length of 1.250 inches and test 21.0 lbs. when compressed to a length of 1.087 inches.

Refer to paragraph 17 for method of checking the steering system operating pressure.

Series 424-444-2424-2444

34. **R&R AND OVERHAUL.** To remove the flow divider valve, first drain the hydraulic system, then disconnect the pump pressure hose, hitch supply pressure pipe, power steering supply pressure pipe and the relief valve return pipe. Unbolt and remove the flow divider valve assembly.

Remove plug (3—Fig. 26) and spool (2), then unscrew slotted plug (8) and withdraw relief valve spring (6) and relief valve (5).

Inspect all parts for scratches, scoring and undue wear. Relief valve spring free length should be 1.238 inches and should test 20-20.4 lbs. when compressed to a length of 63/64-inch. If

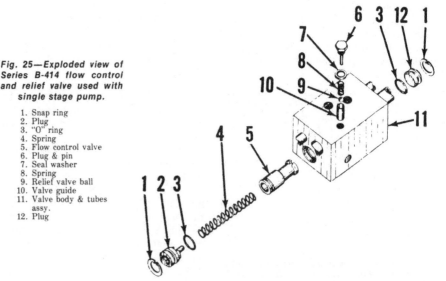

spool (2) or spool bore in valve body (1) is defective, renew complete valve assembly as spool and valve body are not available separately. Spool should move freely in its bore with no binding.

Coat all parts with Hy-Tran fluid, or its equivalent, and reassemble valve. Reinstall valve by reversing removal procedure, then fill and bleed power steering system as outlined in paragraph 14.

To check the relief valve operating pressure proceed as follows: Disconnect a steering cylinder hose, then using a tee connector, install a gage capable of registering 3000 psi in this circuit. Start engine and run at 2000 rpm. Turn steering wheel in the direction needed to pressurize the steering line in which gage is installed. When wheels reach stop, continue to apply steering effort to steering wheel and note reading on the gage. The gage should read 1500-1600 psi. A too high reading indicates a stuck relief valve while a too low reading could be caused by a weak or broken relief valve spring or a worn or scored relief valve.

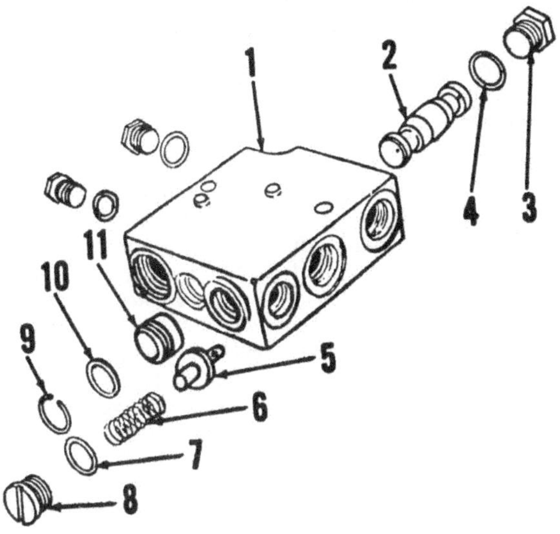

Fig. 26—Exploded view of the proportional type flow divider valve used on Series 424, 444, 2424 and 2444 tractors. Note the power steering relief valve (5).

1. Valve body
2. Spool
3. Plug
4. "O" ring
5. Relief valve
6. Relief valve spring
7. "O" ring
8. Plug
9. Snap ring
10. "O" ring
11. Plug

ENGINE AND COMPONENTS

R&R ENGINE WITH CLUTCH

Series B-275-B-414-354-364-384 Diesel

35. To remove the engine and clutch as an assembly, first drain cooling system, remove air cleaner cap and muffler, if equipped with vertical exhaust, then unlatch and raise hood. On Series B-275 and B-414 remove cotter pins from aft ends of the stay rod slides and move hood at a vertical position. Disconnect headlight wires at junction on front support, then remove hood pivot bolts and hood. Do not lose the two hood pivot bolt spacers. On Series 354, 364 and 384, remove the hood and radiator grille, disconnect the wiring from horn, disconnect ground lead and move the harness back on engine. On all series, disconnect drag link or power steering cylinder, if so equipped, from left steering arm. Disconnect upper and lower radiator hoses. Disconnect fuel filter bracket from radiator on Series B-275. Support tractor under clutch housing, then unbolt front support from engine and roll the front support, axle and radiator assembly away from tractor.

On Series B-275 and B-414 remove battery shield, battery hold-down and batteries. Remove battery carrier. On all series, if tractor is equipped with down swept exhaust, either remove the exhaust pipe from exhaust manifold or disconnect manifold from cylinder head. Disconnect tachometer cable, wires from generator, starter switch and number four glow plug, unclip wiring loom and lay wires rearward. Disconnect fuel shut-off rod at injection pump and remove rod. Close the fuel shut-off valve and disconnect fuel supply line from fuel pump. Disconnect oil pressure gage line from cylinder block and the temperature sending unit from cylinder head. Disconnect fuel return line from fuel tank. Disconnect the inlet and pressure line from hydraulic pump, if so equipped. Attach hoist to engine then unbolt and separate engine from clutch housing.

NOTE: When rejoining engine to clutch housing, time can often be saved, particularly on dual clutch models, if the following procedure is used.

Unbolt clutch assembly from flywheel and place same on transmission input shaft. Move sections together until clutch (input) shaft pilot enters pilot bearing and flywheel butts against clutch cover. Now bolt clutch to flywheel and complete the mating of engine and clutch housing by pulling sections together with the retaining cap screws.

Series B-414-354 Non-Diesel

36. To remove the engine and clutch as an assembly, first drain cooling system, remove air cleaner cap and muffler if equipped with vertical exhaust. Then on Series B-414, unlatch and raise hood, remove cotter pins from aft ends of stay rod slides and move hood to a vertical position. Disconnect headlight wires at junction on front support, then remove hood pivot bolts and hood. Do not lose the two hood pivot bolt spacers. On Series 354 remove the hood, radiator grille and air cleaner hose, disconnect the wiring from horn and battery ground cable, then move the harness back on the engine. On either series disconnect drag link, or power steering cylinder if so equipped, from left steering arm. Disconnect upper and lower radiator hoses. Support tractor under clutch housing, then unbolt front support from engine and roll the front support, axle and radiator assembly away from tractor.

On Series B-414, remove battery shield, battery hold-down and battery. Remove battery carrier. On either series, if tractor is equipped with a down swept exhaust, either remove the exhaust pipe from exhaust manifold or disconnect manifold from cylinder head. Disconnect wires from generator, starter switch and ignition coil, unclip wiring loom and lay wires rearward. Disconnect battery cable from starter and operating rod from starter switch. Disconnect choke control from carburetor. Disconnect and remove the governor to bellcrank rod. Close fuel shut-off and disconnect fuel supply line from fuel pump. Disconnect oil pressure line from cylinder block and temperature sending unit from cylinder head. Disconnect inlet and pressure lines from hydraulic pump. Attach hoist to engine, then unbolt and separate engine from clutch housing.

NOTE: When rejoining engine to clutch housing, time can often be saved, particularly on dual clutch models, if the following procedure is used.

Unbolt clutch assembly from flywheel and place same on transmission input shaft. Move sections together until clutch (input) shaft pilot enters pilot bearing and flywheel butts against clutch cover. Now bolt clutch to flywheel and complete mating of engine and clutch housing by pulling sections together with the retaining cap screws.

Series 424-444-2424-2444 Diesel

37. To remove the engine and clutch as an assembly, first drain the cooling system and disconnect battery cables. Remove the pre-cleaner, muffler (if equipped with vertical exhaust), hood and side panels. On tractors equipped with mechanical steering, unbolt the steering shaft front and rear bearing brackets from the fuel tank support and clutch housing. Slide the steering shafts forward and out of the master splined yoke. On tractors equipped with power steering, remove the steering tube clip from engine and disconnect the tubes from the power steering control valve. Plug and cap all openings to prevent dirt from entering steering system.

Remove the radiator hoses, disconnect headlight wires and unbolt the radiator brace and fan shroud from radiator. Place wood blocks between steering gear housing and axle to prevent tipping. Using a suitable jack under the clutch housing, raise front of tractor to remove most of the weight from the front tires. Unbolt the stay rod bracket from clutch housing and the steering gear housing from front of engine. Raise tractor engine until crankshaft pulley will clear steering gear housing, then roll front end assembly from tractor.

Remove the temperature indicator bulb, fuel shut-off cable and throttle rod. Disconnect wiring harness terminal blocks located under instrument panel and work the harness forward until it is clear of clutch housing. Remove the battery cable from the cranking motor solenoid switch. Disconnect the fuel lines from tank and plug openings. Unbolt the air cleaner bracket from fuel tank support and disconnect the tachometer drive cable. Place a wood block between fuel tank and clutch housing, then disconnect fuel tank support from engine.

Drain hydraulic system and disconnect the hydraulic pump pressure hose from the flow divider valve. Loosen hose clamps and slide the hose coupling forward on the hydraulic pump suction pipe. Plug and cap all openings. Unbolt and remove cranking motor. If tractor is equipped with an underslung muffler, the exhaust pipe can now be removed. Remove clutch housing front dust cover.

Attach a hoist to the engine, unbolt

and separate engine from clutch housing.

When rejoining engine to clutch housing, unbolt clutch assembly from flywheel and place same on the pto driving shaft (two stage clutch) and transmission input shaft. Bolt engine to clutch housing, then working through the opening in bottom of clutch housing, bolt clutch assembly to flywheel.

Complete the engine installation by reversing the removal procedure. Vent the fuel system and bleed the power steering system.

Series 424-444-2424-2444 Non-Diesel

38. To remove the engine and clutch as an assembly, first drain the cooling system and disconnect battery cables. If tractor is equipped with vertical exhaust, remove the muffler, then remove the pre-cleaner, hood and side panels. On tractors equipped with mechanical steering, unbolt the steering shaft front and rear bearing brackets from the fuel tank support and clutch housing. Slide the steering shafts forward and out of the master splined yoke. On tractors equipped with power steering, remove the steering tube clip from engine and disconnect the tubes from the power steering control valve. Plug and cap all openings to prevent dirt from entering steering system.

Remove the radiator hoses, temperature indicator bulb, governor control rod and the tank to filter fuel line. Disconnect the headlight wires, choke cable and tachometer drive cable. Unbolt the radiator brace and fan shroud from radiator, then place wood blocks between steering gear housing and axle to prevent tipping. Using a suitable jack under clutch housing, raise front of tractor to remove most of the weight from front tires. Unbolt stay rod bracket from clutch housing and the steering gear housing from front of engine. Raise tractor engine until crankshaft pulley will clear steering gear housing, then roll front end assembly from tractor.

Disconnect wiring harness terminal blocks located under instrument panel and work the harness forward until it is clear of clutch housing. Remove the battery cable from the cranking motor solenoid switch, then unbolt and remove cranking motor. Unbolt the air cleaner bracket from fuel tank support, place a wood block between fuel tank and clutch housing and then unbolt the fuel tank support from engine. If tractor is equipped with an underslung muffler, remove exhaust pipe.

Drain the hydraulic system and disconnect the hydraulic pump pressure hose from the flow divider valve.

Loosen hose clamps and slide the hose coupling forward on the hydraulic pump suction pipe. Plug and cap all openings.

Attach a hoist to the engine, unbolt and separate engine from clutch housing.

When rejoining the engine to clutch housing, unbolt clutch assembly from flywheel and place same on the pto driving shaft (two stage clutch) and the transmission input shaft. Bolt engine to clutch housing, then working through the opening in bottom of clutch housing, bolt clutch assembly to flywheel.

Complete the engine installation by reversing the removal procedure. Refill hydraulic reservoir and bleed the power steering system.

CYLINDER HEAD

NOTE: In all engines except Series 424, 444, 2424 and 2444 non-diesel, the cylinder bolt bores in cylinder block are fitted with Heli-Coil inserts which can be removed and renewed using the tools shown in Fig. 27, or their equivalents. New Heli-Coil inserts are installed with the driving lug on bottom and are screwed into the tapped hole until top side of 0.100-0.125 below edge of bore.

Series B-275-B-414-354-364-384 Diesel

39. To remove cylinder head, first drain cooling system and on Series B-275 and B-414, remove air cleaner cap and muffler, if equipped with vertical exhaust. Then, unlatch and raise hood. Remove the radiator brace, then disconnect the hood bracket from tappet cover and place hood in a vertical positon. Remove the hood guide to valve tappet cover brace. Remove battery shield, battery hold-down and batteries. Disconnect fuel return line at tank and at front of hood guide. Dis-

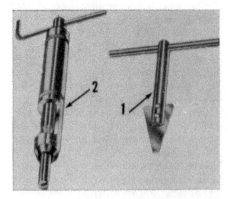

Fig. 27—Tools used to remove and install the Heli-Coil inserts.

1. Extracting tool   2. Inserting tool

connect air cleaner inlet hose from inlet manifold, then unbolt and remove hood guide and air cleaner as a unit. On Series 354, 364 and 384, unbolt and remove hood and disconnect air cleaner inlet hose from inlet manifold. On all series, disconnect wires from alternator or generator and disconnect loom from thermostat housing. Disconnect alternator or generator adjusting strap, remove drive belt, then unbolt and remove mounting bracket and alternator or generator. Disconnect upper hose from radiator, unbolt water outlet elbow and remove outlet and upper hose. Disconnect lead wire from number four glow plug. Remove the temperature indicating sending unit from cylinder head. Remove the breather line which runs from injection pump drive gear cover to intake manifold. On Series B-275 remove lines from fuel filter, disconnect fuel filter bracket from fan shroud and remove bracket and fuel filter. On Series B-414, 354, 364 and 384 disconnect lines from fuel filter but leave filter attached to inlet manifold. On all series, disconnect the pressure lines from injectors, then remove the leak-off (excess) fuel line from the top of each injector. Remove inlet (with fuel filter attached on B-414, 354, 364 and 384 tractors) and exhaust manifolds from cylinder head. Remove tappet lever cover and gasket. Unbolt and remove the tappet levers and shaft assembly.

NOTE: When removing the tappet levers and shaft assembly, keep pressure on outer ends to prevent the two piece shaft assembly from flying apart.

Identify push rods and lift out same, then remove balance of cylinder head bolts and remove cylinder head.

40. Reinstall cylinder head by reversing the removal procedure and when renewing cylinder head gasket, be sure to use the latest type which has ferrules in the two holes shown in Fig. 28. Position cylinder head gasket with word "TOP" upward, use guide studs in the two ferruled gasket holes, then install cylinder head.

Tighten cylinder head bolts in the sequence shown in Fig. 29 and to a torque of 70-75 ft.-lbs. for the B-275, or 75-80 ft.-lbs. for the B-414, 354, 364 or 384 diesel engines. Tighten all manifold stud nuts to 25-30 ft.-lbs. Adjust tappet gap to 0.020 hot on all diesel engines and bleed fuel system as outlined in paragraph 91.

Series B-414-354 Non-Diesel

41. To remove cylinder head, first drain cooling system, and on Series B-414 remove air cleaner cap and muffler, if equipped with vertical exhaust,

then unlatch and raise hood. Remove radiator brace, then disconnect the hood bracket from tappet cover and place hood in a vertical position. Remove the hood guide to tappet cover brace. Remove battery shield, battery hold-down and battery. Disconnect air cleaner inlet hose from carburetor, then unbolt and remove hood guide and air cleaner as a unit. On Series 354 unbolt and remove hood, disconnect air cleaner inlet hose from inlet manifold. Then, on both series, disconnect wires from generator and disconnect loom from thermostat housing. Disconnect generator adjusting strap, remove drive belt from generator, then unbolt mounting bracket and remove bracket and generator. Disconnect upper hose from radiator, unbolt water outlet elbow and remove elbow and upper hose. Remove temperature sending unit from cylinder head. Remove manifold and carburetor assembly from cylinder head. Remove tappet lever cover and gasket. Unbolt and remove tappet levers and shaft assembly.

NOTE: When removing the tappet levers and shaft assembly, keep pressure on outer ends to prevent assembly from flying apart.

Identify push rods and lift out same, then remove balance of cylinder head bolts and remove cylinder head.

42. Reinstall cylinder head by reversing removal procedure and when renewing cylinder head gasket, be sure to position cylinder head gasket with word "TOP" upward. Tighten cylinder

head bolts to a torque of 75-80 ft.-lbs. in the sequence shown in Fig. 30. Tighten all manifold stud nuts to a torque of 25-30 ft.-lbs. and adjust tappets to 0.020 hot.

Series 424-444-2424-2444 Diesel

43. To remove the cylinder head, first drain cooling system, remove precleaner and muffler, if equipped with vertical exhaust. Remove hood and side panels, then disconnect battery ground cable, generator or alternator wires, oil pressure switch wire, headlight wire and lead wire from number four glow plug. Disconnect generator or alternator adjusting strap, remove drive belt, then unbolt mounting bracket and remove bracket and generator or alternator. Disconnect upper hose from radiator, loosen upper clamp on by-pass hose, then unbolt and remove thermostat housing from cylinder head.

Remove the temperature indicator bulb, the crankcase breather tube and the ground wire from number one glow plug. Disconnect the air cleaner pipe and unbolt and remove the intake manifold. Unbolt and remove the air cleaner and heat shield assembly and after first removing the nuts, lift off the exhaust manifold. Disconnect the pressure lines from injectors and remove the leak-off (excess) fuel line from top of each injector.

CAUTION: Cap or plug openings in injectors and lines immediately.

Unbolt and remove the radiator

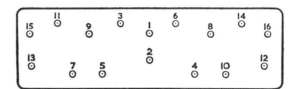

Fig. 28—View showing the two bolt holes fitted with ferrules in the diesel engine cylinder head gaskets.

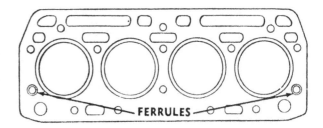

Fig. 29—When reinstalling diesel engine cylinder head, tighten cylinder head bolts in the sequence shown.

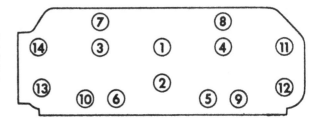

Fig. 30—When reinstalling Series B-414 or 354 non-diesel engine cylinder head, tighten cylinder head bolts in the sequence shown.

brace, rocker arm cover and the rocker arm and shaft assembly.

NOTE: When removing the rocker arm and shaft assembly, keep pressure on outer ends to prevent the two-piece shaft assembly from flying apart.

Lift out push rods, remove balance of cylinder head retaining cap screws and lift off cylinder head.

44. Reinstall cylinder head by reversing the removal procedure and when renewing cylinder head gasket, position gasket with word "TOP" facing upward. Use guide studs in the two ferruled gasket holes, shown in Fig. 28 to align the cylinder head and gasket.

Tighten cylinder head bolts in the sequence shown in Fig. 29 and to a torque of 75-80 ft.-lbs. Tighten all manifold stud nuts to a torque of 25-30 ft.-lbs. Adjust tappet gap to 0.020 hot on both intake and exhaust valves. Bleed fuel system as outlined in paragraph 91.

Series 424-444-2424-2444 Non-Diesel

45. To remove the cylinder head, first drain cooling system, remove precleaner and muffler, if equipped with vertical exhaust. Remove hood and side panels, then disconnect battery ground cable, oil pressure switch wire and ignition wires from coil and resistor. Unclip wiring harness and lay it along left side of engine. Remove upper radiator hose, radiator brace, temperature indicator bulb and loosen upper hose clamp on thermostat by-pass hose. Unbolt and remove air cleaner and pipe assembly. Disconnect choke cable, governor rod and fuel line from carburetor. Then, unbolt and remove the manifold and carburetor assembly. Disconnect spark plug wires and remove the governor control rod clip from right side of cylinder head. Unbolt and remove the rocker arm cover and rocker arms and shaft assembly. Lift out push rods, remove cylinder head retaining cap screws and lift off cylinder head.

46. Reinstall cylinder head by reversing the removal procedure. When renewing cylinder head gasket, apply a light coating of lubricant to top face of cylinder block. The head gasket is marked for correct installation.

Tighten cylinder head bolts in the sequence shown in Fig. 31 and to a torque of 80-90 ft.-lbs. Tighten manifold stud nuts to a torque of 33-37 ft.-lbs. Adjust tappet gap to 0.014 hot for the intake valves and 0.020 hot for exhaust valves.

VALVES AND SEATS

All Models

47. Intake and exhaust valves are not interchangeable and on all series except 424, 444, 2424 and 2444 non-diesels, seat directly in the cylinder head. Series 424, 444, 2424 and 2444 nondiesel exhaust valves seat on renewable inserts which are available in standard size as well as oversizes of 0.015 and 0.030. Valves have a face and seat angle of 45 degrees. Seat width should be 0.070-0.080 and total runout must not exceed 0.002. Adjust valve tappet gap on all models except 424, 444, 2424 and 2444 non-diesels to 0.020 (hot) for both intake and exhaust valves. On Series 424, 444, 2424 and 2444 non-diesels, adjust valve tappet gap to 0.014 (hot) for intake valves and 0.020 (hot) for exhaust valves.

Use the chart shown in Fig. 32 for valve tappet gap adjusting procedure. Four valves are adjusted when No. 1 piston is at TDC (compression) and the remaining four are adjusted when No. 4 piston is at TDC (compression).

Stem diameter of valves used in Series 424, 444, 2424 and 2444 nondiesel engines is 0.3405-0.3415 and normal operating clearance in guides is 0.0015-0.0035. Valve stem diameter on all other series is 0.341-0.342 and normal operating clearance in guides is 0.002-0.004.

When removing exhaust valve seat inserts from a Series 424, 444, 2424 or 2444 non-diesel cylinder head, use the proper puller. Do not attempt to drive a chisel under seat insert as counterbore will be damaged. Chill new

seat insert with dry ice or liquid Freon and when insert is properly bottomed, it should be 0.008 to 0.030 below edge of counterbore. After installation, peen the cylinder head material around the complete outer circumference of the valve seat insert.

VALVE GUIDES AND SPRINGS

All Models

48. The shouldered valve guides used in diesel engines are not interchangeable. The inside diameter of all guides is 0.344-0.345 which provides a normal operating clearance of 0.002-0.004 for the valves. Press out old guides from bottom of cylinder head. Press new guides in top of cylinder head until shoulder of guide bottoms against cylinder head. When correctly installed, guide will protrude 0.938 from top of cylinder head. Guides are pre-sized; however, they should be reamed after installation, if necessary, to obtain the 0.344-0.345 inside diameter.

49. The shouldered valve guides used in B-414 and 354 series non-diesel engines are not interchangeable and can be pressed out from bottom of cylinder head. Press new guides in top of cylinder head until shoulder of guide bottoms against cylinder head. When correctly installed, top of intake guide is 0.828 inch; and top of exhaust guide is 0.984 inch, above spring recess of cylinder head. Guides are pre-sized; however, they should be reamed after installation, if necessary, to obtain an inside diameter of 0.344-0.345.

50. The valve guides used in Series 424, 444, 2424 and 2444 non-diesel

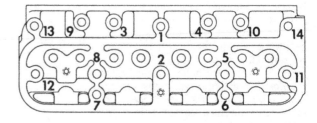

Fig. 31—When installing Series 424, 444, 2424 and 2444 non-diesel cylinder head, tighten cylinder head bolts in the sequence shown.

WITH	ADJUST VALVES (Engine Warm)							
No. 1 Piston at T.D.C. (Compression)	1	2	3		5			
No. 4 Piston at T.D.C. (Compression)				4		6	7	8

Numbering sequence of the valves which correspond to the chart.

Fig. 32—Chart shows the valve tappet gap adjusting procedure.

engines are interchangeable. Valve guides are not shouldered and should be pressed in cylinder head until top of guides are 13/16-inch above surface of cylinder head. Inside diameter of valve guides are pre-sized to 0.343-0.344 and if not distorted during installation, will require no final sizing.

51. On B-275 series diesel engines, intake and exhaust valve springs are interchangeable. Renew springs which are rusted, discolored or do not meet the following specifications:

Free length . . . . 2.531 inches
Test load . . . . . 42.5 lbs. at 1.703 inches

On Series 424, 444, 2424, 2444, B-414, 364 and 384 diesel engines, each valve is fitted with two valve springs. Renew springs which are rusted, discolored or do not meet the following specifications:

Inner spring,
    Free length . . . . 2.125 inches
    Test load . . . . 12.4-13.7 lbs.
                   at 1.653 inches
Outer spring,
    Free length . . . . 2.550 inches
    Test load . . . . 29.5-32.6 lbs.
                   at 1.870 inches

On Series 354 diesel engine intake and exhaust valve springs are interchangeable. Renew springs which are rusted, discolored or do not meet the following specifications:

Free length . . . . 2.48-2.58 inches
Test load . . . . 28.7-31.7 lbs.
                   at 1.922 inches

On B-414 and 354 series non-diesel engines, intake and exhaust valve springs are interchangeable. Renew springs which are rusted, discolored or do not meet the following specifications:

Free length . . . . 2.085 inches
Test load . . . . 71.6-84.0 lbs.
                   at 1.346 inches

On Series 424, 444, 2424 and 2444 non-diesel engines, exhaust valves are equipped with positive type valve rotators ("Rotocaps"). Therefore, exhaust and intake valve springs are not interchangeable. Renew springs which are rusted, discolored or do not meet the following specifications:

Intake valve spring,
    Free length . . . . 2.734 inches
    Test load . . . . 49-54.6 lbs.
                   at 1.683 inches
Exhaust valve spring (with rotocap),
    Free length . . . . 2.240 inches
    Test load . . . . 80.1-87.1 lbs.
                   at 1.456 inches

VALVE TAPPETS (CAM FOLLOWERS)

All Models

52. The mushroom type tappets operate directly in the unbushed crankcase bores and can be removed after removing the camshaft as outlined in paragraph 66 and 67. Normal operating clearance of tappets in crankcase bores is 0.0005-0.003.

VALVE TAPPET LEVERS (ROCKER ARMS)

All Models

53. The valve levers and lever shaft assembly are lubricated via drilled passages in cylinder block and cylinder head. Replacement levers with bushings are available for all engines and bushings are available separately for all levers except those used in Series 424, 444, 2424 and 2444 non-diesel engines.

Procedure for removal of the valve levers assembly is evident after removing the tappet lever cover.

NOTE: When removing the tappet levers and shaft assembly from all engines, except Series 424, 444, 2424 and 2444 non-diesel engines, keep pressure on outer ends to prevent the two-piece shaft assembly from flying apart.

When renewing bushings in valve levers (all engines except Series 424, 444, 2424 and 2444 non-diesel), be sure

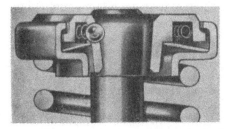

Fig. 33—Cut-away view showing typical installation of a valve rotator.

Fig. 34—View showing timing marks on timing gear train of early B-275 tractor equipped with injection pump having pneumatic governor. Refer also to Fig. 35.

  1. Crankshaft gear
  2. Camshaft gear
  3. Idler gear
  4. Injection pump gear

oil hole in bushing aligns with oil hole in valve lever and ream bushings after installation to an inside diameter of 0.751-0.752.

On all models, the end of the valve lever which contacts the valve can be refinished, if necessary, providing the original contour is carefully maintained. Check the valve levers and valve lever shafts against the values which follow:

Valve lever bore 0.751-0.752
Lever shaft diameter 0.748-0.749
Lever diametral
    clearance 0.002-0.004

VALVE ROTATORS

Series 424-444-2424-2444-354 Non-Diesel

54. Positive type valve rotators ("Rotocaps") are factory installed on the exhaust valves in these engines.

Normal servicing of the valve rotators consists of renewing the units. It is important, however, to observe the valve action after the engine is started. Rotator action can be considered satisfactory if the valve rotates a slight amount each time the valve opens. A cut-away view of a typical "Rotocap" installation is shown in Fig. 33.

VALVE TIMING

All Models

55. Valves are properly timed when the single punch marked tooth on the camshaft gear is meshed with the single punch marked tooth space on crankshaft gear as shown in Fig. 34, 36 and 40.

TIMING GEAR COVER

Series B-275-B-414-354-364-384 Diesel

56. To remove the timing gear cover, first remove the front support, axle and radiator assembly as follows: Drain cooling system, then on Series B-275

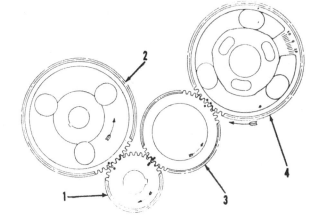

and B-414, raise hood and disconnect headlight and horn wires at junction on front support, then unclip loom from radiator. Remove cotter pins from ends of stay rod slides, remove hood pivot bolts and remove hood. Do not lose the pivot bolt spacers. Disconnect radiator brace and on Series B-275, disconnect the fuel filter bracket from radiator. On Series 354, 364 and 384, remove hood and radiator grille. Disconnect ground lead, horn wire and move the harness back to the engine, then disconnect air cleaner hose. On all models, disconnect upper and lower hoses from radiator. Disconnect the drag link or power steering cylinder, if so equipped, from left steering arm. Support tractor, attach hoist to the front support, axle and radiator assembly, then unbolt front support and move assembly away from tractor.

Remove alternator or generator adjusting strap, then remove belt from alternator or generator and let belt hang on water pump. Loosen water pump pulley flange or the belt adjusting pulley on models so equipped, and remove water pump belt from crankshaft pulley. Unbolt and remove water pump. Remove the breather line which runs from injection pump gear cover to inlet manifold, then unbolt and remove injection pump gear cover.

NOTE: Notice the size and location of the cap screws as they are removed.

Remove crank nut, crankshaft pulley, woodruff key and two front oil pan cap screws. Unbolt and remove front cover as shown in Fig. 35.

When installing new oil seal in timing gear cover, install same with lip facing inward.

Series B-414-354 Non-Diesel

57. To remove the timing gear cover, first remove the front support, axle and radiator assembly as follows: Drain cooling system, then on Series B-414, raise hood and disconnect headlight wires and horn wire (if so equipped) at junction on front support, then unclip wiring loom. Remove cotter pins from ends of stay rod slides, remove hood pivot bolts and remove hood. Do not lose the pivot bolt spacers. Disconnect upper radiator brace. On Series 354, remove hood and radiator grille. Disconnect ground lead, horn wire and move the harness back to the engine, then disconnect air cleaner hose. On both series, disconnect upper and lower radiator hoses from radiator. Disconnect drag link, or power steering cylinder if so equipped, from left steering arm. Support tractor, attach hoist to front support, axle and radiator assembly, then unbolt front support and

move assembly away from tractor.

Remove fan blades, loosen water pump pulley adjusting flange and remove water pump drive belt. Unclip wiring loom from water pump. Remove crankshaft pulley and the two front oil pan bolts. Unbolt and remove timing gear cover.

NOTE: Identify cap screws as they are removed. Refer to Fig. 36.

When installing new oil seal in timing gear cover, install same with lip facing inward.

Series 424-444-2424-2444 Diesel

58. To remove the timing gear cover, first remove the front support, axle and radiator assembly as follows: Drain cooling system, remove hood and disconnect radiator hoses. Disconnect the headlight wires and unbolt the radiator brace and fan shroud from radiator. On tractors equipped with mechanical steering, drive roll pin from forward yoke of front steering shaft universal and remove cap screws from front steering shaft support. Drive front universal from steering worm

shaft. If tractor is equipped with hydrostatic steering, disconnect the steering cylinder hoses, then cap and plug all openings to prevent dirt from entering system.

Then, on all tractors, place wood blocks between steering gear housing and axle to prevent tipping. Using a suitable jack, support tractor under clutch housing. Unbolt stayrod bracket from clutch housing and steering gear housing from front of engine. Raise engine until crankshaft pulley will clear steering gear housing, then roll assembly from tractor.

Remove generator or alternator adjusting strap, then remove drive belt from pulley and let belt hang on water pump. Loosen water pump belt adjusting pulley and remove water pump belt from crankshaft pulley. Unbolt and remove water pump. Remove the breather line which runs from injection pump gear cover to intake manifold, then unbolt and remove injection pump gear cover. Remove crankshaft pulley nut, pulley and Woodruff key. Unbolt and remove timing gear cover.

When installing new oil seal in tim-

Fig. 35—View showing timing gear train of late B-275 and all B-414, 424, 444, 2424, 2444, 354, 364 and 384 tractors equipped with injection pump having mechanical governor. Timing marks are similar to those shown in Fig. 34.

1. Crankshaft gear
2. Camshaft gear
3. Idler gear
4. Injection pump gear
5. Hydraulic pump gear

Fig. 36—View of B-414 and 354 non-diesel engine gear train. Note timing marks on gears (1) and (2).

S. Governor housing screws
1. Crankshaft gear
2. Camshaft gear
3. Idler gear
4. Governor gear
5. Hydraulic pump gear

ing gear cover, install same with lip facing inward.

Series 424-444-2424-2444 Non-Diesel

59. To remove the timing gear cover, first remove the front support, axle and radiator assembly as follows: Drain cooling system, remove hood and disconnect radiator hoses. Disconnect the headlight wires and unbolt the radiator brace and fan shroud from radiator. On tractors equipped with mechanical steering, drive roll pin from forward yoke of front steering shaft universal and remove cap screws from front steering shaft support. Drive front universal from steering worm shaft.

If tractor is equipped with hydrostatic steering, disconnect the steering cylinder hoses, then cap and plug all openings to prevent dirt from entering system.

Then, on all tractors, place wood blocks between steering gear housing and axle to prevent tipping. Support tractor under clutch housing with a suitable jack. Unbolt stay rod bracket from clutch housing and steering gear housing from front of engine. Raise engine until crankshaft pulley will clear steering gear housing, then roll assembly from tractor.

Remove generator or alternator adjusting strap, then remove the drive belt. Unbolt and remove water pump and governor housing. After first removing the crank nut, attach a suitable puller and remove crankshaft pulley and Woodruff key. Unbolt and remove the timing gear cover from engine.

Extra care must be taken when installing the oil seal in timing gear cover, so as not to distort or bend the cover. Install seal with lip of same facing inward toward timing gears.

NOTE: A wear ring is furnished with the new oil seal. Ring must be pressed on sealing surface of crankshaft pulley.

When reassembling, leave the cover retaining cap screws loose until crankshaft pulley has been installed. This will facilitate centering the seal with respect to the pulley.

TIMING GEARS

All Series Except 424-444-2424-2444 Non-Diesel

60. **CAMSHAFT GEAR.** To remove the camshaft gear, first remove timing gear cover as outlined in paragraphs 56, 57 or 58. Remove tappet cover, valve levers and shaft assembly and push rods. Drain and remove oil pan, then unbolt and remove oil pump. Check camshaft end play which should be 0.008-0.017. If end play is excessive, renew thrust plate after camshaft gear is removed. Turn camshaft until holes in gear align with thrust plate cap screws, then remove the cap screws. Carefully pull camshaft from cylinder block and catch each tappet as the camshaft is removed. Press camshaft out of gear and remove Woodruff key and thrust plate.

When installing camshaft gear, press same on camshaft until it bottoms against shoulder on camshaft. Install camshaft and align timing marks as shown in Fig. 34 or 36.

61. **CRANKSHAFT GEAR.** To remove the crankshaft gear, first remove timing gear cover as outlined in paragraphs 56, 57 or 58. The crankshaft gear has two tapped holes to provide for removal and can be removed at this time by using a suitable puller.

When reinstalling crankshaft gear, align timing marks as shown in Fig. 34. or 36.

62. **IDLER GEAR.** The idler gear (3—Fig. 34 or 36) can be removed after the timing gear cover has been removed as outlined in paragraphs 56, 57 or 58. Refer to Fig. 37 for a view of the removed gear and shaft. Prior to installing gear and shaft, inspect the Heli-Coil insert and renew if necessary.

On diesel engines, align timing marks on all gears as shown in Fig. 34 when installing idler gear. When installing idler gear shaft, be sure dowel pin (P—Fig. 37) in shaft enters hole (H—Fig. 38) in cylinder block. Tighten the retaining cap screw to a torque of 75 ft.-lbs. Operating clearance of gear on shaft is 0.0015-0.0028. If clearance is excessive, renew gear and/or shaft.

63. **INJECTION PUMP GEAR.** Refer to Fig. 35 or 39 for views of the two types of injection pump drive gears

Fig. 37—Removed idler gear and shaft. Pin (P) enters oil hole in cylinder block shown in Fig. 38. Diesel engine gear is shown, however, B-414 and 354 non-diesel gear is similar.

Fig. 38—Hole (H) accepts the pin which is on rear side of idler gear shaft. Refer to Fig. 37.

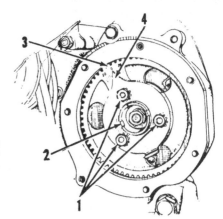

Fig. 39—Injection pump drive gear used on early B-275 tractors equipped with injection pump having pneumatic governor.

1. Cap screws
2. Hub groove
3. Pump gear
4. Timing pointer

that have been used. In both cases, it will be necessary to remove the timing gear cover as outlined in paragraphs 56 and 58 before gear can be removed.

When reinstalling the gear and timing pointer shown in Fig. 39, refer to paragraph 103 for injection pump timing procedure.

Series 424-444-2424-2444 Non-Diesel

64. **CAMSHAFT GEAR.** To remove the camshaft gear, first remove timing gear cover as outlined in paragraph 59. Remove the valve cover, valve levers and shaft assembly, push rods, oil pan and oil pump. Push tappets up into their bores. Check camshaft end play which should be 0.003-0.012. If end play is excessive, renew thrust plate after camshaft gear is removed. Working through openings in camshaft gear, remove the cap screws retaining the shaft thrust plate to crankcase. Carefully withdraw camshaft and gear assembly and catch each tappet as shaft is removed. Gear can now be removed from camshaft by using a suitable press.

When reassembling, mesh the single punch marked tooth on camshaft gear with single punch marked tooth space on crankshaft gear and the double punch marked tooth on camshaft gear with the double punch marked tooth space on the governor and ignition unit drive gear. Refer to Fig. 40.

65. **CRANKSHAFT GEAR.** To remove the crankshaft gear, first remove engine from tractor as outlined in paragraph 38. Then, remove oil pan, oil pump, timing gear cover, clutch, flywheel and rear oil seal assembly. Remove connecting rod caps and main bearing caps and lift crankshaft from cylinder block. The gear can now be removed from crankshaft by using a suitable press.

When reinstalling crankshaft in cylinder block, align timing marks as shown in Fig. 40.

CAMSHAFT

All Series Except 424-444-2424-2444 Non-Diesel

66. If the camshaft only is to be removed, same can be done without removing engine from tractor and the procedure for doing so is given in paragraph 60. However, if service is required on the camshaft bushings and/or expansion plug (4—Fig. 41) located at the rear of camshaft, remove the engine as outlined in paragraphs 35, 36 or 37.

With the engine removed, remove camshaft as outlined in paragraph 60. Then, unbolt clutch from flywheel and flywheel from crankshaft. Bump out expansion plug.

NOTE: Camshaft bushings are furnished semi-finished for service and must be align reamed after installation.

When installing the bushings, be sure oil holes in same align with oil holes in cylinder block. Install the front and rear bushings with the "FRONT" marking toward front of engine and be sure front of rear bushing is flush with the front of its bore in order to allow room for the expansion plug installation.

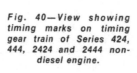

Fig. 40—View showing timing marks on timing gear train of Series 424, 444, 2424 and 2444 non-diesel engine.

Specifications for the camshaft and camshaft bushings are as follows:

Camshaft Journal Diameter,
  Front 1.811-1.812
  Center 1.577-1.578
  Rear 1.499-1.500
Camshaft Bushing (Reamed I.D.),
  Front 1.8135-1.8145
  Center 1.5795-1.5805
  Rear 1.5015-1.5025
Camshaft End Play 0.008-0.017
Gear Backlash 0.0025-0.0045
Journal Operating
  Clearance 0.0015-0.0035

When installing the rear expansion plug, use sealing compound on plug and seat. Install camshaft and gear with the

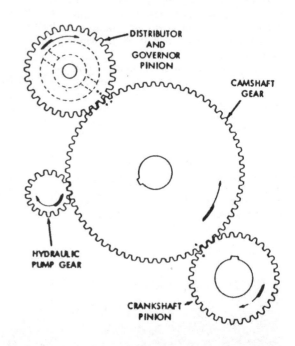

Fig. 41—Crankshaft rear oil seal and retainer installed. Note camshaft expansion plug.

1. Crankshaft
2. Oil seal
3. Retainer
4. Expansion plug

timing marks aligned as shown in Fig. 34 or 36.

Series 424-444-2424-2444 Non-Diesel

67. If camshaft only is to be removed, same can be done without removing engine from tractor and the procedure for doing so is given in paragraph 64. The camshaft front journal rides in a renewable bushing while the intermediate and rear camshaft journals ride directly in the crankcase bores. The bushing can be renewed after removing the camshaft.

However, if the expansion plug (4—Fig. 41) located at rear of camshaft is to be renewed, remove engine as outlined in paragraph 38.

With engine removed, remove camshaft as outlined in paragraph 64. Then, unbolt and remove the clutch, flywheel and engine rear support plate. Bump out expansion plug.

When installing the rear expansion plug, use sealing compound on plug and seat.

The camshaft front bushing is pre-sized and when correctly installed, no final sizing is required. When installing the bushing, make certain that oil holes in bushing align with oil holes in cylinder block.

Specifications for the camshaft are as follows:

Camshaft Journal Diameter,
Front 1.811-1.812
Center 1.577-1.578
Rear 1.499-1.500
Camshaft End Play 0.003-0.012
Journal Operating
Clearance 0.0009-0.0054

Install camshaft and gear with timing marks aligned as shown in Fig. 40.

ROD AND PISTON UNITS

All Models

68. Connecting rod and piston units are removed from above after cylinder head, oil pan and oil pump have been removed. Cylinder numbers are marked on rods and caps. When reassembling, make certain that numbers are in register and on all except Series 424, 444, 2424 and 2444 non-diesel engines, numbers are on the side opposite camshaft. On Series 424, 444, 2424 and 2444 non-diesel engines, rod and cap numbers are on camshaft side of engine. Torque rod bolts to 30-35 ft.-lbs. on Series B-275 and secure with lock wire. Rod bolts on all Series B-414, 354, 364 and 384 are self-locking and should be torqued to 40-45 ft.-lbs. Series 424, 444, 2424 and 2444 non-

diesel rod bolts are self-locking and should be tightened to a torque of 43-49 ft.-lbs. Tighten the self-locking rod bolts used in Series 424, 444, 2424 and 2444 diesel engines to a torque of 40-45 ft.-lbs.

PISTONS, SLEEVES AND RINGS

All Series Except 424-444-2424-2444 Non-Diesel

69. Pistons and sleeves are available as individual parts or as a matched set. Recommended clearance of new pistons in new sleeves is 0.0048-0.0056 for B-275 diesel engines, or 0.0031-0.0039 for 364, 384, 424, 2424 and 2444 diesel engines and Series B-414 and 354 diesel and non-diesel engines, when measured between piston skirt and cylinder sleeve at 90 degrees to piston pin.

NOTE: As individual replacements, fit pistons to a clearance of 0.0031-0.0047.

70. The wet type cylinder sleeves should be renewed when out-of-round exceeds 0.008 and/or taper exceeds 0.005.

Special pullers are available to remove the wet type sleeves from above after the pistons have been removed. Before installing sleeves, check to make certain that the counterbore at top and sealing ring groove at the bottom are clean and free from foreign material. All sleeves should enter crankcase bores full depth and should be free to rotate by hand when tried in bores without sealing rings. After making trial installation without sealing rings, remove the sleeves, wet new sealing rings and end of sleeves with a thick soap solution or equivalent and install sleeves. If sealing ring is in place and not pinched, very little hand pressure is required to press the sleeve completely into place. Normally, the top of the sleeves will extend 0.003-0.007 above the machined top surface of the cylinder block. If sleeve stand out is excessive, check for foreign material under the sleeve flange.

NOTE: The cylinder head gasket forms the upper cylinder sleeve seal, and excessive sleeve stand out will result in coolant leakage.

To test lower sealing rings for proper installation, fill crankcase (cylinder block) water jacket with cold water and check for leaks near bottom of sleeves.

71. Diesel engine pistons are fitted with five rings; three compression and two oil control rings. Gasoline engine pistons are fitted with four rings; three compression and one oil control ring.

Specifications are as follows:

Ring Width (B-275 Diesel)
Compression rings 0.0930-0.0935
Top oil control 0.1875
2nd oil control 0.1860-0.1865
Ring Width (B-414-424-444-2424-2444-354-364-384 Diesel)
Compression rings 0.0927-0.0937
Top oil control 0.1865-0.1875
2nd oil control 0.1865-0.1875
Ring Width (B-414-354 Non-Diesel)
Compression rings 0.0930-0.0935
Oil control 0.1860-0.1865
Ring End Gap (B-275 Diesel)
Compression rings 0.012-0.018
Top oil control 0.015-0.045
2nd oil control 0.012-0.018
Ring End Gap (B-414-424-444-2424-2444-354-364-384 Diesel)
Compression rings 0.010-0.015
Top oil control 0.010-0.015
2nd oil control 0.010-0.015
Ring End Gap (B-414-354 Non-Diesel)
Compression rings 0.012-0.018
Oil control 0.012-0.018
Ring Side Clearance (B-275 Diesel)
Top compression 0.0028-0.0039
2nd compression 0.0032-0.0043
3rd compression 0.0024-0.0035
Top oil control 0.0015-0.0045
2nd oil control 0.0012-0.0023
Ring Side Clearance (B-414-424-444-2424-2444-354-364-384 Diesel)
Top compression 0.0035-0.0055
Other compression 0.0028-0.0048
Top oil control 0.0012-0.0028
2nd oil control 0.0012-0.0028
Ring Side Clearance (B-414-354 Non-Diesel)
Compression rings 0.0018-0.0033
Oil control 0.0025-0.0040

When installing piston rings, be sure that the largest diameter of the stepped top compression ring is on bottom side. Position rings so that end gaps are 90 degrees from thrust side of piston and are 180 degrees from one another.

PISTONS AND RINGS

Series 424-444-2424-2444 Non-Diesel

72. The cam ground aluminum alloy pistons operate directly in the block bores and are available in standard size as well as oversizes of 0.010, 0.020, 0.030 and 0.040. With pistons removed from engine, measure cylinder bores both parallel and at right angle to the crankshaft centerline. If taper from top of cylinder to bottom of piston travel exceeds 0.006, or if out-of-round more than 0.006, rebore cylinder to next larger size.

NOTE: When reboring, bore cylinder to within approximately 0.001 of desired

size to allow finish honing.

To fit pistons in bores, attach a 0.001 ribbon gage (½-inch wide) to a spring scale, then invert piston and position feeler ribbon at 90 degrees from the piston pin hole. Insert piston and feeler ribbon into cylinder bore until piston is about 3 inches below top of cylinder block. Keep piston pin hole parallel with crankshaft. Now withdraw the feeler ribbon by pulling straight up on the spring scale and note reading on scale as feeler ribbon is being withdrawn. Pistons are correctly fitted to the normal 0.001-0.002 clearance when the spring scale pull reads 2-6 lbs.

73. Pistons are fitted with two compression rings and one oil control ring. The two compression rings have an end gap of 0.010-0.020 and the oil control ring end gap should be 0.018-0.028. Side clearance of rings in piston grooves is 0.003-0.0045 for the top compression ring, 0.0015-0.003 for the second compression ring and 0.002-0.0035 for the oil control ring.

Piston rings are available in standard size as well as oversizes of 0.010, 0.020, 0.030 and 0.040.

PISTON PINS

All Models

74. Piston pins are retained in piston bosses by snap rings and are available in standard size and 0.005 oversize. The 0.005 oversize pin is identified by a plus 5 marking on one end.

75. On Series 424, 444, 2424 and 2444 non-diesel engines standard piston pin diameter is 0.8591-0.8593. Piston pin should have a diametral clearance of 0.0002 in piston bosses and 0.0004 in connecting rod bushing. Maximum allowable clearance of pin in piston is 0.0025 and in rod is 0.003. Total clearance between end of piston pin and snap ring is 0.005-0.055.

76. On all other series engines, standard piston pin diameter is 1.1021-1.1024 and should have a diametral clearance of 0.0005-0.0008 in the connecting rod bushing and be a hard hand push fit in the piston at a room temperature of 68 degrees F. Fit the oversize pins in the same manner. Total clearance between end of piston pin and snap ring is 0.012-0.020.

CONNECTING RODS AND BEARINGS

All Series Except 424-444-2424-2444 Non-Diesel

77. Connecting rod bearings are of the slip-in, precision type, renewable from below after removing the oil pan and connecting rod bearing caps. When installing new inserts, make certain that the projections on same engage slots in connecting rod and cap and that the cylinder identifying numbers on rod and cap are in register and face opposite camshaft side of engine. Connecting rod bearings are available in standard size as well as undersizes of 0.015 and 0.030. Bearing inserts should have a running clearance of 0.001-0.0029 on the 1.7495-1.750 diameter crankshaft crankpins. Rod side play is 0.003-0.010.

Piston pin bushing is furnished semi-finished and must be reamed after installation to provide 0.0005-0.0008 clearance for the piston pin. Be sure oil holes in bushing and connecting rod align after bushing is installed.

Torque the connecting rod bolts to 30-35 ft.-lbs. for Series B-275 and secure with lockwire. Rod bolts, for all Series B-414, 354, 364 and 384 engines and Series 424, 444, 2424 and 2444 diesels, are self-locking and should be torqued to 40-45 ft.-lbs.

Series 424-444-2424-2444 Non-Diesel

78. Connecting rod bearings are of the slip-in, precision type, renewable from below after removing the oil pan and connecting rod bearing caps. When installing new inserts, make certain that the projections on same engage slots in connecting rod and cap and that cylinder identifying numbers on rod and cap are in register and face the camshaft side of engine. Connecting rod bearings are available in standard size as well as undersizes of 0.002, 0.010, 0.020 and 0.030. Bearing inserts should have a running clearance of 0.0009-0.0039 on the 2.059-2.060 standard diameter crankshaft crankpins. Rod side play should be 0.005-0.014.

Piston pin bushing is furnished semi-finished and must be reamed after installation to provide 0.0004 clearance for piston pin. Be sure oil holes in bushing and connecting rod align after bushing is installed.

Tighten the self-locking rod bolts evenly to a torque of 43-49 ft.-lbs.

CRANKSHAFT AND MAIN BEARINGS

All Series Except 424-444-2424-2444 Non-Diesel

79. The crankshaft is supported in five slip-in, precision type main bearings, renewable from below after removing the oil pan, oil pump and main bearing caps. Normal crankshaft end play of 0.004-0.008 is controlled by the flanged rear main bearing inserts. Excessive crankshaft end play is cor-

rected by renewing the inserts. Main bearings are available in standard size as well as undersizes of 0.015 and 0.030.

Crankshaft removal requires removal of engine from tractor as outlined in paragraphs 35, 36 or 37, then remove oil pan, oil pump, timing gear cover, clutch, flywheel and rear oil seal. Remove connecting rod caps and main bearing caps and lift crankshaft from cylinder block.

NOTE: Check main bearing caps and if they are not identified, do so before removing.

Check the crankshaft and bearings against the values which follow:

Crankpin diameter 1.7495-1.750
Rod bearing clearance 0.001-0.0029
Main journal diameter 2.124-2.125
Main bearing clearance 0.002-0.004
Crankshaft end play 0.004-0.008
Journal max. allowable
taper (per inch) 0.0015
Journal max. out-of-round 0.005
Crankshaft max. run-out
(center main) 0.0008

Connecting rod and main bearing bolt torque values are as follows: Series B-275 rod bolts, 30-35 ft.-lbs.; main bearing bolts, 70-75 ft.-lbs. All Series B-414, 354, 364 and 384 and Series 424, 444, 2424 and 2444 diesel rod bolts, 40-45 ft.-lbs.; main bearing bolt torque is as follows:

Pitch bolt . . . . .70-75 ft.-lbs.
Place bolt . . . . .80-85 ft.-lbs.

NOTE: Refer to paragraph 80 for bolt identification.

80. On some series of engines, two types of main bearings bolts are used. The PITCH bolt has a standard bolt head with a washer face. The thread diameter is larger than the shank. This type attains its tension by stretching of the shank and should be torqued to 70-75 ft.-lbs. The PLACE bolt has a head that is either notched or concave and the shank and thread diameter are nearly the same. This type attains its tension by bending the bolt head and should be torqued to 80-85 ft.-lbs.

Series 424-444-2424-2444 Non-Diesel

81. The crankshaft is supported in three slip-in, precision type main bearings, renewable from below after removing the oil pan, rear oil seal retainer plate and main bearing caps. Normal crankshaft end play of 0.004-0.010 is controlled by the flanged rear main bearing inserts. Excessive end play is corrected by renewing the inserts. Main bearings are available in standard size as well as undersizes of

0.002, 0.010, 0.020 and 0.030.

Crankshaft removal requires removal of engine from tractor as outlined in paragraph 38. Then, remove the oil pan, oil pump, timing gear cover, clutch, flywheel, engine rear support plate and rear oil seal. Remove connecting rod caps and main bearing caps and lift crankshaft from cylinder block.

Check the crankshaft and bearings against the values which follow:

Crankpin diameter 2.059-2.060
Rod bearing clearance 0.0009-0.0039
Main journal diameter 2.6235-2.6245
Main bearing clearance 0.009-0.0039
Journal max. allowable
  taper 0.003
Journal max. out-of-
  round 0.003
Crankshaft end play 0.004-0.010

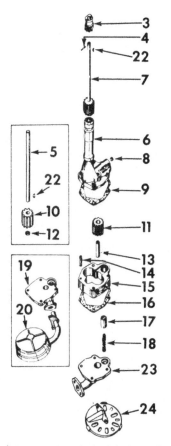

Fig. 42—Exploded view of typical oil pump used on all series except 424, 444, 2424 and 2444 non-diesel engines. Items 5, 10, 12, 19, 20 and 22 are used on early type pump. NOTE: On early B-275 oil pump, a ball was used in place of plunger (17) to regulate oil pressure. Diesel and non-diesel pumps are basically similar.

3.	Pinion	15.	Gearcase
4.	Roll pin	16.	Gasket
5.	Shaft	17.	Pressure regulator
6.	Body		valve
7.	Shaft assy.	18.	Pressure regulator
8.	Plug		spring
9.	Gasket	19.	Pump cover
10.	Gear	20.	Screen assy.
11.	Gear	22.	Woodruff key
12.	Retainer	23.	Pump cover
13.	Idler shaft	24.	Screen assy.
14.	Roll pin		

Connecting rod bolts and main bearing bolts should be tightened evenly to a torque of 43-49 ft.-lbs. for the rod bolts and 75-80 ft.-lbs. for the main bearing bolts.

CRANKSHAFT REAR OIL SEAL

All Series Except 424-444-2424-2444 Non-Diesel

82. The lip type oil seal is contained in a one-piece retainer which is doweled and bolted to the rear of the cylinder block as shown in Fig. 41.

The procedure for renewing the seal is evident after splitting the engine from clutch housing and removing the clutch and flywheel.

When installing seal, press same into retainer until it bottoms against shoulder in retainer. Use a sealant on retainer and lubricate lip of seal prior to installing. The bottom face of retainer must register within 0.020 with face of crankcase.

Series 424-444-2424-2444 Non-Diesel

83. The lip type oil seal is contained in a one-piece retainer which is bolted to the rear of the crankcase.

To renew the oil seal, first split the engine from the clutch housing, then remove the clutch and flywheel. Unbolt and remove the oil seal and retainer assembly.

To install the oil seal and retainer, International Harvester recommends the following procedure: Apply sealer on gasket and retainer and using oil seal driver (IH tool No. FES 6-15) to line up retainer with crankshaft oil seal surface, install retainer to the crankcase. With the driver remaining on the crankshaft and in the retainer, tighten the cap screws in sequence (one across from the other) rotating the driver in retainer at the same time. If binding or driver occurs during the tightening of the cap screws, loosen cap screws and then repeat the tightening procedure. Remove the seal driver after all cap screws have been tightened.

Lubricate the oil seal, crankshaft flange and seal bore in the retainer. Install seal on crankshaft flange, then push seal forward by hand until the sealing lip on seal outer diameter has entered the chamfer on the retainer around the entire circumference of the seal. Position oil seal driver on crankshaft flange and drive seal forward in retainer until the shoulder of the driver contacts the rear surface of crankshaft flange.

NOTE: If special seal driver (IH tool No. FES 6-15) is not available, seal may be installed as follows: Before remov-

ing the old seal from retainer, note the depth of old seal in retainer. Install new seal in retainer in same position as old seal. Apply sealer to gasket and retainer, then install seal and retainer on crankshaft flange. Slide assembly forward on crankshaft flange and install cap screws. Center oil seal to crankshaft flange and tighten cap screws.

FLYWHEEL

All Models

84. The flywheel can be removed after splitting engine from clutch housing and removing the clutch. To install the flywheel ring gear, heat same to approximately 500 degrees F. On Series 424, 444, 2424 and 2444 non-diesel engines, tighten flywheel bolts to a torque of 45-52 ft.-lbs. Flywheel bolts on all other series engines should be tightened to a torque of 65-70 ft.-lbs.

OIL PUMP AND RELIEF VALVE

All Models

85. The gear type oil pump is gear driven from a pinion on the camshaft and removal is evident after removing the oil pan. Overhaul of the pump is obvious after reference to Figs. 42 and 43 and to the specifications which follow:

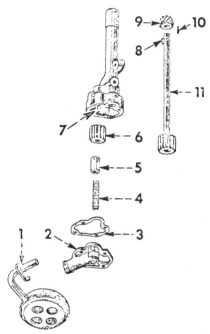

Fig. 43—Exploded view of oil pump used in Series 424, 444, 2424 and 2444 non-diesel engines.

1.	Screen assy.	7.	Pump body
2.	End plate	8.	Woodruff key
3.	Gasket	9.	Drive pinion
4.	Relief valve spring	10.	Pin
5.	Relief valve	11.	Gear & shaft
6.	Idler gear		assy.

Series 424-444-2424-2444 non-diesel:

Gear diametral
clearance 0.0068-0.0108
Gear to end plate end
play 0.0035-0.006
Gear backlash 0.004-0.006
Drive shaft operating
clearance 0.001-0.0035
Idler gear to shaft
clearance 0.0015-0.003
Relief valve spring free
length 2.398
Relief valve opening
pressure . . . . .45-55 psi

All other series:

Gear diametral
clearance 0.0053-0.0083
Gear to end plate end
play 0.0035-0.006
Gear backlash 0.003-0.006
Drive pinion to
camshaft backlash 0.008-0.012
Drive shaft operating
clearance 0.002-0.0035
Idler gear to shaft
operating clearance 0.0015-0.0035
Relief valve spring free
length (ball type) . . . . .2-11/32 in.
Relief valve spring free
length (plunger type) . . . . . .2-9/16 in.
Relief valve opening
pressure . . . . .30-35 psi

GOVERNOR
(NON-DIESEL)

Series B-414-354

The Series B-414 and 354 non-diesel
tractors are equipped with a flyweight
type governor which is mounted on left
front of engine as shown in Fig. 44.

86. **SPEED ADJUSTMENT.** Prior to
making any speed adjustments, check
all operating linkage for lost motion or
binding and correct any defects which
may be present.

Start engine and bring to operating
temperature, place throttle lever in the
high idle position and check the engine
high idle speed which should be 2200
rpm. Move the throttle lever to the low
idle position and check the engine low
idle speed which should be 500-525
rpm.

If engine speeds are not as stated,
remove side cover from governor housing and adjust screw (H—Fig. 44) to
correct engine high idle rpm and/or
screw (L) to correct engine low idle
rpm.

No surge adjustment is provided on
governor assembly.

87. **R&R AND OVERHAUL.** To
remove the governor assembly, it is
first necessary to remove the timing
gear cover as outlined in paragraph 57.

With timing gear cover off, remove
drive gear retaining nut and drive
gear. Disconnect throttle control rod
and carburetor rod from governor
levers, then remove the cap screws
(S—Fig. 36) and pull governor assembly from engine front plate. Bearing
(3—Fig. 45), weight carrier, weights,
shaft and thrust bearing assembly can
be removed from housing after removing snap ring (3). Rockshaft lever (23)
and spring lever (25) and their shafts
can be removed after loosening the
clamping cap screws. Any further disassembly required will be obvious.

Oil seals (17) are installed with lips
toward inside. Use sealant on outer
edge of expansion plug (21) when
renewing.

Series 424-444-2424-2444
Non-Diesel

The centrifugal flyweight type governor is mounted on the right front face
of engine and is driven by the engine
timing gear train. Before attempting
any governor adjustments, check the
operating linkage and remove any
binding or lost motion.

88. **ADJUSTMENT.** To adjust the
governor proceed as follows: With engine stopped, place the speed change
lever in wide open position and remove
clevis pin (3—Fig. 46) from the governor rockshaft arm. Hold the rockshaft
arm and carburetor throttle rod (1) as
far towards carburetor as they will go.
If pin holes are not in alignment, adjust
the length of rod (1) until pin (3) will
slide freely into place. Then, lengthen
rod one full turn and install clevis pin.

With engine running and speed control lever in wide open position, adjust
the high idle adjusting screw (16) to
obtain a high idle speed of 2200 rpm
and then, lock the screw in place with
the jam nut.

With engine running at low idle
speed (425 rpm), quickly move the
speed control lever to high idle position. If the engine surges more than
twice, adjust the governor bumper
spring (25) as follows: Stop engine and
remove acorn nut from bumper spring
adjusting screw (26). Loosen jam nut

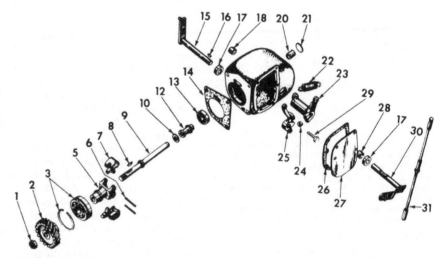

Fig. 45—Exploded view of the governor used on Series B-414 and 354 non-diesel engines.

1. Nut	9. Shaft	17. Oil seal	25. Spring lever
2. Drive gear	10. Thrust washer	18. Rockshaft brg.	26. Gasket
3. Snap ring & bearing	12. Sleeve	20. Bushing	27. Housing cover
4. Governor carrier	13. Thrust bearing	21. Expansion plug	28. Rockshaft brg.
5. Governor carrier	14. Gasket	22. Governor spring	29. Governor spring bolt
6. Pin	15. Speed change lever	23. Rockshaft lever	30. Rockshaft
7. Governor weight	16. Woodruff key	24. Key	31. Control rod
8. Woodruff key			

*Fig. 44—Series B-414 and 354 non-diesel
governor which mounts on left side of
engine is shown with side cover off. Note
high idle (H) and low idle (L) adjusting
screws.*

and turn screw (26) in just enough to stop excessive surging. When the bumper spring screw is properly adjusted, lock it in place with jam nut and install acorn nut.

89. R&R AND OVERHAUL. Remove distributor cap and mark location of rotor. Then, unbolt and remove the distributor and drive housing assembly and mark location of slots in the ignition unit drive coupling on governor gear (19—Fig. 46) in relation to crankcase. This procedure will facilitate reinstallation of governor gear in proper timing mesh with camshaft gear if position of crankshaft is not changed while governor is removed. Remove the governor speed control rod and carburetor throttle rod. Remove the grille side panels and loosen fan belt, then unbolt and remove the governor assembly from timing gear cover.

To disassemble the governor, first remove the gear and flyweight assembly. Remove the nut and lock washer from the speed control lever and shaft (6), then withdraw shaft (6) from the governor spring lever (9) and low speed spring (17). Unhook high speed spring (10) from the rockshaft fork (4) and remove governor spring lever, low speed spring and high speed spring from governor housing (15). Remove the fork (4) from rockshaft (5) and slide the rockshaft from the housing. The two needle bearings (14 and 22), "Oilite" bushing (13), and oil seals (12 and 7) can now be renewed if necessary. Flyweights and pins should be renewed when excessive wear is evident.

NOTE: A governor overhaul service package (part No. 391021R93) is available from International Harvester Co.

When reassembling the high speed spring and low speed spring to the governor spring lever, hook the upper end of high speed spring through both holes in low speed spring as shown in Fig. 47. Then, hook the lower end of the high speed spring in hole (A) of rockshaft lever (fork). Install the spring lever and low speed spring on the lever shaft. The open ends of the high speed spring hooks must be toward the center of the governor housing.

The bushing in the crankcase which supports the governor and ignition unit drive gear hub can be renewed when governor and ignition unit are off. The I&T recommended clearance of gear hub in bushing is 0.0015-0.002.

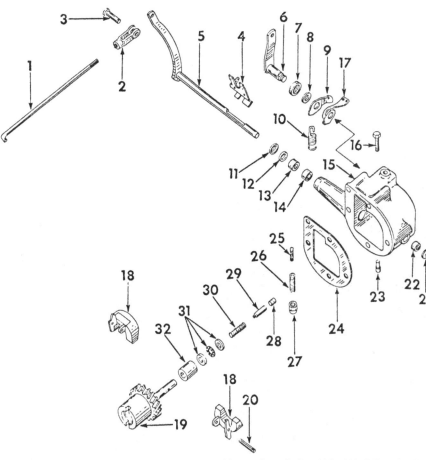

Fig. 46—Exploded view of the governor assembly used on Series 424, 444, 2424 and 2444 non-diesel engines.

1. Throttle rod
2. Clevis
3. Clevis pin
4. Rockshaft fork
5. Rockshaft
6. Speed change lever & shaft
7. Oil seal
8. Washer
9. Spring lever
10. High speed spring
11. Retainer
12. Oil seal
13. Bushing
14. Needle bearing
15. Housing
16. Speed adjusting screw
17. Low speed spring
18. Flyweights
19. Governor gear & shaft assembly
20. Flyweight pin
21. Plug
22. Needle bearing
23. Fork retainer pin
24. Gasket
25. Bumper spring
26. Bumper spring adjusting screw
27. Acorn nut
28. Shaft stop pin
29. Shaft spring pin
30. Shaft spring
31. Thrust bearing
32. Thrust sleeve

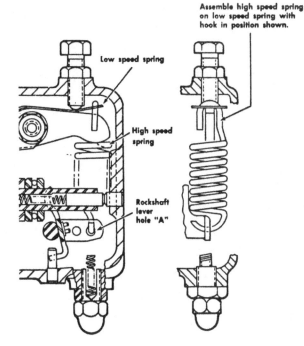

Fig. 47—View showing correct installation of high speed spring in Series 424, 444, 2424 and 2444 non-diesel governor.

DIESEL
FUEL SYSTEM

Two types of injection pumps have been used on the B275 tractor. On early model tractors (prior eng. ser. no. BD-144/17289-A), a C.A.V. multiple plunger pump, fitted with a pneumatic governor was used. On later Model B-275 tractors (eng. ser. no. BD-144/17289-A and up), a C.A.V. distributor type injection pump having a mechanical type governor is used.

Series 424, 444, 2424, 2444, B-414, 354, 364 and 384 tractors also use the C.A.V. distributor pump.

When servicing any unit associated with the fuel system, the maintenance of absolute cleanliness is of utmost importance. Of equal importance is the avoidance of nicks or burrs on any of the working parts.

Probably the most important precaution that service personnel can impart to owners of diesel powered tractors, is to urge them to use an approved fuel that is absolutely clean and free from foreign material. Extra precaution should be taken to make certain that no water enters the fuel storage tanks. This last precaution is based on the fact that all diesel fuels contain some sulphur. When water is mixed with sulphur, sulphuric acid is formed and the acid will quickly erode the closely fitting parts of the injection pump and nozzles.

90. **QUICK CHECK—UNITS ON TRACTOR.** If the diesel engine does not start or does not run properly, and the diesel fuel system is suspected as the source of trouble, refer to the following list of troubles and their possible causes:

1. Sudden Stopping of Engine.
   a. Lack of fuel.
   b. Clogged fuel filter and/or lines.
   c. Faulty injection pump.
   d. Broken spring in by-pass valve.
2. Lack of Power.
   a. Improper injection pump timing.
   b. Inferior fuel.
   c. Faulty injection pump.
   d. Clogged fuel filter and/or lines.
   e. Weak or broken transfer pump plunger spring.
3. Engine Hard to Start.
   a. Inferior fuel.
   b. Clogged fuel filter and/or lines.
   c. Improper injection pump timing.
   d. Faulty injection pump.
4. Irregular Engine Operation.

a. Weak or broken governor springs.
b. Clogged fuel filter and/or lines.
c. Faulty nozzle.
d. Improper injection pump timing.
e. Faulty injection pump.
f. Air leak in venturi vacuum pipe or governor diaphragm.
5. Engine Smokes or Knocks.
   a. Improper injection pump timing.
   b. Faulty nozzle.
   c. Inferior fuel.
6. Excessive Fuel Consumption.
   a. Improper injection pump timing.
   b. Faulty nozzle.

Many of the problems are self-explanatory; however, if the difficulty points to the fuel filter, injection nozzles and/or injection pump, refer to the appropriate paragraphs 91 through 122.

FILTER AND BLEEDING

All Models

91. When fuel lines have been disconnected or the fuel flow interrupted, bleed trapped air from the system as follows:

On early B-275 model tractors having the pneumatic governor, be sure there is sufficient fuel in tank, loosen the sediment bowl and allow same to fill, then tighten bowl. Loosen bleed screw (I—Fig. 48) on top of fuel filter and operate hand primer (P) until bubble free fuel flows, then tighten bleed screw. Loosen bleed screw (A) on injection pump and operate hand primer until bubble free fuel flows, then tighten bleed screw.

On late Model B-275 tractors with the mechanical governor and all 424, 444, 2424, 2444, B-414, 354, 364 and 384 tractors follow the same procedure except loosen both bleeder screws (A and B—Fig. 49) on the injection pump starting with the bottom one first. Primer is incorporated in fuel pump

which is mounted on right side of engine.

NOTE: Models 424, 444, 2424 and 2444 diesel engines do not use the auxiliary fuel pump. Therefore, gravity flow must be used to vent the system.

On all models, loosen injector pressure lines at injectors, make sure shut-

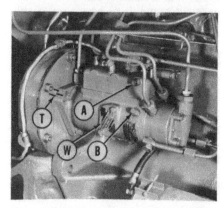

Fig. 49—Bleeder screws (A & B) on injection pump having mechanical governor.

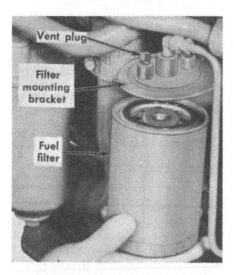

Fig. 50—When installing the diesel fuel filter on Series 424, 444, 2424 or 2444 tractor, rotate filter element until rubber seal contacts filter mounting bracket, then tighten element ¼- to ½-turn.

Fig. 48—Hand primer and bleed screw location on injection pump having pneumatic governor.

A. Bleed screw
D. Breather
G. Governor
I. Bleed screw
P. Hand primer
V. Venturi

off control is in the operating position and turn engine over with starting motor until fuel escapes from line ends. Tighten the pressure line connections.

INJECTOR NOZZLES

All Models

WARNING: Fuel leaves the injection nozzles with sufficient pressure to penetrate the skin. When testing, keep your person clear of the nozzle spray.

92. **TESTING AND LOCATING A FAULTY NOZZLE.** If the engine does not run properly, and a faulty injector is suspected, locate the faulty unit as follows:

If one engine cylinder is misfiring, it is reasonable to suspect a faulty injector. Generally, a faulty injector can be located by loosening the high pressure line fitting on each nozzle holder in turn, thereby allowing fuel to escape at the union rather than enter the cylinder. As in checking spark plugs in a spark ignition engine, the faulty unit is the one which, when its line is loosened, least affects the running of the engine.

93. Remove the suspected injector from the engine as outlined in paragraph 99. If a suitable nozzle tester is available, check the unit as outlined in paragraphs 94, 95, 96, 97 and 98. If a tester is not available, reconnect the fuel line to the injector and with the nozzle tip directed where it will do no harm, crank the engine with the starting motor and observe the nozzle spray pattern.

If the spray patterns are ragged, unduly wet, streaky and/or not symmetrical or, if nozzle dribbles, the nozzle valve is not seating properly and same should be cleaned and/or overhauled.

94. **NOZZLE TESTER.** A complete job of testing and adjusting the nozzle requires the use of a special tester such as that shown in Fig. 51. The nozzle should be tested for opening pressure, seat leakage, back leakage and spray pattern.

Operate the tester lever until oil flows, then attach the nozzle and holder assembly.

NOTE: Only clean approved oil should be used in the tester tank.

Close the tester valve and apply a few quick strokes to the tester lever. If undue pressure is required to operate the tester, the nozzle is plugged and same should be serviced as outlined in paragraph 100.

95. OPENING PRESSURE. While operating the tester lever, observe the gage pressure at which the spray occurs. The gage pressure should be 2130-2205 psi for models with pneumatic governor, or 2350-2425 for models with mechanical governor. On Series B-275 or B-414 injectors, if pressure is not as specified remove cap nut (3—Fig. 52) and loosen locknut (5). Turn adjusting screw (6) either way, as required, to correct opening pressure. Refer also to Fig. 53.

On Series 354, 364 and 384 injectors, add or remove shims (6—Fig. 54) as required to obtain an opening pressure of 2350 psi.

96. SEAT LEAKAGE. To check seat leakage, operate tester until gage pressure is 150 psi below nozzle operating

pressure and hold this pressure for 10 seconds. Examine orifice and if drops of fuel collect at pressures below those specified, the nozzle valve is not seating properly and should be serviced as in paragraph 100.

97. BACK LEAKAGE. Test specifications for used nozzles may vary, however, if nozzle will pass the following test it may be considered satisfactory.

Pump up pressure on tester until gage registers at least 1500 psi, then as pressure starts to drop, observe the time it takes for the gage pressure to drop from 1500 to 1100 psi. A nozzle in good condition should not lose the given amount of pressure in less than 10 seconds at 60 degrees F. However, bear in mind that higher temperatures may give a time of less than the 10 seconds.

If nozzle fails to meet the foregoing test, service same as outlined in paragraph 100.

98. SPRAY PATTERN. Operate the tester handle at approximately 100 strokes per minute and observe the nozzle spray pattern. If the spray pattern is unduly wet, streaky and/or ragged, service the nozzle as outlined in paragraph 100.

99. **REMOVE AND REINSTALL.** Raise or remove hood, then before loosening any fuel lines, wash the

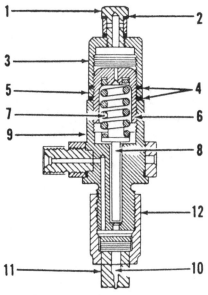

Fig. 53—Cross-sectional view of injector nozzle and holder assembly. Refer to Fig. 52 for an exploded view.

1. Banjo bolt	7. Spring
2. Washer	8. Spindle
3. Cap nut	9. Holder
4. Washers	10. Valve
5. Locknut	11. Valve holder
6. Adjusting screw	12. Nozzle cap nut

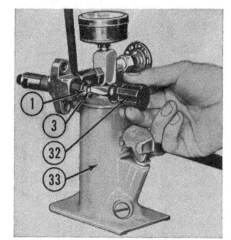

Fig. 51—Typical tester used to check and adjust injector nozzles.

1. Locknut	32. Screwdriver
3. Adjusting screw	33. Nozzle tester

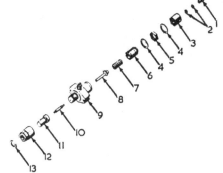

Fig. 52—Exploded view of injector nozzle and holder. Refer to Fig. 53 for a cross-sectional view.

1. Banjo bolt	8. Spindle
2. Washers	9. Holder
3. Holder cap nut	10. Valve
4. Washers	11. Valve holder
5. Locknut	12. Nozzle cap nut
6. Adjusting screw	13. Washer
7. Spring	

nozzle holder and connections with clean diesel fuel. After disconnecting the high pressure and leak-off lines, cover open ends of connections with composition caps to prevent the entrance of dirt or other foreign material. On Series B-414, 354, 364 and 384, remove fuel filter from inlet manifold if number two or three injector is to be removed. Remove the nozzle retaining nuts and carefully withdraw the nozzle from cylinder head, being careful not to strike the tip end of the nozzle against any hard surface. Use a suitable puller if necessary.

Thoroughly clean the nozzle recess in cylinder head before reinserting the nozzle and holder assembly. It is important that the seating surfaces of recess be free of even the smallest particles of carbon which could cause the unit to be cocked and result in blowby of hot gases. No hard or sharp tools should be used for cleaning. A piece of wood dowel or brass stock properly shaped is very effective. Do not reuse the copper ring gasket located between nozzle and precombustion chamber holder, always install a new one. Tighten the nozzle holder stud nuts to a torque of 40-50

ft.-lbs. for Series B-275; or 30-35 ft.-lbs. for Series B-414, 424, 444, 2424, 2444, 354, 364 and 384.

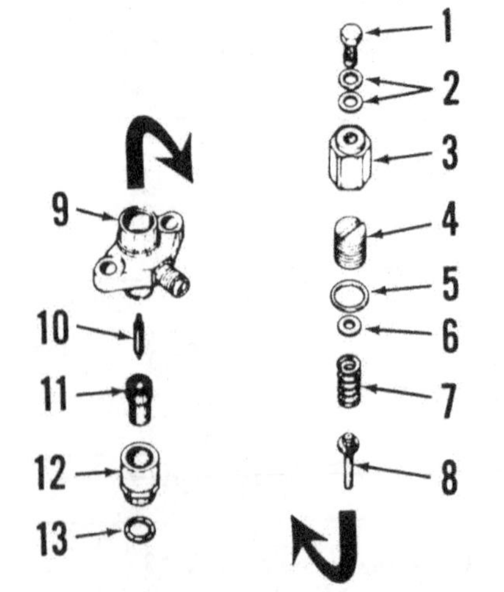

100. MINOR OVERHAUL (CLEANING) OF NOZZLE VALVE AND BODY. Hard or sharp tools, emery cloth, crocus cloth, grinding compounds or abrasives of any kind should NEVER be used in the cleaning of nozzles.

Wipe all dirt and loose carbon from the nozzle and holder assembly with a clean, lint free cloth. Carefully clamp nozzle holder assembly in a soft jawed vise and remove the cap nut (3—Fig. 52 or 54). On Series B-275, B-414, 424, 444, 2424 and 2444, loosen jam nut (5—Fig. 52) and back-off the adjusting screw (6) enough to relieve load from spring (7). On Series 354, 364 and 384, back-off spring cap (4—Fig. 54) to relieve load from spring (7).

On all series remove the nozzle cap nut (12—Fig. 52 or 54) and nozzle body (11). Normally, the nozzle valve (10) can be easily withdrawn from the nozzle body. If the valve cannot be easily withdrawn, soak the assembly in fuel oil or carbon solvent to facilitate removal. Be careful not to permit the valve or body to come in contact with any hard surface.

Examine the nozzle body and remove any carbon deposits from exterior surfaces using a brass wire brush. The nozzle body must be in good condition and not blued due to overheating. All polished surfaces should be relatively bright, without scratches or dull patches. Pressure surfaces (A, B and J—Fig. 56) must be absolutely clean and free from nicks, scratches or foreign material, as these surfaces must register together to form a high pressure joint.

Clean out the small fuel feed channels (C), using a small diameter wire. Insert a suitable groove scraper into nozzle body until nose of scraper locates in fuel gallery (F); then, press nose of scraper hard against side of cavity and rotate scraper to clean all carbon deposits from the gallery. Clean all carbon from valve seat (G), using a suitable seat scraper.

Use a pintle hole cleaning probe of appropriate size and pass the probe

down the bore of the nozzle body until probe protrudes through the orifice; then, rotate the probe until all carbon is cleared.

Examine the pintle and seat end of the nozzle valve and remove any carbon deposits using a brass wire brush. Use care, however, as any burr or small scratch may cause valve leakage or spray pattern distortion. If valve seat (M—Fig. 56) has a dull circumferential ring indicating wear or pitting or if valve is blued, the valve and body should be turned over to an official diesel service station for possible overhaul.

Before reassembling, thoroughly rinse all parts in clean diesel fuel and make certain that all carbon is removed from the nozzle holder nut. Install nozzle body and holder nut, making certain that the valve stem is located in the hole of the holder body. Tighten the holder nut to a torque of 50 ft.-lbs.

NOTE: Over-tightening may cause distortion and subsequent seizure of the nozzle valve.

Test the injector as in paragraphs 94, 95, 96, 97 and 98. If the nozzle does not leak and if the spray pattern is satisfactory, the nozzle is ready for use. If the nozzle will not pass the leakage and spray pattern tests, renew the nozzle valve and seat, which are available only in a matched set; or, send the nozzle and holder assembly to an official diesel

Fig. 54—Exploded view of injector nozzle and holder, used on 354, 364 and 384 diesel.

1. Banjo bolt
2. Washers
3. Holder cap nut
4. Spring cap
5. Washer
6. Shim
7. Spring
8. Spindle
9. Holder
10. Valve
11. Valve holder
12. Nozzle cap
13. Washer

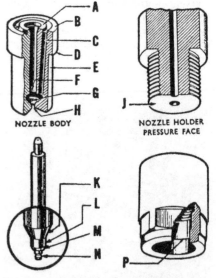

NOZZLE BODY   NOZZLE HOLDER PRESSURE FACE

Fig. 56—Views of nozzle and holder showing various points for detailed cleaning and inspection.

A. Nozzle body pressure face
B. Nozzle body pressure face
C. Fuel feed hole
D. Shoulder
E. Nozzle trunk
F. Fuel gallery
G. Valve seat
H. Pintle orifice
J. Holder pressure face
K. Valve cone
L. Stem
M. Valve seat
N. Pintle
P. Nozzle retaining shoulder

Fig. 55—View showing injector assembly, precombustion chamber holder, precombustion chamber and glow plug removed from engine.

1. Banjo bolt
13. Washer
14. Injector
15. Precombustion chamber holder
16. Washer
18. Glow plug
19. Precombustion chamber

service station for a complete overhaul which includes reseating the nozzle valve cone and seat.

101. OVERHAUL OF NOZZLE HOLDER. On Series B-275, B-414, 424, 444, 2424 and 2444, refer to Fig. 52 and remove cap (3). Remove jam nut (5) and adjusting screw (6). Withdraw spring (7) and spindle (8). On Series 354, 364 and 384 refer to Fig. 54 and remove cap (3). Remove spring cap (4) and shims (6). Withdraw spring (7) and spindle (8).

On all models, thoroughly wash all parts in clean diesel fuel and examine the end of the spindle which contacts the nozzle valve stem for any irregularities. If the contact surface is pitted or rough, renew the spindle. Renew any other questionable parts.

Reassemble the nozzle holder and leave the adjusting screw locknut loose until after the nozzle opening pressure has been adjusted as outlined in paragraph 95.

PRE-COMBUSTION CHAMBERS

All Models

102. The necessity for cleaning the pre-combustion chambers is usually indicated by excessive smoking or when fuel economy drops.

To remove the precombustion chambers, first remove injector nozzles as outlined in paragraph 99. Remove glow plug wires and glow plugs. Precombustion chambers can now be removed.

NOTE: In cases where precombustion chambers are stuck extremely tight, it may be necessary to remove cylinder head as outlined in paragraphs 39 and 43.

When reinstalling precombustion chambers, use new gaskets and be sure glow plug bores are aligned. Misaligned bores could cause the glow plug to contact the precombustion chamber and result in a short circuit.

INJECTION PUMP

All Models

The subsequent paragraphs will outline ONLY the injection pump service work which can be accomplished without the use of special, costly pump testing equipment. If additional service work is required, the pump should be turned over to an offical Diesel service station for overhaul. Inexperienced service personnel should never attempt to overhaul a Diesel injection pump.

103. **TIMING TO ENGINE, (PNEU-MATIC GOVERNOR).** To check and adjust the injection pump timing on models equipped with pump having pneumatic governor, proceed as follows: Disconnect the number one or number four injector line from pump, remove the delivery valve holder, withdraw its valve and spring, then reinstall the delivery valve holder. Attach a gooseneck or bent tube to the delivery valve holder. Remove the pump side cover, place speed control lever in maximum speed position and check to be sure rack is in delivery position. Turn engine in direction of normal rotation, until pump plunger starts to lift, then operate primer pump to maintain fuel pressure in fuel gallery. Continue to turn engine in same direction until fuel flow from spill line decreases. Stop engine rotation at this instant.

NOTE: Spill cut-off point is when fuel flow decreases to two to four drops of fuel per second at full fuel gallery pressure.

Measure the distance around the circumference of the crankshaft pulley from marker in pulley to timing pointer. This distance should be 1-1/8 inches which equals 20 degrees BTDC.

NOTE: While 20 degrees BTDC is considered optimum, timing can be retarded as far as 16 degrees BTDC in cases where engine noise is critical and power requirements are not exacting. However, timing should not be advanced more than 22 degrees BTDC or maximum power will be reduced and glow plug failure will be increased. If necessary, the timing angle can be found as follows: (A) Multiply the distance measured on crankshaft pulley by 360. (B) Multiply pulley diameter by 3.14. (C) Divide (B) into (A).

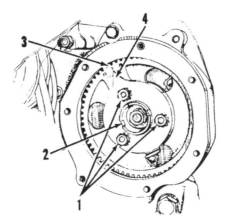

Fig. 57—Injection pump drive gear used on early B-275 tractors equipped with injection pump having pneumatic governor.

1. Cap screws	3. Pump gear
2. Hub groove	4. Timing pointer

To change the injection pump timing, first remove the injection pump drive gear cover from timing gear cover. Refer to Fig. 57 and loosen the three cap screws (1). Turn indicator (4) clockwise to advance timing, or counterclockwise to retard. One division on the injection pump drive gear equals four degrees on the engine crankshaft.

Reassemble by reversing disassembly procedure and torque the delivery valve holder to 30 ft.-lbs. Bleed the fuel system as in paragraph 91.

104. **TIMING TO ENGINE (ME-CHANICAL GOVERNOR).** The injection pump drive shaft and the drive gear adapter are equipped with a master spline. As long as the pump drive gear is in proper relation to the engine timing gear train as shown in Fig. 35, the pump may be installed at any time without regard to crankshaft or timing mark location. When a new injection pump, pump drive gear or drive gear adapter is installed, or when incorrect pump timing is suspected, the pump timing can be checked as follows: Shut off fuel and remove injection pump timing window and clutch housing dust cover. Turn engine in direction of normal rotation until No. 1 piston is coming up on compression stroke and continue turning engine until the "TDC" mark on flywheel aligns with the scribe mark located on left front flange of clutch housing. Check the timing marks on engine front end plate and pump mounting flange as shown at (T) in Fig. 49. If alignment is required, loosen nuts on pump mounting studs, align the mark on pump mounting flange midway between the two marks on the engine front end plate and retighten nuts. At this time the "E" scribed line on injection pump rotor should align with the scribed lines at the lower hole of the pump snap ring.These marks are visible after removing cover (W).

NOTE: Movement of the pump between the two scribed lines on the engine front end plate gives a variation of approximately three degrees.

105. **REMOVE AND REINSTALL (PNEUMATIC GOVERNOR).** Before removing injection pump, thoroughly wash pump and all connections with clean diesel fuel. Shut off fuel and disconnect the fuel supply line at the primary pump. Disconnect the fuel pump to fuel filter line at fuel pump. Disconnect fuel filter to injection pump line at injection pump. Disconnect and remove injector lines.

NOTE: Use plastic caps and plugs to seal all openings of lines and pump to prevent entry of foreign material.

Remove the two lines between injection pump governor and intake venturi. Disconnect stop control rod at injection pump. Remove the injection pump drive gear from the timing gear cover, then remove the cap screws which retain timing pointer and drive gear to injection pump hub. Rotate injection pump gear until holes in same align with mounting cap screws and remove cap screws. Remove cap screws from injection pump mounting flange, then remove the injection pump.

Reinstall by reversing the removal procedure and bleed fuel system as in paragraph 91.

106. REMOVE AND REINSTALL (MECHANICAL GOVERNOR). Before removing injection pump, thoroughly wash pump and all connections with clean diesel fuel. Remove the breather line between injection pump drive gear cover and inlet manifold, then remove the injection pump drive gear cover from timing gear cover. Disconnect control rod and fuel shut-off rod from pump. Shut off fuel and disconnect injection pump supply line at injection pump. Disconnect excess fuel line from injection pump and pressure lines from injectors.

NOTE: Use plastic caps and plugs to seal all openings of lines and pump to prevent entry of foreign material.

Unbolt injector pump drive gear from pump hub, then unbolt and remove injection pump with pressure lines attached. Pressure lines can now be removed, if necessary.

NOTE: On Series B-414, 354, 364 and 384, injection pump removal will be eased if fuel filter is removed from inlet manifold.

Reinstall by reversing the removal procedure and bleed fuel system as in paragraph 91. Check timing marks as outlined in paragraph 104.

107. GOVERNOR (PNEUMATIC). The pneumatic governor is actuated by vacuum in the venturi of the engine air induction system.

108. ADJUSTMENT. Recommended governed speeds are as follows:

Pneumatic Governor

Engine high idle rpm	2075
Engine rated rpm	1875
Engine low idle rpm	570
Belt pulley high idle rpm	1479
Belt pulley rated rpm	1336
Belt pulley low idle rpm	407
Pto high idle rpm	614
Pto rated rpm	555
Pto low idle rpm	169

To adjust the governor, first make sure the air cleaner is properly serviced and that no leaks are present in the venturi, governor housing or piping. Check travel of venturi cross shaft and be sure it contacts the limit stops at the extremes of hand lever positions. Adjust link rod length if necessary, to obtain correct travel.

Loosen the idling valve locknut (11—Fig. 58), remove pitot line connections and screw idle valve (7) out until the stem is completely clear of the diaphragm. Tighten locknut. Start engine and run same at high idle for approximately 15 minutes and when both the temperature and rpm have stabilized, check the idle valve for leakage by placing finger over end of valve. If the engine rpm decreases, the valve is leaking and same should be removed, cleaned and inspected to determine the cause.

With idle valve installed and the engine temperature and high idle rpm stabilized, check the engine high idle which should be 2075 rpm. If high idle rpm is not as specified, turn adjusting screw (H—Fig. 59) as required. Tighten locknut and recheck.

109. IDLE (DAMPER) VALVE. Although this valve can be adjusted at low idle, the following procedure will preclude the possibility of over adjustment which can cause poor governor action.

With pitot line removed, start engine and run same at high idle until both engine temperature and rpm have stabilized. Now with engine running at

high idle, slowly turn idle valve inward until engine rpm decreases 20 to 30 rpm. Tighten locknut and connect pitot line.

110. To check the engine low idle rpm, place the hand lever in low idle position and check to see that the venturi cross shaft lever is against the stop. Check the engine low idle speed which should be 570 rpm. If engine rpm is not as specified, turn screw (L—Fig. 59) as required. Tighten locknut and recheck.

111. GOVERNOR DIAPHRAGM. The governor diaphragm is made from specially prepared leather and any pin holes or fractures in same will result in faulty governor action.

To test the diaphragm, proceed as follows: With engine stopped, disconnect both lines from the governor unit. Move the stop control lever to the stop position, place a finger over each of the two holes from which the lines were removed, then release the stop lever. The control rod should remain stationary and any movement of the control rod indicates leakage either at diaphragm, housing or housing cover joints.

To renew the governor diaphragm, disconnect pitot line (8—Fig. 58) from governor housing cover, then remove governor housing cover and spring. Pull diaphragm rim from its recess and pull as far from governor housing as possible. Push the control rod as far back as it will go, then turn diaphragm 90 degrees, slide toward top of housing and remove.

When reassembling, tighten the governor housing screws securely. Check and adjust if necessary, the governed speeds as outlined in paragraphs 108, 109 and 110.

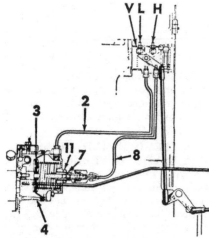

Fig. 58—Schematic view showing governor assembly, venturi and the connecting pipes used on early B-275 tractors.

H. High idle adjusting screw	
L. Low idle adjusting screw	3. Diaphragm
V. Venturi	4. Housing
2. Vacuum line	7. Idle valve
	8. Pitot line
	11. Locknut

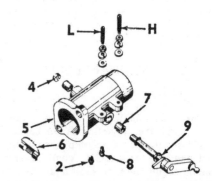

Fig. 59—Exploded view of the venturi and component parts used on early B-275 tractors fitted with injection pump having pneumatic governor.

H. High speed adjusting screw	
L. Low speed adjusting screw	5. Housing
2. Vacuum tube	6. Throttle valve
4. Plug	7. Bushing
	8. Pitot tube
	9. Spindle

112. GOVERNOR (MECHANICAL). The mechanical governor is an integral part of the injection pump and if service is required, the pump should be turned over to an official diesel service station.

113. ADJUSTMENT. Recommended governed speeds are as follows:

Series B-275

Engine high idle rpm	2000
Engine rated rpm	1900
Engine low idle rpm	530-580
Pto high idle rpm	595
Pto rated rpm	563
Pto low idle rpm	158-176

Series B-414-424-444-2424-2444

Engine high idle rpm	2200
Engine rated rpm	2000
Engine low idle rpm,	
B-414-424-2424	550
444-2444	650
Pto high idle rpm	600
Pto rated rpm	545
Pto low idle rpm,	
B-414-424-2424	150
444-2444	177

Series 354

Engine high idle rpm	2075
Engine rated rpm	1900
Engine low idle rpm	550
Pto high idle rpm	615
Pto rated rpm	556
Pto low idle rpm	163

Series 364-384

Engine high idle rpm	2310
Engine rated rpm	2100
Engine low idle rpm	550
Pto high idle rpm	634
Pto rated rpm	573
Pto low idle rpm	150

To adjust the governor, first start engine and bring to normal operating temperature. Move the speed control hand lever to the high idle position at which time the engine high idle should be 2000 rpm for Series B-275, 2200 rpm for Series 424, 444, 2424, 2444 and B-414, 2075 rpm for Series 354 and 2310 for 364 and 384. If engine rpm is not as specified, loosen locknut and turn the high idle adjusting screw (H—Fig. 60) as required. Tighten locknut and recheck.

With engine high idle rpm properly adjusted, move the speed control hand lever to the low idle position at which time the engine low idle should be 530-580 rpm on Series B-275, 550 rpm on Series B-414, 354, 364, 384, 424 and 2424 or 650 rpm on Series 444 and 2444. If engine rpm is not as specified, loosen locknut and turn low idle adjusting screw (L—Fig. 60) as required. Tighten locknut and recheck.

VENTURI

Series B-275

114. R&R AND OVERHAUL. To remove the venturi, raise hood and loosen clamp which retains air cleaner hose to venturi. Remove air cleaner cup, then unbolt air cleaner clamp and remove air cleaner. Disconnect control rod from venturi cross-shaft, then unbolt and remove venturi from inlet manifold.

Further disassembly and/or overhaul is obvious after an examination of the unit and reference to Fig. 59.

FUEL PUMP

Series B-275-B-414-354-364-384

115. When tractor is fitted with an injection pump having a pneumatic governor a fuel pump such as that shown in Figs. 61 and 62 is used. On tractors which are fitted with the injection pump having a mechanical governor a pump such as that shown in Figs. 63 and 64 is used.

116. TESTING. The fuel pump used in conjunction with the pneumatic governed injection pump can be tested by mounting injection pump and fuel pump on test stand and proceeding as follows:

117. SUCTION TEST. With injection pump running at 700 rpm, fuel pump should deliver fuel in 60 seconds through dry hoses from a 2-foot suction head.

118. DELIVERY TEST. At 700 injection pump rpm, fuel pump should deliver a minimum of 300 cubic centimeters of fuel in 30 seconds from a 2-ft. suction head.

119. MAXIMUM PRESSURE TEST. With injection pump running at 700

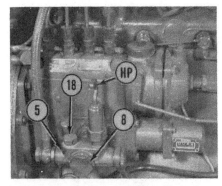

Fig. 61—When tractor is fitted with an injection pump having a pneumatic governor, the fuel pump is bolted to the side of the injection pump as shown.

5. Housing	18. Valve guide
8. Plug	HP. Hand primer

Fig. 60—On late B-275 and all B-414, 354, 364, 384, 424, 444, 2424 and 2444 tractors having injection pump fitted with mechanical governor the high (H) and low (L) idle speed adjusting screws are located as shown.

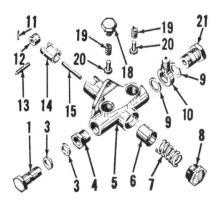

Fig. 62—Exploded view of fuel pump shown in Fig. 61. Hand primer lever is not shown.

1. Outlet stud	11. Retaining pin
3. Washers	12. Tappet roller
4. Nipple	13. Roller pin
5. Housing	14. Tappet guide
6. Plunger	15. Tappet spindle
7. Spring	18. Valve guide
8. Plug	19. Springs
9. Washers	20. Valves
10. Threaded connection	21. Inlet stud

rpm, fuel pump should maintain an outlet pressure of 5-7 psi.

120. VALVES TEST. With maximum pump pressure built up and pump stopped, the maximum pressure should not decrease by more than 0.5 psi in 20 seconds.

121. **R&R AND OVERHAUL.** To remove and overhaul the fuel pump used on models with pneumatic governed injection pump, proceed as follows:

Shut off fuel and disconnect fuel inlet and outlet lines from pump body. Refer to Fig. 62 and remove hand primer and

valve guide (18). Lift out springs (19) and valves (20). Remove plug (8) and spring (7). The pump housing can now be removed from injection pump.

NOTE: Above disassembly should be done prior to removing pump housing from injection pump as pump housing can be easily damaged if clamped in a vise.

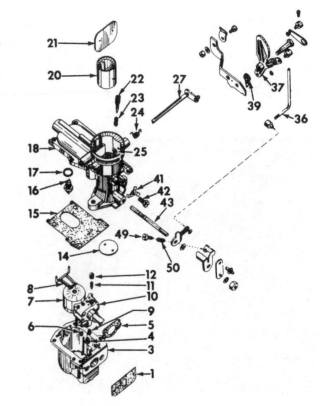

Fig. 64—View of B-275 tractor showing the auxiliary fuel pump (AP) used when tractor is equipped with an injection pump having a mechanical governor. B-414, 354, 364 and 384 tractors are similar.

With pump housing removed, invert same and shake out plunger (6) and tappet spindle (15). If necessary, the tappet roller (12) and tappet guide (14) can be disassembled by driving out pin (11).

Clean all parts and inspect valves and their seats in the pump housing. All seating surfaces should be clean and smooth. Renew any pitted, worn or damaged valves. Inspect all springs and renew those which are rusted, distorted or show signs of being fractured. Inspect tappet assembly and pay particular attention to the roller and its pin. If wear is excessive, renew both parts. Inspect plunger (6) for scoring, pitting, wear or hammering from spindle (15) and if excessive wear or damage is found, it is recommended that both plunger (6) and housing (5) be renewed. Inspect the condition of the tappet spindle (15) gland washer, which is located in the pump housing, and renew same if necessary.

Reassemble by reversing the disassembly procedure and if necessary, test pump as outlined in paragraphs 117 through 120.

122. The fuel pump used when tractor is equipped with an injection pump having a mechanical governor is conventional and the removal and/or overhaul is obvious after an examination of the unit and reference to Figs. 63 and 64.

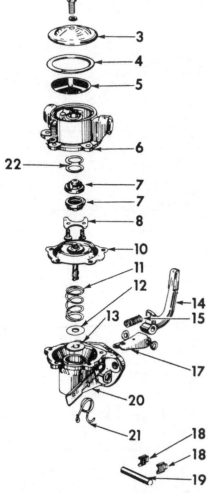

Fig. 63—Exploded view of fuel pump used on late Series B-275 and all Series B-414, 354, 364 and 384 tractors. Diesel and non-diesel pumps are similar.

3. Cover	13. Washer (fabric)
4. Gasket	14. Rocker arm
5. Filter screen	15. Spring
6. Valve body	17. Link
7. Valves	18. Pin retainers
8. Valve retainer	19. Rocker arm pin
10. Diaphragm assy.	20. Body & hand primer
11. Spring	21. Spring
12. Washer (metal)	22. Valve gasket

Fig. 65—Exploded view of the Zenith downdraft carburetor used on early Series B-414 and 354 non-diesel tractors.

1. Gasket
3. Bowl
4. Main jet
5. Gasket
6. Bleed jet
7. Float
8. Float arm & pivot
9. Compensating jet
10. Emulsion block
11. Idle jet
12. Plug
14. Throttle plate
15. Gasket
16. Float needle
17. Washer
18. Barrel assembly
20. Choke tube
21. Choke plate
22. Idle mixture adjusting screw
23. Spring
24. Spring
27. Choke shaft & lever
36. Interconnection rod
37. Choke control lever
39. Spring
41. Tab washer
42. Choke tube retaining screw
43. Throttle shaft
49. Stop screw
50. Spring

FUEL SYSTEM (NON-DIESEL)

CARBURETOR

Series B-414-354

123. The early production B-414 non-diesel tractors (prior to engine serial No. BC-144/3525) and all 354 non-diesel tractors were equipped with an English built, Model VNN, Zenith downdraft carburetor. On later production B-414 tractors, the carburetor used is the Model VNP Zenith downdraft which is equipped with an accelerator pump.

124. R&R AND OVERHAUL. Removal of the carburetor is obvious upon examination of the unit.

With carburetor removed, separate the bowl cover and barrel from bowl, then remove float and emulsion block from bowl. Jets can now be removed from emulsion block and the fuel inlet needle valve assembly removed from bowl cover. Remove choke plate and shaft, then remove set screw and pull choke tube from barrel.

On Model VNP carburetor, remove piston retaining screw and withdraw accelerator piston and spring. Then, unscrew and remove the accelerator pump check valve.

Any further disassembly required on either model will be obvious upon examination of the unit and reference to Fig. 65 or 66.

Reassemble by reversing the disassembly procedure. Float level is non-adjustable.

Series 424-444-2424-2444

125. The carburetor used on Series 424, 444, 2424 and 2444 tractors is the Marvel-Schebler Model TSX 896. The 1-1/8-in. updraft carburetor is equipped with a fuel solenoid shut-off valve which prevents engine from "dieseling" when ignition is switched to off position.

126. R&R AND OVERHAUL. Removal of the carburetor is obvious upon examination of the unit.

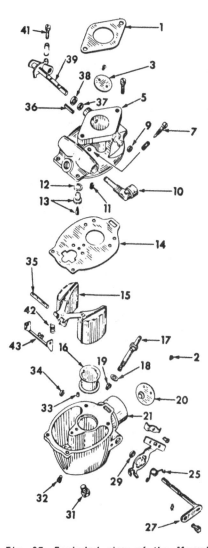

Fig. 67—Exploded view of the Marvel-Schebler Model TSX896 carburetor used on Series 424, 444, 2424 and 2444 tractors.

1. Gasket	21. Fuel bowl
3. Throttle plate	25. Spring
5. Throttle body	27. Choke shaft
7. Idle mixture needle	29. Packing
9. Throttle shaft cup	31. Drain plug
10. Fuel inlet screen	32. Plug
11. Idle jet	33. Vent
12. Gasket	34. Choke shaft cup
13. Fuel inlet valve	35. Float axle
14. Gasket	36. Throttle stop
15. Float	37. Packing
16. Venturi	38. Retainer
17. Main nozzle	39. Throttle shaft
18. Gasket	41. Throttle stop screw
19. Power jet	42. Float spring
20. Choke plate	43. Spring bracket

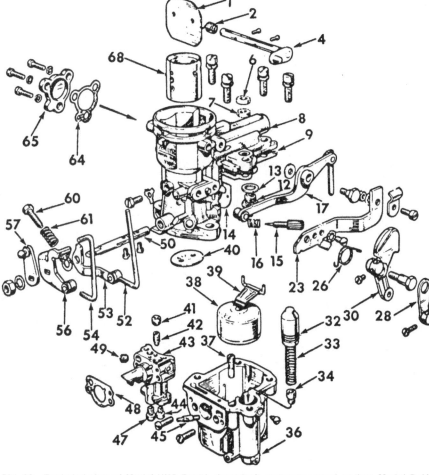

Fig. 66—Exploded view of Model VNP Zenith downdraft carburetor used on late Model B-414 tractors.

1. Choke plate	16. Spring	38. Float	50. Throttle shaft
2. Spring	17. Pump lever	39. Float arm	52. Rod
4. Choke shaft	23. Bracket	40. Throttle plate	53. Lever
6. Retainer	26. Spring	41. Plug	54. Pump link
7. Felt washer	28. Choke link	42. Idle jet	56. Throttle stop
8. Carburetor barrel	30. Choke lever	43. Emulsion block	57. Throttle lever
9. Gasket	32. Pump piston	44. Compensating jet	60. Throttle stop screw
12. Inlet needle & seat	33. Pump spring	45. Pump jet	61. Spring
13. Washer	34. Check valve	47. Main jet	64. Gasket
14. Gasket	36. Carburetor bowl	48. Gasket	65. Blanking plate
15. Idle mixture needle	37. Discharge valve	49. Ventilating screw	68. Choke tube

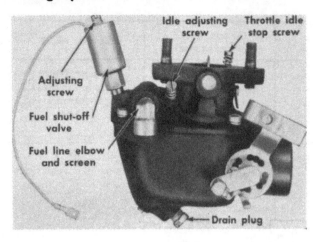

Fig. 68—View showing adjusting screws on Marvel-Schebler Model TSX896 carburetor. Main fuel adjusting screw is on fuel solenoid shut-off valve.

With carburetor removed, first remove the fuel solenoid shut-off assembly, then unbolt and separate the throttle body from fuel bowl. Remove float and fuel inlet needle valve assembly from the throttle body. Jets can now be removed from throttle body and fuel bowl. Any further disassembly required will be obvious upon examination of the unit and reference to Figs. 67 and 68.

Reassemble by reversing the disassembly procedure. Float setting and parts data are as follows:

Float setting . . . . ¼-in.
Repair kit .286-1465
Gasket set 16-613
Inlet needle and seat 233-608
Idle jet 49-345
Nozzle 47-277
Power jet 49-188

The power adjusting needle on fuel solenoid shut-off valve should be adjusted to four turns off its seat for correct adjustment. The initial setting for the idle mixture needle is one full turn open. Adjust the throttle stop screw to obtain the correct engine low idle speed of 425 rpm.

FUEL PUMP

Series B-414-354

127. The fuel pump used on Series B-414 and 354 non-diesel tractors is a conventional diaphragm type. Removal, installation and overhaul of the pump will be obvious upon examination and reference to Fig. 63.

Series 424-444-2424-2444

128. An "Autopulse" electric fuel pump (IH part No. 391586R91) was used on early production Series 424 and 2424 non-diesel tractors. Late production Series 424 and 2424 and all Series 444 and 2444 non-diesel tractors are equipped with a "Bendix" electric fuel pump (IH part No. 394327R91). Removal of either pump is obvious upon examination of the unit.

COOLING SYSTEM

RADIATOR

Series B-275-B-414-354-364-384

129. To remove radiator, drain cooling system and on Series B-275 and B-414, lift hood, remove cotter pins from ends of stayrod channels and move hood to a vertical position. Disconnect radiator brace and on Series B-275, the fuel filter brace from radiator. On Series 354, 364 and 384 remove hood and grille support, then remove air cleaner hose from air cleaner and disconnect the horn wire. On all series, disconnect upper and lower hoses from radiator, then unbolt radiator from front support and lift from tractor. Fan shroud can now be removed, if necessary.

Radiator is of one-piece construction with non-detachable upper and lower tanks.

Fig. 69—Exploded view showing component parts of the water pump assembly and thermostat housing used on Series 354, 364, 384 and late B-414. Items 16, 17, 18 and 19 are used on Series B-275 and early B-414.

1. Fan
2. Pulley flange
3. Pulley hub
4. Shaft & brg. assy.
5. Set screw
6. Housing
7. Gasket
8. Flinger
9. Seal
10. Impeller
11. Cover thermostat housing
12. Gasket
13. Thermostat
14. Gasket
15. Thermostat housing
16. Retaining ring
17. Thermostat
18. Gasket
19. Thermostat housing

Reinstall by reversing the removal procedure.

Series 424-444-2424-2444

130. To remove radiator, drain cooling system and remove hood and side panels. Disconnect radiator hoses and unbolt the radiator brace. Unbolt the fan shroud and lay same back over fan. Remove nuts from lower support bolts and lift radiator from tractor.

Radiator is of one-piece construction with non-detachable upper and lower tanks.

Reinstall by reversing the removal procedure.

THERMOSTAT

All Models

131. Thermostat is located in the water outlet elbow and the removal procedure is evident. Standard thermostat for all B-275 and early production B-414 tractors begins to open at 176 degrees F and is fully open at 190 degrees F. On late production B-414 tractors starting with engine serial numbers BC-144/6029 and BD-154/32795 and all 354, 364, 384, 424, 444, 2424 and 2444 tractors, standard thermostat begins to open at 170 degrees F and is fully open at 199 degrees F.

WATER PUMP

Series B-275-B-414-354-364-384

132. **R&R AND OVERHAUL.** To remove the water pump, first remove the radiator as outlined in paragraph 129, then loosen the water pump pulley flange, or bracket on those models with belt adjusting attachment, and remove

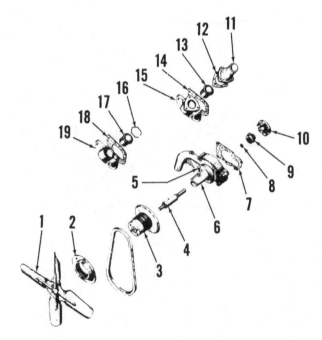

fan belt. Loosen alternator or generator adjustment strap and remove alternator or generator belt. Unbolt and remove water pump.

To disassemble water pump, refer to Fig. 69 and remove fan blades (1) and pulley flange (2). Pulley hub (3) can now be pressed from pump shaft (4). With pump hub removed, remove set screw (5), then press shaft, bearing, seal and impeller rearward out of housing. Shaft can now be pressed from impeller.

Inspect all parts for excessive rust or scale and for undue wear or damage. Both impeller and pulley hub must fit pump shaft with no less than 0.001 interference fit. Pump shaft bearing fit in housing is from 0.0003 loose to 0.0008 tight. Renew any parts which do not meet the above conditions.

Reassemble pump by reversing the disassembly procedure. Longest section of pump shaft is toward impeller end of housing. Install new seal (9). Press impeller and pulley hub on pump shaft until they are flush with ends of shaft and when pressing on pulley hub be sure that impeller end of shaft is supported. Do not subject pump housing to the pressing pressure.

After assembly, turn pump by hand to see that it rotates freely. As a further check for correct assembly, measure the distance between fan mounting surface of pulley hub and mounting face of pump housing. This distance should be 5-17/32 inches.

Use new gasket and reinstall pump by reversing removal procedure.

NOTE: In some cases, where overheating has become a problem, a belt pulley attachment such as that shown in Fig. 70 is available for Series B-275. The attachment consists of an additional pulley mounted on the injection pump gear cover which provides adjustment and a fixed diameter water pump pulley. In order to accommodate the attachment, a new injection pump gear cover is required and the crankcase breather pipe has been shortened and relocated.

Series 424-444-2424-2444 Diesel

133. R&R AND OVERHAUL. To remove the water pump, first drain cooling system, then remove hood and side panels. Loosen fan belt adjuster and generator or alternator belt adjuster and remove belts. Unbolt fan and lay same forward in shroud. Remove hoses then unbolt and lift out water pump.

To disassemble water pump, use a suitable puller and remove pulley (6—Fig. 71) from shaft (7). Remove set

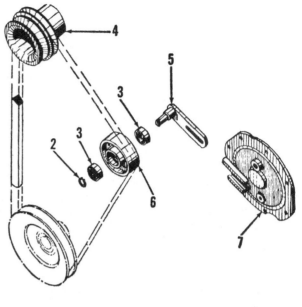

Fig. 70—View showing component parts which make up the fan belt adjusting pulley attachment for Series B-275.

　2. Retainer
　3. Bearings
　4. Pulley hub
　5. Bracket
　6. Adjusting pulley
　7. Injection pump gear cover

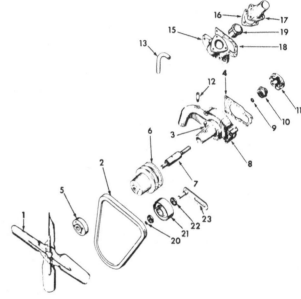

Fig. 71—Exploded view of the water pump used on Series 424, 444, 2424 and 2444 diesel tractors.

　1. Fan blade
　2. Fan belt
　3. Set screw
　4. Gasket
　5. Spacer
　6. Pulley
　7. Shaft & bearing assy.
　8. Pump body
　9. "O" ring
　10. Seal
　11. Impeller
　12. By-pass tube
　13. By-pass hose
　15. Thermostat housing
　16. Gasket
　17. Water outlet
　18. Gasket
　19. Thermostat
　20. Snap ring
　21. Belt adjuster pulley
　22. Bearing
　23. Bracket

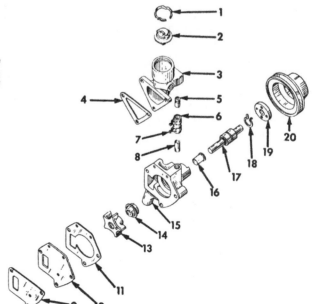

Fig. 72—Exploded view of the water pump used on Series 424, 444, 2424 and 2444 non-diesel tractors.

　1. Retainer
　2. Thermostat
　3. Water outlet elbow
　4. Gasket
　5. Nipple
　6. By-pass hose
　7. Clamp
　8. Nipple
　9. Gasket
　10. Plate
　11. Gasket
　13. Impeller
　14. Seal
　15. Pump body
　16. Slinger
　17. Shaft & bearing assy.
　18. Snap ring
　19. Hub
　20. Pulley

screw (3), then press shaft, bearing, seal and impeller rearward out of pump body. Shaft can now be pressed from impeller.

The shaft and bearing (7) are available only as a pre-assembled unit. When renewing seal (10), press only on outer diameter. Both impeller and pulley hub must fit pump shaft with not less than 0.001 interference fit. Pump shaft bearing fit in housing is from 0.0003 loose to 0.0008 tight. After the seal and bearing and shaft unit are installed and set screw tightened, press impeller and pulley hub on pump shaft. Impeller must be pressed on flush with end of shaft. When pressing on pulley hub, be sure impeller end of shaft is supported. Do not subject pump housing to the pressing pressure.

Use new gasket and reinstall pump by reversing removal procedure.

Series 424-444-2424-2444 Non-Diesel

134. **R&R AND OVERHAUL.** To remove water pump, first drain cooling system, then remove hood skirts and side panels. Unbolt and remove the generator or alternator. Disconnect the lower hose and by-pass hose. Unbolt fan blades and pulley from water pump and let fan blades rest in shroud. Unbolt water pump from engine and withdraw pump from left side of tractor.

To disassemble pump, use a suitable puller and remove hub (19—Fig. 72) from shaft (17). Remove plate (10) and snap ring (18), then support pump body and press shaft and bearing assembly (17) from impeller (13) and pump body. Remove seal (14).

The shaft and bearing are available only as a pre-assembled and pre-lubricated unit. Water pump overhaul package (IH part No. 391028R92) is available from International Harvester Co. When renewing seal (14), press only on outer diameter.

Reassembly is evident, keeping in mind that vanes of impeller (13) and flange of hub (19) are toward front of tractor. Press impeller and hub on shaft until bottom of chamfer at each end shows. Clearance from face of body to face of impeller hub should be 0.031.

ELECTRICAL SYSTEM

Series B-275-B-414-354-364-384

135. Lucas generators, alternators, regulators and starting motors are used and the specifications are as follows:

Generator (B-275-B-414-354-364)
Model C-39-P2
Brush spring tension (oz.) 22-25
Field resistance (ohms) 6.1
Output
  Hot or cold .... Cold
  Amperes 11.0
  Volts 13.5
  Rpm 1450-1700

Alternator (364-384)
Make .... Lucas
Model 15ACR
Output
  Rpm 3000
  Amperes 28

Regulator (B-275-B-414-354-364)
Model RB108
Cut-out relay
  Air gap 0.025-0.040
  Point gap 0.010-0.020
  Closing voltage range 12.7-13.3
Voltage regulator,
  Air gap .... See note
  Setting (volts) range
    at 68°F 16.0-16.5

Fig. 73—Field winding continuity and resistance test. Refer to text.

Fig. 74—Field winding insulation check. Refer to text.

Fig. 75—Stator winding continuity test. Refer to text.

Ground polarity .... Negative

NOTE: 0.015 for those with disc or wire winding core; 0.021 for those with square winding core.

Starting Motor (Series B-275-B-414-354-364-384 Diesel)
Model M45G
Volts 12.0
Brush spring tension 30-40

No Load Test
  Volts 12.0
  Amperes 70.0
  Rpm 8000

Lock Test
  Volts 5.2
  Amperes 930
  Torque (Ft.-Lbs.) 29

Starting Motor (Series B-414-354 Non-Diesel)
Model M418G
Volts 12.0
Brush spring tension (oz.) 30-40

No Load Test
  Volts 12.0

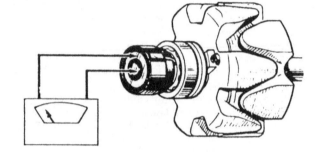

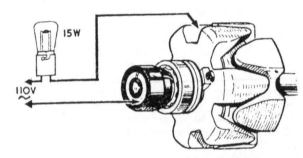

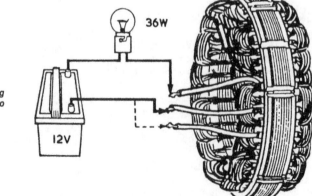

Amperes 45.0
Rpm .7400-8500

Lock Test

Volts 7.0-7.4
Amperes 440-460
Torque (Ft.-Lbs.) 17

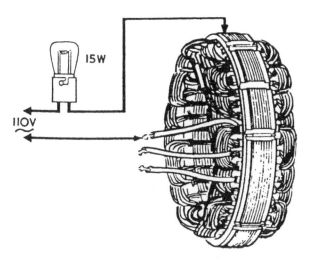

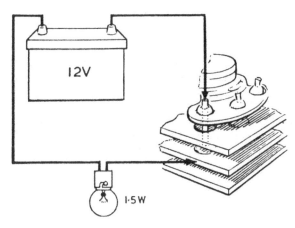

135A. ALTERNATOR AND REGU-LATOR. Late Series 364 and all Series 384 are equipped with Lucas alternators. The following component testing may be accomplished with minimum disassembly. See paragraph 135B for disassembly procedures.

Remove slip ring end cover. Note position of stator winding connections and unsolder connections from rectifier. Remove brush and regulator assembly. Renew brushes if overall length is less than 5/16-inches.

Check field winding continuity and resistance simultaneously by connecting a battery operated ohmmeter as shown in Fig. 73. The ohmmeter should read 4.3 ohms for 15 ACR rotor with pink winding or 3.3 ohms for 15 ACR rotor with purple winding.

To check field winding insulation, connect a 110 AC-15 watt test lamp as shown in Fig. 74. The lamp should not light.

Inner stator winding short-circuiting can be indicated by signs of burning of the insulation varnish covering. If this is obvious renew stator assembly. To check the continuity of the stator windings, connect any two of the three stator winding leads in series with a 12V battery and 36 watt test lamp (Fig. 75). Lamp should light. Transfer test lamp lead to third stator lead (Fig. 75). Lamp should light.

Check insulation of stator windings by connecting 110 AC-15 watt test lamp between lamination and any one of the three stator leads (Fig. 76). The lamp should not light.

To test the rectifier diodes connect a 12V battery and 1.5 watt bulb in series as shown in Fig. 77. Lamp should light during one-half of test only. If one diode is unsatisfactory renew rectifier assembly.

135B. ALTERNATOR OVERHAUL. Refer to Fig. 78. Scribe a mark across

Fig. 76—Stator winding insulation check. Refer to text.

Fig. 77—Testing rectifier diodes. Refer to text.

1. Moulded cover
2. Brush box assembly
3. Regulator
4. Brush & spring assy.
5. Regulator grounding screw & brush box screw
6. Rectifier
7. Slip ring end bracket
8. Stator winding assy.
9. Slip ring moulding
10. Ball bearing
11. Rotor & field winding
12. Woodruff key
13. Bearing assy.
14. Drive end bracket
15. Fan & pulley

REGULATOR

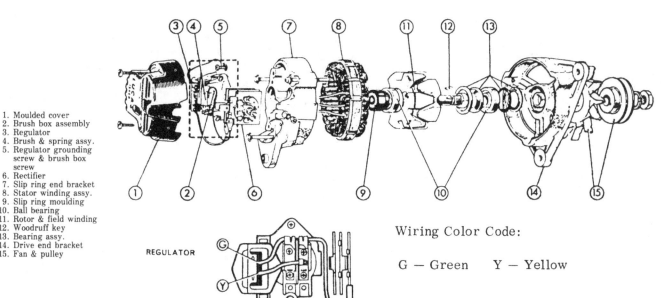

Wiring Color Code:

G — Green   Y — Yellow

Fig. 78—Lucas alternator used on late Series 364 and all Series 384.

alternator halves. Remove mounting bolts and separate alternator. Separate stator from slip ring end bracket and rotor assembly from drive end bracket. Remove drive pulley, fan and shaft key. Then press rotor shaft from front bearing. Unsolder field winding connections and remove slip ring. Then press off rear bearing. To reassemble, reverse disassembly process. Use only "M" grade 45-55 resin-cored solder to attach stator wire to diode pins (Fig. 79). Torque alternator mounting bolts to 55 in.-lbs.

Series B-414 Non-Diesel

136. DISTRIBUTOR AND IGNITION TIMING. Lucas distributor is used and specifications are as follows:

Model D3A4/LT
Breaker contact gap (in.) 0.014
Breaker arm spring
 pressure (oz.) 28-34
Advance Data*
 Start advance . . . . 5 at 500
 Maximum advance . . . . . . 19.5 at 1100
*Advance data is in distributor degrees and rpm. Double the listed values for flywheel degrees and rpm.

Static timing should occur at 5 degrees BTDC. The rear flange of the crankshaft pulley is marked with a TDC mark (notch) and may or may not have an IGN. (5 degrees) mark (notch). If pulley does not have the IGN. mark, add such a mark 9/32-in. before the TDC mark.

NOTE: If advance timing is to be checked with a timing light, add an additional mark (notch) 2-7/32 inches ahead of the TDC mark at this time. This mark will indicate the 39 degrees BTDC advance timing specified for the engine.

To set static timing, proceed as follows: Check to see that distributor breaker points are in good condition and set at 0.014, then turn engine until number one piston is coming up on compression stroke. Continue to turn engine until the 5 degree mark on crankshaft pulley is aligned with timing pointer. Loosen distributor mounting, pull spark plug wire from number one spark plug and turn on ignition switch. Hold disconnected spark plug wire about 1/8-in. from a clean point on engine, then rotate distributor body clockwise until a spark jumps between the spark plug wire and engine. Tighten distributor mounting at this point.

NOTE: Always approach the number one cylinder firing position by rotating distributor body in the direction oppo-

site to breaker cam rotation (clockwise). If necessary, turn distributor body counter-clockwise past timing point, then set timing by turning distributor body clockwise.

Turn ignition switch off and reinstall the number one spark plug wire.

If running timing is to be checked with a timing light, proceed as follows: Be sure the advance timing mark (2-7/32 in. before TDC) has been added to crankshaft pulley. Attach timing light, start engine and direct timing light at timing pointer. Timing pointer should be aligned with the advance timing mark (39 degrees) when engine is operating at high idle (2200 rpm).

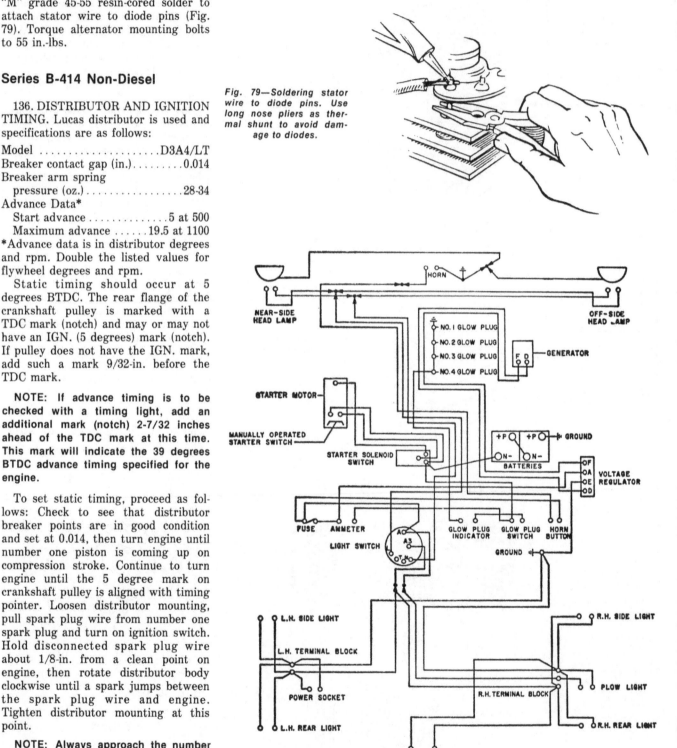

Fig. 79—Soldering stator wire to diode pins. Use long nose pliers as thermal shunt to avoid damage to diodes.

Fig. 80—Wiring diagram of the International B-275 tractor. Some of the units above may not be included.

Series 354 Non-Diesel

137. DISTRIBUTOR AND IGNITION TIMING. Model D204 Delco-Remy distributor is used and specifications are as follows:

Breaker contact gap (in.) 0.020
Breaker arm spring
   pressure (oz.) 19-23
Cam dwell angle . . . . 35°-37°
Advance Data*
   Start advance . . . . 1 at 300
   Intermediate advance . . . . . 7.5 at 600
   Maximum advance . . . . . 16.5 at 1100

*Advance data is in distributor degrees and rpm. Double the listed values for flywheel degrees and rpm.

To set static timing proceed as follows: Check to see that distributor breaker points are in good condition and set at .020. Turn engine until the notch in the crankshaft pulley is in line with the pointer on the timing cover. Remove the distributor cap and check that the points are just opening (0.003 maximum). If the points are not in this position loosen the mounting bolts and turn the distributor clockwise to open and counter-clockwise to close the points. Tighten the mounting bolts and install the distributor cap.

Slight advance or retard may be required to obtain smoothest operation. Turning the distributor clockwise will advance the timing and counter-clockwise will retard the timing.

Series 424-444-2424-2444

138. Delco-Remy electrical units are used and the specifications are as follows:

Generators 1100409 and 1100422
Brush spring tension (oz.) 28
Field draw:
   Volts 12
   Amperes 1.58-1.67
Cold output:
   Volts 14
   Amperes 25
   Rpm 3040

Regulator 1119270E
Ground polarity . . . . Negative
Cut-out relay:
   Air gap 0.020
   Point gap 0.020
   Closing voltage, range 11.8-13.5
     Adjust to 12.6

Voltage regulator:
   Air gap 0.060
   Voltage setting at degrees F.,
     14.4-15.4 at 65°
     14.2-15.2 at 85°
     14.0-14.9 at 105°
     13.8-14.7 at 125°
     13.5-14.3 at 145°
     13.1-13.9 at 165°

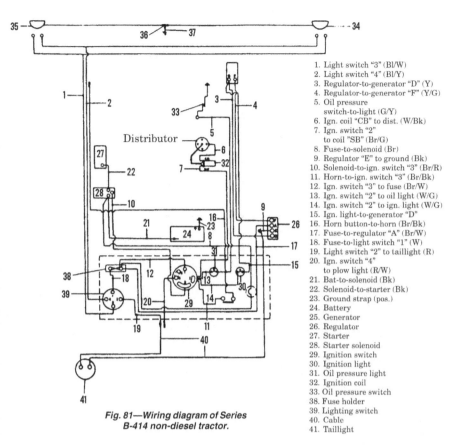

Distributor

1. Light switch "3" (Bl/W)
2. Light switch "4" (Bl/Y)
3. Regulator-to-generator "D" (Y)
4. Regulator-to-generator "F" (Y/G)
5. Oil pressure
   switch-to-light (G/Y)
6. Ign. coil "CB" to dist. (W/Bk)
7. Ign. switch "2"
   to coil "SB" (Br/G)
8. Fuse-to-solenoid (Br)
9. Regulator "E" to ground (Bk)
10. Solenoid-to-ign. switch "3" (Br/R)
11. Horn-to-ign. switch "3" (Br/Bk)
12. Ign. switch "3" to fuse (Br/W)
13. Ign. switch "2" to oil light (W/G)
14. Ign. switch "2" to ign. light (W/G)
15. Ign. light-to-generator "D"
16. Horn button-to-horn (Br/Bk)
17. Fuse-to-regulator "A" (Br/W)
18. Fuse-to-light switch "1" (W)
19. Light switch "2" to taillight (R)
20. Ign. switch "4"
   to plow light (R/W)
21. Bat-to-solenoid (Bk)
22. Solenoid-to-starter (Bk)
23. Ground strap (pos.)
24. Battery
25. Generator
26. Regulator
27. Starter
28. Starter solenoid
29. Ignition switch
30. Ignition light
31. Oil pressure light
32. Ignition coil
33. Oil pressure switch
38. Fuse holder
39. Lighting switch
40. Cable
41. Taillight

Fig. 81—Wiring diagram of Series B-414 non-diesel tractor.

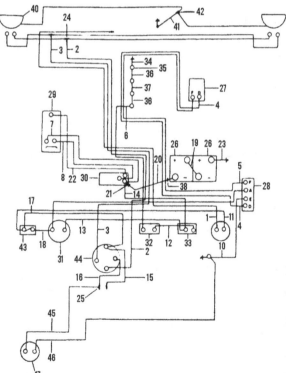

1. Horn button (Br/Bk)
2. Light switch "3" to headlight (Bl/W)
3. Light switch "4" to headlight (Bl/Y)
5. Regulator "D" to generator (Y)
6. Glow plug indicator (Br/G)
7. Solenoid-to-start switch (R/G)
8. Solenoid-to-start switch (R/Bk)
9. Regulator "A" to ammeter (Br/W)
10. Regulator "E" to ground (Bk)
11. Fuse-to-horn button (Br/Bk)
12. Glow plug switch-to-indicator
   (Br/Y)
13. Glow plug switch-to-ammeter
   (Br/G)
14. Solenoid-to-glow
   plug switch (Br/R)
15. Light switch "1" to
   plow light (R/W)
16. Light switch "2" to taillight (R)
17. Light switch "1" to fuse (W)
18. Ammeter-to-fuse (Br/Bl)
19. Battery cable
20. Battery-to-solenoid
22. Solenoid-to-starter (Bk)
23. Ground strap (pos.)
24. Connector
25. Connector
26. Batteries
27. Generator
28. Voltage regulator
29. Starter
30. Starter solenoid
31. Ammeter
32. Resistor
33. Glow plug switch
35. Glow plug
36. Wires
43. Fuse housing
44. Light switch
47. Taillight

Fig. 82—Wiring diagram of Series B-414 diesel tractors. Ignition switch is not shown.

Current regulator:
  Air gap 0.075
  Current setting at degrees F.,
    25.0-30 at 65°
    24.5-29 at 85°
    23.5-28 at 105°
    23.0-27 at 125°
    21.5-25.5 at 145°
    20.5-24.5 at 165°
    19.5-23.5 at 185°

Starting Motors 1107364 and 1108323
Volts 12
Brush spring tension (oz.) 35
No-load test:
  Volts 9.0
  Amperes (min.)................50*
  Amperes (max.) 60*
  Rpm (min.) 6000
  Rpm (max.) 8500
Resistance test:
  Volts 4.3
  Amperes (min.) 270
  Amperes (max.) 310
*Includes solenoid.

Starting Motor 1107585
Volts 12
Brush spring tension (oz.) 35
No-load test:
  Volts 9.0
  Amperes (min.)................50*
  Amperes (max.) 80*
  Rpm (min.) 5500
  Rpm (max.) 9000
Resistance test:
  Volts 3.5
  Amperes (min.) 330
  Amperes (max.) 395
*Includes solenoid.

139. ALTERNATOR AND REGULA-TOR. Series 444 and 2444 are equipped with a "DELCOTRON" generator (alternator) and a double contact voltage regulator.

CAUTION: Because certain components of the alternator can be damaged by procedures that will not affect a D.C. generator, the following precautions MUST be observed:

a. When installing batteries or connecting a booster battery, the negative post of battery must be grounded.

b. Never short across any terminal of the alternator or regulator.

c. Do not attempt to polarize the alternator.

d. Disconnect all battery ground straps before removing or installing any electrical unit.

e. Do not operate alternator on an open circuit and be sure all leads are properly connected before starting engine.

Specification data for alternator and regulator are as follows:

Alternator 1100805
Field current at 80° F,
  Amperes 2.2-2.6
  Volts 12.0
Cold output at specified voltage,
  Specified volts 14.0
  Amperes at rpm ....21 at 2000
  Amperes at rpm ....30 at 5000

Rated output hot,
  Amperes 32.0

Regulator 1119516
Ground polarity ....Negative
Field relay,
  Air gap 0.015
  Point opening................0.030

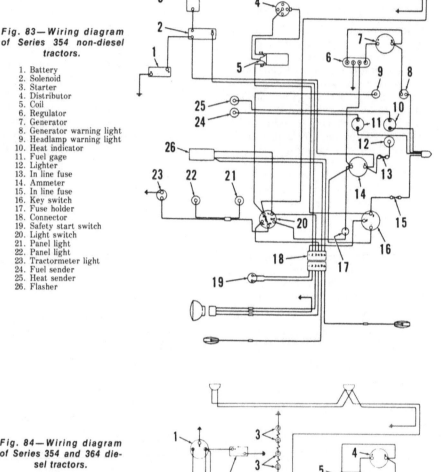

Fig. 83—Wiring diagram of Series 354 non-diesel tractors.
1. Battery
2. Solenoid
3. Starter
4. Distributor
5. Coil
6. Regulator
7. Generator
8. Generator warning light
9. Headlamp warning light
10. Heat indicator
11. Fuel gage
12. Lighter
13. In line fuse
14. Ammeter
15. In line fuse
16. Key switch
17. Fuse holder
18. Connector
19. Safety start switch
20. Light switch
21. Panel light
22. Panel light
23. Tractormeter light
24. Fuel sender
25. Heat sender
26. Flasher

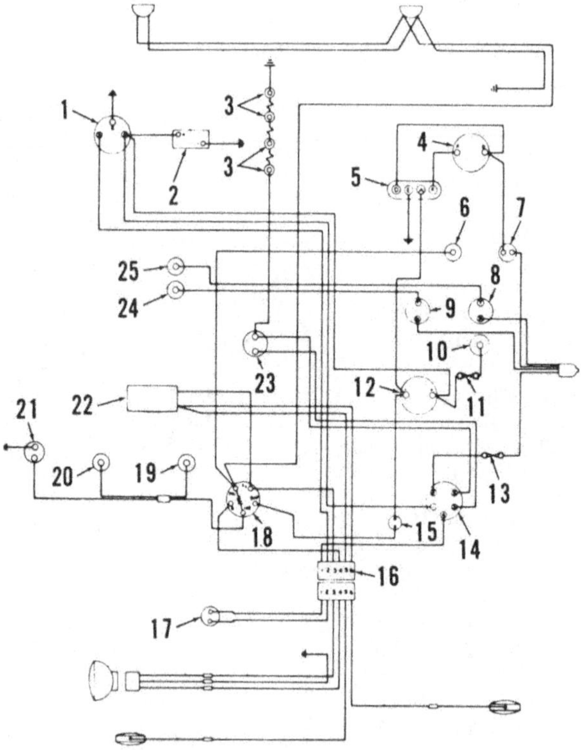

Fig. 84—Wiring diagram of Series 354 and 364 diesel tractors.
1. Starter
2. Battery
3. Glow plugs
4. Generator
5. Regulator
6. Headlamp warning light
7. Generator warning light
8. Heat indicator
9. Fuel gage
10. Lighter
11. In line fuse
12. Ammeter
13. In line fuse
14. Key switch
15. Fuse holder
16. Connector
17. Safety start switch
18. Light switch
19. Panel light
20. Panel light
21. Tractormeter light
22. Flasher
23. Flow plug indicator
24. Fuel sender
25. Heat sender

Closing voltage range 1.5-3.2
Voltage regulator,
Air gap (lower points
closed) 0.067 (1)
Upper point opening
(lower points closed) 0.014
Voltage setting at degrees F,
13.9-15.0 at 65°
13.8-14.8 at 85°
13.7-14.6 at 105°
13.5-14.4 at 125°
13.4-14.2 at 145°
13.2-14.0 at 165°
13.1-13.9 at 185°

(1) When bench tested, set air gap at 0.067 as a starting point, then adjust air gap to obtain specified difference between voltage settings of upper and lower contacts. Operation on lower contacts must be 0.05-0.4 volt lower than on upper contacts. Voltage setting may be increased up to 0.3 volt to correct chronic battery under-charging or decreased up to 0.3 volt to correct battery over-charging. Temperature (ambient) is measured ¼-inch away from regulator cover and adjustment should be made only when regulator is at normal operating temperature.

140. ALTERNATOR TESTING AND OVERHAUL. The only test which can be made without removal and disassembly of alternator is output test. Output should be approximately 32 amperes at 5000 alternator rpm.

To disassemble the alternator, first place match marks (M—Fig. 85) on the two frame halves (6 and 16), then remove the four through-bolts. Pry frame apart with a screwdriver between stator frame (11) and drive end frame (6). Stator assembly (11) must remain with slip ring end frame (16) when unit is separated.

NOTE: When frames are separated, brushes will contact rotor shaft at bearing area. Brushes MUST be cleaned of lubricant if they are to be re-used.

Clamp the iron rotor (12) in a protected vise only tight enough to permit loosening of pulley nut (1). Rotor and end frame can be separated after pulley is removed. Check bearing surfaces of rotor shaft for visible wear or scoring. Examine slip ring surfaces for scoring or wear and windings for overheating or other damage. Check rotor for grounded, shorted or open circuits using an ohmmeter as follows:

Refer to Fig. 86 and touch the ohmmeter probes to points (1-2 and 1-3); a reading near zero will indicate a ground. Touch ohmmeter probes to the two slip rings (2-3); reading should be 4.6-5.5 ohms. A higher reading will indicate an open circuit and a lower reading will indicate a short. If windings are satisfactory, mount rotor in a lathe and check runout at slip rings using a dial indicator. Runout should not exceed 0.002. Slip ring surfaces can be trued if runout is excessive or if surfaces are scored. Finish with 400 grit or finer polishing cloth until scratches or machine marks are removed.

Disconnect the three stator leads and separate stator assembly (11—Fig. 85) from slip ring end frame assembly. Check stator windings for grounded or open circuits as follows: Connect ohmmeter leads successively between each pair of stator leads. A high reading would indicate an open circuit.

NOTE: The three stator leads have a common connection in the center of the windings. Connect ohmmeter leads

between each stator lead and stator frame. A very low reading would indicate a grounded circuit. A short circuit within the stator windings cannot be readily determined by test because of the low resistance of the windings.

Three negative diodes (19) are located in the slip ring end frame (16) and three positive diodes (20) in heat sink (15). Diode should test at or near infinity in one direction when tested with an ohmmeter, and at or near zero when meter leads are reversed. Renew any diode with approximately equal meter readings in both directions. Diodes must be removed and installed using an arbor press or vise and suitable tool which contacts only the outer edge of the diode. Do not attempt to drive a faulty diode out of end frame or heat sink as shock may cause damage to the other good diodes. If all diodes are being renewed, make certain the positive diodes (marked with red printing) are installed in the heat sink and negative diodes (marked with black printing) are installed in the end frame.

Brushes are available only in an assembly which includes brush holder (13). Brush springs are available for service and should be renewed if heat damage or corrosion is evident. If brushes are reused, make sure all grease is removed from surface of brushes before unit is reassembled. When reassembling, install brush springs and brushes in holder, push brushes up against spring pressure and

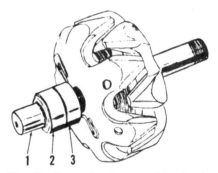

Fig. 86—Removed rotor assembly showing test points when checking for grounds, shorts and opens.

Fig. 87—Exploded view of brush holder assembly. Insert wire in hole (W) to hold brushes up. Refer to text.

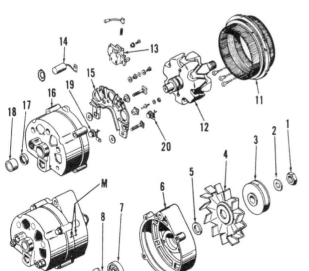

Fig. 85—Exploded view of "DELCOTRON" alternator used on Series 444 and 2444.

1. Pulley nut
2. Washer
3. Drive pulley
4. Fan
5. Spacer
6. Drive end frame
7. Ball bearing
8. Gasket
9. Spacer
10. Bearing retainer
11. Stator assembly
12. Rotor assembly
13. Brush holder
14. Capacitor
15. Heat sink
16. Slip ring end frame
17. Felt seal & retainer
18. Needle bearing
19. Negative diode (3 used)
20. Positive diode (3 used)

insert a short piece of straight wire through hole (W—Fig. 87) and through end frame (16—Fig. 85) to outside. Withdraw the wire only after alternator is assembled.

Capacitor (14) connects to the heat sink and is grounded to the end frame. Capacitor protects the diodes from voltage surges.

Remove and inspect ball bearing (7). If bearing is in satisfactory condition, fill bearing ¼-full with Delco-Remy lubricant No. 1960373 and reinstall. Inspect needle bearing (18) in slip ring end frame. This bearing should be renewed if its lubricant supply is exhausted; no attempt should be made to relubricate and reuse the bearing.

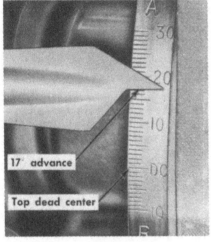

Fig. 88—View showing timing marks on Series 424, 444, 2424 and 2444 non-diesel engine.

Press old bearing out towards inside and press new bearing in from outside until bearing is flush with outside of end frame. Saturate felt seal with SAE 20 oil and install seal and retainer assembly.

Reassemble alternator by reversing the disassembly procedure. Tighten pulley nut to a torque of 45 ft.-lbs.

NOTE: A battery powered test light can be used instead of ohmmeter for all electrical checks except shorts in rotor winding. However, when checking diodes, test light must not be of more than 12 volts.

141. **DISTRIBUTOR.** Identification and tune-up data for the Series 424, 444, 2424 and 2444 distributor is as follows:

Make .... IH
Identification symbol............. AK
Breaker contact gap............. 0.020
Breaker arm spring
  pressure .... 21-25 oz.
Advance Data (degrees at flywheel rpm)
  Start advance.............. 0 at 375
  Maximum advance........ 17 at 2200

142. IGNITION TIMING. If the distributor has been removed for service, reinstall same as follows: Crank engine until No. 1 (front) piston is coming up on compression stroke and continue cranking slowly until TDC (0 degree) mark on crankshaft pulley is aligned with the pointer extending from front

face of the timing gear cover. Turn distributor shaft until rotor arm is in the No. 1 firing position and install distributor.

To set static timing, proceed as follows: Make certain that distributor breaker points are in good condition and set at 0.020. Loosen the distributor mounting cap screws and retard distributor about 30 degrees by turning distributor assembly in same direction as breaker cam rotates. Turn engine until No. 1 piston is coming up on compression stroke, then continue cranking slowly until TDC mark on crankshaft pulley is in register with the timing pointer. Disconnect coil secondary wire from distributor cap and hold free end of cable 1/16 to 1/8 inch from distributor primary terminal. Advance the distributor by turning distributor body in opposite direction from breaker cam rotation until a spark occurs between cable and primary terminal. Tighten distributor mounting cap screws at this point. Assemble spark plug cables to the distributor cap in the proper firing order of 1-3-4-2.

To set running timing, attach a timing light and start engine. Adjust engine low idle speed to 375 rpm. At this time, the timing light should show the TDC mark in register with the timing pointer. Then with engine speed set at 2200 rpm check to see that the 17 degree advance mark is aligned with timing pointer as shown in Fig. 88. After the preceding checks and adjustments have been made, readjust engine low idle speed to 425 rpm.

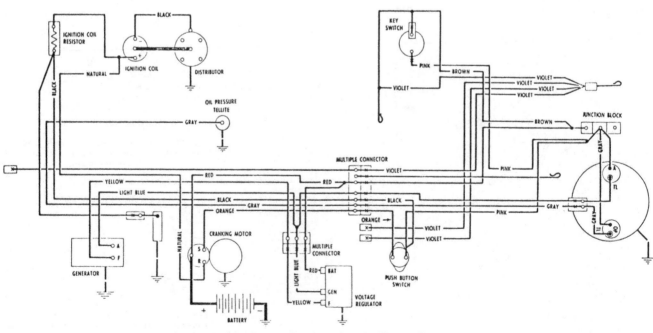

Fig. 89—Wiring diagram of Series 424 and 2424 non-diesel tractors.

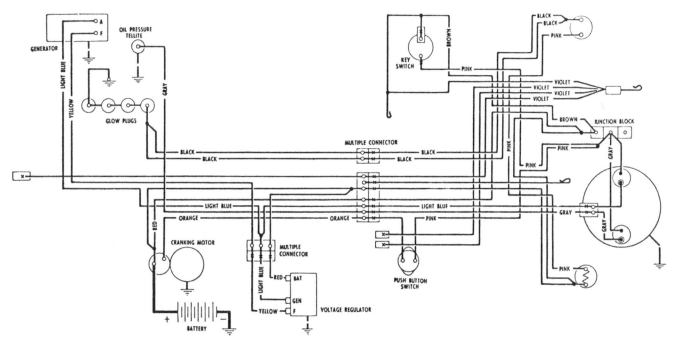

Fig. 90—Wiring diagram of Series 424 and 2424 diesel tractors.

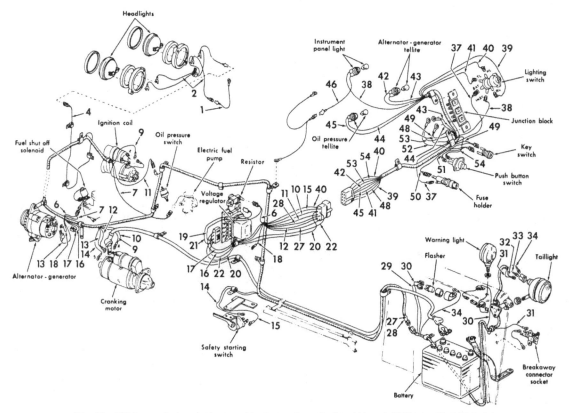

Fig. 91—Wiring and electrical components used on Series 444 and 2444 non-diesel tractors.

1. Pink	11. Gray	19. Light green
2. Red	12. Red	20. Light green
4. Violet	13. Light blue	21. Light blue
6. Light green	14. Orange w/black	22. Light blue
7. Light green w/black	tracers	27. White w/black
tracers	15. Orange	tracers
9. White	16. Yellow	28. Black
10. Natural w/orange &	17. Brown	29. Black
purple tracers	18. Pink	

30. Brown	39. White	45. Gray
31. Black w/white	40. Violet	46. Pink
tracers	41. Black w/white	48. Red
32. Black	tracers	49. Red
33. White	42. Light blue	50. Red
34. Pink	43. White w/black	51. Light green
37. Light green	tracers	52. Light green
38. Tan	44. Black w/white	53. Light green
	tracers	54. Orange

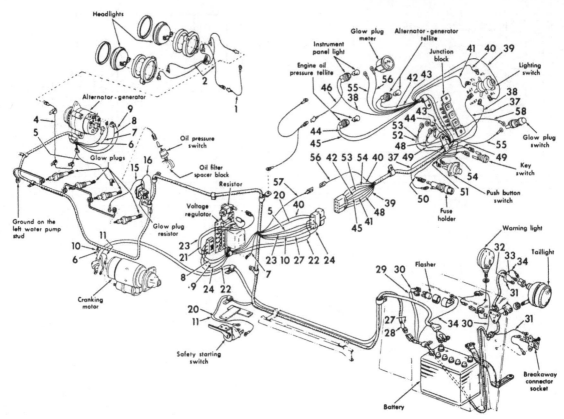

Fig. 92—Wiring and electrical components used on Series 444 and 2444 diesel tractors.

1. Pink
2. Red
4. Violet
5. Gray
6. Light blue
7. Pink
8. Brown
9. Yellow
10. Red

11. Orange w/black tracers
16. Black
20. Orange
21. Light green
22. Light green
23. Light blue
24. Light blue

27. White w/black tracers
28. Black
29. Black
30. Brown
31. Black w/white tracers
32. Black

33. White
34. Pink
37. Light green
38. Tan
39. White
40. Violet
41. Black w/white tracers

42. Light blue
43. White w/black tracers
44. Black w/white tracers
45. Gray
46. Pink
48. Red
49. Red

50. Red
51. Light green
52. Light green
53. Light green
54. Orange
55. Pink
56. Black
57. Black
58. Red

CLUTCH

Series B-275-B-414-354-364-384

143. Tractor may be equipped with a single plate or a dual plate clutch. Series B-414, 364 and 384 have an 11-inch single plate clutch and Series 354 has a standard 10-inch or optional 11-inch heavy duty single plate clutch. All models are available with the dual plate clutch which has an 11-inch main driven member with a 9-inch pto driven member.

144. **ADJUST.** Adjustment to compensate for lining wear is accomplished by adjusting the clutch pedal height and free travel.

To adjust clutch pedal height, position clutch pedal by turning adjusting screw (S—Fig. 93) until bottom of clutch pedal measures 6-7/8 inches from foot plate for Series B-275, B-414, 354, 364 and 384 tractors equipped with dual plate clutch. With single plate

clutch, the measurement should be 5-13/16 inches for Series B-414, 7 inches on 364 and 384 and 7 inches for 10 inch clutch or 4-9/16 inches for 11

Fig. 93—Clutch pedal of B-275, B-414, 354, 364 or 384 tractor with dual plate clutch. Refer to text for adjustment procedure and note that on Series B-414 with single plate clutch, pedal height is adjusted to 5-13/16 inches. On Series 354, pedal height is 7 inches for 10-inch single plate clutch or 4-9/16 inches for 11-inch single plate clutch. Pedal height on 364 and 384 with single plate 11-inch clutch is 7 inches.

inch clutch used on Series 354. The pedal free travel of ¾-inch (dual plate clutch) and 1-7/8 inch on 11 inch or 7/8-inch on 10 inch (single plate clutch) is obtained by loosening retaining bolts (B—Fig. 93) and moving pedal around clutch cross-shaft. Clutch cross-shaft may be held by using a wrench on flats of cross-shaft outer end.

145. **REMOVE AND REINSTALL.** The procedure for removing and reinstalling the clutch is evident after splitting the engine from the clutch housing as outlined in the following paragraph.

146. TRACTOR SPLIT. To separate engine from clutch housing, first raise or remove hood. On Series B-275 and B-414, remove battery shield, battery hold-down and battery, or batteries. Remove battery tray and throttle bellcrank. On all series, if tractor is equipped with a down swept exhaust, disconnect exhaust pipe at manifold,

then remove exhaust pipe and muffler. Disconnect wires from alternator or generator, starter switch and ignition coil (non-diesel models), or number four glow plug (diesel models). Disconnect battery cable from starter and operating rod from starter switch. On diesel models, disconnect fuel shut-off rod at injection pump and remove rod. On non-diesel models, disconnect choke control from carburetor. Close fuel shut-off valve and disconnect fuel supply line from fuel pump. On diesel models, disconnect fuel return line from fuel tank. On all models, disconnect oil pressure gage line from cylinder block. On Series B-275 and B-414, drain cooling system and remove temperature sending unit from cylinder head. On Series 354, 364 and 384 disconnect temperature gage wire and tachometer cable. On all series, disconnect the pressure and inlet lines from hydraulic pump. Disconnect drag link from steering gear drop arm on manual steering models; or on models with power steering, disconnect hoses from power steering cylinder, cylinder from reaction bracket and drag link from steering gear drop arm. Support clutch housing, attach hoist to engine, then unbolt and separate engine from clutch housing.

NOTE: When rejoining engine to clutch housing, time can often be saved, particularly on dual plate clutch models, if the following procedure is used. Place clutch assembly on input shafts (Fig. 94), then move sections together until transmission input shaft pilot enters pilot bearing and flywheel butts against clutch cover. Now bolt clutch to flywheel and complete the mating of engine and clutch housing by pulling sections together with the retaining cap screws.

147. OVERHAUL (DUAL PLATE).

With tractor split as outlined in paragraph 146, proceed as follows: Remove the cap screws retaining the clutch cover assembly to flywheel and withdraw the cover assembly and 11-inch lined disc. Examine the clutch shaft pilot bearing in crankshaft and renew the bearing if same is damaged or worn.

To disassemble the clutch cover assembly, proceed as follows: Carefully place punch marks on all of the components so they can be reassembled in the same relative position. Place the assembly in a press with the cover assembly up. Place a bar across the cover and compress the assembly until the release levers are just free. Loosen the locknuts and completely unscrew the finger adjusting screws (41—Fig. 95). Release the compressing pressure and disassemble the remaining parts.

Inspect all parts and renew those showing damage or wear. New facings can be riveted to the driven discs if discs are otherwise in good condition.

To reassemble the unit, refer to Fig. 95 and proceed as follows:

NOTE: It is important that a new 11-inch disc be used during adjustment of clutch fingers and pto adjustment screws. The original 11-inch disc can be reinstalled after adjustment is complete providing it is in serviceable condition.

Place the 11-inch pressure plate (32) on work bench with friction face down. Install the red marked Belleville washer (35) on pressure plate with convex side up and outer edge in counterbore of pressure plate. Align the previously affixed punch marks and install flywheel plate (36). Position the 9-inch driven disc (24) on flywheel plate with hub of same upward; then, again aligning punch marks, install the 9-inch pressure plate (37). Install second (blue marked) Belleville washer (35) on

pressure plate (37) with concave side up. Align punch marks and place clutch bracket (cover) (38) in position, using caution to see that outer edge of Belleville washer is in counterbore of bracket (cover) (38). To facilitate starting of clutch finger adjusting screws, drop three cap screws through bracket (38) and flywheel plate (36), equidistant around outer bolt circle. Start clutch finger adjusting screws (41) and tighten only enough to hold clutch parts in position.

Install a NEW 11-inch driven disc in counterbore of flywheel, with hub pointing away from engine. Position clutch assembly on flywheel and start the retaining cap screws. Install aligning tool (pilot) and tighten cap screws. See Fig. 96.

To adjust clutch, refer to Fig. 97 for tools required and proceed as follows: With clutch finger height gage in position as shown in Fig. 98 or 98A and while holding slightly on finger to take out the slack, turn finger adjusting screw until the 2-21/32 inches dimension for Series B-275 and B-414 is

Fig. 96—Using aligning tool when installing dual plate clutch assembly to flywheel.

Fig. 94—Two concentric shafts are used when tractors are equipped with dual plate clutches. Outer shaft (2) drives pto, inner shaft (1) drives transmission.

Fig. 95—Exploded view of Series B-275, B-414, 354, 364 and 384 dual clutch and related parts.

24. 9" driven disc
32. 11" pressure plate
34. Cap screw (special)
35. Belleville washer
36. Flywheel plate
37. 9" pressure plate
38. Bracket (cover)
39. Spring
40. Release lever
41. Adjusting screw
42. Lock pin
43. Pivot pin
44. Rod end
45. Pin
46. 11" driven disc

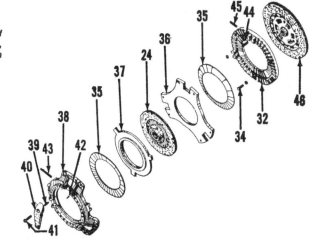

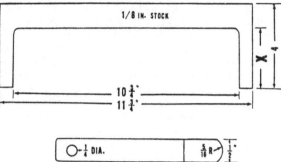

Fig. 97—Tools used to adjust the dual plate clutch. Tools can be made locally, using the dimensions shown. Dimension "X" is 2-21/32 inches for Series B-275 and B-414 and 2-25/32 inches for Series 424, 444, 2424, 2444, 354, 364 and 384.

normal range of travel several times and recheck adjustment.

With finger adjustment made, refer to Fig. 99 and set the clearance between the 9-inch pressure plate and adjusting cap screws. Loosen locknuts and turn adjusting screw until the 0.080 thickness gage can just be inserted between head of adjusting screw and 9-in. pressure plate. Tighten locknuts.

148. OVERHAUL (SINGLE PLATE). With tractor split as outlined in paragraph 146, refer to Fig. 100 and proceed as follows: Unbolt and remove cover assembly from flywheel. Disengage release lever plate retainers and remove release lever plate. Place clutch assembly in a press and apply pressure until release levers are relieved of tension. Remove release lever pins, then release press pressure and separate clutch assembly. Further disassembly is obvious.

Component parts of clutch assembly are catalogued separately, and in addition, a facing package for the driven disc is available for service.

Clutch pressure springs should have a free length of 2.68 inches and should test 120-130 lbs. when compressed to a length of 1.69 inches. Adjust clutch release fingers to a height of 1.955 inches, measured from contacting surface of fingers to friction face of pressure plate.

Fig. 98—Using special tool to adjust the dual plate clutch release lever height. Adjustment must be made with a NEW 11-inch driven disc installed.

obtained. On Series 354, 364 and 384, adjustment is 2-25/32 inches, measured with release lever plate installed. See Fig. 98A. When adjustments are obtained, tighten locknut, repeat same procedure on other fingers. It is IMPORTANT that clutch fingers be adjusted to within 0.015 of each other. Use a box end wrench, or similar tool, and actuate each finger through its

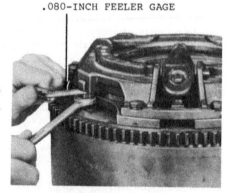

.080-INCH FEELER GAGE

Fig. 99—Using special tool to adjust the 0.080 clearance between the 9-inch pressure plate and the releasing cap screws.

Series 424-444-2424-2444

149. Tractors may be equipped with a 11-inch single plate clutch or a dual plate clutch with a 11-inch main driven member and a 9-inch pto driven member. The single plate clutch is used in tractors equipped with transmission driven pto and the dual stage clutch is used in tractors equipped with a continuous running pto.

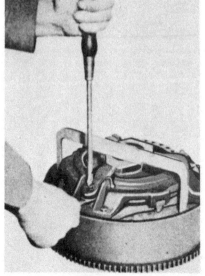

Fig. 98A—Using special tool to adjust the dual plate clutch release lever height on 354, 364 and 384. Adjustment must be made with a NEW 11-inch driver disc and RELEASE LEVER PLATE installed.

Fig. 100—Exploded view of the single plate clutch used on B-414, 354, 364 and 384 tractors without constant running pto.

1. Release bearing & cup
2. Retainer spring
3. Lever plate
4. Retainer spring
5. Clutch cover
6. Anti-rattle spring
7. Clutch spring
8. Strut
9. Release lever
10. Pressure plate
11. Clutch plate

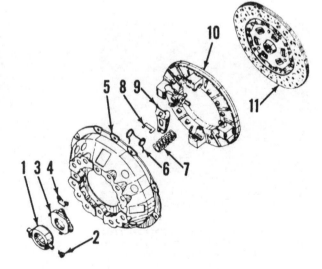

150. ADJUST. Adjustment to compensate for lining wear is accomplished by adjusting the clutch pedal. First, adjust set screw shown in Fig. 101 until pedal height is 6 inches (measured from the foot rest to highest point on pedal flange). Then, loosen pedal clamp bolts and rotate the clutch pedal on clutch release shaft until ¾-inch free travel is obtained. Clutch cross-shaft may be held by using a wrench on flats of cross-shaft outer end. Tighten the pedal clamp bolts securely.

151. REMOVE AND REINSTALL. The procedure for removing the clutch is evident after splitting the engine from the clutch housing as outlined in the following paragraph.

152. TRACTOR SPLIT. To separate the engine from clutch housing, first remove hood and drain cooling system. Disconnect battery cables and wiring harness multiple plug under instrument panel, then work harness forward until it is clear of clutch housing.

On tractors equipped with mechanical steering, unbolt the steering shaft front and rear bearing brackets from the fuel tank support and clutch housing. Slide steering shaft forward and out of the master splined yoke.

On tractors equipped with power steering, remove the steering tube clip from engine and disconnect the tubes from the power steering control valve. Plug and cap all openings to prevent dirt from entering steering system.

Then, on all tractors, disconnect the fuel supply line, choke cable or fuel shut-off cable, engine speed control rod and tachometer cable. Remove the temperature indicator bulb and disconnect battery cable from the cranking motor solenoid switch. Unbolt and remove the cranking motor. Unbolt the air cleaner bracket from fuel tank support, place a wood block between fuel tank and clutch housing and then, unbolt the fuel tank support from engine. Remove the underslung exhaust pipe, if so equipped.

Drain the hydraulic system and disconnect the hydraulic pump pressure and suction lines. Plug and cap all openings.

On diesel models, remove the clutch housing front dust cover.

Then, on all tractors, unbolt stay-rod bracket from clutch housing and place wood blocks between steering gear housing and front axle to prevent tipping. Support clutch housing, attach a hoist to engine, then unbolt and separate engine from clutch housing. Unbolt and remove clutch from flywheel.

When rejoining engine to clutch housing, place clutch assembly on the pto driving shaft (two stage clutch) and transmission input shaft. Bolt engine to clutch housing, then working through the opening in bottom of clutch housing, bolt clutch assembly to flywheel.

153. OVERHAUL (DUAL PLATE). With tractor split as outlined in paragraph 152, proceed as follows: Remove cap screws retaining the clutch assembly to flywheel and withdraw the cover assembly and 11-inch lined disc. Examine the clutch shaft pilot bearing in flywheel and renew the bearing if same is damaged or worn.

To disassemble the clutch cover assembly, proceed as follows: Carefully place punch marks on all of the components so they can be reassembled in the same relative position. Place the assembly in a press with the cover assembly up. Place a bar across the cover and compress the assembly until the release fingers are just free. Loosen locknuts and completely unscrew the finger adjusting screw (3—Fig. 102). Release the compressing pressure and disassemble the remaining parts.

Inspect all parts and renew those showing damage or wear.

To reassemble the unit, refer to Fig. 102 and proceed as follows: Place the 11-inch pressure plate (15) on work bench with friction face down. Install the purple paint marked Belleville washer (14) on pressure plate with convex side up and outer edge in counterbore of pressure plate. Align the previously affixed punch marks and install flywheel plate (13). Position the 9-inch (pto) driven disc (12) on flywheel plate with hub of same upward. Then, again aligning punch marks, install the 9-inch pressure plate (11). Install second (green paint marked) Belleville washer (10) on pressure plate (11) with concave side up. Align punch marks and place clutch cover (9) in position, making certain that outer edge of Belleville washer is in counterbore of cover. Start clutch finger adjusting screws (3) and tighten only enough to hold clutch parts in position.

Install a NEW 11-inch driven disc in flywheel, with hub pointing away from engine. Position clutch assembly on flywheel and start the retaining cap screws. Install aligning tool (pilot) and tighten cap screws.

NOTE: It is important that a new 11-inch driven disc be used during adjustment of clutch fingers and pto adjustment screws. The original 11-inch disc can be reinstalled after adjustment is complete providing it is in serviceable condition.

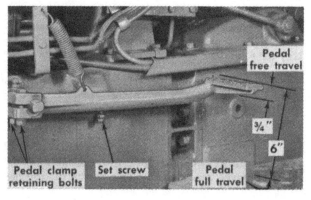

Fig. 101—View showing clutch pedal adjustment on Series 424, 444, 2424 and 2444 tractors.

Pedal free travel

Pedal clamp retaining bolts   Set screw   Pedal full travel   ¾"   6"

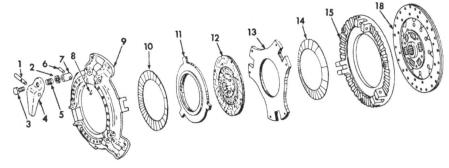

Fig. 102—Exploded view of Series 424, 444, 2424 and 2444 dual plate clutch assembly.

1. Pivot pin	9. Clutch cover
2. Locknut	10. Belleville washer
3. Adjusting screw	11. Pto pressure plate
4. Clutch finger	12. Pto driven disc
5. Spring	13. Flywheel plate
6. Pivot block pin	14. Belleville washer
7. Pivot block	15. Pressure plate
8. Retainer	18. Main driven disc

To adjust clutch, refer to Fig. 97 for tools required and proceed as follows: With clutch finger height gage in position as shown in Fig. 98 and while holding down slightly on finger to take out the slack, turn finger adjusting screw until the 2-25/32 inch dimension (X—Fig. 97) is obtained and tighten locknut. Repeat same procedures on other fingers. It is IMPORTANT that clutch fingers be adjusted to within 0.015 of each other. Use a box wrench or similar tool, and actuate each finger through its normal range of travel several times and recheck adjustment.

With finger adjustment completed, refer to Fig. 99 and set the clearance between the 9-inch pressure plate and adjusting cap screws. Loosen locknuts and turn adjusting screw until the 0.080 thickness gage can just be inserted between head of adjusting screw and 9-inch pressure plate. Tighten locknuts.

154. OVERHAUL (SINGLE PLATE). With tractor split as outlined in paragraph 152, unbolt and remove clutch assembly from flywheel. Disassembly procedure is obvious upon examination of the unit and reference to Fig. 103.

Clutch pressure springs should have a free length of 3.852 inches and should test 152 lbs. when compressed to a length of 2.062 inches. Adjust clutch release fingers to a height of 2.569 inches, measured from contacting surface of fingers to friction face of pressure plate. Clutch cover to pressure plate friction face distance should be 1.006 inches.

The "Dyna-Life" clutch driven disc (11—Fig. 103) is available as a unit only.

RELEASE BEARING

All Models

155. To remove the release bearing first split tractor as outlined in paragraph 146 or 152. Disconnect the clutch pedal return spring and remove snap ring or cotter pin and washer from right hand end of clutch cross-shaft. Remove bolts and key from release fork then tap cross-shaft toward left side of clutch housing until release fork comes off cross-shaft. Release fork, bearing and sleeve can now be removed by removing the three retaining cap screws. Any further disassembly and/or overhaul is obvious.

NOTE: On Series 424, 444, 2424, 2444, B-414, 354, 364 and 384, inserts (sleeves) are used on the lugs of the release bearing sleeve; when reassembling be sure flange of insert is between release fork and bearing sleeve.

CLUTCH SHAFT

All Models

156. The clutch shaft for models not equipped with forward and reverse transmission is the transmission input shaft. For information concerning this shaft, refer to paragraphs 161 and 162.

The clutch shaft for models equipped with forward and reverse transmission is the input shaft for the forward and reverse transmission. For information concerning this shaft refer to paragraph 167.

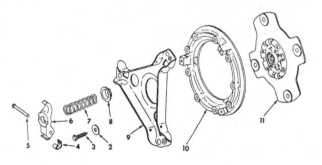

Fig. 103—Exploded view of the single plate clutch assembly used on Series 424, 444, 2424 and 2444 tractors without constant running pto.

2. Washer
3. Adjusting screw
4. Return clip
5. Pivot pin
6. Clutch finger
7. Pressure spring
8. Spring cup
9. Clutch cover
10. Pressure plate
11. "Dyna-Life" driven disc

TRANSMISSION

All Models

157. **MAJOR OVERHAUL.** Data for removing and overhauling the various transmission components are as follows:

158. SHIFTER RAILS AND FORKS. To remove the transmission top cover on Series B-275, disconnect hose from hydraulic system suction line and catch the hydraulic oil. Unbolt hydraulic control valve lever quadrant from reservoir, then remove the suction filter. Disconnect hydraulic pressure line from control valve, then unbolt and remove hydraulic control valve. Place both shifter levers in neutral, then unbolt and remove transmission top cover. See Fig. 104.

159. To remove the transmission top cover on Series 424, 444, 2424, 2444, B-414, 354, 364 and 384, disconnect hose from hydraulic system filter and catch oil, then remove the suction filter. Disconnect the pressure line from cylinder head of control valve. Unbolt hydraulic lift housing from rear frame and either remove the housing, or block it up high enough to allow removal of transmission top cover.

Place both shifter levers in neutral position, then unbolt and remove transmission top cover. See Fig. 104.

160. With transmission top cover removed, secure same in a vise, straighten cap screw locks (28—Fig. 105), then unbolt and remove guide brackets (27). Shifter forks, rail guides and detents can now be removed. Further disassembly, if required, is obvious.

161. DRIVING (INPUT) SHAFT. To remove the transmission driving (input) shaft, the transmission case must be separated from clutch housing as follows: Disconnect hydraulic system suction line from lower right front of

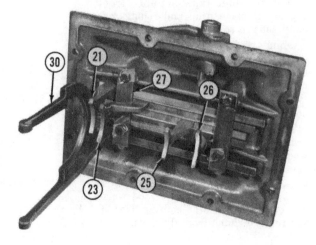

Fig. 104—View of the assembled gear shifting mechanism contained in the transmission top cover. Mechanism will differ on 424, 444, 2424, 2444, B-414, 364 and 384 tractors equipped with forward and reverse transmission. Refer to Fig. 105 for legend.

hydraulic reservoir and drain system. Disconnect hydraulic pressure line from control valve. Remove lower instrument plate, disconnect rear side light wire and remove clip which retains hydraulic lines to fuel tank support. Unbolt front of left step plate from clutch housing then, if equipped with downswept exhaust, either disconnect muffler from rear frame or exhaust pipe from manifold. On Series 354, 364 and 384, disconnect foot accelerator if so equipped, disconnect wiring harness at connector under fuel tank, then unbolt footplate from transmission case and clutch housing. On all series, drain oil from transmission and on dual plate clutch models, drain clutch housing. Remove all covers from bottom of clutch housing and if tractor is a dual plate clutch model remove the plug and "O" ring from clutch housing at front of pto driven shaft. Remove snap ring and washer or nut and washers from end of shaft. If tractor is equipped with forward and reverse transmission, disconnect shifter lever link, remove shaft retaining plate and shims, then pull shifter shaft connector shaft outward as far as it will go. Wedge front axle to prevent tipping, support both sections of tractor, then unbolt and separate

transmission from clutch housing. See Fig. 106.

NOTE: On tractors equipped with transmission driven pto, six mounting cap screws are located inside the clutch housing. On tractors equipped with constant running pto, four mounting cap screws are located inside the clutch housing.

162. With transmission separated from clutch housing, remove the hydraulic control valve and the transmission top cover as outlined in paragraphs 158 or 159. Engage pto shifter lever, then unbolt retainer (2, 17, 24 or 31—Fig. 107) and remove the pto rear shaft assembly.

NOTE: If it is desired to remove clutch (10) it will be necessary to remove snap ring (6, 21, 27 or 34) before clutch will pass through bearing bore.

On models with dual plate clutches, and no forward and reverse transmission, remove the front pto shaft (29—Fig. 106) by sliding same rearward after disengaging clutch and either removing same from housing or allowing it to remain on floor of housing. On

dual plate clutch models, remove countershaft nut (40—Fig. 108) then remove bearing retainer (41). On single plate clutch models, remove retainer (41—Fig. 109), then remove countershaft retainer (45).

On models with direction reverser, remove retainer (46—Fig. 108) and loosen nut (40). Loosen direction reverser housing from face of transmission housing and pull direction reverser unit as far forward as possible, then complete removal of nut (40) and lockwasher (39). Slinger (45) must be removed prior to removal of front pto shaft.

On all models, remove bearing (38) from front end of countershaft.

NOTE: It is highly recommended that International Harvester tool FES 44-3 be used to remove bearing and snap ring assembly (38). Bearing is a tight fit on countershaft and if wedged off by using pinch bars, damage to snap ring and/or bearing could result.

On all models with no direction reverser, remove cap screws from input shaft bearing retainer (25). On models with direction reverser, complete removal of direction reverser housing retaining cap screws, if necessary.

Push down on forward end of countershaft to provide clearance for teeth of input shaft and withdraw input shaft or reverser unit.

Any further disassembly of the input shaft used on models with no direction reverser is obvious. For information on the input shaft used in models with direction reverser, refer to paragraph 167.

163. SPLINE SHAFT. Removal of the spline shaft requires removal of the hydraulic lift attachment (or the housing rear top cover), in addition to split-

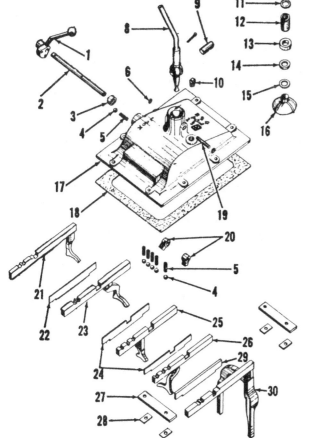

Fig. 105—Exploded view of the gear shifting mechanism.

1. High-Low range shifter
2. Shifter shaft
3. Oil seal
4. Detent ball
5. Detent spring
6. Expansion plug
7. Gearshift lever
8. Gearshift lever
9. Large swivel shaft
10. Filler plug
11. Spring stop
12. Spring
13. Retainer
14. Felt washer
15. Plain washer
16. Shield
17. Transmission top cover
18. Gasket
19. Small swivel shaft
20. High-Low shifter cam
21. High speed shifter fork
22. Left rail guide
23. 4th speed shift fork
24. Inner rail guides
25. 2nd & 3rd shift fork
26. 1st & rev. shift fork
27. Fork & guide bracket
28. Lock
29. Right rail guide
30. Low speed shift fork

Fig. 106—Transmission assembly shown separated from clutch housing. Tractor shown is a B-275 dual plate clutch model, however, B-414, 424, 444, 2424, 2444, 354, 364 and 384 series tractors are similar. Single plate clutch models will not have front pto shaft (29) and will have a cover type retainer instead of retainer (41) shown.

ting the tractor and removing the driving (input) shaft or direction reverser as outlined in paragraphs 161 and 162.

With tractor split and the driving (input) shaft or direction reverser removed, disconnect lift arms from rockshaft, then unbolt and lift off hydraulic lift assembly. Unbolt the spline shaft bearing retainer (6—Fig. 108 or 109) and withdraw spline shaft as a unit. See Fig. 110.

NOTE: On Series 354, 364 and 384, the spline shaft and bevel ring gear have been cut using either the Oerlikon or Gleason system of gear cutting, resulting in different tooth curvature. The spline shaft with Gleason cut teeth can be removed without disturbing the bevel ring gear. On tractors having the Oerlikon cut teeth, remove right brake housing and loosen right bull pinion bearing cage cap screws to provide the necessary clearance to remove the spline shaft assembly.

On all series, after removing gears from spline shaft, disassembly can be completed as follows: Unstake lockwasher (10—Fig. 108 or 109) and remove locknuts (9). Use a suitable press and press spline shaft out of tapered bearings. Save spacer (4) and shims present for subsequent reinstallation. Bearing cups can be driven from retainer.

164. If original parts are to be used, reassemble components to shaft by reversing disassembly procedure. However, if incorrect bearing adjustment is suspected, or new parts are installed, check the bearing preload as follows: Assemble bearings, spacer, shims, and bearing retainer to spline shaft and tighten locknuts securely. Clamp spline shaft in a vise, wrap a length of cord around bearing retainer and check the amount of pull on a spring scale required to keep the bearing retainer rotating. This measurement should be 3-6 lbs. on the spring scale which is equivalent to 5-15 in.-lbs. preload. If preload is not as specified, vary shims between spacer (4—Fig. 108 or 109) and bearings until preload is correct. Shims are available in thicknesses of 0.002, 0.004 and 0.010. Stake lockwasher when adjustment is completed. Check and adjust, if necessary, the cone center depth as outlined in paragraph 170.

165. COUNTERSHAFT. Removal of transmission countershaft can be accomplished after spline shaft is removed as outlined in paragraph 163.

With spline shaft removed, remove snap ring (27—Fig. 108 or 109) and drive bearing (28) from countershaft

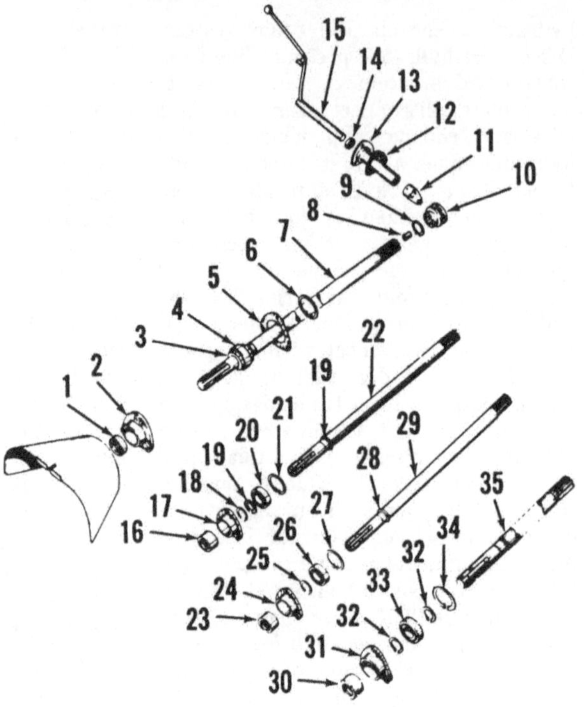

Fig. 107—Exploded view of various pto rear (output) shafts. Items 1 through 7 are used on Series 354, 364 and 384, items 16 through 22 are used on 424, 2424, 444 and 2444, items 23 through 29 are used on B-414 and items 30 through 35 are used on B-275.

1. Oil seal
2. Retainer
3. Collar
4. Bearing
5. Gasket
6. Snap ring
7. Shaft
8. Bushing
9. Snap ring
10. Clutch
11. Shifter
12. Gasket
13. Bracket
14. Oil seal
15. Shift lever
16. Oil seal
17. Retainer
18. Snap ring
19. Collars
20. Bearing
21. Snap ring
22. Shaft
23. Oil seal
24. Retainer
25. Snap ring
26. Bearing
27. Snap ring
28. Collar
29. Shaft
30. Oil seal
31. Retainer
32. Snap ring
33. Bearing
34. Snap ring
35. Shaft

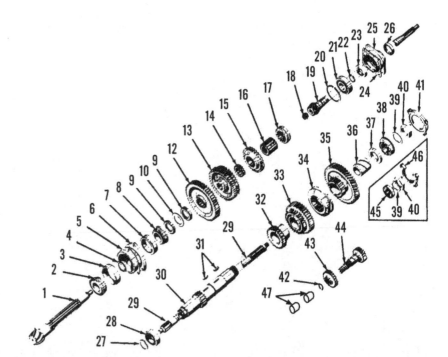

Fig. 108—Exploded view of transmission shafts, gears and component parts used when tractor is equipped with constant-running pto (dual plate clutch). Items shown in the inset are used when equipped with forward and reverse transmission. Refer also to Fig. 109.

1. Spline shaft
2. Bearing cone
3. Bearing cup
4. Spacer
5. Shims
6. Bearing retainer
7. Bearing cup
8. Bearing cone
9. Locknut
10. Lockwasher
12. 1st & rev. sliding gear
13. 2nd & 3rd sliding gear
14. 4th & direct gear coupling
15. 4th & direct sliding gear coupling
16. Sliding quill gear
17. Coupling
18. Pilot bearing
19. Driving (input) shaft
20. Snap ring
21. Ball bearing
22. Snap ring
23. Driving shaft oil seal
24. Gasket
25. Bearing retainer
26. Pto driving shaft oil seal
27. Snap ring
28. Ball bearing
29. Front pto shaft
30. Countershaft
31. Woodruff keys
32. 2nd speed drive gear
33. 3rd & 4th speed driving gear
34. Coupling
35. Constant mesh gear
36. Bushing (cast iron)
37. Spacer
38. Ball bearing & snap ring
39. Lockwasher
40. Locknut
41. Bearing retainer
42. Snap ring
43. Rev. idler driving gear
44. Rev. idler gear & shaft
45. Bushing
46. Bearing retainer
47. Bushings

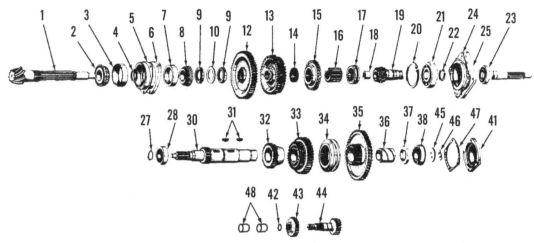

Fig. 109—Exploded view of the transmission shafts, gears and component parts used when tractor is equipped with transmission driven pto (single plate clutch). Note the solid countershaft (30). Refer also to Fig. 108.

1. Spline shaft	10. Lockwasher	32. 2nd speed
2. Bearing cone	12. 1st & rev.	drive gear
3. Bearing cup	sliding gear	33. 3rd & 4th speed
4. Spacer	13. 2nd & 3rd	driving gear
5. Shims	sliding gear	34. Coupling
6. Bearing retainer	14. 4th & direct	35. Constant mesh gear
7. Bearing cup	gear coupling	36. Bushing (cast iron)
8. Bearing cone	15. 4th & direct	37. Spacer
9. Locknut	sliding gear	38. Ball bearing & snap

16. Sliding quill gear	23. Oil seal	41. Bearing retainer
17. Coupling	24. Gasket	42. Snap ring
18. Pilot bearing	25. Bearing retainer	43. Rev. idler driving gear
19. Driving (input) shaft	27. Snap ring	44. Rev. idler gear
20. Snap ring	28. Ball bearing	& shaft
21. Ball bearing	30. Countershaft	45. Retaining washer
22. Snap ring	31. Woodruff keys	46. Lock plate
		47. Gasket
		48. Bushings

using a soft drift positioned on inner race of bearing. Lift out countershaft.

Slide off constant mesh gear (35) and bushing (36). Gears (32 and 33) can now be pressed from shaft. See Fig. 111.

NOTE: When reinstalling countershaft, bear in mind that the front end

Fig. 110—Removed spline shaft showing gears and pilot bearing (18). Refer to Fig. 108 or 109 for legend.

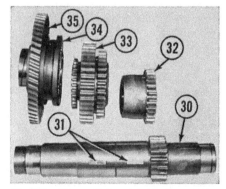

Fig. 111—Dismantled countershaft from tractor having dual plate clutch. Note the hollow shaft. Refer to Fig. 108 for legend.

will have to be pushed downward before driving (input) shaft can be installed.

Install the countershaft as follows: Place gears on shaft and install assembly in housing. Install front bearing and bearing retainer to position shaft, then install rear bearing and snap ring. Remove front bearing retainer and bearing and install spline shaft and input shaft, or direction reverser shaft. Reinstall front bearing, bearing retainer and nut (or retaining washer) and tighten nut (40—Fig. 108) to a torque of 30-35 ft.-lbs. Check end play of constant mesh gear (35) which should be 0.005-0.010. If end play is insufficient, drive countershaft forward and retorque nut or retaining washer cap screws.

NOTE: Insufficient end play of constant mesh gear can only be caused by incorrect assembly of countershaft components.

166. REVERSE IDLER. To remove the reverse idler assembly, first remove the driving (input) shaft or direction reverser, spline shaft and countershaft as outlined in paragraph 161 through 165, then remove snap ring (42—Fig. 108 or 109) and remove reverse idler assembly from housing.

NOTE: Some mechanics prefer to remove the right bull gear to provide easier access to snap ring.

Check reverse idler gear shaft and bushings against the values which follow:

Shaft front diameter 1.249-1.250
Shaft rear diameter 0.991-0.992
Front bushing inside
   diameter 1.251-1.253
Rear bushing inside
   diameter 0.993-0.995
Operating clearance 0.001-0.004

When renewing reverse idler bushings, install bushings with the split at the bottom. Remove expansion plug at front of transmission case and ream bushings to specified size. Use sealing compound and renew expansion plug.

FORWARD AND REVERSE TRANSMISSION

Series B-414-424-444-2424-2444-364-384

Series B-414, 364, 384, 424, 444, 2424 and 2444 tractors are available with a forward and reverse transmission which provides eight speeds in reverse as well as the eight forward speeds. Transmission is a self-contained unit mounted on front face of main transmission housing and is controlled by a lever mounted on left side of clutch housing. See Fig. 112.

167. **R&R AND OVERHAUL.** To remove the forward and reverse trans-

mission, split tractor between clutch housing and transmission as outlined in paragraph 161.

With tractor split, remove bearing retainer (46—Fig. 108), and loosen nut (40) until it reaches cluster gear of direction reverser. Loosen (or remove) the nuts and cap screw which retain direction reverser to transmission housing and pull unit forward as far as possible. Complete removal of nut (40) and lockwasher (39). Remove bearing and snap ring assembly (38). Push down on forward end of countershaft to provide clearance and withdraw the forward and reverse unit.

NOTE: It is highly recommended that International Harvester tool FES 44-3 be used to remove bearing and snap ring (38). Bearing is tight fit on countershaft and if wedged off by using pinch bars, damage to snap ring and/or bearing could result.

With unit removed, refer to Fig. 113, remove lockbolt (23), then drive roll pin (28) from countershaft (26) and remove countershaft, spacer (30) and cluster gear (33). The two needle bearings (32) can now be removed from cluster gear, if necessary.

Remove lockbolt (23) and snap ring (41) from idler shaft, then slide thrust washer (40) rearward and remove the small locking pin from its bore in idler shaft. Idler gear shaft (29), idler gear (36), needle bearing (38), bearing sleeve (39) and thrust washers (34) can now be removed. Needle bearing can now be removed from idler gear, if necessary.

Remove shifter rail (24), detent ball (11) and detent spring (10), then remove snap ring (4) and bump input shaft (3) and bearing assembly (5) from housing. Bearing can be removed from input shaft after removing snap ring (6).

Remove snap ring (12) and input gear (13) from aft end of clutch shaft (17), remove fork shoes (8), then bump clutch shaft (17) forward and out of bearing. Remove snap ring (14) and bump bearing (15) rearward out of housing. Oil seal (21) and pilot bearing (16) can now be removed, if necessary.

Drive roll pins from shifter fork (9) and shaft (20) and remove fork and shaft.

Inspect all gears for chipped, broken or unduly worn teeth. Check shifter fork shoes in groove of output gear and renew shoes and/or gear if excessive clearance is present. Check condition of thrust washers and needle bearings of cluster gear and idler gear, and pay particular attention to bearing sleeve (39). Be sure the small sealing ball (25) is securely in place in forward end of

Fig. 112—Forward and reverse transmission mounts on front face of transmission housing as shown.

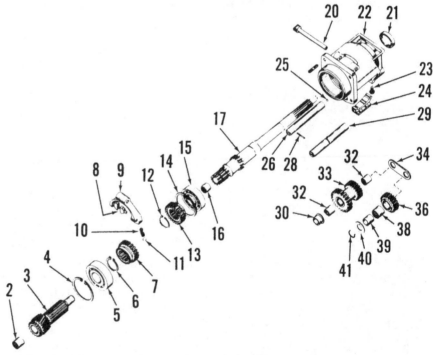

Fig. 112A—Forward and reverse transmission with cluster gear and idler gear removed.

Fig. 113—Exploded view of the forward and reverse transmission which is available for the Series 364, 384, 424, 444, 2424, 2444 and B-414 tractors.

2. Pilot bearing	29. Idler shaft
3. Transmission input shaft	30. Spacer
4. Snap ring	32. Needle bearing
5. Ball bearing	33. Cluster gear
6. Snap ring	34. Thrust washer
7. Output gear	36. Idler gear
8. Shifter fork shoes	38. Needle bearing
9. Shifter fork	40. Thrust washer
10. Poppet spring	41. Snap ring
11. Poppet ball	
12. Snap ring	
13. Input gear	
14. Snap ring	
15. Ball bearing	
16. Needle bearing	
17. Clutch shaft	
20. Shifter fork shaft	
21. Oil seal	
22. Housing	
23. Locking bolt	
24. Shifter rail	
25. Steel ball	
26. Countershaft	
28. Groove pin	

cluster gear shaft. Ball bearings should turn freely with no lumpy (tight) spots.

Reassemble by reversing the disassembly procedure. Oil seal (21) is installed with lip facing toward rear of housing. Large gear of countershaft cluster gear (33) is toward rear. Shifter groove of output gear (7) is toward rear. The small locking pin in idler gear shaft fits in notch of thrust washer (40).

Reinstall by reversing the removal procedure and tighten the countershaft nut (40—Fig. 108) to a torque of 30-35 ft.-lbs.

MAIN DRIVE BEVEL GEARS
AND DIFFERENTIAL

RENEW BEVEL GEARS

All Models

168. To renew the bevel pinion, follow the procedure for overhaul of spline shaft given in paragraphs 163 and 164. To renew the main drive bevel ring gear, follow the procedure for overhaul of differential in paragraph 173.

Neither bevel pinion, nor ring gear, are available separately and renewal of one necessitates the renewal of the other. After renewal of gears, check and adjust, if necessary, the carrier bearing adjustment, the mesh position and backlash as follows:

169. **DIFFERENTIAL BEARING PRELOAD.** Prior to making either the mesh or backlash adjustments, the preload on the differential carrier bearings should be established. To adjust the taper roller bearings (7—Fig. 115), first install more than enough shims (2) under each of the differential bearing cages so that differential has a slight amount of end play and make certain there is some backlash between the ring gear and pinion. Wind a cord around the machined diameter of left differential case half, attach spring scale to end of cord and check the amount of pull required to keep the differential rolling once it has started. Record this reading.

Now remove an equal amount of shims (2) from each side until the bearing preload requires 6 to 10 pounds more than the previously recorded pull on the spring scale to keep the differential rolling at a constant speed.

170. **MESH POSITION.** The mesh position (cone point distance) is controlled by shims (5—Fig. 108 and 109) located between spline shaft bearing retainer and wall of transmission case. The correct cone point distance will be found etched on the machined aft end of spline shaft and represents the distance from machined end of shaft to center line of differential. The nominal radius of the left hand differential case

half is 3.938 inches at its machined surface, however, it should be measured at several points to insure accuracy. Mesh position is correct when the distance between end of spline shaft and machined surface of left half of differential case is equal to the difference between the etched cone point distance and the radius of the differential case. For example: If the etched cone center distance was 4.423 and the differential case radius 3.938, the difference would be 0.485 which is the measureable distance between end of spline shaft and the differential case. Shims are available in thicknesses of 0.0016, 0.006 and 0.0108.

NOTE: Do not confuse cone center point markings with the gear matching numbers.

171. **BACKLASH ADJUSTMENT.** After the carrier bearings are adjusted as outlined in paragraph 169, and the mesh position, adjusted as outlined in paragraph 170, the backlash can be adjusted as follows: Transfer shims (2—Fig. 115) from under one bearing cage to the other to provide correct backlash between teeth of the main drive bevel pinion and bevel ring gear. The backlash value will be etched on the outer machined face of the ring gear and may vary between gear sets. To decrease backlash, remove shim, or shims, from cage on bevel ring gear side of housing and install same under cage on opposite side. Only transfer shims, do not remove shims or the previously determine carrier bearing adjustment will be changed.

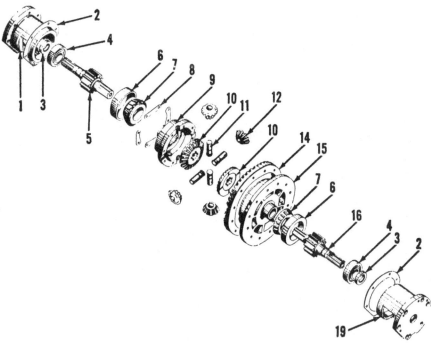

Fig. 115—Exploded view of differential, bull pinion shafts and associated parts. Shims (2) control carrier bearing preload and gear backlash adjustments.

1. L.H. bearing cage	
2. Shims	6. Bearing cup
3. Oil seal	7. Bearing cone
4. Ball bearing	8. Lock plates
5. L.H. bull pinion	9. Left differential half
10. Differential side gear	15. Right differential
11. Pinion shafts	half
12. Differential pinions	16. R.H. bull pinion
14. Bevel ring gear	19. R.H. bearing cage

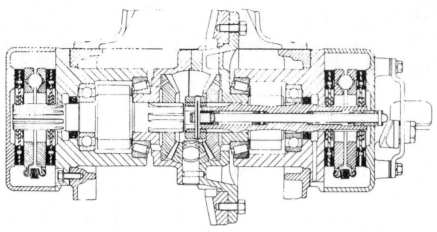

Fig. 116—Sectional view of differential, bull pinions and brakes. Note differential lock which operates through hollow right hand bull pinion shaft.

172. TOOTH CONTACT. If desired, the bevel pinion and bevel ring gear adjustment can be finally checked as follows: Paint several teeth at 90 degree intervals with Prussian Blue or red lead and rotate pinion in both directions. Correct adjustment will result in patterns similar to those shown in Figs. 117 and 118.

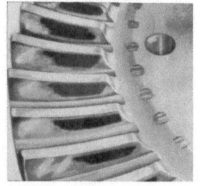

Fig. 117—Correct mesh and backlash will leave pattern shown when pinion shaft is turned to produce forward motion. Refer to text.

Fig. 118—Correct mesh and backlash will leave pattern shown when pinion shaft is turned to produce reverse motion. Refer to text.

DIFFERENTIAL AND CARRIER BEARINGS

All Models

173. R&R AND OVERHAUL. To remove the differential first drain transmission case, then either remove differential top cover, if so equipped, or drain and remove the hydraulic lift unit as outlined in paragraph 207 or 209. Support rear of tractor in a raised position and remove rear wheels. Wedge front axle as an aid to prevent tractor from tipping. Remove cap screws from bearing retainer of rear pto shaft and remove shaft and bearing assembly.

On Series B-275, B-414, 354, 364 and 384 unbolt right hand step plate (platform) and hand brake from right fender. Remove cap screw (1—Fig. 121) from inner end of axle shaft, unbolt axle carrier (8) and remove axle, axle carrier and fender as a unit.

On Series 424, 444, 2424 and 2444, unbolt and remove the right fender. Remove snap ring (1—Fig. 122) and unbolt bearing retainer (10), then

withdraw axle shaft (11).

NOTE: The axle and axle carrier can be removed as a unit to renew axle carrier gasket (3).

On all models where axle is a tight fit in the bull gear hub, a cap screw and nut can be used in conjunction with a short piece of pipe as a pusher.

On Series 424, 444, 2424 and 2444 remove the battery, then on all models unbolt and remove the left fender. Follow same procedure as outlined before and remove left hand axle and axle carrier. Lift bull gears from rear frame.

With axle and axle carrier assemblies removed, remove the differential lock thrust plate and lever from right brake housing. Depress "O" ring retainer (8—Fig. 119) and remove the retaining ring.

On Series B-275 and early production B-414 tractors, loosen jam nut at brake operating rod and unscrew brake rod assembly from yoke.

On Series 354, 364, 384, 424, 444, 2424, 2444 and late production B-414 tractors, disconnect the brake pedal return springs and remove the brake adjusting rod pin.

Then, on all models, unbolt brake housing and remove housing, outer brake disc, actuating assembly and inner brake disc. Unbolt and remove bull pinion shaft bearing cage and bull pinion shaft. Bearing cages can usually be withdrawn with no difficulty, however, should a tight fit be encountered, tapped holes are provided for puller screws. Use of puller screws usually causes shim damage which requires renewal of shims.

Removal of left brake, bull pinion shaft bearing cage and bull pinion shaft is accomplished in a like manner except that no differential lock is involved.

Differential assembly can be lifted from rear frame after bull pinion shafts are removed.

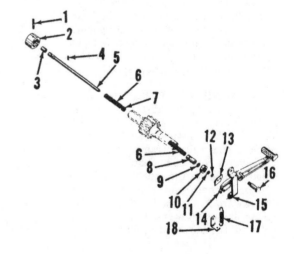

Fig. 119—Exploded view of the differential lock assembly.

1. Groove pin
2. Clutch
3. Spacer
4. Dowel pin
5. Shaft
6. Spring
7. Collar
8. "O" ring retainer
9. Inner "O" ring
10. Oil seal
11. Outer "O" ring
12. Retaining ring
13. Cam plate
14. Lever
15. Thrust plate
16. Pivot bolt
17. Return spring
18. Spring anchor

174. To inspect and/or overhaul removed differential, straighten lock plates (8—Fig. 115), unbolt and separate the differential case halves. Pinion shafts (11) can be driven off pins at this time. Any further disassembly and/or overhaul is obvious after an examination of the unit.

When reinstalling differential assembly, adjust preload on carrier bearings as outlined in paragraph 169. Adjust mesh position as outlined in paragraph 170 and backlash as outlined in paragraph 171.

DIFFERENTIAL LOCK

All Models

All tractors are equipped with a differential lock which operates through the right hand bull pinion shaft and is controlled by a foot pedal mounted on the right hand brake housing.

Operating the differential lock connects the two bull pinion shafts together which causes the differential to act as a solid hub and the driving wheels to operate as though they were on a common shaft.

Refer to Fig. 119 for an exploded view of the differential lock.

175. **R&R AND OVERHAUL.** Removal and overhaul of the differential lock requires removal and separation of the differential assembly as outlined in paragraphs 173 and 174.

Overhaul of the unit is obvious. Springs (6—Fig. 119) can be checked by using the following specifications:

Free length . . . . .3-5/8 in.
Test length . . . . .2-1/8 in.
Test load . . . . .54.8 lbs.

When reassembling, adjust differential lock thrust plate by means of adjustment washers until there is 1/32 to 1/16 inch clearance between cam plate and lock shaft as shown in Fig. 120.

FINAL DRIVE

The final drive consists of two bull pinions and integral brake shafts and two bull gears which are splined to the inner ends of the rear axle shafts. Each axle shaft is carried in a sleeve which is bolted to the side of the transmission case.

All Models

176. **R&R WHEEL AXLE SHAFT.** To remove either wheel axle shaft, proceed as follows: Drain transmission housing. Remove rear frame top cover, if so equipped, or drain hydraulic lift reservoir and remove same as outlined in paragraph 207 or 209. Wedge front axle to aid in preventing tractor from tipping, then raise and support rear of tractor. Remove rear wheel and tire assembly. Disconnect rear light wires.

On Series B-275, B-414, 354, 364 and 384 remove cap screw (1—Fig. 121), unbolt axle carrier (8) and withdraw axle, axle carrier and fender assembly.

On Series 424, 444, 2424 and 2444, remove snap ring (1—Fig. 122), unbolt axle carrier (6) and withdraw axle, axle carrier and fender assembly.

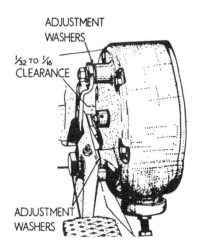

Fig. 120—Use adjustment washers as shown to adjust cam plate clearance.

Fig. 121—Exploded view of axle, axle carrier and component parts used on Series B-275, B-414, 354, 364 and 384.

1. Cap screw
2. Retainer
3. Bull gear
4. Gasket
5. Snap ring
6. Ball bearing
7. Collar
8. Axle carrier
9. Ball bearing
10. Oil seal
11. Felt washer
12. Gasket
13. Bearing retainer
14. Axle shaft

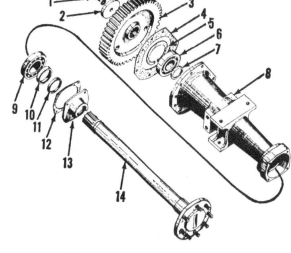

Fig. 122—Exploded view of axle, axle carrier and component parts used on Series 424, 444, 2424 and 2444.

1. Snap ring
2. Bull gear
3. Gasket
4. Snap ring
5. Ball bearing
6. Axle carrier
7. Ball bearing
8. Oil seal
9. Gasket
10. Bearing retainer
11. Axle shaft

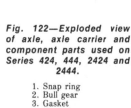

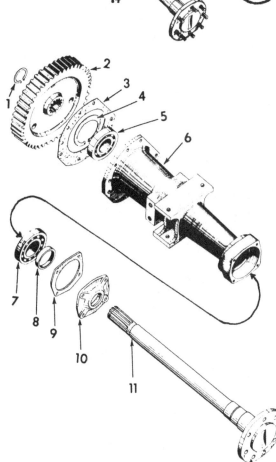

NOTE: Although the axle carriers and fenders can be removed as a unit, some mechanics prefer to remove the fenders before removing the axle carriers.

Use a cap screw and nut in conjunction with a short piece of pipe if a pusher is required to push axle from bull gear.

With axle and carrier removed, unbolt bearing retainer, bump axle on inner end and drive axle out of axle carrier. Remove snap ring and drive out axle inner bearing. Any further disassembly required is obvious.

Use new oil seals and gaskets and reassemble by reversing the disassembly procedure.

177. R&R BULL GEARS. To remove the bull gears, first remove the wheel axle shafts as outlined in paragraph 176. Unbolt the rear pto shaft bearing retainer and remove the rear pto shaft. Bull gears can now be removed, however, the left bull gear must be out of housing before enough clearance is available to lift out right bull gear. Bear this is mind when occasions arise which dictate work on right bull gear.

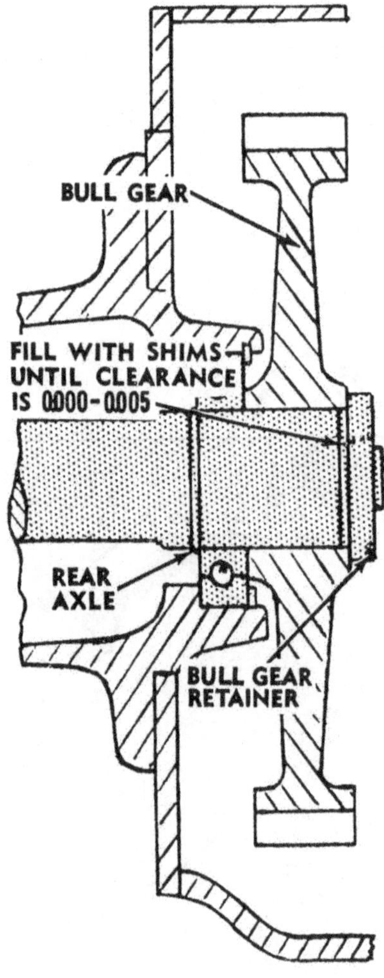

NOTE: In a few isolated cases a clicking noise has occurred in the axle and bull gear assembly of the Series B-275 tractors and the cause has been determined by the manufacturer to be excessive clamping between bull gear and axle. This noise is not conducive to failures, however, it may tend to be objectionable and may be eliminated as follows:

Remove rear frame top cover, if so equipped, or the hydraulic lift reservoir, then remove the cap screw and retaining washer from inner end of axle. Place a straightedge across hub of bull gear and measure distance between end of axle and straightedge. If this distance is excessive, add shims (IHC part No. 3042081R1) as shown in Fig. 123 until clearance is 0.000-0.005. Reinstall cap screw and retainer and hydraulic lift reservoir.

178. R&R BULL PINION SHAFTS. To remove either bull pinion shaft, first remove bull gear as outlined in paragraph 177. Remove right brake as follows: Remove differential lock thrust plate (15—Fig. 119) and lever (14). Push "O" ring retainer (8) inward and

remove retainer (12). Disconnect the brake operating rod, then unbolt and remove brake housing, actuating assembly and inner disc. Left brake can be removed after brake operating rod is disconnected as no differential lock mechanism is involved.

Remove the cap screws which retain bull pinion shaft bearing cage to rear frame and withdraw same. Bearing cages usually are withdrawn with no difficulty, however, if difficulty is encountered tapped holes are provided in bearing cages to permit use of puller screws.

NOTE: The use of puller screws usually results in damage to shims (2—Fig. 115) and when renewing be sure to install the same thickness of shims as were removed in order to preserve the carrier bearing preload. Shims are available in thicknesses of 0.004, 0.007, 0.015 and 0.032.

Disassembly of the bull pinion shaft and bearing cage assembly is obvious after an examination of the unit and reference to Fig. 115. Bearing cup, shaft, ball bearing and oil seal are removed from inner end of bearing cage.

BRAKES

All Models

179. Brakes are double disc, self-energizing type and are splined to the outer ends of the bull pinion shafts. The moulded type linings are riveted to the discs and brake lining service packages are available for renewing the linings.

Series B-275 and early production B-414 tractors are equipped with the 5-3/8 inch brakes shown in Fig. 125. B-414 tractors having serial number 31374 and up (diesel) or 4360 and up (non-diesel) and all Series 354, 364, 384, 424, 444, 2424 and 2444 tractors are equipped with 6½ inch brakes. See Fig. 127.

180. **ADJUST.** To adjust the brakes on Series B-275 and early B-414 tractors, loosen jam nut (9—Fig. 124) and turn rod (3) either way as required to obtain a pedal free travel of ¾-inch. Tighten jam nut.

On Series 354, 364, 384, 424, 444, 2424, 2444 and late B-414 tractors, loosen the locknut and turn adjusting nut either way as required until 1¾-inches pedal free travel is obtained. Tighten locknut. Refer to Fig. 126 for location of adjusting nut.

To equalize the brakes on all models, first block up rear of tractor securely, start engine and shift transmission into

Fig. 124—Left brake of B-275 tractor showing brake rod and points of adjustment. Step plate has been removed for illustrative purposes. Early Series B-414 tractors are similar. Refer to Fig. 125 for legend.

Fig. 123—Schematic view showing position of shims used to eliminate axle noises.

BULL GEAR

FILL WITH SHIMS UNTIL CLEARANCE IS 0.000-0.005

REAR AXLE

BULL GEAR RETAINER

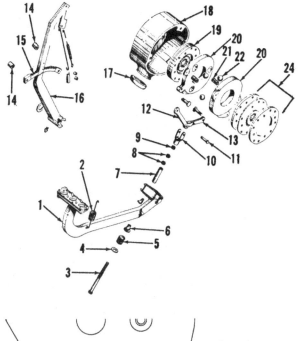

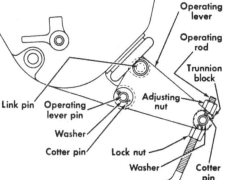

Fig. 125—Exploded view showing component parts of Series B-275 right hand disc brake assembly. Early Series B-414 are similar.

1. R.H. pedal
2. Return spring
3. Brake rod
4. Washer
5. Rod spring
6. Rod ball
7. Spacer
8. Jam nuts
9. Jam nut
10. Rod yoke
11. Pin
12. Male link
13. Female link
14. Spacers
15. Quadrant (rack)
16. Hand lever
17. Boot
18. Brake housing
19. Outer brake disc
20. Actuating disc
21. Extension spring
22. Steel ball
24. Inner brake disc

Fig. 126—View showing brake adjuster on Series 354, 364, 384, 424, 444, 2424, 2444 and late B-414 tractors.

Fig. 127—Exploded view of the 6½-inch disc brake used on Series 354, 364, 384, 424, 444, 2424, 2444 and late B-414 tractors.

1. Brake housing
2. Brake disc
5. Actuating disc
6. Steel ball
7. Disc return spring
8. Actuating link
9. Tension link
10. Rubber boot
12. Washer
13. Operating lever pin
14. Pivot pin
15. Operating lever
16. Adjusting nut
17. Trunnion block
18. Operating rod
19. Bushing
20. Brake pedal R.H.
22. Operating rod pin

Disassembly of brakes is obvious upon examination of the units and reference to Figs. 125 and 127. Wash all parts except the linings in solvent. Inspect all parts and renew defective or worn parts as necessary.

On Series B-275 and early B-414 tractors, if brake operating rod is disassembled, reset the spring preload by turning down first jam nut (8—Fig. 125) until spring (5) is compressed solid, then back off nut three turns. Install second jam nut (8) and tighten against first to hold this setting.

On Series 424, 444, 2424, 2444 and late B-414 tractors, pin the brake operating rod in the front (19-1 ratio) hole in foot pedal for normal operation and in operations where increased braking is desired, install pin in rear (25-1 ratio) hole.

POWER TAKE-OFF

All Models

182. Tractors are available either with a transmission driven or a constant-running power take-off. Refer to paragraphs 187 through 189 for dual speed constant running pto used on some Series 354, 364 and 384 tractors.

When tractor is equipped with a transmission driven pto, the output shaft is driven by a solid transmission

third or fourth gear. With the pedals latched together, apply the brakes. If both wheel stop at the same time, the adjustment is satisfactory. If not, loosen the adjustment slightly on the tight brake until equalization is obtained.

181. **R&R AND OVERHAUL.** To remove right brake, proceed as follows: Remove differential lock thrust plate and pedal. On Series B-275 and early B-414 tractors, loosen jam nut (9—Fig. 125) and unscrew brake operating rod from yoke. On Series 354, 364, 384, 424, 444, 2424, 2444 and late B-414 tractors equipped with 6½-inch brakes, disconnect brake pedal return springs and remove the brake operating rod pin. Then, on all models, unbolt brake housing and remove housing, outer brake disc, actuating assembly and inner brake disc.

Left brake can be removed after disconnecting brake operating rod and unbolting brake housing from transmission housing.

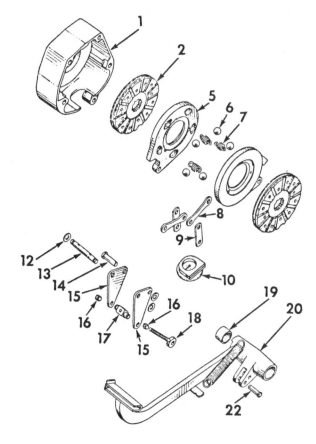

countershaft. When tractor is equipped with constant-running pto, the pto driving shaft and transmission countershaft are both hollow. In both cases power is directed to the output shaft through a

splined clutch (collar) located in the differential section of the transmission housing.

Refer to Fig. 128 for an illustration showing arrangement of shafts for the

constant-running pto.

Occasions for overhauling the complete power take-off will be infrequent. Usually any failed or worn part will be so positioned that localized repairs can be made as outlined in the following paragraphs:

183. REAR (OUTPUT) SHAFT. To remove and/or overhaul the pto rear shaft, first drain transmission housing, then unbolt bearing retainer and shield and pull shaft from housing. Keep pto shift lever in engaged position so clutch (collar) will remain on aft end of transmission countershaft or front pto shaft. If desired, the clutch can be removed by disengaging snap ring from bearing bore at rear of transmission case, pushing shift lever forward and taking clutch out through bearing bore.

Overhaul is obvious after an examination of the unit and reference to Fig. 107. Oil seal in bearing retainer is installed with lip facing front of tractor.

NOTE: During reinstallation of rear shaft it may be necessary to remove shifter lever bracket to insure mating of clutch shifter with groove of clutch.

184. COUNTERSHAFT OR FRONT PTO SHAFT. On those models equipped with transmission driven pto, the pto output (rear) shaft is driven from a spline on the aft end of a solid transmission countershaft. Refer to paragraph 165 for information pertaining to this shaft.

185. On models equipped with a constant-running pto, the output shaft is driven by a front mounted shaft which runs through a hollow transmission countershaft and is splined to a gear which is driven from the hollow pto driving shaft. See Fig. 128.

The front pto shaft can be removed from the rear after removing bottom cover from clutch housing, nut or snap ring from front of shaft and the pto output shaft as outlined in paragraph 183.

186. DRIVING SHAFT AND DRIVEN GEAR. Due to the fit of bearings in their bores and on driving shaft (4—Fig. 129) and driven gear (5), plus the inaccessibility of the front of shaft and gear, it is usually necessary to perform the clutch split as outlined in paragraph 146 or 152 and the transmission split as outlined in paragraph 161. With the clutch housing removed as a unit, remove snap ring (9—Fig. 129) and press driven gear (5) from bearing (8). Remove "O" ring (13) and snap ring (7) and push bearing from its bore.

Remove snap ring (1) and press driving shaft (4) out rear of clutch housing.

Fig. 128—Illustration showing shaft arrangement when tractor is equipped with single speed constant-running pto. Refer to Fig. 129 for legend.

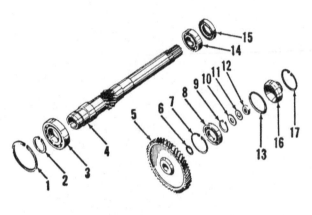

Fig. 129—Exploded view of pto driving shaft, driven gear and their component parts.

1. Snap ring
2. Snap ring
3. Ball bearing
4. Driving shaft
5. Driven gear
6. Snap ring
7. Snap ring
8. Ball bearing
9. Snap ring
10. Retaining washer
11. Lockwasher
12. Nut
13. "O" ring
14. Ball bearing
15. Oil seal
16. Cap (plug)
17. Snap ring

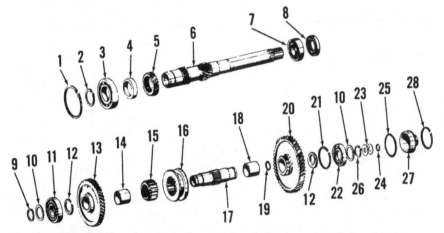

Fig. 130—Exploded view of dual speed pto driving shaft, pto countershaft and their component parts used on some Series 354, 364 and 384 tractors.

1. Snap ring	8. Front seal	15. Clutch body	22. Front bearing
2. Snap ring	9. Snap ring	16. Coupling	23. Retaining washer
3. Rear bearing	10. Retaining washer	17. Countershaft (pto)	24. Snap ring
4. Spacer	11. Countershaft rear brg.	18. Bushing	25. "O" ring
5. Driving gear	12. Spacer	19. Snap ring	26. Snap ring
6. Driving shaft	13. Driven gear	20. Driven gear	27. Bearing cap
7. Front bearing	14. Bushing	21. Snap ring	28. Snap ring

Remove snap ring (2) and pull bearing (3) from shaft. Remove oil seal (15) and if bearing (14) has remained in its bore, remove same. Oil seal for aft end of driving shaft is contained in the transmission input shaft bearing retainer.

Dual Speed Constant Running Pto

187. Some Series 354, 364 and 384 tractors are equipped with a dual speed pto, which is similar to single speed constant-running pto except that the hollow driving shaft has two gears. See Fig. 130. These gears are in constant mesh with two gears (13 and 20) that are free wheeling on the pto countershaft (17). The countershaft is splined to the pto front shaft.

There is a sliding coupling on the pto countershaft which is operated by a lever on the right side of clutch housing. You may lock either of the free wheeling gears to the pto countershaft giving you either 540 rpm or 1000 rpm output shaft speed. Refer to paragraph 183 for R&R and overhaul procedure on rear (output) shaft.

188. **R&R DRIVING SHAFT AND COUNTERSHAFT.** To remove the pto countershaft first perform the clutch split as outlined in paragraph 146 or 152 and transmission split as outlined in paragraph 161. With housing removed as a unit, remove pto countershaft rear bearing snap ring (9—Fig. 130), retainer washer (10) and "O" ring (25). Drive the countershaft and front bearing forward into clutch housing compartment. Gears and coupling will be removed as the countershaft is removed. Remove snap ring (26) and press bearing (22) from countershaft. Remove bushing (18) from countershaft and snap ring (19) from inside of countershaft.

Procedure for removal of driving shaft is same as for single speed pto

outlined in paragrah 184.

189. **R&R DUAL SPEED SHIFTER.** To remove the pto dual speed shifter, remove countershaft as outlined in paragraph 188. Loosen pinch bolt and remove operating lever (7—Fig. 131) from shaft. Remove bracket (4), shifter

fork (1) and "O" ring (2). Reinstall by reversing the removal procedure.

NOTE: To adjust the shifter, loosen the pinch bolt, position shifter in center location hole in clutch housing. Move the clutch (16—Fig. 130) to neutral position, tighten pinch bolt.

BELT PULLEY

All Models

190. **R&R AND OVERHAUL (Clockwise Rotation).** The belt pulley unit bolts to the rear of the tractor rear frame and is driven by the power take-off. Removal of the unit from the tractor is obvious.

To disassemble, remove pulley, bearing retainer (19—Fig. 132), withdraw pinion shaft (12) and bearings, then remove spacer (16) from housing. Remove cover (1), using two of the bolts in extractor holes, bevel gear assembly and oil seal (8) from housing. Any further disassembly required will be obvious.

To reassemble, position bevel gear retaining bolts in the drive shaft (6—Fig. 132), press inner bearing (7) on the shaft and install bevel gear (4)

using original shims (5). Measure from back face of bearing (7) to the front face of the gear (teeth side). The dimension should be between 3.554 to 3.564 inch for 550 rpm unit or 3.399 to 3.409 for 700 rpm unit. Add or remove shims (5) as required. Install the drive shaft assembly into the housing. Press the outer bearing (2) into the cover (1) and install the cover on the drive shaft until bearing contacts shoulder on the shaft. Measure the gap between the cover (1) and housing (9) and select shims (3) to give a total thickness greater than the gap. At least one shim will be required. Remove the drive shaft assembly. Press the oil seal (8) into housing, with the lip facing in toward bearing, until the back face of the oil seal is 1.88 inch from the mounting flange. Assemble the drive shaft

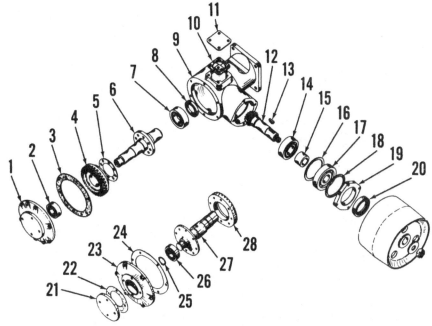

Fig. 132—Exploded view of the belt pulley unit. Drive is taken from pto shaft. Items 4 through 6 are used for clockwise rotation and items 27 and 28 are used for counterclockwise rotation. Items 21 through 25 are used on early models.

1. Cover	8. Oil seal	15. Spacer	22. Gasket
2. Bearing	9. Housing	16. Spacer	23. Cover
3. Shim	10. Gasket	17. Bearing	24. Shim
4. Bevel gear	11. Plate	18. Gasket	25. Snap ring
5. Shim	12. Pinion shaft	19. Bearing retainer	26. Bearing
6. Drive shaft	13. Woodruff key	20. Oil seal	27. Drive shaft
7. Bearing	14. Bearing	21. Bearing retainer	28. Bevel gear

Fig. 131—Exploded view of the dual speed shifter located on right side of the clutch housing.

1. Shifter fork	5. Sleeve
2. "O" ring	6. Spring
3. Gasket	7. Lever
4. Bracket	8. Lever knob

into housing take care not to damage oil seal (8). Install cover so that the extraction holes are at 5 and 11 o'clock positions.

Reverse the removal procedure for the pinion shaft using original spacers. Remove the inspection cover (11) and check backlash using a dial indicator. Backlash should be 0.005 to 0.007 inch. Add or remove spacers (16) to obtain correct backlash.

191. R&R AND OVERHAUL (Counter-clockwise Rotation). To disassemble, remove pulley, bearing retainer (19—Fig. 132), withdraw pinion shaft (12) and bearings, then remove spacer (16) from housing. Remove cover (1), using two of the bolts in the extractor holes, bevel gear assembly and oil seal (8). Any further disassembly required will be obvious.

To reassemble, press the oil seal (8) with the lip facing inward, toward the bevel gear. Position bevel gear (28) on drive shaft (27). Press inner and outer bearings on drive shaft, install snap ring (25). Press the outer bearing and shaft assembly into the cover (23). Install original shims (24) on cover, then install cover and shaft assembly into housing, taking care not to damage oil seal (8). Position cover so that the extraction holes are at 5 and 11 o'clock positions. Reverse the removal procedure for the pinion shaft using original spacers. Remove the inspection cover (11) and check backlash using a dial indicator. Backlash should be at 0.005 to 0.007 inch. Add or remove spacers (16) to obtain correct backlash.

HYDRAULIC LIFT SYSTEM

The hydraulic lift system is composed of three basic units: A pressure loaded gear type pump which is driven from the engine timing gear train; a control valve mounted on front side of reservoir and the reservoir which incorporates the work cylinder, rockshaft and operating linkage.

The maintenance of absolute cleanliness of all parts is of the utmost importance in the operation and servicing of the hydraulic system. Of equal importance is the avoidance of nicks and burrs on any of the working parts.

LUBRICATION AND BLEEDING

All Models

192. Capacity of hydraulic reservoir

is 3.0 gallons and IH "Hy-Tran" fluid is recommended. If "Hy-Tran" fluid is not available, the following oils may be used. Below 32 degrees F, SAE-10 engine oil; 32 degrees F to 80 degrees F, SAE-20 engine oil and above 80 degrees, SAE-30 engine oil.

To bleed system, fill reservoir to bottom of filler hole then start engine and cycle system until all air is bled from system. Recheck reservoir fluid level and add fluid as required.

TROUBLESHOOTING

Series B-275

193. The following troubleshooting chart lists troubles which may be encountered in the operation and servicing of the Series B-275 hydraulic lift system. The procedure for correcting many of the causes of trouble is obvious, however, for those not so obvious, refer to the appropriate subsequent paragraphs.

Install a gage capable of registering at least 2500 psi in the lower plug (15—Fig. 138) hole before starting system check.

A. System unable to lift load. High gage pressure. Could be caused by:
　1. Isolating valve closed.
　2. System overloaded.
　3. Linkage restricted.
B. System unable to lift load or load raises slowly. Gage shows little or no pressure. Could be caused by:
　1. Suction filter clogged.
　2. Insufficient fluid in system.
　3. Faulty pump.
　4. Main valve not seating.
C. Erratic or sluggish operation. Could be caused by:
　1. Insufficient fluid in system.
　2. Suction filter clogged.
D. System will not hold load in raised position with control lever at neutral. Could be caused by:
　1. Incorrect non-return valve tappets setting.
　2. Leakage past piston or through thermal relief valve.
　3. "O" ring on isolating valve damaged.
　4. Oil leaking through non-return valve.
E. System will not lower. Could be caused by:
　1. Incorrect non-return valve tappet setting.
　2. Linkage restricted.
F. Excessive operating pressure. Could be caused by:
　1. Faulty or malfunctioning main valve.

Series B-414-424-444-2424-2444-354-364-384

194. The following troubleshooting chart lists troubles which may be encountered in the operation and servicing of the 424, 444, 2424, 2444, B-414, 354, 364 and 384 hydraulic lift system. The procedure for correcting many of the causes of trouble are obvious, however, for those not so obvious, refer to the appropriate subsequent paragraphs.

A. System will not lift with either draft or position control.
　1. System overloaded.
　2. Faulty relief valve.
　3. Regulator piston stuck open.
　4. Speed control piston stuck open.
　5. Faulty hydraulic pump.
B. System lifts slowly.
　1. System overloaded.
　2. Faulty hydraulic pump.
　3. Suction filter plugged.
　4. Speed control piston stuck in slow position.
　5. Faulty relief valve.
　6. Low oil level in reservoir.
C. System will not hold load.
　1. Piston seal leaking.
　2. Thermal relief valve in piston leaking (if so equipped).
　3. Control valve or cylinder "O" rings leaking.
　4. Internal parts of control valve leaking.

NOTE: Turn isolator valve in and if lift arms stop settling, leak is in control valve. If lift arms continue to settle, leak is in work cylinder.

D. System will lift with one control lever but not the other.
　1. Faulty or maladjusted control linkage.
　2. Position control plunger binding.
E. System noisy.
　1. Air in system.
　2. Suction filter plugged.
　3. Oil level in reservoir low.
F. System overheats.
　1. Relief valve operating continuously.
　2. High pressure line restricted.
　3. Air in system.
　4. Operating fluid contaminated.
G. Oil discharges through relief valve with lift arms fully raised.
　1. Regulator piston stuck in closed position.
　2. Control lever stop incorrectly adjusted.
　3. Orifice or orifice filter plugged.
H. Lift arms creep up when operating external cylinder.
　1. Isolator valve not fully closed.
　2. Isolator valve leaking.

TESTING

Series B-275

195. When testing, all connections and mounting surfaces should be clean and free of dirt or other foreign material. Remove filler plug and check fluid level. Fluid should reach bottom of filler hole. Refill if necessary. If fluid is excessively low, check level of oil in crankcase to see if there is leakage from pump into crankcase.

Remove lower plug from left end of control valve and install a gage capable of registering at least 2500 psi. Connect lower links to a weight of at least 1250 pounds and measure to see that the lower link ends are approximately 10 inches from the ground. Start engine, place control lever in raise position and check the time required for the system to reach full lift. This time should be approximately 1½ seconds.

With engine running, control lever in raised position and with load fully raised, the pressure gage should read 2050-2250 psi. A lower reading indicates an inefficient pump or a malfunctioning pressure relief valve.

With system in raised position, place control lever in "Hold" position and let unit set for about 3 minutes. During this 3 minutes the lower links should not drop more than 1/16-inch when measured at extreme ends. If the 1/16-inch drop is exceeded, oil may be leaking past the piston or through the non-return valve. Close the isolating valve and recheck. If lower links continue to fall, oil is leaking past the piston, the non-return valve "O" ring or the thermal relief valve which is incorporated into the piston. If the lower links remain stationary, oil is leaking through the non-return valve.

Start engine and place control lever in raise position. Allow engine to run at full throttle for about 3 minutes and check all connections and mounting surfaces for leaks. Return control lever to neutral and stop engine. Recheck reservoir level and crankcase level for evidence of leakage into engine crankcase. Remove pressure gage and reinstall lower plug in control valve.

Series B-414-424-444-2424-2444-354-364-384

196. When testing, all connections and mounting surfaces should be clean and free of dirt or other foreign material. Remove filler plug and check fluid level. Fluid should reach bottom of hole. Refill if necessary. If fluid is excessively low, check level of oil in engine crankcase to see if there is leakage of oil from pump into crankcase.

Prior to testing lift system, connect at least 1250 pounds of weight to hitch.

Open isolating valve and set the draft control (outside) lever to the deepest (full forward) position. Place speed control knob in the "Slow" lift position, slowly raise lift with position control lever until the distance between centerline of top pin on hitch lower link attaching plate and centerline of lift arm pin is 24½ inches, then move position control lever stop until it firmly contacts rear of lever and tighten stop in this position. Check this setting by lowering and raising the lift arms again with the position control lever. It should not be possible to raise the hitch an additional ½-inch by moving the draft control lever to the shallowest (rear) position. Return draft control lever to deepest position.

Install a gage capable of registering at least 3000 psi in the accessory plug hole (right hand plug) in the front side of control valve cylinder head. Now run engine at rated rpm and be sure position control lever is in full lower (forward) position. Close the isolator valve, then move the draft control lever to the shallowest position and note the gage reading as the relief valve opens. This reading should be 2300-2400 psi.

CAUTION: Do not maintain this test for more than 20 seconds as damage to hydraulic system could result.

If the observed gage reading is not as specified and the relief valve is operating, a faulty relief valve indicated and relief valve spring should be checked.

NOTE: Late production relief valves are shim adjusted. Installing one additional shim behind relief valve spring will increase relief pressure approximately 50 psi.

If the observed gage reading is not as specified and the pressure relief valve is not operating, either the flow control valve spool is stuck in the fully opened position or the hydraulic pump is faulty. Before removing gage, make a back pressure test as follows: Place draft control lever in deepest position and the position control lever in full down position and observe gage which should have a zero reading. If condition is not as stated, check for bent, damaged, worn or maladjusted linkage.

Hydraulic system operation can be further checked as follows: With isolator valve open and hitch in fully lowered position, place speed control knob in "Fast" position and move position control lever to lift position. Hitch should raise to full lift in not more than two seconds. Move the position control handle slowly forward and check to see that it is possible to control the rate of drop so that at least five seconds are

required for a full drop. Move the speed control knob to "Slow" position, move the position control lever to lift position and check the time required for the hitch to raise to full lift. This time should be three to four seconds.

Raise hitch to full lift position and close the isolator valve. Move the position control lever to lower position and check drop of hitch. The hitch should not drop more than 0.005 in thirty seconds. Open isolator valve.

Again place hitch in full lift position, if necessary, then move position control lever to mid-position. Check the hitch drop which should be not more than 0.005 in thirty seconds. Move the position control lever to lower position and check height of lower link ends above the ground. Lower link ends must be not more than five inches from ground.

Secure the position control lever stop about mid-way of the quadrant, fully raise hitch, then move position control lever forward until it contacts stop. Measure and record the height of lower link ends from ground. Now fully lower hitch, then raise it until the position control lever is in the same position and against the same side of the stop as it was when hitch was lowered. Again measure height of lift link ends from ground and if this measurement differs more than one inch from the first measurement, plunger spring (17—Fig. 143) is faulty and should be renewed.

PUMP

Series B-275

197. **R&R AND OVERHAUL.** To remove the hydraulic pump, remove the Allen screws from suction and pressure line manifolds. Remove the cap screws retaining pump to crankcase front cover plate and remove pump.

To disassemble the removed pump, straighten tab washer (2—Fig. 133), remove nut (1) and using a suitable puller, or a soft faced hammer, remove gear from shaft (14). Remove the eight cap screws and lift cover assembly (7) from pump body (16). At this time, note the location of pressure relief plate (10). Prior to removing same, remove "O" ring (8). Remove snap ring (3), oil seal washer (5) and oil seal (6).

NOTE: Bearings (12) may stay with cover or may remain in pump body. In either case they may be withdrawn by moving one out slightly in advance of the other. Be sure not to lose the small locking wires.

Drive shaft (14) and driven shaft (13) can now be removed. Disassembly can be completed on pumps as shown in Fig. 133 by removing name plate (19), snap rings (18), sealing plugs (17) and bumping out bearings (15). On those pumps which do not include items (17) and (18), bearings can be removed using a hooked tool.

With pump disassembled, inspect all parts for burrs, scoring, wear or other damage. Bearing faces can be refinished providing they are only lightly scored.

When reassembling, use new drive shaft oil seal and "O" rings. Install bearings as follows: Hold bearing together with flats in engagement and locking wires in position. Start bearings into their bores with one bearing slightly in advance of the other. Push bearings into bores until first bearing bottoms, rotate bearings in the direction of pump rotation, then press second bearing into bore until it bottoms. Be sure pressure relief plate and relief plate "O" ring are positioned on inlet side of pump body and prior to installing pump cover, place a straightedge across machined face of pump body and measure the distance between straightedge and pressure relief plate. This distance should not be less than 0.003 nor more than 0.0055. Various sized pressure relief plates are available to obtain this distance.

Balance of reassembly is obvious. Reinstall pump, then fill and bleed system as outlined in paragraph 192.

Series B-414-354-364-384 And Series 424-444-2424-2444 Diesel

198. **R&R AND OVERHAUL.** To remove the hydraulic pump, remove the Allen screws from suction and pressure line manifolds. Unbolt pump from engine front cover and remove pump.

To disassemble the removed pump, straighten tab washer (1—Fig. 134), remove nut and using a suitable puller, or a soft faced hammer, remove gear from shaft (8). Place a scribe line across length of pump, remove the four through-bolts and separate pump. The "O" rings (5) and (6) can now be removed from cover (12) and flange (4). Bearings (7) and the drive and driven gears and shafts (8 and 9) can now be removed from pump body (11) by gently bumping ends of shafts.

NOTE: When removing the bearing assemblies (7) be sure to note which side the cut-outs in bearings are positioned and be sure bearings are in the same position when reinstalling. The smallest cut-out must be on the pressure side of the pump.

With pump disassembled, inspect all parts for burrs, scoring, wear or other damage. Bearing faces can be refinished providing they are only slightly scored. Use all new "O" rings and drive shaft oil seal when reassembling.

Reassemble by reversing the disassembly procedure and be sure to align the previously affixed scribe line. Refer to Fig. 135 and note how "O" rings are installed in pump cover. "O" rings in pump mounting flange are installed in a similar manner. Reinstall pump, then fill and bleed the hydraulic system as outlined in paragraph 192.

199. Series 354, 364, 384 and late Series B-414 tractors, serial number 21196 and later, equipped with power steering, use the dual pump shown in Fig. 136. The removal and disassembly of this pump is similar to the procedure outlined in paragraph 198 for the single pump.

Series 424-444-2424-2444 Non-Diesel

200. **R&R AND OVERHAUL.** To remove the hydraulic pump, drain the hydraulic system and remove the two bolts securing the manifold flange to pump. Unbolt pump from engine crankcase and remove pump.

200A. To disassemble the Cessna pump, first remove nut, drive gear and key from pump shaft. Place a scribe line across length of pump to assure proper reassembly. Remove the two remaining cap screws, then bump pump shaft against a wood block to separate front plate (9—Fig. 137) from rear housing (1). Withdraw gear and shaft assemblies (3 and 12) and thrust plate (2) from rear housing. Using a sharp tool, pry diaphragm (4) from diaphragm seal (7). Identify the check ball hole and remove spring (11) and check ball (10). Lift nylatron gasket (5), protector gasket (6) and diaphragm seal (7) from front plate (9). Remove pump shaft oil seal (8).

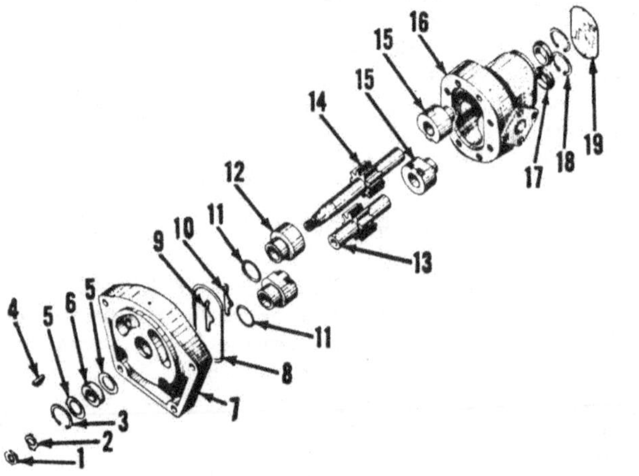

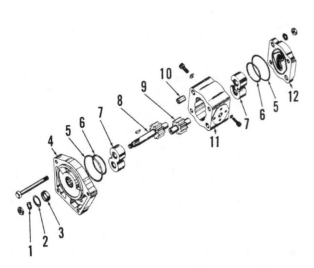

Fig. 133—Exploded view of the Series B-275 engine driven hydraulic pump. Some pumps do not include items (17) and (18).

1. Nut
2. Lockwasher
3. Snap ring
4. Woodruff key
5. Seal washer
6. Oil seal
7. Pump cover
8. "O" ring
9. "O" ring
10. Pressure relief plate
11. "O" rings
12. Bearings
13. Driven gear
14. Drive gear
15. Bearings
16. Pump body
17. Sealing plugs
18. Snap rings
19. Name plate

Fig. 134—Exploded view of hydraulic pump used on Series B-414, 354, 364 and 384 tractors not equipped with power steering and Series 424, 444, 2424 and 2444 diesel tractors.

1. Lockwasher
2. Snap ring
3. Oil seal
4. Pump flange
5. "O" ring
6. "O" ring
7. Bearing
8. Drive gear & shaft
9. Driven gear & shaft
10. Dowel
11. Pump body
12. End cover

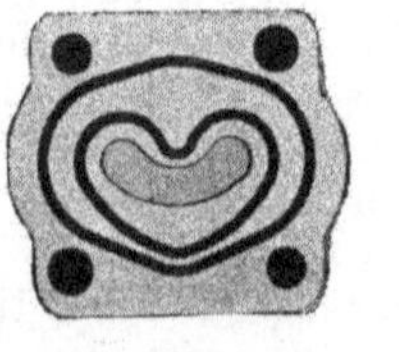

Fig. 135—Hydraulic pump cover showing installation of "O" rings. Pump flange will be similar.

Check all parts for burrs, scoring, wear or other damage.

4.5 GPM Pump

O.D. of shafts at
bushings . . . . .0.5605 min.
I.D. of bushings in body
and cover . . . . .0.5655 max.
Gear width . . . . .0.376 min.
I.D. of gear pocket in body . .1.404 max.

9 GPM Pump

O.D. of shafts at
bushings . . . . .0.5605 min.
I.D. of bushings in body
and cover . . . . .0.5655 max.
Gear width . . . . .0.748 min.
I.D. of gear pocket in
body . . . . .1.404 max.

When reassembling, use new diaphragm, nylatron gasket, protector gasket, diaphragm seal, thrust plate and shaft oil seal. These parts are available as a service package (IH part number 381002R92). With open part of diaphragm seal (7) towards front plate (9), work same into grooves of front plate using a dull tool. Press the protector gasket (6) and nylatron gasket (5) into the relief in the diaphragm seal. Install check ball (10) and spring (11) in front plate, then install diaphragm (4) with bronze face toward gears. Entire diaphragm must fit inside raised rim of diaphragm seal. Dip gear and shaft assemblies in oil and install them in front plate. Position thrust plate (2) in rear housing (1) with the bronze side toward gears and the half moon cut-out on inlet side of pump. Install rear housing over pump gears, then install cap screws. Lubricate the shaft seal (8) and carefully work seal over drive gear shaft. Seat the shaft seal by tapping with a plastic hammer. Check the pump rotation. Pump should have a slight amount of drag but should rotate evenly.

After reinstalling the pump drive gear, reinstall pump on tractor, then fill and bleed the hydraulic system as outlined in paragraph 192.

200B. To disassemble the Thompson pump, first remove the nut, drive gear and key from pump shaft. Place a scribe line across length of pump to assure proper reassembly. Remove the two remaining cap screws, then bump the pump shaft against a wood block to separate front cover from pump body. Remove gears and shafts, bearings and bearing spring plate, identifying all parts so they can be reassembled in the same position. Any further disassembly will be evident after examination of unit. Small nicks and/or scratches can be removed from body, cover, shafts, gears and bearings by using crocus cloth or a fine oil stone.

Check pump for excessive wear using the following specifications:

9 GPM Pump

O.D. of shafts at
bearings . . . . .0.624 min.
I.D. of bearing bore . . . . . . .0.632 max.
I.D. of gear pocket in
body . . . . .2.651 max.
Gear width . . . . .0.599 min.

When reassembling, lubricate all parts with clean Hy-Tran fluid, or its equivalent, and renew all seals and "O" rings. A hydraulic pump "O" ring and gasket package (IH part number 368634R92) is available for this pump.

Reinstall pump on tractor, then fill and bleed the hydraulic system as outlined in paragraph 192.

CONTROL VALVE

Series B-275

201. **R&R AND OVERHAUL.** To remove the control valve, drain the hydraulic system reservoir, then remove the quadrant strip. Disconnect

pressure line, and if so equipped, the remote control line, from the control valve. Unbolt and remove control valve. If necessary, use a screwdriver to separate control valve from lift cylinder as it is being removed.

Disassemble the removed control valve as follows: Refer to Fig. 138, remove cap screws from isolating valve (4) stop plate and unscrew isolating valve. Remove plug (6) or remote control line coupling, if so equipped. Remove plug (20), washer (19), spring (18) and dashpot piston (17). Remove plugs (15 and 31) and their washers. Loosen jam nut (35) on intercouple lever adjusting screw (34) and remove adjusting screw from intercouple lever. Place control lever in lift position so that intercouple lever is clear of non-return valve (10). Place a screwdriver through non-return valve hole, lift intercouple lever and at the same time place control lever in lower position. This will withdraw relief valve (27) from intercouple lever and allow intercouple lever to be removed. Remove the intercouple lever spring (13) then

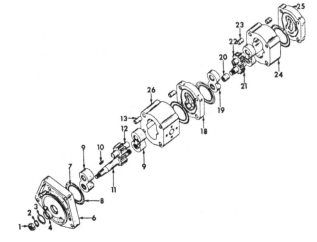

Fig. 136—Exploded view of dual hydraulic pump used on late Series B-414, 354, 364 and 384 tractors equipped with power steering.

1. Nut
2. Lock
3. Snap ring
4. Oil seal
6. Pump flange
7. "O" ring
8. "O" ring
9. Bearing
10. Key
11. Drive gear
12. Driven gear
13. Dowel pin
18. Center plate
19. Bearing
20. Coupling
21. Drive gear
22. Driven gear
23. Dowel pin
24. Rear body
25. Rear cover
26. Front body

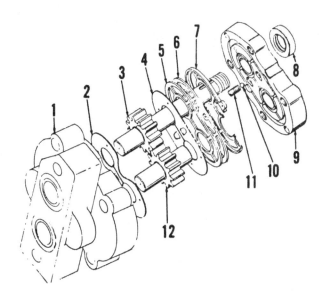

Fig. 137—Exploded view of the Cessna hydraulic pump used on Series 424, 444, 2424 and 2444 non-diesel tractors.

1. Rear housing
2. Thrust plate
3. Drive gear
4. Diaphragm
5. Nylatron gasket
6. Protector gasket
7. Diaphragm seal
8. Oil seal
9. Front plate
10. Steel ball
11. Spring
12. Idler gear

unscrew the non-return valve seat (12), valve (10), washer (11), spring (9) and stop plate (8). Remove end cover (21) and gasket (22), then remove pin (2) and withdraw control lever (1) from cam (23).

CAUTION: During the removal and disassembly of the relief valve assembly, care must be exercised to prevent bodily injury. Should the spring stirrup (25) spread open and release the relief valve, the valve will fly out with considerable force due to the spring

compression. DO NOT stand in line with relief valve assembly when working on same.

Pull relief valve assembly out of housing far enough to remove cam (23) from cam stirrup (24), then either install a 1¼-in. I.D. safety collar over spring stirrup (25) or wrap same with wire to insure against spring stirrup spreading and allowing relief valve to fly. Now pull relief valve assembly from housing.

Do not disassemble the relief valve

assembly unless necessary. However, should it prove to be necessary, proceed as follows: Support closed end of spring stirrup in a press, place a hollow mandrel or a piece of pipe over stem of relief valve and take up slack. Slide safety collar up over mandrel, or loosen wire wrapping, then depress spring and spread stirrup. Slowly release pressure and separate relief valve assembly.

Use a tool similar to that shown in Fig. 139 and remove relief valve seat (28—Fig. 138) and washer (29).

202. Thoroughly clean all parts and inspect valves and seats for scoring, pitting or undue wear. Inspect all springs for rust, distortion or fractures. Inspect all other parts for undue wear or damage. Relief valve spring can be checked against the following specifications.

Free length . . . . .3-7/8 in.
Test length . . . . .2¾ in.
Test load at 2¾ in. . . . . . . .231-245 lbs.

Approximate free lengths of the remaining springs shown in Fig. 138 are as follows: Intercouple lever spring (13), 7/8-inch; non-return valve spring (9), 11/16-inch; dashpot spring (18), 13/16-inch.

203. Reassemble by reversing the disassembly procedure, however, keep the following points in mind: If new valves and/or seats are being installed the mating faces should be lightly lapped to insure against leakage. Renew seat washers (11 and 29—Fig. 138) and all "O" rings. Cam (23) is installed with "V" cut-out toward outside of housing as shown in Fig. 140.

With unit assembled and prior to installing plug (15—Fig. 138), adjust intercouple lever adjusting screw (34) a as follows: Place control lever in neutral position, turn adjusting screw in until it just touches the non-return valve, then back-off screw ⅓-turn and tighten jam nut (35). Install plug (15).

204. Reinstall control valve to reser-

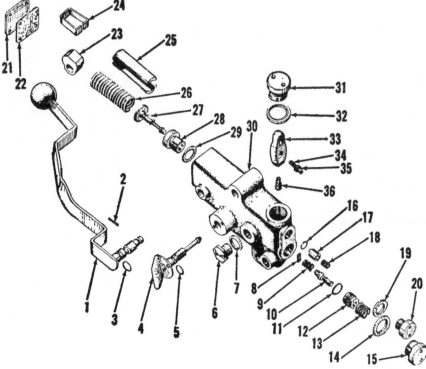

Fig. 138—Exploded view of the B-275 series hydraulic control valve. A remote coupling attachment may be installed instead of plug (6).

1. Control lever & shaft	10. Non-return valve	19. Washer	28. Valve seat
2. Pin	11. Copper washer	20. Housing plug	29. Copper washer
3. "O" ring	12. Valve seat	21. End cover	30. Valve body
4. Isolating valve	13. Spring	22. Gasket	31. Plug
5. "O" ring	14. Washer	23. Cam	32. Washer
6. Plug	15. Plug	24. Cam strirrup	33. Intercouple lever
7. Washer	16. Snap ring	25. Spring stirrup	34. Socket set screw
8. Valve stop	17. Dashpot piston	26. Relief valve spring	35. Jam nut
9. Spring	18. Spring	27. Relief valve	36. Ball end

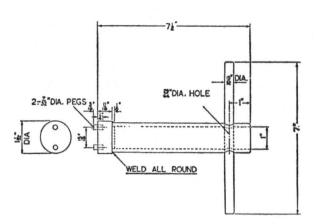

WELD ALL ROUND

Fig. 139—View showing dimensions of special tool used to remove relief valve seat of the control valve shown in Fig. 138.

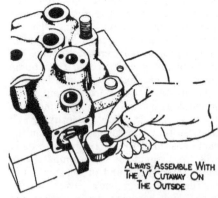

ALWAYS ASSEMBLE WITH THE "V" CUTAWAY ON THE OUTSIDE

Fig. 140—When installing cam in cam stirrup, position "V" cutout as shown.

voir, then fill and bleed system as outlined in paragraph 192.

CONTROL VALVE CYLINDER HEAD

Series B-414-424-444-2424-2444-354-364-384

205. **R&R AND OVERHAUL.** To remove the control valve cylinder head, either remove the complete hydraulic housing as outlined in paragraph 209, or drain housing, remove retaining cap screws, then raise front of housing far enough for control valve cylinder head to clear transmission top cover and support in this position with wood blocks. Remove the cap screws which retain the cylinder head and control valve to lift housing and separate cylinder head from control valve. Catch spacer (9—Fig. 143) and spring (10) as the units are separated.

With cylinder head removed, refer to Fig. 141 and proceed as follows: Remove adapter pressure line (coupling) and spacer (14). Remove snap ring (28) and pull plug (25) and "O" ring (26), speed control piston spring (22) and speed control piston (21) from bore. Unscrew latch (29) and remove speed control spool assembly (items 23, 24, 27, 29 and 30). Remove plug (3) and washer (4), then remove orifice and screen (5) and "O" ring (6).

NOTE: Orifice and screen assembly will generally remain in its position in the control valve flange.

Remove cap screws from isolating valve stop plate (8), then unscrew and remove isolating valve assembly (items 7 through 13). Auxiliary plug (1) need not to be removed unless cleaning is indicated. Any further disassembly required will be obvious, however, be sure to note and identify "O" rings (17, 18, 19 and 20) and their locations.

Inspect all parts for burrs, scoring, wear or other damage. Speed control piston should be a snug fit in its bore yet slide freely. Speed control piston spring should have a free length of 2.51 inches and should test 10 lbs. when compressed to a length of 1.625 inches.

Use all new "O" rings and reassemble by reversing the disassembly procedure. Reinstall cylinder head to control valve and hydraulic lift housing to rear frame, then fill and bleed the hydraulic system as outlined in paragraph 192.

CONTROL VALVE

Series B-414-424-444-2424-2444-354-364-384

206. **R&R AND OVERHAUL.** To

remove the control valve, either remove the complete hydraulic housing as outlined in paragraph 209 or drain housing, remove hydraulic housing retaining cap screws, then raise front of

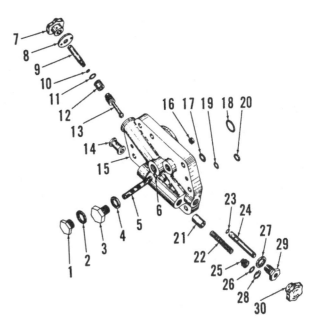

Fig. 141—Exploded view of hydraulic control valve cylinder head.

1. Plug
2. Seal washer
3. Plug
4. Seal washer
5. Orifice assembly
6. "O" ring
7. Knob
8. Stop plate
9. Shaft
10. "O" ring
11. "O" ring
12. Bushing
13. Isolator valve
14. Spool (spacer)
15. Cylinder head
16. Plug
17. "O" ring
18. "O" ring
19. "O" ring
20. "O" ring
21. Speed control piston
22. Piston spring
23. "O" ring
24. Speed control spool
25. Plug
26. "O" ring
27. Seal washer
28. Snap ring
29. Spool latch
30. Knob

Fig. 142—Control valve assembly and draft plunger linkage after removal from hydraulic lift housing.

housing and support with a wood block. Remove auxiliary valve cover plate from left front of lift housing and pull the return tube. Tube is internally threaded and a cap screw can be used to assist in pulling tube. Disconnect valve operating link from draft control shaft crankshaft. Disconnect draft link plunger return spring and on models so equipped, disconnect draft control spool safety chain. Remove the retaining cap screws and pull control valve, cylinder head and cylinder assembly from housing. See Fig. 142. Remove the valve operating link, radius arm and the draft link plunger assembly from control valve. Remove the two small cap screws and separate cylinder head from control valve which will release spacer (9—Fig. 143) and spring (10). Remove spring seat and lowering spool (11) from control valve flange (body). Remove plunger (19) and plunger spring (17). Remove the internal snap ring (25) and draft spool (24).

NOTE: To remove snap ring (25), disengage and work ends into bore which will cause snap ring to extend above end of bore where it can be grasped.

Remove snap ring (35) and withdraw plug (33) and "O" ring (34). Withdraw piston (31) and "O" ring (32). Use International Harvester tool FES 10-28, or its equivalent, and remove valve seat (29) and "O" ring (30). Remove flow control ball (28), ball rider (27) and spring (26). Unscrew relief valve spring housing (46) and remove relief valve spring (45), ball rider (44) and relief valve ball (43). Remove relief valve ball seat (41) and "O" ring (42), spacer (39), spring (40), non-return ball (38), ball seat (37) and copper washer (36).

At this point, sleeve (12), gland (14), sleeve (18), plug (22) and sleeve (23) are still in the valve body bores. They can be pressed from the bores if necessary. Sleeve (18) is shouldered and must be removed toward front of valve body. Be sure to renew all "O" rings if these parts are removed.

Cylinder (2) can be pressed from flange of valve body, if necessary. An International Harvester crankshaft rear oil seal driver (FES 6-15) or a piece of pipe of proper diameter can be used for this operation, however, take care not to damage the machined surface of valve body flange.

Clean all parts in a suitable solvent and blow out all oil passages. Use compressed air to dry parts. DO NOT use rags as lint could clog filters or lodge between valves and seats causing them to malfunction.

Inspect all parts for nicks, burrs or scoring. Nicks and burrs may be dressed with a fine stone, however, if any dressing is done, be extremely careful not to remove any of the sharp edges of spools or valve seats. No adjustments are available for this valve. Renew any parts which are the least bit doubtful and refer to the

following table for the spring specifications.

Lubricate parts prior to assembly, then reassemble and reinstall control valve by reversing the disassembly and removal procedures. Reinstall hydraulic lift housing, then fill and bleed hydraulic system as outlined in paragraph 192.

RESERVOIR AND COMPONENTS

Series B-275

The hydraulic lift system reservoir is fitted over the differential portion of the transmission case (rear frame) and

Spring	Call-out Fig. 143	Free Length	Test Lbs. at In.
Position plunger	17	1.038	7-9 at ¾
Flow control	26	1.439	15.8-17.5 at 1-1/32
Non-return valve	40	1-1/8 to 1¼	4.74 at 5/8
Position spool (B-414)	10	2-5/32	24-29.4 at 1-3/16
Position spool (all other models)	10	2.53	37 at 2
Relief valve	45	2.88	85.5-94.5 at 2.568

forms the top cover of that portion. Incorporated within the reservoir are the rockshaft assembly, work cylinder and piston, a spring loaded depth control, a by-pass filter and a suction filter. In addition, the hydraulic control valve is bolted to the left front face of the reservoir directly over the work cylinder. See Fig. 144 for an exploded view of reservoir and rockshaft assembly.

207. REMOVE AND REINSTALL. To remove the hydraulic reservoir unit, proceed as follows: Drain reservoir by disconnecting suction line hose. Disconnect quadrant strip from control lever

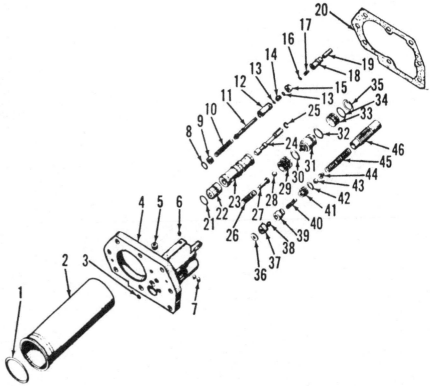

Fig. 143—Exploded view of the B-414, 354, 364, 384, 424, 444, 2424 or 2444 series hydraulic lift system control valve. Spacer (9) used on early Series B-414 units.

1. "O" ring	13. "O" ring	25. Snap ring
2. Cylinder	14. Spool gland	26. Spring
3. "O" ring	15. Plug	27. Ball rider
4. Valve body	16. "O" ring	28. Flow control valve ball
5. Pipe plug	17. Spring	29. Seat
6. Pipe plug	18. Sleeve	30. "O" ring
7. Pipe plug	19. Plunger	31. Piston
8. "O" ring	20. Gasket	32. "O" ring
9. Spacer	21. "O" ring	33. Plug
10. Spring	22. Plug	34. "O" ring
11. Lowering spool	23. Sleeve	35. Snap ring
12. Sleeve	24. Draft spool	

36. Copper washer	
37. Seat	
38. Non-return valve ball	
39. Sleeve	
40. Spring	
41. Seat	
42. "O" ring	
43. Relief valve ball	
44. Ball rider	
45. Relief valve spring	
46. Spring housing	

quadrant. Disconnect pressure line and if so equipped, the remote line coupling, from control valve. Unbolt and remove control valve and if necessary, use a screwdriver to separate control valve from work cylinder. Remove seat. Disconnect lift linkage from rockshaft arms and reservoir, then unbolt and remove the reservoir and rockshaft unit.

208. OVERHAUL. To disassemble and/or overhaul the reservoir and rockshaft unit after it has been removed as outlined in paragraph 207, proceed as follows: Unbolt and remove control lever quadrant, then pull suction filter and attached pipe from housing. Use a cap screw, if necessary, and pull the by-pass filter from housing, then remove piston (27—Fig. 144) from cylinder (19). Remove cylinder from housing. Unbolt depth control support (38) from housing, turn unit until key on spring housing aligns with slot in housing, then remove unit. Remove cap screws and retainers (15), then remove lift arms (9) from rockshaft (6). Turn unit over so cover plate (2) is on top side, then remove cover plate and gasket. Pull cotter pin from connecting rod pin (4) and remove pin and connecting rod (17). Remove clamping bolts from depth control cam (5) and rocker arm (3), then using a lead hammer, drive rockshaft from housing.

NOTE: During removal of the rockshaft, the oil seal (14), washer (12) and in most cases, bushing (11) will be pushed out as rockshaft emerges. Oil seal, washer and bushing can be removed from opposite side after rockshaft is out.

Any further disassembly required will be evident upon examination of the units. If thermal relief valve is removed from piston, be extremely careful not to lose the small steel ball (30). Free length of spring (28) is 11/16-inch and spring should test 12.4 lbs. when compressed to a length of ½-inch.

Clean all parts and examine for any undue wear or damage. It is always good policy to renew bushings (11) when housing is stripped as these bushings receive the greatest amount of wear and are the least accessible.

When reassembling the reservoir and rockshaft assembly, use all new "O" rings, gaskets and oil seals. The only seal that should be considered for reuse is the piston seal (13) and then only if it shows no signs of wear or damage. To install oil seals (14) use special tool provided by International Harvester, or a piece of shim stock and be sure seals fit over major diameter of rockshaft.

Reassemble reservoir and rockshaft

unit by reversing disassembly procedure, reinstall on tractor, then fill and bleed system as outlined in paragraph 192.

Series B-414-424-444-2424-2444-354-364-384

The hydraulic reservoir (Fig. 145) is fitted over the differential portion of the transmission case (rear frame) and forms the top cover of that portion.

Incorporated within the reservoir are the rockshaft assembly, work cylinder and piston, the control valve assembly, the position and draft control operating and sensing linkage and a suction filter. See Fig. 146 for a view of the position and draft control operating and sensing linkage.

209. REMOVE AND REINSTALL. To remove the rockshaft and reservoir

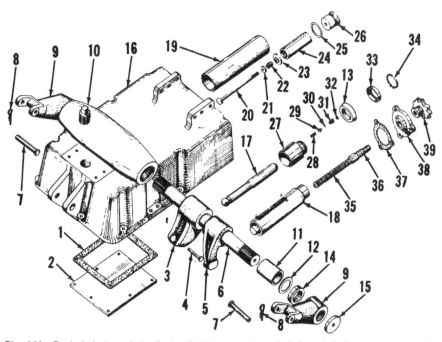

Fig. 144—Exploded view of the Series B-275 reservoir, rockshaft and their component parts. Items 28 through 32 comprise the thermal relief valve.

1. Gasket	11. Bushing	20. Bolt	30. Ball
2. Cover plate	12. Washer	21. Washer	31. Plug
3. Rocker arm	13. Piston seal	22. Spring	32. "O" ring
4. Pin	14. Oil seal	23. Spring seat	33. Seal retainer
5. Depth control cam	15. Retainer	24. By-pass filter	34. Snap ring
6. Rockshaft	16. Housing	25. "O" ring	35. Screw
7. Pin	17. Connecting rod	26. Filter body	36. Seal ring
8. Cotter pin	18. Depth control spring	27. Piston	37. Gasket
9. Lift arm	assembly	28. Spring	38. Support
10. Filler plug	19. Cylinder	29. Ball rider	39. Hand wheel

Fig. 145—Typical hydraulic lift reservoir used on Series B-414, 354, 364, 384, 424, 444, 2424 and 2444. Refer to Fig. 146 for control linkage and Fig. 148 for the rockshaft assembly.

1. Suction filter
2. Connector
3. Sealing washer
4. "O" ring
5. Lift housing (reservoir)
6. Gasket
7. Top cover
8. Filler & level plug
9. "O" ring
10. Return tube
11. Gasket
12. Auxiliary valve cover
13. Gasket
14. Bottom cover
15. Gasket
16. Filter retainer
17. Control valve assembly

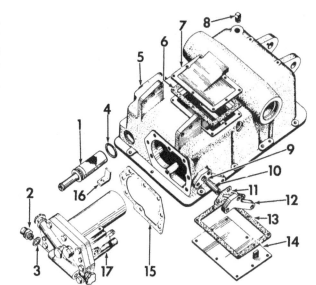

(lift unit) assembly, first drain reservoir by removing drain plug at right rear of housing, or by loosening clamps of suction line coupling hose and sliding hose forward on suction line. Disconnect pressure line, and if so equipped, the remote cylinder line from the control valve cylinder head. Disconnect lift links from rockshaft lift arms and the upper link from unit control bellcrank. Remove seat. Remove the top cover plate (7—Fig. 145) and disconnect the safety chain (46—Fig. 146) on models so equipped. Remove reservoir retaining cap screws, attach hoist to unit and lift same from rear frame.

When placing unit on work bench, be sure draft spring assembly is beyond edge of bench, or support unit with blocks.

210. **OVERHAUL.** Overhaul of all control and sensing linkage except the draft control spindle and spring assembly can be accomplished without removing the hydraulic lift unit from tractor.

To remove the rockshaft and/or rockshaft bushings, rocker arm or rockshaft cam, refer to paragraph 213.

For information relating to control valve cylinder head, refer to paragraph 206.

211. QUADRANT, LEVERS AND SHAFTS. The quadrant, control levers and control lever shafts can be removed and serviced as follows: Drain reservoir and remove seat. Remove housing top cover. Disconnect position control link (21–Fig. 146) from position control sleeve (16), then unbolt quadrant bracket from housing and pull quadrant and control levers and shafts from housing.

To disassemble unit, remove quadrant strip from quadrant and nut from outer end of draft control shaft. Remove levers (3 and 7), friction discs (4 and 6) and locating washer (5). If necessary, remove quadrant from bracket. Remove snap ring (8) from outer end of position control sleeve, pull shaft assembly from bracket, then remove "O" ring (12) from inner bore of bracket. Use a press, or a valve spring compresser, and pressing on end of draft control shaft (27), compress spring (22) enough to allow removal of snap ring (23). Carefully release pressure and pull draft control shaft from position control sleeve.

The position control tube (19) can be removed from housing after removal of the cross tube pivots (17).

Any further disassembly required will be obvious. Renew any damaged, worn or bent parts.

Use new "O" rings and gaskets and reassemble by reversing the disassembly procedure. Be sure to mate the correlation marks of control levers and shafts when installing levers on shafts

and enter pin on draft control crankshaft in its hole in the draft link (29) when installing quadrant assembly.

212. DRAFT CONTROL LINKAGE. To completely service the draft control linkage will require removal of lift unit as outlined in paragraph 209.

Remove lock plate (31–Fig. 146), then pull pivot pin (32) and remove bellcrank (33). Remove locknut from top of trunnion (34), then unscrew spindle (41) from trunnion. All springs and spring seats are now free. Remove the spindle oil seal from bore in bottom of housing.

Inspect all parts for damage or undue wear and renew as necessary. Spring and pivot pin specifications are as follows: Upper draft control spring has a free length of 2.170 inches and should test 1000 lbs at 1.937 inches. Lower draft control spring has a free length of 5-3/16 inches and should test 3100-3500 lbs. when compressed to a length of 4-11/16 inches. Pivot pin (32) has a diameter of 0.996-1.000 and pivot pin hole in bellcrank has an inside diameter of 1.002-1.005 which gives the pivot pin a normal operating clearance of 0.002-0.009.

Reinstall and adjust the draft control spring assembly as follows. Install new

oil seal in spindle bore in bottom of housing. Place spring seat (39), spring (35) and trunnion (34) in their position on housing, then insert spindle (41) and screw it into trunnion until shoulder on spindle contacts spring seat (39) and all slack is taken out of assembly but DO NOT compress the spring. Install and tighten locknut. Install top spring seat (42), spring (43), bottom spring seat and retaining nut. Tighten nut until all slack is taken out of assembly but DO NOT compress spring. Install cotter pin.

NOTE: It is important that all slack be removed from spring assemblies but the springs must not be compressed.

Mate slots of bellcrank and pins of trunnion and install bellcrank pivot pin and lock plate.

With unit assembled as outlined, place draft control lever in its deepest position and measure distance between end of draft link plunger (38) and operating pad of bellcrank. This distance should be 3/8 to 18/32-inch as shown in Figure. 147. If measurement is not as stated, bring to correct dimension by adding flat washers as shown.

Reinstall life unit on tractor by reversing the removal procedure.

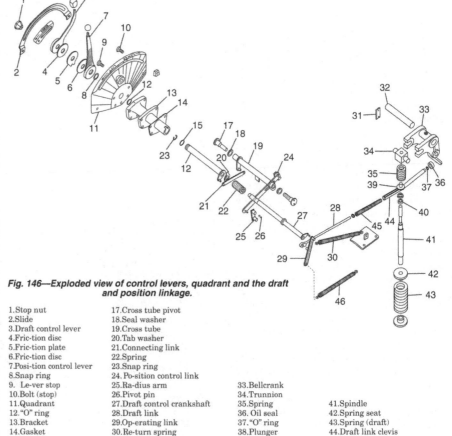

Fig. 146—Exploded view of control levers, quadrant and the draft and position linkage.

1.Stop nut	17.Cross tube pivot	
2.Slide	18.Seal washer	
3.Draft control lever	19.Cross tube	
4.Fric-tion disc	20.Tab washer	
5.Fric-tion plate	21.Connecting link	
6.Fric-tion disc	22.Spring	
7.Posi-tion control lever	23.Snap ring	
8.Snap ring	24. Po-sition control link	
9. Le-ver stop	25.Ra-dius arm	33.Bellcrank
10.Bolt (stop)	26.Pivot pin	34.Trunnion
11.Quadrant	27.Draft control crankshaft	35.Spring	41.Spindle
12."O" ring	28.Draft link	36.Oil seal	42.Spring seat
13.Bracket	29.Op-erating link	37."O" ring	43.Spring (draft)
14.Gasket	30.Re-turn spring	38.Plunger	44.Draft link clevis
15."O" ring	31.Lock plate	39.Spring seal	45.Draft link spring
16. Po-sition control linkage	32.Bellcrank pivot pin	40. Oil seal	46.Safety chain

213. ROCKSHAFT. To remove the rockshaft (6—Fig. 148), remove the lift unit as outlined in paragraph 209. Remove both lift arms from rockshaft. Turn unit bottom side up, remove housing bottom cover and disconnect draft plunger link return spring. On models so equipped, disconnect draft control spool safety chain. Remove the connecting rod. Remove lock bolt from rocker arm, then bump rockshaft from left to right and remove rockshaft from right side of housing. Rocker arm (7) and rocker cam (8) can be lifted from housing.

NOTE: Rockshaft right hand bushing, washer and oil seal will come out with rockshaft and left hand bushing and oil seal assembly can be driven out after rockshaft is out.

Inspect all parts for damage and undue wear. Renew parts as necessary. Rockshaft and rockshaft bushings specifications are as follows: Rockshaft bushing journal diameter, 2.248-2.250. Rockshaft bushings inside diameter, 2.251-2.253. This provides a normal operating clearance of 0.001-0.005 for

the rockshaft.

If necessary, the connecting rod bushing (9), located in the rocker arm, can be renewed at this time.

When reassembling, coat parts with oil, start rockshaft into housing at right side and align master spline of rockshaft with master splines of rocker arm and rocker cam. Rocker cam must be installed with chamfered bore side next to rocker arm. Install bushings and oil seals after rockshaft is positioned and use a seal driver when installing seals to insure against seal damage.

Reinstall unit on tractor by reversing the removal procedure. Fill and bleed hydraulic system as outlined in paragraph 192.

214. WORK CYLINDER AND PISTON. To service the work cylinder and piston, remove the control valve as outlined in paragraph 206. With control valve out, remove valve cylinder head and bump piston from cylinder head. On Series B-414 and early Series 424 and 2424, remove snap ring, seal retainer and seal from piston. On late Series 424 and 2424 and all Series 354, 364, 384, 444 and 2444, an "O" ring (12—Fig. 148) and a back-up ring (13) are used for the piston seal.

NOTE: On all tractors except Series B-414 having serial numbers 727 to 2160 (non-diesel) and 8106 to 16650 (diesel), a thermal relief valve assembly was incorporated in the work piston.

Removal of the relief valve assembly is obvious, however, be extremely careful not to lose the small 3/32-inch steel ball. Free length of relief valve spring is 11/16-inch and spring should test 12.4 lbs. when compressed to a length of ½-inch.

The cylinder can be pressed from flange of control valve, if necessary, by placing rear of cylinder on a block and using an International Harvester crankshaft rear oil seal driver (FES 6-15), or a piece of pipe of proper diameter, positioned on valve flange over cylinder bore. Use caution during this operation not to mar the machined surface of the valve flange. Cylinder inside diameter is 3.000-3.006 for new cylinder.

REMOTE CONTROL VALVE

Series B-275-B-414

The Series B-275 and B-414 tractors have available a double acting remote control valve which mounts on the scuttle panel (lower instrument panel) and provides hydraulic control for trailed implements.

215. **R&R AND OVERHAUL.** Removal of the remote control valve is obvious.

To overhaul the removed valve, refer to Fig. 149 and proceed as follows: Disconnect and remove link (12) from valve spool and lever, then remove lever pivot pin and separate lever from valve housing. Remove cap (1), snap ring (2) and stop disc (3). Pull spool and centering spring assembly from valve body. Remove screw (3A) and lock washer, stop collar (4), centering spring (5) and stop washer (6) from end of valve spool, then remove spool seals (7) from spool bore. Unscrew relief valve body (15) and remove body, relief valve spring (18), spring guide (19) and ball (20) from valve body. DO NOT lose shims (17) which are located between outer end of relief valve spring and bottom of spring bore in relief valve body. Unscrew relief valve seat (21) and remove seat, "O" ring (22) and

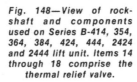

Fig. 147—After assembly, draft control spring and spindle must be adjusted as shown.

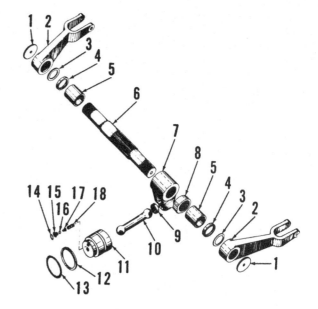

Fig. 148—View of rockshaft and components used on Series B-414, 354, 364, 384, 424, 444, 2424 and 2444 lift unit. Items 14 through 18 comprise the thermal relief valve.

1. Retaining plate
2. Lift arm
3. Washer
4. Oil seal
5. Bushing
6. Rockshaft
7. Rocker arm
8. Cam
9. Bushing
10. Connecting rod
11. Piston
12. "O" ring
13. Back-up ring
14. "O" ring
15. Relief plug
16. Steel ball
17. Ball rider
18. Spring

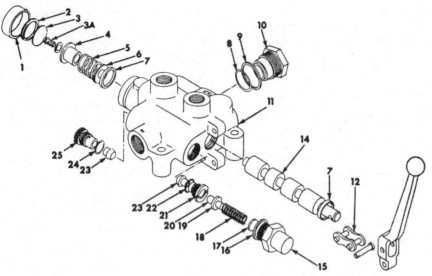

1. Bonnet cap	14. Valve spool
2. Snap (stop) ring	15. Relief valve body
3. End cap	16. Gasket
4. Collar stop	17. Shim
5. Centering spring	18. Spring
6. Stop washer	19. Spring guide
7. Spool seal	20. Check ball
8. Seal	21. Relief valve seat
9. Gasket	22. "O" ring
10. Sleeve	23. Poppet
11. Valve body	24. "O" ring
12. Connecting link	25. Plug

Fig. 149—Exploded view of the remote control valve which can be mounted on the scuttle plate (lower instrument panel) on Series B-275 and early Series B-414 tractors.

follows: Remove plug (23), then using a valve spring compressor, compress centering spring (20) and drive out roll pin (9A). Remove centering spring and spring seats then withdraw spool assembly from top of valve body. Drive out roll pin (9) and remove ball (4), rider (5), spring (6), spring seat (10) on double acting valves only (Fig. 150), and spool end (8) from spool (3). Unscrew relief valve housing (16) and remove relief valve spring (14), rider (13), ball (12) and seat (11). Using a screwdriver, pry out seal (1).

Clean and inspect all parts for excessive wear, scoring or scratches. If spool (3) or bore in valve body shows evidence of scoring or excessive wear, both parts must be renewed as they

poppet (23). Unscrew plug (25) and remove plug, "O" ring (24) and poppet (23). Any further disassembly is obvious.

Wash all parts in a suitable solvent and inspect. Renew parts as necessary. Spool (14) and valve body (11) are available only as a matched set.

Relief valve spring should have a free length of 2.135 inches and should test 225-275 lbs. when compressed to a length of 1¾ inches.

Lubricate parts and reassemble by reversing disassembly procedure. Relief valve pressure of 2050-2250 psi for Series B-275 or 2200-2400 psi for Series B-414 is adjusted by varying the number of shims (17). Relief pressure can be checked by installing a pressure gage in one of the break-away couplings and directing oil to the gage by the control valve.

Series 424-444-2424-2444-B-414-354-364-384

216. **R&R AND OVERHAUL.** Removal of the auxiliary valve is obvious.

To overhaul the removed double or single acting control valves, refer to Fig. 150 and 151 and proceed as

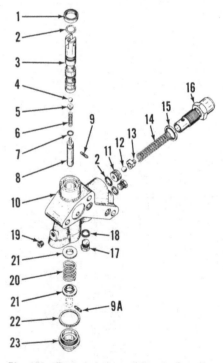

1. Oil seal	12. Steel ball
2. "O" ring	13. Ball rider
3. Valve spool	14. Relief valve spring
4. Steel ball	15. "O" ring
5. Ball rider	16. Relief valve housing
6. Spring	17. Washer
7. "O" ring	18. Spring seat
8. Spool end	19. Spring seat
9. Roll pin	20. Return spring
9A. Roll pin	21. "O" ring
10. Spring seat	22. Valve body
11. Relief valve seat	23. Plug

Fig. 150—Exploded view of the double acting auxiliary control valve used on Series 354, 364, 384, 424, 444, 2424, 2444 and late Series B-414 tractors.

1. Oil seal	12. Ball
2. "O" ring	13. Ball rider
3. Valve spool	14. Relief valve spring
4. Ball	15. Washer
5. Ball rider	16. Spring housing
6. Spring	17. Screw
7. "O" ring	18. Washer
8. Spool end	19. Pipe plug
9. Roll pin	20. Return spring
9A. Roll pin	21. Spring seat
10. Valve body	22. Fibre washer
11. Relief valve seat	23. Plug

Fig. 151—Exploded view of the single acting independent auxiliary valve used on Series 384.

are available only as a matched set.

Relief valve spring free length is 2-27/32 inches and should exert 85-95 lbs. pressure when compressed to a length of 2.568 inches.

When reassembling, dip all parts in clean Hy-Tran oil and renew seal (1) and all "O" rings. The spool (3) must be installed in top of valve body to prevent damage to "O" ring (2). Install seal (1) after spool is installed in valve body.

Reinstall valve on tractor, then fill and bleed hydraulic system as outlined in paragraph 192.

NOTES

INTERNATIONAL HARVESTER

Models ■ 330 ■ 340 ■ 504 ■ 2504

Previously contained in I&T Shop Manual No. IH-23

SHOP MANUALS

SHOP MANUAL
INTERNATIONAL HARVESTER

SERIES 330-340-504-2504

Engine serial number is stamped on left side of engine crankcase. Engine serial number will be preceeded by engine model number. Suffix letters to engine serial number are as follows:

U. High Altitude Engine

V. Exhaust Valve Rotators

Tractor serial number is stamped on name plate attached to right side of clutch housing. Suffix letters to tractor serial numbers indicate following attachments:

J. Rockford Clutch

P. Independent PTO Drive

R. "Torque Amplifier" With Provision for Transmission Driven PTO

S. "Torque Amplifier" With Provision for Independent PTO

T. Cotton Picker Mounting Attachment (Low Drum)

U. High-Altitude

V. Exhaust Valve Rotators

W. Forward and Reverse Drive

X. High Speed Low and Reverse

Y. Hydraulic Power Supply (12 gpm pump)

Z. Hydraulic Power Supply (17 gpm pump)

FF. Hydraulic Power Supply (4.5 gpm pump)

GG. Hydraulic Power Supply (7.0 gpm pump)

INDEX (By Starting Paragraph)

CONDENSED SERVICE DATA

	Series 330-340 Non-Diesel	Series 340 Diesel	Series 504, 2504 Non-Diesel	Series 504, 2504 Diesel
GENERAL				
Engine Make	Own	Own	Own	Own
Engine Model	C-135	D-166	C-153	D-188
Number of Cylinders	4	4	4	4
Bore—Inches	$3\,^1/_4$	$3\,^{11}/_{16}$	$3\,^3/_8$	$3\,^1/_{16}$
Stroke—Inches	$4\,^1/_{16}$	$3\,^7/_8$	$4\,^1/_4$	$4\,^{25}/_{64}$
Displacement—Cubic Inches	135	166	153	188
Pistons Removed From	Above	Above	Above	Above
Main Bearings, Number of	3	3	3	3
Cylinder Sleeves	Wet	Dry	None	Dry
Forward Speeds, With T.A.	10	10	10	10
Forward Speeds, Without T.A.	5	5	5	5
Generator and Starter Make	---------Delco-Remy---------			
TUNE-UP				
Firing Order	1-3-4-2	1-3-4-2	1-3-4-2	1-3-4-2
Valve Tappet Gap (Hot)	*	0.027	*	0.027
Inlet and Exhaust Valve Seat Angle (Degrees)	45	45	45	45
Ignition Distributor Make	IH	...	IH	...
Ignition Distributor Symbol	------See Paragraph 130------			
Breaker Gap	0.020	...	0.020	...
Timing, Retard	TDC	...	TDC	...
Distributor Timing, Advanced	30° BTDC	...(2)	...	
Magneto Timing, Advanced	35° BTDC	...	...	
Timing Mark Location	-------Crankshaft Pulley------			
Spark Plug Electrode Gap	0.025	...	0.023G, 0.015LP	...
Battery Terminal Grounded	Pos.	Neg.	Neg.	Neg.
Engine Low Idle rpm	425	650	425	700-750
Engine High Idle rpm No Load	2200	2180	2500	2400
Engine Rated rpm	2000	2000	2200	2200

SIZES—CAPACITIES—CLEARANCES
(Clearances in thousandths)

	Series 330-340 Non-Diesel	Series 340 Diesel	Series 504, 2504 Non-Diesel	Series 504, 2504 Diesel
Camshaft Journal Diameter, No. 1	1.8115	2.1095	1.8115	2.1095
Camshaft Journal Diameter, No. 2	1.5775	2.0895	1.5775	2.0895
Camshaft Journal Diameter, No. 3	1.4995	2.0695	1.4995	2.0695
Camshaft Diametral Clearance	.9-5.4	.5-5	.9-5.4	.5-5
Camshaft End Play	3-12	2-10	3-12	2-10
Crankpin Diameter	(1)	2.3735	2.0595	2.3735
Rod Bearing Diametral Clearance	(3)	.9-3.4	(3)	.9-3.4
Crankshaft Main Journal Diameter	(4)	2.7485	(4)	2.7485
Main Bearing Diametral Clearance	.9-3.9	1.2-4.2	.9-3.9	1.2-4.2
Crankshaft End Play	**	5-13	6-10	5-13
Piston Skirt Diametral Clearance	1.1-1.9	4-5.6	1-2	4-5.6
Piston Pin Diameter	0.859	1.1248	0.859	1.1248
Valve Steam Diameter	0.341	0.372	0.341	0.372
Cooling System Capacity, Qts.	14 (330) 15 (340)	18 $^1/_2$ ...	15 ...	18.6 ...
Crankcase Oil, Qts.	6 (330) 5 (340)	7 ...	5 ...	7 ...
Transmission and Differential, Gals	7 (330) 10 (340)	10 ...	12 ...	12 ...
PTO Rear Unit, Qts.— Planetary Type	2	2	...	...
Clutch Type	$^1/_2$	$^1/_2$	2	2

TIGHTENING TORQUES—Ft.-Lbs.

	Series 330-340 Non-Diesel	Series 340 Diesel	Series 504, 2504 Non-Diesel	Series 504, 2504 Diesel
Cylinder Head Bolts	80-90	110-120	70-75	110-120
Connecting Rod Nuts	43-49	...	...	...
Connecting Rod Bolts	40-45	40-50	40-45	40-50
Main Bearing Bolts	75-80	75-85	75-80	75-85
Camshaft Nut (Diesel)	...	110-120	...	110-120
Injection Nozzle Hold Down Screws	...	20-25	...	20-25
Injection Nozzle Fitting	...	45-50	...	45-50

(1) 1.809-1.810 for C-135 engines below engine serial number 100501; 2.059-2.060 for C-135 engines serial number 100501 and up. (2) 22 degrees for gasoline engines prior to engine number 4484, 20 degrees for gasoline engines 4484 and up, 25 degrees for LP-Gas engines. (3) 0.0009-0.0034 for C-135 engines prior engine number 100501, 0.0009-0.0039 for C-135 engine 100501 and up and all C-153 engines. (4) 2.244-2.245 for C-135 engines prior to engine number 100501, 2.6235-2.6245 for C-135 engines number 100501 and up and all C-153 engines. *See paragraph 50. ** See paragraph 66.

FRONT SYSTEM TRICYCLE TYPE

SINGLE WHEEL
Farmall 340-504

1. The single front wheel is mounted in a fork which is retained to the steering worm wheel shaft by four stud nuts.

On series 340 and 504 having a 7.50 x 10 tire, male and female wheel halves (Fig. IH1300) are used to accommodate the tire and wheel halves are fitted with bushings (8). Replacement wheel halves are factory fitted with bushings; or, the bushings are available separately as repair parts. Ream bushings after installation, if necessary, to provide a free fit for the axle (10). Side play of wheel is adjusted by nuts (2) and nuts are locked in adjustment by locks (3).

On series 504 fitted with a 7.50 x 16 tire, a solid, drop center type wheel (Fig. IH1300A) is used and hub (12) is fitted with tapered roller bearings. Wheel bearings (8) should be adjusted to a slight preload using adjusting nut (15). Lock adjusting nut with jam nut (17) after adjustment is complete.

Fig. IH1300 — Exploded view of the single front wheel unit which is available on Farmall 340 and 504 tractors.

1. Fork
2. Adjusting nuts
3. Lock washers
4. Dust shields
5. Felt washers
6. Oil seals
8. Bushings
10. Axle

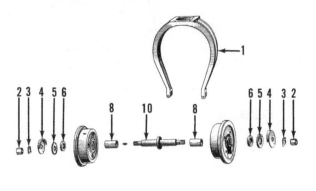

Fig. IH1300A — Farmall 504 tractors have available a solid, drop center, single front wheel as shown.

3. Jam nut
4. Lock washer
5. Shield
6. Axle
7. Oil seal
8. Bearing cone
9. Oil seal retainer
10. Bearing cup
11. Grease retainer
12. Hub
15. Adjusting nut
16. Spacer
17. Jam nut
18. Shield

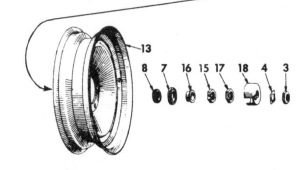

DUAL WHEELS
Farmall 340-504

2. The bolster for tricycle type dual front wheels is retained to the steering worm wheel shaft by four stud nuts. The bolster (2—Fig. IH1301) and axle (1) are available as a pre-riveted assembly; or, the bolster and axle are available as individual repair parts. Wheel bearings (5, 7, 9 & 10) should be adjusted to a slight preload by turning the bearing adjusting nut (12).

Fig. IH1301 — Exploded view of the dual front wheel unit which is available on Farmall 340 and 504 tractors.

1. Axle
2. Bolster
3. Dust shield
4. Oil seal
5. Bearing cone
6. Oil seal wear ring
7. Bearing cup
8. Hub
9. Bearing cup
10. Bearing cone
11. Washer
12. Adjusting nut
13. Gasket
14. Cap

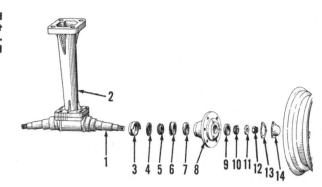

FRONT SYSTEM AXLE TYPE

AXLE MAIN MEMBER
Farmall 340-504

3. On Farmall 340 and 504 tractors equipped with an adjustable wide tread front axle, refer to Fig. IH1302. The axle main member (20) pivots on pin (22) which is pinned in the pivot bracket (17). The pivot pin bushing (19), which is pressed into the main member, should be reamed after in-

stallation, if necessary, to provide a free fit for the pivot pin. On Farmall 504 models with adjustable wide tread front axle, disconnect power steering cylinder prior to removing front axle.

International 330-340-504-2504

4. International 330, 340, 504 and 504 High-Clearance tractors are available with an adjustable front axle (Fig.

IH1303). A non-adjustable axle (Fig. IH1304) is optionally available for International 330, 340 and 504 tractors but is standard equipment on 2504 tractors. On both types of axles, the stay rod is welded to the axle main member and is not available separately. On series 330 and 340, the 1.240-1.242 diameter front axle pivot pin (5—Fig. IH1303 or 1304) is car-

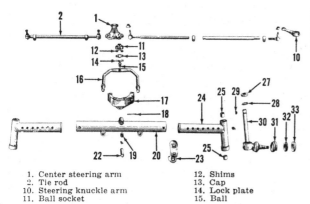

16. Radius (stay) rod
17. Axle pivot bracket
18. Pin
19. Bushing
20. Axle main member
22. Pivot pin
23. Clamp
24. Axle extension
25. Bushings
27. Thrust bearing
28. Felt washer
29. Woodruff key
30. Steering knuckle
31. Dust shield
32. Dust seal
33. Washer

1. Center steering arm
2. Tie rod
10. Steering knuckle arm
11. Ball socket
12. Shims
13. Cap
14. Lock plate
15. Ball

tween the pivot ball and socket can be adjusted by the addition or subtraction of shims (12). The radius rod (16) can be detached from the axle main member (20).

International 330-340-504-2504

6. The radius rod pivot ball is not available for service; however, the clearance between the pivot ball and socket can be adjusted by the addition or subtraction of shims (3-Fig. IH1303 or 1304). The radius rod is welded to the axle main member and is not available for service.

TIE RODS AND DRAG LINKS
All Models

7. Procedure for removing the tie rod or rods, tie rod tubes, tie rod ends and/or drag links is self-evident after an examination of the unit. The tie rod and drag link ends on all models are of the automotive type and are not adjustable for wear.

Adjust the toe-in of the front wheels to $\frac{3}{16}$-⅜-inch for International 330, 340, 504 and 2504 tractors, ⅛-¼-inch for 340 and 504 Farmall tractors. Adjustment is made by varying the length of the tie rods or drag links.

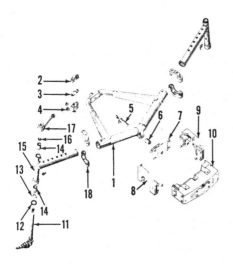

ried in the 1.2505-1.2545 diameter (presized) bushing (6). Pin (5) is bolted to the front axle pivot bracket (7).

On series 504 and 2504, the 1.748-1.750 diameter front axle pivot pin is carried in a 1.751-1.755 diameter (presized) bushing. Pivot pin is retained in front bolster by a lock plate bolted to front axle pivot bracket.

RADIUS (STAY) ROD AND PIVOT
Farmall 340-504

5. The radius rod pivot ball (15-Fig. IH1302) is available for service as an individual part. Clearance be-

Fig. IH1303—Exploded view of the adjustable front axle which is available on 330 and 340 International tractors. Except for front bolster and brackets, series 504 International tractors are similar.

1. Axle main member
2. Bracket cap
3. Shims
4. Pivot bracket
5. Pivot pin
6. Bushing
7. Pivot bracket
8 & 9. Bolster support brackets
10. Front bolster
11. Steering knuckle
12. Felt washer
13. Thrust bearing
14. Bushing
15. Axle extension
16. Snap ring
17. Steering knuckle arm
18. Clamp

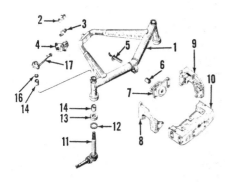

Fig. IH1304—Exploded view of the non-adjustable front axle which is available on 330 and 340 International tractors. Except for front bolster and brackets, series 504 and 2504 International tractors are similar. Refer to Fig. IH1303 for legend.

Fig. IH1305—Exploded view of the manual steering gear unit used on Farmall 340 tractors. Farmall 504 tractors are basically similar.

5. Rear steering shaft
6. Woodruff keys
7. Bearings
8. Upper support
9. Snap ring
10. Lower support
11. Universal joint yoke
13. Center steering shaft
14. Universal joint yoke
16. Retainer
17. Bearing
18. Bearing support
19. Retainer
20. Front steering shaft
21. Universal joint yoke
23. Front bolster
25. Starting crank bracket
27. Retainer
28. Steering worm nut
29. Ball bearing
30. Worm and shaft
31. Bushing
32. Oil seal
33. Worm wheel shaft washer and stud
34. Seal
35. Ball bearing
36. Steering worm wheel
37. Gasket
38. Bearing cage
39. Ball bearing
40. Gasket
41. Oil seal
42. Bearing cage cover
43. Steering worm wheel shaft
44. Studs (4 used)

On some 330 and 340 International tractors, alignment marks are provided on steering arms and axle (or axle extensions) as well as on Pitman arms and gear housing to facilitate making the toe-in adjustment. Drag links are properly adjusted if all marks align at the same time.

STEERING KNUCKLES
All Models

10. Procedure for removing knuckles from axle or axle extensions is evident after an examination of the unit and reference to Fig. IH1302, 1303 or 1304.

New knuckle bushings should be installed with the outer ends flush with the counter bores and oil hole in bushing aligned with oil hole in axle or axle extension. Ream the bushings after installation, if necessary, to provide a diametral clearance of 0.001-0.006 for the steering knuckles. Steering knuckle diameter for series 330 and 340 is 1.422-1.423; for series 504 and 2504, 1.671-1.672.

MANUAL STEERING SYSTEM

ADJUSTMENTS

Farmall 340-504

12. On Farmall 340 and 504 tractors, the manual steering gear unit is non-adjustable; however, excessive backlash between the worm and worm wheel can be partially corrected by changing the position of the worm wheel (36—Fig. IH1305) on the worm wheel shaft splines so as to bring un-worn teeth of the worm wheel into mesh with the steering worm.

International 330 and 340

13. The manual steering gear unit is provided with the following adjustments: The cam (worm) shaft end play, the mesh position between the lever shaft stud and cam, and the end play of the gear shaft.

Before attempting to make any adjustments, first make certain that the gear housing is properly filled with lubricant, then disconnect the drag links from the steering (Pitman) arms.

14. **CAM (WORM) SHAFT END PLAY.** To check and/or adjust the steering cam shaft end play, loosen both locks (25 & 26—Fig. IH1306 or 1307) and back-off the adjusting screws (24 & 27) three or four turns. Pull up and push down on the steering wheel to detect any end play of the cam shaft (12). Adjustment is correct when no end play exists and a barely perceptible drag is felt when turning the steering wheel with thumb and forefinger. If adjustment is not as specified, unbolt the upper cover (7) from gear housing and vary the number of shims (8) until desired adjustment is obtained. Shims are available in thicknesses of 0.002, 0.003 and 0.010. Tighten the two adjusting screws (24 & 27) as outlined in paragraphs 15 and 16.

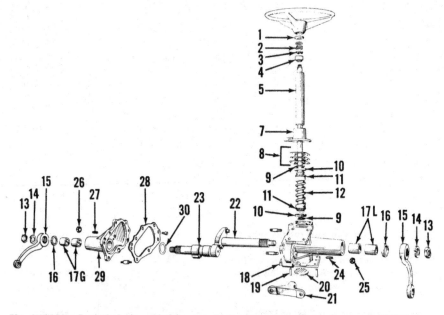

Fig. IH1306—Exploded view of the manual steering gear unit used on early 330 International tractors. Refer to Fig. IH1307 for an exploded view of the steering gear unit used on later 330 International and all 340 International tractors. Legend also applies to Fig. IH1307.

1. Dust seal	9. Cam retaining
2. Spring	rings (2 used)
3. Spring seat	10. Bearing cup
4. Bearing	(2 used)
5. Steering tube	11. Bearing balls
6. Governor control	(14 per race)
shaft pilot bushing	12. Cam (worm) and
7. Cover	shaft
8. Shims (0.002,	13. Nuts
0.003 and 0.010)	14. Lock washer

15. Steering (Pitman)	21. Steering gear
arm	bracket
16. Oil seal	22. Lever shaft
17G. Gear shaft bush-	23. Gear shaft
ings (2 used)	24. Adjusting screw
17L. Lever shaft	25. Lock nut
bushings (2 used)	26. Lock nut
18. Steering gear	27. Adjusting screw
housing	28. Gasket
19. Dowel pins	29. Trunnion cover
(2 used)	30. Thrust washer
20. Expansion plug	

15. **LEVER SHAFT STUD MESH.** With the cam shaft end play adjusted as in paragraph 14, turn the steering gear to the mid or straight ahead position and turn adjusting screw (27—Fig. IH1306 or 1307), located in side (trunnion) cover (29) on right side of housing, until a slight drag is felt when turning the steering gear through the mid or straight ahead position. Tighten the adjusting screw lock (26).

16. **GEAR SHAFT END PLAY.** With the cam shaft end play and the lever shaft mesh adjusted as outlined in paragraphs 14 and 15, turn the steering gear to the mid or straight

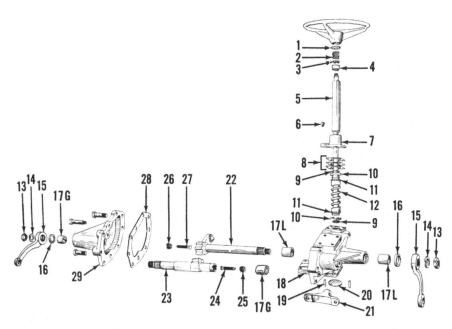

Fig. IH1307—Exploded view of the manual steering gear unit used on late 330 International and all 340 International tractors. Refer to Fig. IH1306 for legend.

ahead position and turn adjusting screw (24—Fig. IH1306 or 1307) to remove all end play from gear shaft (23) without increasing the amount of pull required to turn the steering gear through the mid or straight ahead position.

REMOVE AND REINSTALL

Farmall 340-504

18. To remove the steering gear housing and front axle and support, bolster or wheel fork as an assembly, proceed as follows: Remove hood, grille, radiator and on series 340, the generator. Disconnect the steering shaft front universal joint, and pry universal joint rearward and off the steering worm shaft. Place jack under torque tube (clutch housing) and remove weight from front wheels. Support steering gear housing and remove side rail retaining cap screws and lower dust cover on series 504. On series 340, remove housing to engine retaining bolts. On series 340, jack up tractor high enough for crankshaft pulley to clear steering gear housing and move assembly away from tractor. On series 504, pull steering gear and front bolster and axle assembly from side rails.

International 330 and 340

19. To remove the steering gear unit first drain cooling system and disconnect drag links from the steer-

ing (Pitman) arms. Remove hood, battery and starting motor. Disconnect the heat indicator sending unit, fuel lines, oil pressure gage line, wiring harness and controls from engine and engine accessories. Disconnect head light and rear light wires. Unbolt steering gear housing and fuel tank from tractor and using a hoist, lift the fuel tank, instrument panel and steering gear housing assembly from tractor.

Remove the steering wheel retaining nut and using a suitable puller, remove the steering wheel. Unbolt and remove the instrument panel assembly and fuel tank.

Fig. IH1308—Farmall 340 and 504 steering worm wheel and shaft assembly removed from the steering gear housing. The unit can be disassembled after removing stud (33). See legend for Fig. IH1305.

OVERHAUL
Farmall 340-504

21. The steering gear unit can be overhauled as follows without removing the assembly from tractor.

22. To remove the steering worm, proceed as follows: Remove grille and drain steering gear housing. Disconnect the steering shaft front universal joint, and slide universal joint rearward and off the steering worm shaft. Remove Woodruff key from worm shaft. On series 340, remove both starting crank bracket front retaining cap screws and block-up between bracket and steering gear housing enough to permit worm to come out. On series 504, remove bolster front cover. Remove steering worm bearing retainer (27—Fig. IH1305), and turn worm forward and out of housing. Worm bushing (31) and worm shaft oil seal (32) can be renewed at this time. Install worm shaft oil seal with lip of seal facing inward toward steering gears. Use a tin sleeve or shim stock when reinstalling worm shaft to prevent damaging the seal. To remove worm shaft ball bearing, remove cotter key and nut, and bump worm shaft out of bearing.

23. To remove the steering worm wheel and shaft assembly, proceed as follows: Remove grille and drain steering gear housing. Jack up tractor under torque tube (clutch housing) and remove bolster or wheel fork on tricycle type models; or, on adjustable axle versions, disconnect the center steering arm from worm wheel shaft. Remove the four bearing cage to steering gear housing retaining cap screws, and bump entire assembly out of steering gear housing. The unit is shown removed in Fig. IH1308. To disassemble the unit, remove nut from stud (33) and pull out the stud; place assembly on a suitable press, and press bearings and worm wheel off worm wheel shaft. At this time, lower bearing and oil seal (41—Fig. IH1305) can be renewed. Install oil seal with lip of seal facing up toward worm wheel. Use a tin sleeve or shim stock when reinstalling worm wheel shaft to prevent damaging the seal.

Reassemble the unit by reversing the disassembly procedure.

International 330 and 340

24. To overhaul the steering gear, first remove the unit from tractor as outlined in paragraphs 19. Remove the steering (Pitman) arm retaining

nuts (13—Fig. IH1306 or 1307) and using a suitable puller, remove the Pitman arms (15) from the lever shaft and gear shaft. Unbolt the side cover (29) from gear housing and remove the side cover and gear shaft (23). Withdraw lever shaft (22). Unbolt and remove the housing upper cover (7) and save shims (8) for reinstallation. Withdraw cam shaft (12) and remove ball cups (10) by removing their retaining snap rings (9). Thoroughly clean and examine all parts for damage or wear. The lever shaft and gear shaft should

be renewed if the spur gear teeth are damaged or worn.

Inside diameter (new) of the lever shaft and gear shaft bushings is 1.374-1.375. Diameter of lever shaft and gear shaft at bearing surfaces is 1.3725-1.3735. Renew the shafts and/or bushings if clearance is excessive.

New bushings are pre-sized and if carefully pressed into position using a suitable piloted arbor until outer ends of bushings are flush with inner edge of chamfered surface in bores, should need no final sizing.

When installing the cam shaft and

jacket tube, vary the number of shims (8) to remove all camshaft end play and provide a barely perceptible drag when turning steering wheel with thumb and forefinger. Install the lever shaft, gear shaft and side cover. On early production 330 International tractors, be sure thrust washer (30—Fig. IH1306) is installed between gear shaft and side cover. Install the steering (Pitman) arms.

Complete the overhaul by adjusting the lever stud mesh as in paragraph 15 and the gear shaft end play as outlined in paragraph 16.

POWER STEERING SYSTEM

Power steering is optional equipment on series 330, 340 and Farmall 504, while International 504 and 2504 series tractors have full time (Hydrostatic) power steering as standard equipment. Refer to Fig. IH1308A for a schematic view showing the general layout of components and tubing for the Hydrostatic steering system.

NOTE: The maintenance of absolute cleanliness of all parts is of utmost importance in the operation and servicing of the hydraulic power steering system. Of equal importance is the avoidance of nicks or burrs on any of the working parts.

LUBRICATION AND BLEEDING

Series 330-340

With "Hydra-Touch"

25. The regular Hydra-Touch system fluid reservoir is the source of fluid supply to the power steering system. Refer to paragraph 184 or 185 concerning the lubrication requirements.

Series 340 Without

"Hydra-Touch"

26. The transmission and differential housing serves as the hydraulic fluid reservoir for the power steering system. The power steering fluid serves also as the lubricant for the "Torque Amplifier," transmission and differential gears. The fluid should be drained and renewed and the filter (Fig. IH1309) should be serviced at least every 1,000 hours or once a year whichever occurs first.

NOTE: Early filters were of the wire screen type and had a by-pass valve (V—Fig. IH1310). Later filters were of the wire screen type but did not have the by-pass

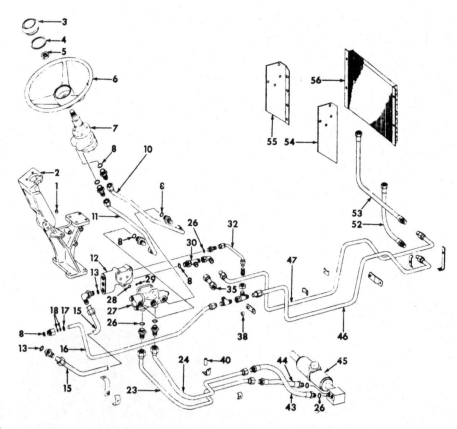

Fig. IH1308A — Schematic view showing the general lay-out of components and tubing for the full time (Hydrostatic) power steering used on International 504 and 2504 tractors.

1. Governor control shaft bushing	17. Retainer	35. Oil cooler relief valve
2. Support	18. "O" ring and back-up washer	38. Spacer
3. Steering wheel cap	23. Cylinder tube	40. Spacer
4. Cap retainer	24. Cylinder tube	43. Cylinder hose
5. Nut	26. "O" ring	44. Cylinder hose
7. Hand pump	27. Control (pilot) valve	45. Steering cylinder
8. "O" ring	28. "O" ring	46. Oil cooler return tube
10. Hand pump tube (LH)	29. "O" ring	47. Oil cooler inlet tube
11. Hand pump tube (RH)	30. Tee	52. Oil cooler outlet hose
12. Flow divider	32. Flow divider return tube	53. Oil cooler inlet hose
13. "O" ring		54. Oil cooler bracket (RH)
15. Flow divider to transfer block tube		55. Oil cooler bracket (LH)
		56. Oil cooler

valve. The latest type (Fig. IH1310A) incorporate the wire screen and two renewable filter elements.

When an early type filter (with the by-pass valve) is encountered, it should be discarded and one of the later types should be installed.

Only IH "Hy-Tran" fluid should be used in the hydraulic system and reservoir fluid level should be maintained at the oil level plug located on the right side of the transmission and differential housing. Whenever the power steering oil lines have been disconnected or fluid drained, cycle the power steering system several times to bleed air from the system; then, refill the reservoir to the lower level of the oil level plug.

Farmall 504

26A. If power steering system has been drained, or any connections disconnected, system should be bled of any air which is present. To bleed power steering on Farmall models, proceed as follows:

On tricycle models, be sure reservoir (rear frame) is at proper level, start engine and run at low idle speed; then cycle system from lock to lock until all air is purged from system. Refill reservoir, if necessary.

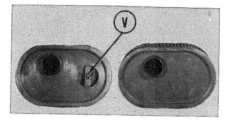

Fig. IH1309—On series 340 tractors with power steering, the system filter should be cleaned in kerosene.

Fig. IH1310—Some very early filter elements were equipped with a by-pass valve (V). A later filter element should be installed in place of the early type element. Refer to Fig. IH1310A for an exploded view of the latest type.

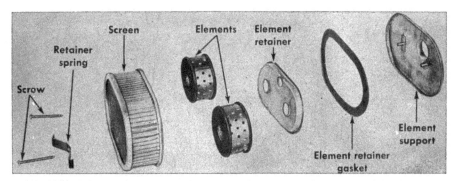

Fig. IH1310A—The latest type filter uses a wire screen and two renewable elements.

On Farmall models equipped with adjustable wide tread front axle, the steering cylinder should be primed, if necessary, to facilitate bleeding of system. Prime steering cylinder as follows: Disconnect the flexible cylinder lines at the aft end, hold lines up and slowly pour oil into both hoses. NOTE: Be sure to pour oil slowly in order to allow air to escape. Reconnect lines as quickly as possible, then check to see that reservoir is at proper level. Start engine and run at low idle speed, then cycle system from lock to lock until all air is purged from system. Refill reservoir (rear frame), if necessary.

International 504-2504

26B. If steering cylinder has been emptied by servicing, or for any other reason, prime cylinder as follows: Disconnect the flexible cylinder lines at the aft end, hold lines up and slowly pour oil into both hoses. NOTE: Be sure to pour oil slowly in order to allow air to escape. Reconnect lines as quickly as possible, then check to see that reservoir is at proper level. Start engine and run at low idle speed, then, rotate steering wheel (hand pump) as rapidly as possible in order to activate the control valve and continue to rotate the steering wheel until the front wheels reach the stop in the direction in which the steering wheel is being turned. Now quickly reverse the direction of steering wheel and follow same procedure until front wheels reach stop in opposite direction. Continue to turn front wheels from lock to lock until steering wheel has no wheel spin (free wheeling) and has a solid feel with no skips or sponginess. Check and add fluid to reservoir (rear frame), if necessary.

TROUBLE SHOOTING

26C. The following table lists some of the troubles, and their causes, which may occur in the operation of

the series 504 and 2504 power steering system. When the following information is used in conjunction with the information contained in the Power Steering Operational Tests section (paragraph 29C through 29J), no trouble should be encountered in locating system malfunctions.

Farmall 504

1. Hard steering.
   a. Oil in system too heavy.
   b. Power unit spool and/or sleeve sticking or binding.
   c. Gerotor set binding.
   d. Damaged thrust bearing.
2. Loss of power or slow steering.
   a. Internal or external leaks in power unit.
   b. Low pump pressure.
   c. Defective or kinked lines.
   d. Faulty flow control valve.
   e. Faulty hydraulic pump relief valve.
3. Creeping.
   a. Weak or broken centering springs in power unit.
   b. Power unit valve spool and/or sleeve sticking or binding.
4. Unequal turning between left and right.
   a. Tie-rods not adjusted equally.
   b. Damaged steering worm wheel gear or steering worm.
5. Steering gear free wheels.
   a. Pin broken in power unit spool and sleeve assembly.
   b. Broken steering worm wheel gear or steering worm.
6. Oil overheating.
   a. Power unit control spool and/or sleeve sticking or binding.
   b. Defective or plugged oil cooler or lines.
   c. Hydraulic pump relief valve stuck partially open.
7. Oil leaks.
   a. Seals in end caps leaking.
   b. "Loctite" on assembly cap screws not sealing screw threads.
   c. Faulty connections or manifold "O" rings.

International 504-2504

1. No power steering or steers slowly.
   a. Binding mechanical linkage.
   b. Excessive load on front wheels and/or air pressure low in front tires.
   c. Steering cylinder piston seal faulty or cylinder damaged.
   d. Control (pilot) valve relief valve setting too low or valve leaking.
   e. Faulty commutator in hand pump.
   f. Flow divider valve spool sticking or leaking excessively.
   g. Control (pilot) valve spool sticking or leaking excessively.
   h. Circulating check ball not seating.
2. Will not steer manually.
   a. Binding mechanical linkage.
   b. Excessive load on front wheels and/or air pressure low in front tires.
   c. Pumping element in hand pump faulty.
   d. Faulty seal on steering cylinder or cylinder damaged.
   e. Pressure check valve leaking.
   f. Control (pilot) valve spool binding or centering spring broken.
   g. Control (pilot) valve relief valve spring broken.
3. Hard steering through complete cycle.
   a. Low pressure from supply pump.
   b. Internal or external leakage.
   c. Line between hand pump and control (pilot) valve obstructed.
   d. Faulty steering cylinder.
   e. Binding mechanical linkage.
   f. Excessive load on front wheels and/or air pressure low in front tires.
   g. Cold hydraulic fluid.
4. Momentary hard or lumpy steering.
   a. Air in power steering circuit.
   b. Control (pilot) valve sticking.
5. Shimmy
   a. Control (pilot) valve centering spring weak or broken.
   b. Control (pilot) valve centering spring washers bent, worn or broken.

OPERATING PRESSURE, RELIEF VALVE AND FLOW CONTROL VALVE

Series 330-340 Except 340 With Transmission Driven Pump

27. Working fluid for the hydraulic power steering system is supplied by the same pump which powers the "Hydra-Touch" system. Interposed between the pump and the "Hydra-Touch" system is a flow control valve

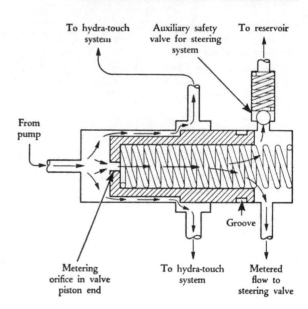

Fig. IH1311 — Schematic illustration of the power steering flow control valve used on all 330 and 340 models except 340 with the transmission driven pump. The flow control valve satisfies the 2½-3 gpm requirement of the power steering system before any oil flows to the "Hydra-Touch" system.

mechanism which is shown schematically in Fig. IH1311. The small metering hole in the end of the flow valve piston passes between 2½ to 3 gallons per minute to the power steering system; but, since the pump supplies considerably more than three gpm, pressure builds up in front of the piston and moves the piston, against spring pressure, until the ports which supply oil to the "Hydra-Touch" system are uncovered. The power steering system, therefore, receives priority and the fluid requirements of the steering system are satisfied before any oil flows to the "Hydra-Touch" system. The auxiliary safety valve for the power steering system maintains a system operating pressure of 1200-1500 psi. The components of the flow control valve are shown exploded from the valve housing in Fig. IH1312.

28. A pressure test of the power steering circuit will disclose whether the pump, safety valve or some other unit in the system is malfunctioning. To make such a test, proceed as follows: Connect a pressure test gage and shut-off valve in series with the line connecting the flow control valve to the steering valves as shown in Fig. IH1313. Notice that the pressure gage is connected in the circuit between the shut-off valve and the flow control valve. Open the shut-off valve and run engine at low idle speed until oil is warmed. Advance the engine speed to the specified high idle rpm, close the shut-off valve, observe the pressure gage reading, then open the shut-off valve. If the gage reading is between 1200 and 1500 psi with the shut-off valve closed, the hydraulic

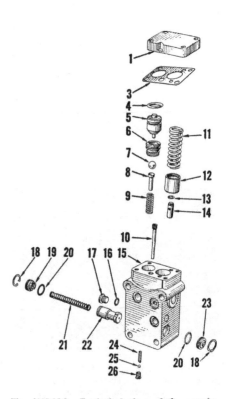

Fig. IH1312—Exploded view of the regulator, safety and flow control valve used on all 330 and 340 models except 340 with the transmission driven pump. The auxiliary safety valve (24, 25 and 26) protects only the power steering system. Valve (22) is not available separately.

1. Cover	14. Safety valve piston
3. Gasket	15. Valve housing
4. Seal ring	16. Seal ring
5. Regulator valve piston	17. Plug
	18. Snap ring
6. Regulator valve seat	19. Retainer
	20. Seal ring
7. Steel ball	21. Flow valve spring
8. Ball rider	22. Flow control valve
9. Ball rider spring	23. Retainer
10. Safety valve orifice screen & plug	24. Auxiliary safety valve spring
11. Safety valve spring	25. Steel ball
12. Spring retainer	26. Plug
13. Snap ring	

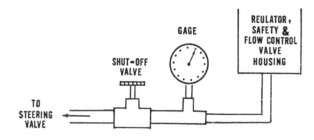

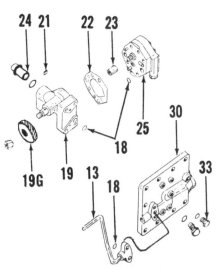

Fig. IH1313 — Shut-off valve and pressure gage installation diagram for trouble shooting the power steering system on all 330 and 340 models except 340 with the transmission driven pump.

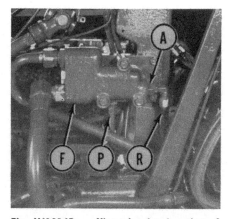

Fig. IH1315 — On 340 tractors equipped with power steering but not a hydraulic system, the power steering pump (25) is mounted on a bracket (19) and a drive coupling (23) is used.

13. Power steering pressure line	23. Coupling
18. "O" rings	24. Suction line
19. Adaptor	25. Power steering pump
19G. Drive gear	30. Pump mounting flange (manifold)
21. Key	33. Plugs
22. Gasket	

pump and auxiliary safety valve are O.K. and any trouble is located elsewhere in the system.

If the gage reading is more than 1500 psi, the auxiliary safety valve may be stuck in the closed position. If the gage reading is less than 1200 psi, renew the auxiliary safety valve spring and recheck the pressure reading. If the gage reading is still less than 1200 psi, a faulty hydraulic pump is indicated.

Series 340 With Transmission Driven Pump

29. Working fluid for the hydraulic power steering system is supplied by a pump contained in the clutch housing. The flow of approximately 2.3-2.7 gpm and pressure of 1200-1500 psi are regulated by spring loaded valves located in the pump body.

The overall condition of the system can be checked by installing a pressure gage of sufficient capacity in place of plug (S—Fig. IH1314). A gage reading of lower than 1100 psi with front wheels in either the extreme left or extreme right turn positions may be caused by one or more of the following: Clogged filter ele-

Fig. IH1314B — View showing location of flow divider valve on 504 Farmall model tractors. Note that valve is early type. Refer to Fig. IH1329C for late type.

A. Adaptor
F. Flow divider valve
P. Pressure line
R. Return line

ment; insufficient amount of oil; leaking lines; leaking seals; sticking or worn pressure relief valve; broken or weak relief valve spring; wear and consequent by-passing of oil in the power steering cylinder or control valve; worn or damaged pump.

To more accurately check the system and partially isolate the cause of trouble, proceed as follows: Connect a pressure gage of sufficient capacity and a shut-off valve in series with the pump pressure line (PL—Fig. IH1314). NOTE: The gage should be installed between the shut-off valve and the pump. Run tractor until

Fig. IH1314—A power steering system pressure test can be made on series 340 with transmission driven pump by installing a gage in place of plug (S).

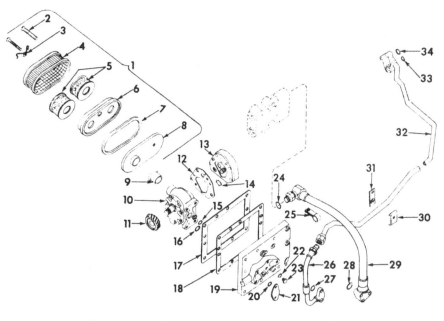

Fig. IH1315A — On 504 and 2504 series tractors with draft and position control and power steering, two pumps are used as shown. On Farmall 504 tractors with no power steering or auxiliary system, an adapter shown in Fig. IH1315C is used in place of pump (10).

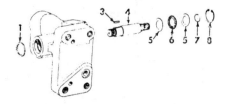

Fig. IH1315B — Early type adapter used on series 340 tractors without "Hydra-Touch".

1. Oil seal	6. Thrust bearing
3. Key	7. Retainer ring
4. Shaft	8. Retainer ring
5. Bearing race	

power steering fluid reaches operating temperature, then close the shut-off valve and observe the gage reading. NOTE: If the shut-off valve is closed for an excessive length of time, damage to the pump may result. If the gage reading is 1200-1500 psi, the filter, pump, relief valve and pump intake seals are O.K. and any trouble is located in the pressure lines, power steering cylinder and/or control valve. Excessive pressure could be caused by the relief valve being stuck in the closed position.

If pressure is insufficient and the filter is known to be O.K., observe the oil. If the oil appears to be milky, an air leak in the suction line can be suspected. Air leakage at the pump intake seal can be corrected by removing the pump (or pumps) and mounting flange and renewing the intake pipe and seal (24—Fig. IH-1316). If renewing the intake pipe and seal does not correct the condition after a reasonable length of time, the tractor should be split and seal (S—Fig. IH1394) renewed.

If pressure is low and the condition of the filter and the oil are known to be O.K., service the pressure relief valve.

The relief valve on both Cessna and

Thompson pumps is pre-set and available only as an assembly. If cleaning the valve does not restore the pressure, the relief valve assembly (11R—Fig. IH1317) or (9—Fig. IH1317C) should be renewed. Recheck pressure and if still insufficient, overhaul the pump.

Farmall 504

29A. Working fluid for the power steering system is supplied by a pump located in the clutch housing. The flow of approximately three gallons per minute and pressure of 1550-1600 psi is controlled by a flow divider valve located on left side of tractor as shown in Fig. IH1314B.

To check power steering operating pressure, connect a gage capable of registering at least 2000 psi in series with the pressure line (P) at bottom of flow control valve cover. With fluid at operating temperature, run engine at high idle rpm and turn front wheels against stop. Continue to apply turning motion to steering wheel and observe reading on gage. Gage should read 1550-1600 psi. A gage reading lower than that stated may be caused by a clogged filter; insufficient amount of oil; defective relief valve in flow divider valve; faulty control valve (power unit or rotary valve) or a faulty power steering pump.

Excessive pressure could be caused by the relief valve being stuck in the closed position.

NOTE: If pressure is not within specifications and pump, filter, flow divider valve and control valve are believed to be satisfactory, observe the oil and if it appears to be milky, remove the pump mounting flange and pump (or pumps) and renew the intake pipe "O" ring and/or intake pipe. If after renewing intake pipe

Fig. IH1316—View of the pumps and mount-flange. The longer smooth section of the suction tube should be pressed in the hydraulic lift pump (or power steering pump drive adaptor) until it bottoms. Although Cessna pumps are shown, Thompson pumps are similar.

"O" ring and/or intake pipe the milky oil condition does not correct itself after a reasonable length of time, split the tractor between clutch housing and rear frame and renew "O" ring on suction pipe. (See S-Fig. IH1394).

International 504-2504

29B. To check power steering operating pressure, refer to paragraph 29J, which gives the method of checking the flow divider relief valve pressure.

POWER STEERING OPERATIONAL TESTS
International 504-2504

The following tests are valid only when the power steering system is completely void of any air. If necessary, bleed system as outlined in paragraph 26B before performing any operational tests.

29C. MANUAL PUMP. With transmission pump inoperative (engine not running), attempt to steer manually in both directions. NOTE: Manual steering with transmission pump not running will require high steering effort. If manual steering can be accomplished with transmission pump inoperative, it can be assumed that the manual pump will operate satisfactorily with the transmission pump operating.

Refer also to paragraph 29E for information regarding steering wheel (manual pump) slip.

29D. CONTROL (PILOT) VALVE. Attempt to steer manually (engine not running). Manual steering will require high steering effort but if steering can be accomplished, control (pilot) valve is working.

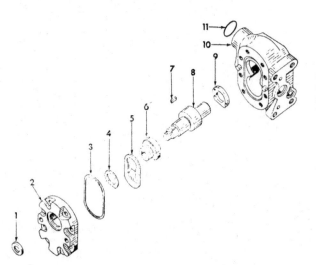

Fig. IH1315C—On Farmall 504 tractors with no power steering, an adapter is used instead of large pump.

1. Oil seal
2. Cover
3. "O" ring
4. Bearing spacer
5. Bearing plate
6. Bearing
7. Key
8. Adapter shaft
9. Bearing
10. Body
11. "O" ring

No steering can be accomplished if control valve is stuck on center. A control valve stuck off center will allow steering in one direction only. If excessive leakage of control valve is suspected, refer to section on Steering Wheel Slip for further checking.

29E. STEERING WHEEL SLIP (CIRCUIT TEST). Steering wheel slip is the term used to describe the inability of the steering wheel to hold a given position without further steering movement. Wheel slip is generally due to leakage, either internal or external, or a faulty hand pump, steering cylinder or control (pilot) valve. Some steering wheel slip, with hydraulic fluid at operating temperature, is normal and permissible. A maximum of four revolutions per minute is acceptable. By using the steering wheel slip test and a process of elimination, a faulty unit in the power steering system can be located.

However, before making a steering wheel slip test to locate faulty components, quick disconnect assemblies should be installed in both steering cylinder lines and both hand pump to control (pilot) valve lines. This is necessary because of the fact that lines have to be disconnected and if it were not for the quick disconnects, air would be introduced into the circuit which in turn would not permit circuit testing having valid results. It is imperative that the complete power steering system be completely free of air before any testing is attempted.

To check for steering wheel slip (circuit test), proceed as follows: Check reservoir (rear frame) and fill to correct level, if necessary. Bleed power steering system, if necessary. Bring power steering fluid to operating temperature, cycle steering system until all components are approximately the same temperature and be sure this temperature is maintained throughout the tests. Remove steering wheel cap (monogram), then turn front wheels until they are against stop. Attach a torque wrench to steering wheel nut. NOTE: Either an inch-pound, or a foot-pound wrench may be used, however, an inch-pound wrench is recommended as it is easier to read. Advance hand throttle until engine reaches rated rpm, then apply 72 inch-pounds (6 foot-pounds) to torque wrench in the same direction as the front wheels are positioned against the stop. Keep this pressure (torque) applied for a period of one minute and count the revolutions of the steering wheel. Use same procedure and check the steering wheel

slip in the opposite direction. A maximum of four revolutions per minute in either direction is acceptable and system can be considered as operating satisfactorily. If, however, the steering wheel revolutions per minute exceed four, record the total rpm for use in checking the steering cylinder, control valve or hand pump.

NOTE: While four revolutions per minute of steering wheel slip is acceptable, it is generally considerably less in normal operation.

29F. STEERING CYLINDER TEST. If steering wheel slip, as checked in paragraph 29E, exceeds the maximum of four revolutions per minute, proceed as follows: Be sure operating temperature is being maintained, then disconnect the steering cylinder lines at quick disconnects. Repeat the steering wheel slip test, in both directions, as described in paragraph 29E. If steering wheel slip is ½ rpm or more, **less** than that recorded in paragraph 29E, overhaul or renew the steering cylinder.

If steering wheel slip remains the same as that obtained in the test outlined in paragraph 29E, check the hand pump and control valve as outlined in paragraph 29G.

29G. MANUAL PUMP AND CONTROL VALVE. To check the hand pump and/or control valve, proceed as follows: Be sure system operating temperature is maintained, then disconnect the quick disconnects in the hand pump to control valve lines. Repeat the steering wheel slip test, in both directions, as described in paragraph 29E.

If steering wheel slip is two rpms, or **more,** greater than those obtained in the circuit test outlined in paragraph 29E, a faulty hand pump is indicated and hand pump should be serviced as outlined in paragraph 46.

If steering wheel slip is two rpms, or **less,** than those obtained in the steering cylinder test outlined in paragraph 29F, a faulty control (pilot) valve is indicated and valve should be serviced as outlined in paragraph 46A.

NOTE: At this time, the safety relief valve in the control (pilot) valve can be checked. To check this valve proceed as follows: Be sure hydraulic fluid is at operating temperature and hand pump to control valve lines are reconnected, then attach a gage, capable of registering at least 3000 psi, in series with one of the steering cylinder lines. Direct oil to the gage by turning the steering wheel in that direction until front wheels engage

stop. With engine running at rated rpm, apply high steering effort to the steering wheel and observe the gage reading. Gage should read 2200 psi. If gage reading is not approximately 2200 psi, adjust safety relief valve by loosening jam nut and turning adjusting screw inward to increase pressure, or outward to decrease pressure. If safety relief valve will not adjust, remove and inspect spring, ball and seat. Renew spring and/or ball if necessary. If seat is damaged, renew complete control valve assembly as seat is not available separately. Recheck and readjust relief pressure. Relief valve adjusting screw is shown at (V—Fig. IH1329).

29H. CONTROL (PILOT) VALVE TEST. Maintain the fluid temperature as outlined in paragraph 29E, then turn front wheels against stop. Disconnect the control valve to steering cylinder line which is furnishing the pressurized oil in the direction of the turn. Repeat slip test of manual pump as outlined in paragraph 29E. If slippage of steering wheel exceeds 2 rpm, or more, than that recorded in paragraph 29E, service the hand pump as outlined in paragraph 46. If steering wheel slip is 2 rpm, or less, than recorded in paragraph 29E, the control valve is faulty and should be serviced as outlined in paragraph 46A.

29J. FLOW DIVIDER VALVE. The flow divider valve as applied to the International model tractors cannot be satisfactorily tested with unit installed in tractor. Therefore, to test the safety relief valve it is recommended that the unit be removed and bench tested. Refer to paragraph 46B.

PUMP
All Models Except 340-540-2504
With Transmission Driven
Pump

30. On all models except 340, 504 and 2504 tractors equipped with the transmission driven pump, the regular "Hydra-Touch" (camshaft driven) pump supplies fluid to both the power steering system and the "Hydra-Touch" system. Refer to paragraphs 192 and 193 or 194 and 195 for R&R and overhaul procedures for this pump.

Series 340-504-2504 With
Transmission Driven Pump

Cessna (Fig. IH1317 or IH1317A) and Thompson (Fig. IH1317C or IH1317D) gear type pumps are used and are interchangeable. On series 340, the small (2.3-2.7 gpm) pump is used for the power steering system and the large pump supplies the "Hydra-Touch" system. On series 504 and 2504, a priority flow of approximately 3 gpm is and 0.021-0.023 for series 340 or 0.011-0.019 and 0.016-0.024 for series 504 and 2504.

Fig. IH1317 — Exploded view of the Cessna power steering pump and flow control and relief valves used on series 340. Shims (17) are used to adjust the volume of flow to 2.3-2.7 gpm.

1. Cover
2. "O" ring
3. "O" ring
4. "O" ring
5. Diaphragm seal
6. Phenolic gasket
7. Diaphragm
8. Pumping gears and shafts
9. Body
10. "O" rings
11. Flow control valve sleeve
11R. Flow control and relief valve
12. Screen (cloth)
13. Screen (brass)
14. Gasket
15. Spring
16. Spring
17. Shims
18. "O" ring
19. Plug

32. OVERHAUL CESSNA PUMP. (SERIES 340). To overhaul the removed pump, refer to Fig. IH1317 and proceed as follows: Remove the two pump body to pump cover cap screws, separate the cover from body and remove gears. Remove diaphragm (7), phenolic (back-up) gasket (6) and diaphragm seal (5) from cover (1).

The bushings are not available separately. The drive and driven gears, shafts and snap rings are available only as a complete set. Pressure relief valve (11R) can be disassembled and cleaned, but if its condition is questionable, renew the unit. Make certain that valve (11R) slides freely in sleeve (11). Shims (17) control the pump output volume of 2.3-2.7 gpm. Be sure to install the same shims that were removed.

O.D. of shafts at bushings . . 0.685 min.

I.D. of bushings in body
and cover . . . . 0.691 max.

Thickness of gears . . . . 0.456 min.

I.D. of gear pockets
in body . . . . 1.719 max.

Depth of gear pockets
in body . . . . 0.4619 max.

When reassembling, use a new diaphragm (7), phenolic gasket (6), diaphragm seal (5) and "O" rings (2, 3, 4, 10 & 18). With open part of diaphragm seal (5) towards cover (1), work same into grooves of cover using a dull tool. Press the phenolic gasket (6) into diaphragm seal (5) and install diaphragm (7) with bronze face towards gears. NOTE. The diaphragm must fit inside the raised rim of the diaphragm seal (5).

taken from the large pump via a flow divider valve and the balance is utilized for the auxiliary hydraulic system, while the small (4½ or 7 gpm) pump supplies the draft and position control system. Also see Fig. IH1475A.

On series 340 tractors which have power steering but no "Hydra-Touch", and Farmall 504 tractors which have draft and position control but no power steering, an adapter such as that shown in Fig. IH1315B (early) or IH1315C (late) is used in place of the large pump and is secured to the pump mounting flange. This adapter is used to provide the drive for the small tandem mounted pump. See Figs. IH1315 and 1315A.

31. REMOVE AND REINSTALL. The pump mounting flange with the pump (or pumps) attached can be removed from left side of tractor after the power steering pressure line and, if so equipped, the hydraulic lift system pressure line are detached from mounting flange and the flange retaining cap screws are removed. The removed flange and pumps are shown in Fig. IH1316.

The small pump can be separated from the large pump or adapter after removing the four attaching cap screws. Take care to prevent damage to the sealing surfaces when separating the pumps. When reinstalling the small pump, tighten the four retaining screws and the two pump

covers to body screws evenly to 25 Ft.-Lbs. of torque.

When reinstalling the pumps and flange on tractor, make certain the longer smooth section of the suction pipe (24—Fig. IH1316) is pressed completely into the large pump or adapter. The lip of the seal on the suction pipe should face toward the left (pump) side of the tractor. Vary the number and thicknesses of the gaskets which are located between pump mounting flange and clutch housing until the backlash between the pump driven gear and the PTO gear is 0.002-0.026. Gaskets are available in two thicknesses; 0.011-0.013

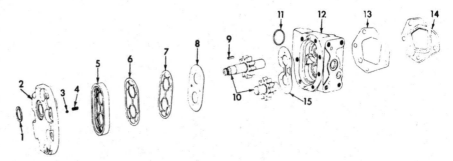

Fig. IH1317A — Exploded view of the Cessna power steering and auxiliary circuit pump used on series 504 and 2504.

1. Oil seal
2. Cover
3. Check ball
4. Spring
5. Diaphragm seal
6. Protector gasket
7. Back-up gasket
8. Pressure diaphragm
9. Key
10. Pump gears and shafts
11. "O" ring
12. Pump body
13. Gasket
14. Rear cover
15. Body diaphragm

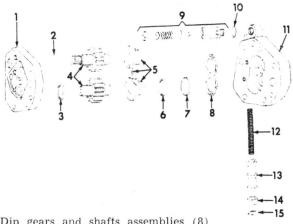

Fig. IH1317C — Exploded
view of the Thompson
power steering pump and
flow control and relief
valves used on series 340.
The relief valve assembly
is shown at (9).

1. Cover
2. "O" ring
3. "O" ring
4. Pumping gears
   and shafts
5. Pressure bearings
6. "O" rings (4 used)
7. Retainers (2 used)
8. Pressure plate spring
9. Pressure relief valve
10. Gasket
11. Body
12. Flow control valve
    spring
13. Flow control valve
14. Plug
15. Snap ring

Dip gears and shafts assemblies (8) in oil and install them in cover. Apply a thin coat of heavy grease on the finished side of pump body (9) and install body over gears and shafts. Check pump rotation. Pump should have very slight amount of drag but should rotate evenly.

32A. OVERHAUL CESSNA PUMP (SERIES 504 AND 2504). To overhaul the removed pump, refer to Fig. IH-1317A and proceed as follows: Remove the four retaining cap screws and separate the small pump from large pump, then remove large pump from pump mounting flange.

NOTE: If tractor is not equipped with draft and position control, cover (14) and gasket (13) will be used on pump body in place of the small pump.

Remove pump drive gear, then unbolt and remove pump cover (2). The remainder of disassembly will be obvious after an examination of the unit and reference to Fig. IH1317A.

Bushings in pump body and cover are not available separately and the pump gears and shafts are available only in sets. Gaskets and "O" rings are also available in a package only.
Pump specifications are as follows:
12 GPM Pump

O. D. of shafts at bushings,
  Minimum....................0.8101
I. D. of bushings in body and
  cover, maximum 0.8161
Thickness (width) of gears,
  minimum...................0.5711
I. D. of gear pockets, max......2.002
Max. allowable shaft to
  bushing clearance 0.006
17 GPM Pump
O. D. of shafts at bushings.....0.8101
I. D. of bushings in body
  and cover 0.8161
Thickness (width) of gears....0.8121
I. D. of gear pockets.........2.002
Max. allowable shaft to
  bushing clearance 0.006
When reassembling, use new diaphragms, gaskets, back-up washers, diaphragm seal and "O" rings. With

open part of diaphragm seal (5) towards cover (2), work same into grooves of cover using a dull tool. Press protector gasket (6) and back-up gasket (7) into the relief of diaphragm seal. Install check ball (3) and spring (4) in cover, then install diaphragm (8) inside the raised lip of the diaphragm seal and be sure bronze face of diaphragm is toward pump gears. Dip gear and shaft assemblies in oil and install them in cover. Position diaphragm (15) in pump body with the bronze side toward pump gears and cut-out portion toward inlet (suction) side of pump. Install pump body over gears and shafts and install retaining cap screws. Torque cap screws to 20 ft.-lbs. for the 12 gpm pump and to 25 ft.-lbs. for the 17 gpm pump.

Check pump rotation. Pump will have a slight amount of drag but should turn evenly.

33. OVERHAUL THOMPSON PUMP (SERIES 340). Disassembly and overhaul procedures will be evident after an examination of the unit and reference to Fig. IH1317C. All

seals and questionable parts should be renewed. If any part of relief valve (9) appears questionable, renew the assembly. Make certain that flow control valve (13) slides freely in its bore.

33A. OVERHAUL THOMPSON PUMP (SERIES 504 AND 2504). To overhaul the removed pump, refer to Fig. IH1317D and proceed as follows: Remove the four retaining cap screws and separate small pump from large pump, then remove large pump from pump mounting flange.

NOTE: If tractor is not equipped with draft and position control, cover (15) and gasket (14) will be used on pump body in place of the small pump.

Remove pump drive gear and key (8), then unbolt and remove pump cover (2). Bearings (7), pressure plate spring (6), "O" ring retainers (5), "O" rings (4), back-up washers (3) and oil seal (1) can now be removed from cover. Note location of bearings (7) so they can be reinstalled in the same position. Remove "O" rings (11 and 12) and the pump gears and shafts (9) from pump body. Bump body on a wood block and remove bearings (10). Note location of bearings so they can be reinstalled in the same position.

Pump gears and shafts, as well as the pump shaft bearings, are available only in sets. Except for suction port "O" ring (12), none of the gaskets or "O" rings are available separately.
Pump specifications are as follows:
12 GPM Pump
O. D. of shafts at bearings..... 0.8120
I. D. of bearings in body
  and cover 0.8155
Thickness (width) of gears.... 0.600
I. D. of gear pockets......... 1.7718
Max. allowable shaft to
  bearing clearance 0.010

Fig. IH1317D — Exploded
view of the Thompson
power steering and aux-
iliary circuit pump used
on series 504 and 2504
tractors.

1. Oil seal
2. Cover
3. Back-up washer
4. "O" ring
5. Retainer
6. Pressure plate spring
7. Pressure bearings
8. Key
9. Pump gears and shafts
10. Body bearings
11. "O" ring
12. "O" ring
13. Body
14. Gasket
15. Rear cover

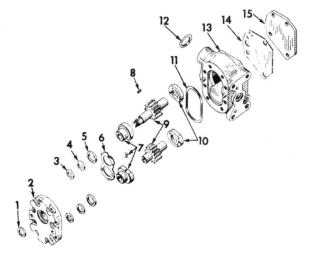

17 GPM Pump

O. D. of shafts at bearings..... 0.8120
I. D. of bearings in body
  and cover 0.8155
Thickness (width) of gears.... 0.720
I. D. of gear pockets.......... 1.7718
Max. allowable shaft to
  bearing clearance 0.010

Lubricate all parts during assembly, use all new gaskets and seals and be sure bearings in body and cover are reinstalled in their original positions if same bearings are being reinstalled. Tighten cover to body cap screws to a torque of 20 ft.-lbs. for the 12 gpm pump, or 30 ft.-lbs. for the 17 gpm pump.

Check pump rotation. Pump will have a slight amount of drag but should turn evenly.

POWER UNIT

Farmall 340-504

34. **REMOVE AND REINSTALL.** To remove the power steering unit on the Farmall 340, first remove hood skirts and hood. Unbolt instrument panel from instrument panel housing cowl and remove the instrument panel housing. Disconnect steering shaft universal from lower end of power steering unit. Disconnect manifold flange from power steering unit and discard the "O" rings. Remove steering wheel, then unbolt power steering unit from steering shaft lower support and remove the power unit.

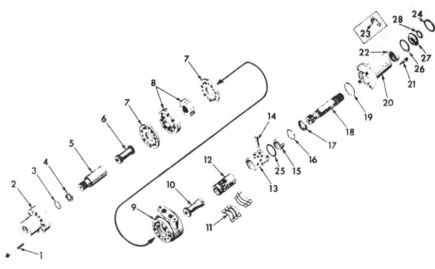

Fig. IH1318B — Exploded view of the power steering unit used on Farmall 340 and 504 tractors. Items (9, 12 and 13), as well as gerotor set (8), are available as units only.

1. Cap screws (7)	8. Gerotor set	15. Bearing race	22. Needle bearing
2. Power end housing	9. Housing	16. Steel balls (14)	23. Oil seal (early)
3. "O" ring	10. Control end drive	17. Bearing race	24. Retainer ring
4. Retainer washer	11. Centering springs	18. Control shaft	25. "O" ring
5. Power end shaft	12. Valve spool	19. "O" ring	26. "O" ring
6. Power end drive	13. Valve sleeve	20. Control end cap	27. Retainer
7. Plate	14. Pin	21. Cap screws (4)	28. "O" ring

On Farmall 504, remove steering wheel and cover for power unit. See Fig. IH1318A. Disconnect steering shaft universal from lower end of power steering unit. Disconnect manifold flange and discard "O" rings, then unbolt unit and remove from steering shaft lower support.

Use new "O" ring seals when reinstalling power steering unit.

35. **OVERHAUL.** With power steering unit removed as outlined in paragraph 34, place a scribe line lengthwise of the unit so it can be reassembled in the same position. Refer to Fig. IH1318B and remove cap screws which retain control end cap (20) to spool and sleeve housing (9). Pull cap, control shaft (18), bearing assembly (15, 16 & 17) and "O" ring (25) from housing (9). Pull sleeve (13) and

spool (12) assembly from housing. Remove pin (14) and centering springs (11), then separate sleeve and spool. Pull control shaft (18) from cap (20). Remove "O" ring (19) from end cap (20). On early type units, remove oil seal (23), if necessary. On later type units, remove retainer ring (24), "O" ring retainer (27) and "O" rings (26 & 28), if necessary.

Remove cap screws, then pull power end housing (2) and power end shaft (5) from spool and sleeve housing. Lift lower plate (7), power end drive (6), gerotor assembly (8) and upper plate (7) from spool and sleeve housing. Remove power end shaft, "O" ring retainer (4) and "O" ring (3) from power end housing.

Wash all parts in suitable solvent and inspect as follows: Inspect center-

Fig. IH1318A — Farmall 504 power steering unit shown with steering wheel and cover removed. Rotary valve used on tractors with adjustable wide front tread axle are similar in appearance except for manifold which has two additional oil tubes.

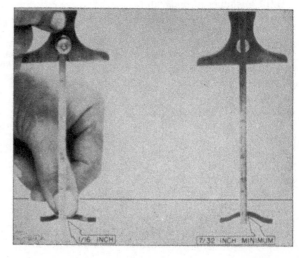

Fig. IH1318C — Centering springs are checked as shown. Refer to text.

ing springs for fractures or distortion. Springs (11) should return to a minimum height of 1/2-inch after being depressed to a height of 1/16-inch. See Fig. IH1318C.

Inspect plates (7—Fig. IH1318B) for wear and scoring. Polish patterns resulting from rotation of inner rotor are normal and care should be taken not to mistake these patterns for wear. Plates are lapped to within 0.0002 of being flat. Do not attempt to polish out any scratches as sealing of pressure depends upon the plate flatness. NOTE: In cases of emergency, plates that are damaged may be turned over so smooth side will be next to gerotor assembly, however, be sure damaged side is lapped until it is flat.

Inspect control valve spool, spool sleeve and valve body for nicks, scoring or undue wear. Damage of any one part will require renewal of all three as they are not available separately. Spool should fit sleeve and sleeve should fit body snugly yet both should move freely. Cross pin (14) should have a snug fit and if pin is bent or has a diameter of less than 0.2498 at its contacting points, renew pin.

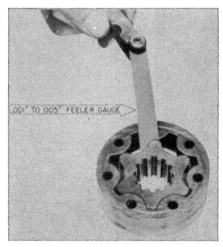

Fig. IH1318D—Measure tooth clearance of rotor and stator with gerotor set positioned as shown.

Fig. IH1318E—When reassembling, pin slot of control end drive must be aligned with any tooth valley of rotor.

Check gerotor assembly as shown in Fig. IH1318D. Clearance between teeth of inner rotor and outer rotor should be 0.001-0.005. Renew rotor assembly if clearance exceeds 0.005.

Inspect output shaft and output shaft housing for wear or damage. Measure outside diameter of output shaft (5—Fig. IH1318B) and inside diameter of housing (2) bore. If bore of housing is 0.006 or more larger than diameter of shaft, renew housing and/or shaft.

Inspect control (input) shaft (18), ball bearings (16), race (17), needle bearing (22) and oil seal (23) (early units) for wear, scoring or other damage and renew parts as necessary. On early units, use a press to install seal in end cap (20); do not drive on seal. When seal bottoms, it will be between flush and 1/64-inch below end of housing. Stake seal in position.

When reassembling, use all new "O" rings and with the exception of needle bearing (22) being coated with Lubriplate, and "O" ring (3) being oiled, all parts are assembled dry. NOTE: "O" ring (3) must be worked into its groove.

Reassemble by reversing disassembly procedure. However, BE SURE the pin slot in control end drive (10) is aligned with the valley of any tooth as indicated in Fig. IH1318E. If pin is not aligned as shown, unit will attempt to operate in reverse when hydraulic pressure is applied. Tighten cap screws retaining end caps evenly and to 12 ft.-lbs. torque.

IMPORTANT: When installing the seven screws which retain end housing (2—Fig. IH1318B) to housing (9), to be sure both cap screws and tapped holes are clean and perfectly dry. Apply one drop of "Loctite" to each cap screw, tighten evenly to a torque of 12 ft.-lbs. and allow to 8 to 12 hours for the "Loctite" to set before attempting to put unit in operation.

ROTARY VALVE

Farmall 504 With Adjustable Front Axle

35A. The rotary valve used on Farmall tractors with adjustable front axle is similar in size and configuration to the power steering unit used on the 504 Farmall tricycle tractors. However, it is not a torque generating unit, but is a self-centering valve which directs pressurized oil to the steering cylinder. Like the torque generating unit, it is mechanically connected to the steering shaft and allows mechanical steering should a hydraulic failure occur. Refer to Fig. IH1318F.

35B. REMOVE AND REINSTALL. Remove steering wheel and the cover for the rotary valve. See Fig. IH1318A. Disconnect steering shaft universal from lower end of rotary valve unit. Disconnect manifold flange and discard "O" rings, then unbolt and remove rotary valve from steering shaft lower support.

Use new "O" rings when reinstalling power steering unit.

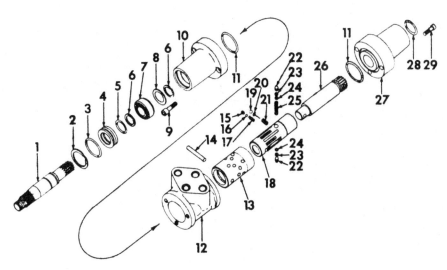

Fig. IH1318F — Exploded view of the power steering rotary valve used on Farmall 504 tractors equipped with adjustable wide tread front axle.

1. Control end (input) shaft
2. Retaining ring
3. "O" ring
4. "O" ring retainer
5. "O" ring
6. Bearing retainer
7. Bearing
8. Shim
9. Cap screw
10. Control end housing
11. "O" ring
12. Housing
13. Valve sleeve
14. Pin
15. Valve seat
16. Steel ball
17. Spring
18. Valve spool
19. Orifice ball
20. Jiggle wire
21. Spring
22. Centering piston
23. Back-up washer
24. "O" ring
25. Centering spring
26. Power end (output) shaft
27. Power end housing
28. "O" ring
29. Cap screw

35C. OVERHAUL ROTARY VALVE. With unit removed as outlined in paragraph 35B, proceed as follows: Remove retaining screws from top end cap (10—Fig. IH1318F) and remove cap and steering wheel (input) shaft (1) from valve body (12) as a unit. Pull sleeve (13), spool (18) and steering (output) shaft (26) from valve body, then remove the cross pin (14) and pull output shaft from sleeve and spool. Start spool out of sleeve and as centering pistons (22) and orifice ball (19) emerge from sleeve, grasp them with fingers to keep parts from flying. Do not attempt to remove check ball assembly (15, 16 & 17) from spool unless the assembly is damaged. If check ball assembly must be renewed, proceed as outlined in paragraph 35D.

35D. Place valve spool in a soft jawed vise with check ball seat upward. Use a 3/64-inch drill with a pin vise and drill through orifice of check valve seat (15). Use caution not to jam check ball into spring as drill nears inner end of valve seat and DO NOT use any other means of drilling except the pin vise. Select 1/16, 5/64, 3/32 and 7/64-inch drills, and in this sequence, drill through the valve seat with each drill. After drilling with the 7/64-inch drill, remove check ball and spring, then final clean the valve seat bore using a ⅛-inch drill in the pin vise. Use extreme care during this final drilling not to enlarge the valve seat bore.

Clean all parts in a suitable solvent and dry with compressed air or lint free wipers. Be sure all chips and other foreign material are removed from bores, holes or grooves of valve spool and sleeve. Small nicks or scratches can be removed with a No. 1 Arkansas stone. Sleeve should fit body and spool should fit sleeve snugly yet both should move freely when assembled. Inspect check valve spring and orifice valve spring for fractures, distortion or other damage. Be sure cross pin is straight and not unduly worn. Seal and bearings in end caps can be renewed and the procedure for doing so is obvious.

If check valve assembly was renewed, proceed as follows: Place check valve seat (15) on a flat surface with ball seat (small) end upward. Place check ball on seat and mate ball and seat by striking ball using a soft brass drift and a small ball peen hammer.

Place valve spool in a soft jawed vise with check valve bore upward. Install check valve spring in bore, then install ball on top of spring. Place check valve ball seat on installing tool (IHC No. FES-64-5) with the large flat end toward shoulder of installing tool, and holding installing tool at a 45-degree angle, drive check ball seat into its bore until outer end is flush. Rotate installing tool when removing. Remove any burrs which

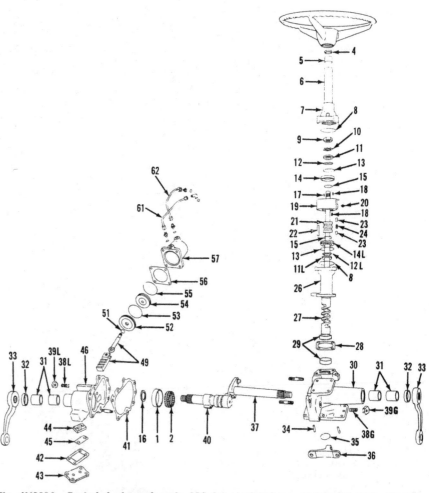

Fig. IH1319—Exploded view of early 330 International power steering gear assembly. The adjusting screw (38L) controls the lever shaft end play; screw (38G) controls the gear shaft end play. The rack adjusting shims (45) are available in thicknesses of 0.003, 0.005 and 0.020. Refer to Fig. IH1320 for unit used on later models.

1. Bearing race	16. Thrust washer	28. Gasket	43. Adjusting pad
2. Gear shaft roller	17. Restricting plug	29. Bushings (2 used)	plate
bearing	(7/64 O.D.)	30. Housing	44. Adjusting pad
4. Oil seal	18. Restricting plugs	31. Bushings (4 used)	45. Shims
5. Needle bearing	(9/64 O.D.)	32. Oil seals (2 used)	46. Cover
6. Tube	(2 used)	33. Steering arms	49. Piston rod and
7. Cover	19. Control valve body	34. Dowel pins	rack
8. "O" rings	20. Check valve	35. Expansion plug	51. "O" ring
9. Bearing adjust-	21. Control valve	36. Bracket	52. Cylinder adapter
ing nut	spool	37. Lever shaft	53. "O" ring
10. Lock washer	22. Inactive plungers	38G. Adjusting screw	54. Piston
11. Thrust bearing	(2 used)	38L. Adjusting screw	55. Piston ring
race	23. Active plungers	39G. Lock nut	56. Gasket
12. Thrust bearing	(6 used)	39L. Lock nut	57. Cylinder
13. "O" rings	24. Springs (3 used)	40. Gear shaft	61. Front pipe
14. Thrust bearing	26. Valve adapter	41. Gasket	62. Rear pipe
race	27. Cam (worm) and	42. Gasket	
15. "O" rings	shaft		

Fig. IH1318G — Oil cooler for 504 and 2504 tractors with power steering is mounted as shown.

may be present from seat, slot and/or bore.

NOTE: Do not use any tool other than IHC installing tool No. FES-64-5 and do not use air, or probe through hole, to unseat check ball as check ball could be forced into spring and the assembly rendered inoperative.

When reassembling, use all new "O" rings and dip all parts in IHC Hy-Tran fluid, or its equivalent. Install "O" rings and back-up washers on centering pistons with "O" rings toward flat end of piston. Place centering spring in bore, then install centering pistons in ends of bore with rounded ends toward outside of spool using care not to cut "O" rings as pistons are installed. Align orifice hole of valve spool with cross pin hole of valve sleeve and while depressing centering pistons, start valve spool into valve sleeve. Push valve spool into sleeve far enough to retain centering pistons in their slots and yet leave orifice hole exposed. Insert orifice wire through its spring, the insert wire and spring in orifice bore and be sure orifice hole wire is in the orifice hole. Place orifice ball in orifice hole, depress same and complete insertion of valve spool in valve sleeve. Install output shaft in lower end of valve spool, align pin holes of output shaft, valve spool and valve sleeve and install valve cross pin. Install lower end cap and seal on valve body, if previously removed, then install valve and output shaft assembly in valve body. Install steering wheel (input) shaft in top end cap and install assembly on valve body. Tighten both end cap retaining cap screws to a torque of 150 in.-lbs.

Complete assembly by reversing the disassembly procedure and bleed steering system as outlined in paragraph 26A.

OIL COOLER AND RELIEF VALVE
Series 504-2504

35E. Models equipped with power steering are fitted with an oil cooler mounted in front of radiator as shown in Fig. IH1318G. This unit is used to cool only the oil used by the power steering system. Also incorporated into the oil cooler system is a relief (by-pass) valve (O-Fig. IH1329) which opens and by-passes the oil to reservoir, should the oil cooler become plugged or if oil is too cold to circulate.

Oil cooler can be unbolted from radiator after removing hood and grille and disconnecting hoses. Relief valve can be removed at any time and procedure for doing so is obvious.

Relief valve can be bench tested and should open at 85 psi. Correct faulty units by renewing same.

STEERING CONTROL VALVE
International 330 and 340

36. R&R AND OVERHAUL. To remove the power steering valves, it is first necessary to remove the complete steering gear unit from tractor as outlined in paragraph 43.

With the gear unit removed from tractor, unbolt and remove the jacket tube and steering valve upper cover assembly. Refer to Fig. IH 1319 or 1320. Unbolt and remove the bearing adjusting nut (9) and lift out the up-

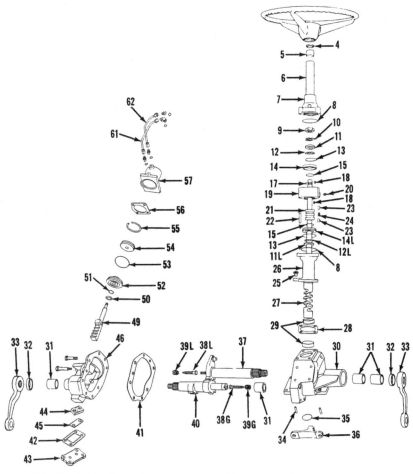

Fig IH1320—Exploded view of 340 and late 330 International power steering gear assembly. Governor control shaft pilot bushing is shown at (25) and back-up ring at (50). Refer to Fig. IH1319 for legend.

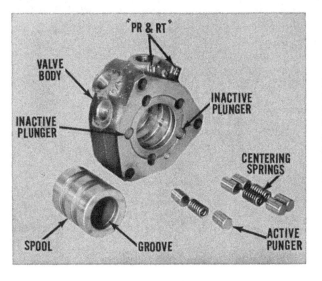

Fig. IH1321 — Series 330 and 340 International power steering control valve, with the control spool, active plungers and centering springs removed. The inactive plungers can be pressed from the valve body.

per thrust bearing (11, 12 & 14). Disconnect the oil lines from valve body and withdraw the valve assembly and lower thrust bearing. Be careful when removing the valve assembly and do not drop or nick any of the component parts.

Carefully slide the spool from the valve body and remove the six active plungers and the three centering springs as shown in Fig. IH 1321. The two inactive plungers can be pressed from the valve body at this time. Thoroughly clean all parts in a suitable solvent and be sure all passages and bleed holes are open and clean.

The valve spool and body are mated parts and must be renewed as an assembly if damaged. Each of the centering springs should have a free length of 0.703 and should test 11-15 pounds when compressed to a height of 0.638 inches. The plungers and springs are available separately.

When reassembling the control valve, be sure to renew all "O" rings and seals and proceed as follows: Install the lower race (11L—Fig. IH-1319 or 1320) of the lower thrust bearing in the adapter casting (26) with ball groove up. Pack the thrust bearing (12L) with wheel bearing grease and install bearing. Then, install upper race of lower thrust bearing with ball groove down. Press the two long inactive plungers into the valve body with one plunger hole between them as shown in Fig. IH1322. Refer to Fig. IH1321 and install the six active plungers and three springs in the remaining three plunger holes in valve body. Install the control spool in the valve body so that the identification groove in I.D. of spool is toward the same side of body as the port identification symbols "PR" and "RT." Refer to Fig. IH1323. Then install the assembled valve body with the symbols "PR" and "RT" up toward steering wheel. Install the upper thrust bearing (11, 12 and 14—Fig IH1319 or 1320) with larger diameter race (14) toward valve body and be sure to pack the bearing with wheel bearing grease. Install the tongued lockwasher and nut (9), tighten the nut just enough to hold the assembly from slipping and center the upper thrust bearing lower race as follows:

Using a pair of dividers or six-inch scale, measure the distance from outer edge of lower race (14) to outside edge of each of the five plunger holes. Shift the bearing race until these five measurements are identical. NOTE: The bearing race must be perfectly centered to prevent subsequent binding of the valve spool. Before tightening the adjusting nut, make certain that the lever shaft stud mesh, gear shaft end play and rack mesh are properly adjusted as in paragraphs 40, 41 and 42.

Install the steering wheel loosely on the shaft serrations and turn the steering wheel to the left, off the mid or straight ahead position, to lift the steering shaft to the upper extreme. Hold the steering wheel in this left turn position and tighten the valve adjusting nut to a torque of 10-12 Ft.-Lbs. If a suitable torque wrench is not available, a pair of ten inch multi-slip joint pliers can be used. Turn the steering wheel to both extreme positions several times and recheck the nut adjustment with the shaft raised or in the left turn position. Back-off the nut 1/12 turn (½ of a hex face) and bend two of the washer prongs against flats of nut.

Install the jacket tube assembly, mount a dial indicator and check the up and down movement of the steering (cam) shaft. When turning from the mid or straight ahead position to the extreme left, the shaft should move upward 0.050-0.055. When turning from the mid or straight ahead position to the extreme right, the shaft should move downward 0.050-0.055. In other words, when turning the steering wheel from one extreme position to the other, the shaft should have a total movement of 0.100-0.110. If the shaft movement is not as specified, it will be necessary to recheck the lever shaft stud mesh position and the valve nut adjustment.

Complete the assembly by installing the oil lines, instrument panel and steering wheel. Tighten the steering wheel retaining nut to a torque of 30-40 Ft.-Lbs.

STEERING CYLINDER
International 330 and 340

37. R&R AND OVERHAUL. To remove the power steering cylinder from an International 330 tractor, first drain the hydraulic system and remove the battery, generator regulator and the hydraulic manifold connecting pump to reservoir. On International 340 tractors, remove the hydraulic lift system control valves rear and side covers. On all models, unbolt and remove the rack adjusting pad and plate from bottom of gear housing. Disconnect oil lines from cylinder, then unbolt and remove cylinder and rack assembly from gear housing.

Using a plastic or lead hammer, bump the cylinder off the cylinder end plate and piston. Remove the nut retaining piston to piston rod, remove piston and withdraw the piston rod from the cylinder end plate. The procedure for further disassembly is evident. Thoroughly clean all parts in a suitable solvent and renew any damaged or worn parts. Be sure to examine rubbing surfaces of the gear rack and adjusting pad for excessive wear. Be sure to renew all rings, seals and gaskets.

When reassembling, lubricate all parts in power steering fluid and proceed as follows: Slide the piston rod

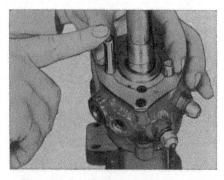

Fig. IH1322—Installing the inactive plungers in series 330 and 340 International power steering valve body. There must be one plunger hole (for active plungers) between the two inactive plungers as shown.

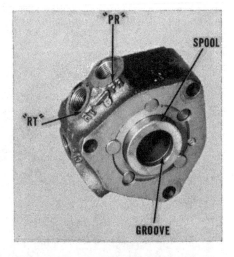

Fig. IH1323—Assembled steering valve for series 330 and 340 International tractors. The spool should be installed so that groove in I.D. of same is toward same side of valve body as the cast in "PR & RT" markings.

through the cylinder end plate and install piston so that cupped end will be toward the retaining nut. Install the self-locking nut and tighten same securely as shown in Fig. IH1324. Install piston, end plate and rod assembly into cylinder. Check the lever stud mesh position as in paragraph 40 and the gear shaft end play as in paragraph 41, then install the assembled cylinder so that the first tooth of the gear sector meshes with the first space in the power rack.

Install the rack adjusting plate pad and vary the number and thickness of shims between the adjusting pad and plate to provide a slight drag between the rack teeth and gear shaft teeth when the pad retaining screws are securely tightened; then, deduct one 0.003 shim, coat gasket with sealer and reinstall the adjusting pad and plate.

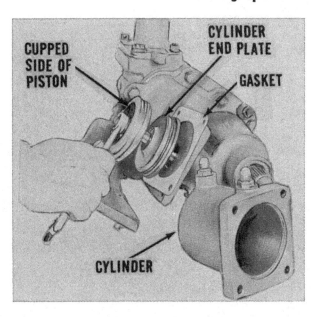

Fig. IH1324—On International 330 and 340 tractors, the power steering system piston should be installed with the cupped side up toward the retaining nut.

F504-I504-I2504

37A. R & R AND OVERHAUL STEERING CYLINDER. To remove the power steering cylinder, disconnect and immediately plug the hydraulic lines. Remove cap screws from the steering cylinder support (steering arm) and disengage from steering cylinder. Remove pin retaining anchor assembly to axle and remove cylinder from tractor.

With cylinder removed, move piston rod back and forth several times to clear oil from cylinder. Refer to Fig. IH 1324A and proceed as follows: Place end of piston rod which has the flats in a vise, then unscrew and remove anchor assembly (15). Remove cylinder head retainer (7) as follows: Lift end of retainer out of slot, then using a pin type spanner, rotate cylinder head (2) and work retainer out of its groove. Cylinder head and the piston rod and piston assembly (10) can now be removed from cylinder. Remove the remaining cylinder head in the same manner. All seals, "O" rings and back-up washers are now available for inspection and/or renewal.

Clean all parts in a suitable solvent and inspect. Check cylinder for scoring, grooving and out-of-roundness. Light scoring can be polished out by using a fine emery cloth and oil providing a rotary motion is used during the polishing operation. A cylinder that is heavily scored or grooved, or that is out-of-round, should be renewed. Check piston rod and cylinder for scoring, grooving and straightness. Polish out very light scoring with fine emery cloth and oil, using a rotary motion. Renew rod and piston assem-

bly if heavily scored or grooved, or if piston rod is bent. Inspect piston seal (9) for frayed edges, wear and imbedded dirt or foreign particles. Renew seal if any of the above conditions are found. NOTE: Do not remove the "O" ring (8) located under the piston seal unless renewal is indicated as it is not necessary to renew this "O" ring unless it is damaged. Inspect balance of "O" rings, back-up washers and seals and renew as necessary. Inspect bores of cylinder heads and renew same if excessively worn or are out-of-round.

Reassemble steering cylinder as follows: Place "O" ring (14), with back-up washer (13) on each side, in groove at inner end of anchor assembly oil tube. Install piston rod "O" ring (11) in groove on piston rod. Install wiper seal (1), back-up washer (3), piston rod "O" ring (4), cylinder head "O" ring (6) and back-up washer (5) to cylinder head, then install the cylinder head assembly over threaded end of piston rod.

Lubricate "O" ring and back-up washers on inner end of anchor assembly oil tube and carefully insert into threaded end of piston rod. Lubricate the piston rod "O" ring (11) and push anchor assembly toward piston rod. As "O" ring on inner end of oil tube approaches the drilled hole (port) in the piston rod (located near piston), use IHC tool FES 65, or equivalent, to depress "O" ring and washers so they will pass the port without being damaged. Screw anchor assembly onto piston rod and tighten to a torque. of 150 ft.-lbs. Lubricate piston seal and cylinder head "O" rings and using a ring compressor, or a suitable hose clamp, install piston and rod assembly into cylinder. Install cylinder head in cylinder so hole in cylinder head will accept nib of retaining ring, then position retaining ring and pull same into its groove by rotating cylinder head. Complete balance of re-assembly by reversing disassembly procedure.

1. Wiper seal
2. Cylinder head
3. Back-up washer
4. "O" ring
5. Back-up washer
6. "O" ring
7. Cylinder head retainer
8. Piston seal "O" ring
9. Piston seal
10. Piston and rod
11. "O" ring
12. Cylinder
13. Back-up washer
14. "O" ring
15. Anchor assembly

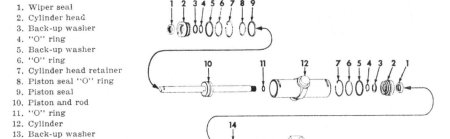

Fig. IH1324A—Exploded view of the power steering cylinder used on 504 Farmall tractors with adjustable wide tread front axle, and the International 504 and 2504 series tractors.

Reinstall unit on tractor, then fill and bleed the power steering system as outlined in paragraph 26A or 26B.

NOTE: Prior to installing steering cylinder on tractor, inspect the bushings in the cylinder support (steering arm) and renew if necessary. Bushings are available separately and renewal procedure is obvious. Farmall series tractors equipped with adjustable front axle are fitted with a one-piece cylinder support (steering arm) and any bushing service is readily accomplished after removing the steering arm clamps which frees the bushings.

GEAR UNIT

Farmall 340-504

38. The gear unit used on Farmall 340 and 504 tractors equipped with power steering is the same as the unit used without power steering. Refer to paragraphs 12, 18 and 21 for adjustment, removal and overhaul.

International 330 and 340

39. **ADJUSTMENT.** The steering gear unit is provided with the following adjustments. The mesh position between the lever shaft stud and cam shaft, the end play of the gear shaft and the power rack mesh position. These adjustments can be accomplished without removing the steering gear unit from tractor. The steering valve thrust bearings are adjusted by tightening the bearing adjusting nut to a torque of 10-12 Ft.-Lbs., but this should be done only when the valve unit is being serviced. Refer to paragraph 36.

Before attempting to make any adjustments, first make certain that the gear housing is properly filled with lubricant, then disconnect the drag links from the steering (Pitman) arms.

40. **LEVER SHAFT STUD MESH.** With the steering gear in the mid or straight ahead position, loosen the lock nut and back-off the gear shaft end play adjusting screw (38G—Fig. IH1319 or 1320) three or four turns. Then loosen the lock nut and turn the lever shaft adjusting screw (38L), located in (trunion) cover on right side of housing, until a slight drag is felt when turning the steering gear through the mid or straight ahead position. Tighten the lever shaft adjusting screw lock nut.

41. **GEAR SHAFT END PLAY.** With the lever shaft stud mesh position adjusted as outlined in para-

graph 40, turn the steering gear to the mid or straight ahead position and turn adjusting screw (38G—Fig. IH-1319 or 1320) to remove all end play from gear shaft without increasing the amount of pull required to turn the steering gear through the mid or straight ahead position.

42. **POWER RACK MESH.** Refer to Fig. IH1319 or 1320. With the rack adjusting plate (43) removed from the

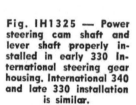

Fig. IH1325 — Power steering cam shaft and lever shaft properly installed in early 330 International steering gear housing. International 340 and late 330 installation is similar.

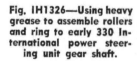

Fig. IH1326—Using heavy grease to assemble rollers and ring to early 330 International power steering unit gear shaft.

gear unit, add enough shims (45) between pad (44) and plate (43) to provide a slight drag between the rack teeth and gear shaft teeth when the pad retaining screws are securely tightened; then, remove the adjusting plate and deduct one 0.003 shim, coat gasket (42) with sealer and reinstall the adjusting plate.

43. **REMOVE AND REINSTALL.** To remove the steering gear unit first

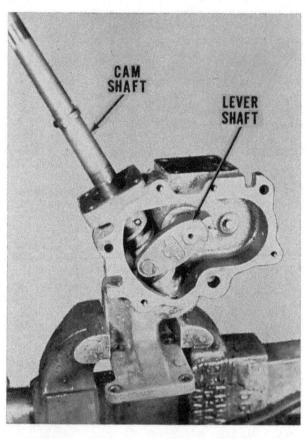

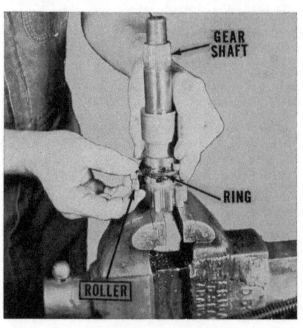

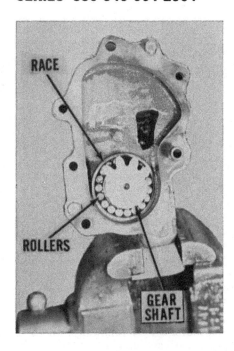

Fig. IH1327 — Early International 330 power steering gear shaft properly installed in housing side cover.

moved, unbolt the side cover from gear housing and remove the side cover and gear shaft. Withdraw lever shaft and cam shaft. Thoroughly clean and examine all parts for damage or wear. The lever shaft and gear shaft should be renewed if the spur gear teeth are damaged or worn. Inspect also the roller bearing at upper end of the jacket tube and on early model 330, gear shaft rollers and race. If any part of the roller bearing (outer race or rollers) is damaged, it will be necessary to renew the complete bearing, as component parts are not catalogued.

Inside diameter (new) of the lever shaft and gear shaft bushings is 1.374-1.375. Diameter of lever shaft and gear shaft at bearing surfaces is 1.3725-1.3735. Renew the shafts and/or bushings if running clearance is excessive. New bushings should be pressed into position with a suitable piloted arbor until outer ends of bushings are flush with inner edge of chamfered surface in bores. Factory recommendations state that bushings should be burnished after installation to an inside diameter of 1.374-1.375.

On all models, inside diameter (new) of the cam shaft bushings (29—Fig. IH1319 or 1320) is 1.6235-1.6250. Diameter of cam shaft at bushing surfaces is 1.620-1.621. Renew the cam shaft and/or bushings if clearance is excessive. Bushings may need to be sized after installation.

On early 330 International tractors, install the lever shaft and cam shaft as shown in Fig. IH1325. Using heavy grease, assemble the rollers and retainer ring to gear shaft as shown in Fig. IH1326. Then, install gear shaft in side cover as shown in Fig. IH1327. Install side cover and the steering (Pitman) arms.

On 340 and late 330 International tractors, install the lever shaft and cam shaft (Fig. IH1325); then, install gear shaft in side cover. Install side cover and the steering (Pitman) arms.

On all models, complete the gear unit overhaul by adjusting the lever stud mesh as in paragraph 40 and the gear shaft end play as outlined in paragraph 41. Then install the power cylinder and rack assembly and adjust the rack mesh position as in para-

Fig. IH1328 — Exploded view of the hand pump used on International 504 and 2504 series tractors.

1. Cap screw	11. Thrust bearing
2. End plate	12. Bearing race
3. Seal retainer	13. Body
4. Seal	14. Needle bearing
5. Pumping element	15. Seal
6. Spacer plate	16. Back-up washer
7. Link pin	17. Spacer
8. Drive link	18. Washer
9. Commutator	19. Retainer
9A. Commutator pin	20. Felt seal
10. Coupling (input)	21. Water seal
shaft	22. Nut

drain cooling system. Disconnect drag links from the steering (Pitman) arms; remove hood, battery and starting motor. Disconnect the heat indicator sending unit, fuel lines, oil pressure gage line, wiring harness and controls from engine and engine accessories. Disconnect head light and tail light wires. Disconnect the oil pressure and return lines from the steering valves. Unbolt steering gear housing and fuel tank, from tractor and using a hoist, lift the fuel tank, instrument panel and steering gear housing assembly from tractor.

Remove the steering wheel retaining nut and using a suitable puller, remove the steering wheel. Unbolt and remove the instrument panel assembly and fuel tank.

44. **OVERHAUL.** To overhaul the steering gear, first remove the unit from tractor as outlined in paragraph 43. Remove the steering (Pitman) arm retaining nuts and using a suitable puller, remove the Pitman arms from the steering lever shaft and gear shaft. Refer to paragraph 36 for removal, reinstallation and overhaul of the steering valves and to paragraph 37 for R&R and overhaul of the steering cylinder.

With the valves and cylinder re-

graph 42. Refer to paragraph 36 when installing the steering valves and adjusting the bearing nut.

HAND PUMP

International 504-2504

45. REMOVE AND REINSTALL. To remove the hand pump used in International 504 and 2504 tractors, remove hood, and steering wheel. NOTE: Use a puller to remove steering wheel, do not bump on upper end of steering wheel shaft. Unbolt instrument panel from instrument panel housing (cowl) and remove housing. Disconnect lines from hand pump, then unbolt and remove hand pump from steering shaft support.

Reinstall by reversing the removal procedure and bleed power steering system as outlined in paragraph 26B.

46. OVERHAUL MANUAL (HAND) PUMP. Remove the manual pump as outlined in paragraph 45. Clear fluid from unit by rotating steering wheel (input) shaft back and forth several times. Place unit in a soft jawed vise with end plate on top side, then remove end plate retaining cap screws and lift off end plate (2-Fig. IH1328).

NOTE: Lapped surfaces of end plate (2), pumping element (5), spacer (6), commutator (9) and pump body (13) must be protected from scratching, burring or any other damage as sealing of these parts depends only on their finish and flatness.

Remove seal retainer (3), seal (4), pumping element (5) and spacer (6) from body (13). Remove commutator (9) and drive link (8), with link pins (7) and commutator pin (9A), from body. Smooth any burrs or nicks which may be present on input shaft (10), wrap spline with masking tape, then remove input shaft from body. Remove bearing race (12) and thrust bearing (11) from input shaft. Remove snap ring (19), washer (18), spacer (17), back-up washer (16) and seal (15). Do not remove needle bearing (14) unless renewal is required. If it should be necessary to renew bearing, press same out pumping element end of body.

Clean all parts in a suitable solvent and if necessary, remove paint from outer edges of body, spacer and end plate by passing these parts lightly over crocus cloth placed on a perfectly flat surface. Do not attempt to dress out any scratches or other defects since these sealing surfaces are lapped to within 0.0002 of being flat. However, in cases of emergency, a spacer that is damaged on one side only may be used if the smooth side is positioned next to the pumping element and the damaged side is lapped flat.

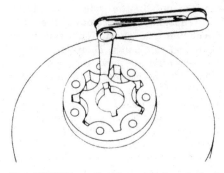

Fig. IH1328A — Position pumping element as shown to check tooth clearance. Refer to text.

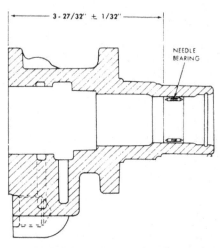

Fig. IH1328B — When renewing needle bearing in body, install same to dimensional shown.

Inspect commutator and housing for scoring and undue wear. Bear in mind that burnish marks may show, or discolorations from oil residue may be present, on commutator after unit has been in service for some time. These can be ignored providing they do not interfere with free rotation of commutator in body.

Check fit of commutator pin in the commutator. Pin should be a snug fit and if bent, or worn until diameter at contacting points is less than 0.2485, renew pin.

Measure inside diameter of input shaft bore in body and outside diameter of input shaft bearing surface. If body bore is 0.006, or more, larger than shaft diameter, renew shaft and/or body and commutator. Note: Body and commutator are not available separately.

Check thrust bearing and race for excessive grooving, flat spots or any other damage and renew bearing assembly if necessary.

Place pumping element on a flat surface and in the position shown in Fig. IH1328A. Use a feeler gage and check clearance between ends of rotor teeth and high points of stator. If clearance exceeds 0.003, renew pumping element. Use a micrometer and measure width (thickness) of rotor and stator. If stator is 0.002 or more wider (thicker) than the rotor, renew the pumping element. Pumping element rotor and stator are available only as a matched set.

Check end plate for wear, scoring and flatness. Do not confuse the polish pattern on end plate with wear. This pattern, which results from rotor rotation, is normal. Renew end plate if worn or scored and is not within 0.00002 of being flat.

When reassembling, use all new seals and back-up washers. All parts, except those noted below, are installed dry. Reassemble as follows: If needle bearing (14) was removed, lubricate with IH Hy-Tran fluid, install from pumping element end body and press bearing into bore until inside end measures $3\frac{13}{16}$-$3\frac{7}{8}$ inches from pumping element end of body as shown in Fig. IH1328B. Lubricate thrust bearing assembly with IH Hy-Tran fluid and install assembly on input shaft with race on top side. Install input shaft and bearing assembly in body and check for free rotation. Install a link pin in one end of the drive link, then install drive link in input shaft by engaging the flats on link pin with slots in input shaft. Use a small amount of grease to hold commutator pin in commutator, then install commutator and pin in body while engaging pin in one of the long slots of the input shaft. Commutator is correctly installed when edge of commutator is slightly below sealing surface of body. Clamp body in a soft jawed vise with input shaft pointing downward. Again make sure surfaces of spacer, pumping element, body and end plate are perfectly clean, dry and undamaged. Place spacer on body and align screw holes with those of body. Put link pin in exposed end of drive link, then install pumping element rotor while engaging flats of link pin with slots in rotor. Position pumping element stator over rotor and align screw holes of stator with those of spacer and body. Lubricate pumping element seal lightly with IH Hy-Tran fluid and install seal in seal retainer, then install seal and retainer over pumping element stator. Install end cap, align screw holes of end cap with those in pumping element, spacer and body, then install cap screws. Tighten cap screws evenly and to a torque of 18-22 ft.-lbs.

NOTE: If input shaft does not turn evenly after cap screws are tightened, loosen and retighten them again. However, bear in mind that the unit was assembled dry and some drag in normal. If stickiness or binding cannot be eliminated, disassemble unit and check for foreign material, nicks or burrs which could be causing interference.

Lubricate input shaft seal with IH Hy-Tran fluid and with input shaft splines taped to protect seal, install seal, back-up washer, spacer, washer and snap ring. The felt washer and water seal may be installed at this time but there will be less chance of loss or damage if installation is postponed until the time the steering wheel is installed.

After unit is assembled, turn unit on side with hose ports upward. Pour unit full of oil and work pump slowly until interior (pumping element) is thoroughly coated. Either plug ports or drain excess oil.

Reinstall unit by reversing the removal procedure and bleed steering system as outlined in paragraph 26B.

CONTROL (PILOT) VALVE

International 504-2504

46A. R&R AND OVERHAUL. Procedure for removal of control valve is obvious after an examination of the unit and reference to Fig. IH1329.

With valve removed, disassemble as follows: Refer to Fig. IH1329A and remove end caps (1) with "O" rings (2). Pull spool and centering spring assembly from valve body. Place a punch or small rod in hole of centering spring screw (3) and remove screw, centering spring (5) and centering spring washers (4) from spool. Remove plug (11), "O" ring (12) and circulating check ball (10). Remove retainer (16), seat (15), pressure check valve (14) and spring (13).

Do not disassemble safety relief valve (items 6 through 10) unless it is deemed absolutely necessary. Valve is set at factory and in normal operation, is seldom actuated. Therefore, it is unlikely that valve will be damaged. However, should it be necessary to disassemble valve, count and record the number of exposed threads on the adjusting screw and be sure to reinstall the adjusting screw to the same position.

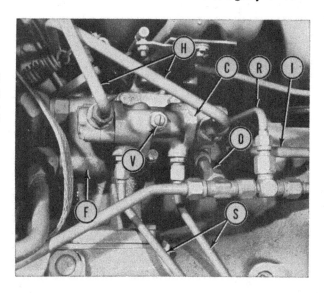

Fig. IH1329 — View of control (pilot) valve and flow divider valve mounting on International 504 and 2504 tractors.

C. Control valve
F. Flow divider valve
H. Hand pump lines
I. Oil cooler inlet line
O. Oil cooler relief valve
R. Return line
S. Steering cylinder lines
V. Safety relief valve

Wash all parts is a suitable solvent and inspect. Valve spool and spool bore in body should be free of scratches, scoring or excessive wear. Spool should fit its bore with a snug fit and yet move freely with no visible side play. If spool or spool bore is defective, renew complete valve assembly as spool and valve body are not available separately.

Inspect pressure check valve and seat. Renew parts if grooved or scored.

If safety relief valve was disassembled, inspect spring for fractures, distortion or lack of tension. Inspect ball for nicks and grooves. Renew parts as necessary. If seat is damaged, it will be necessary to renew complete valve as seat is not available separately. However, very light scoring or nicks can be removed from seat by using a new ball, with fine grinding compound, to lap the seat. Be absolutely certain that valve body is completely cleaned and all traces of grinding compound removed.

Reassembly is the reverse of disassembly and the following points should be observed. Coat all parts with Hy-Tran fluid, or its equivalent, prior to installation. If safety relief valve was disassembled, be sure the adjusting screw is installed with the same number of threads exposed as were exposed prior to removal. Measure distance between gasket surface of circulating check ball plug and inner end of roll pin. This distance should be $\frac{11}{16}$-inch and if necessary, obtain this measurement by adjusting roll pin in or out. Tighten end cap retaining cap screws to a torque of 22-27 ft.-lbs.

Reinstall valve by reversing removal procedure and bleed power steering

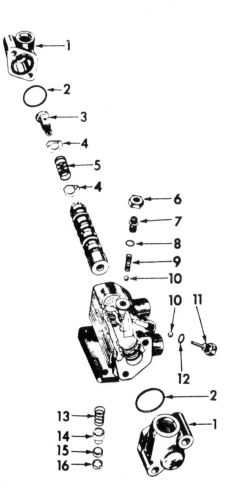

Fig. IH1329A — Exploded view of control (pilot) valve used on International 504 and 2504 tractors.

1. End cap	8. "O" ring
2. "O" ring	9. Spring
3. Centering spring screw	10. Ball
4. Centering spring washer	11. Plug assembly
5. Centering spring	12. "O" ring
6. Jam nut	13. Spring
7. Adjusting screw	14. Check valve
	15. Seat
	16. Retainer

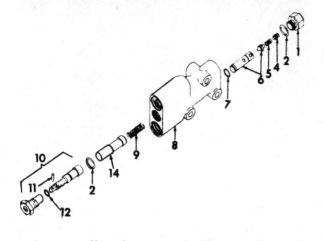

A. Adaptor
B. Valve body
P. Plug
R. Safety (relief) valve
S. Spring
V. Valve spool

Fig. IH1329C—Exploded view of the late type flow divider valve.

1. Adaptor
2. "O" ring
4. Adjusting screw
5. Spring
6. Safety (relief) valve
7. "O" ring
8. Valve body
9. Spring
10. Lock-out assembly
11. Roll pin
12. "O" ring
14. Spool

for scoring and/or excessive wear. Scored or worn parts will require renewal of both parts as they are not available separately. Spring (S) should have a free length of 4 5/64 inches and should test 31-36 lbs. at 2 1/2 inches. Do not attempt to disassemble or adjust safety (relief) valve (R). Faulty relief valves are corrected by renewing the unit.

To overhaul the later type flow divider valves. refer to Fig. IH1329C and proceed as follows: Unscrew gland nut and remove lock-out assembly (10) and "O" ring (2), then pull spool (14) and spring (9) from body (8). Remove valve stop (1) and "O" ring (2), then using a pair of needle-nose pliers, pull safety (relief) valve assembly (items 4, 5 and 6) from valve body. Inspect spool and bore in body for scoring and/or excessive wear. Scored or worn parts will require renewal of both parts as they are not available separately. Spring (9) should have a free length of 4 5/64 inches and should test 31-36 lbs. at 2 1/2 inches. Late type units having the poppet type relief valve as shown have parts catalogued separately.

NOTE: Safety valve can be bench tested providing proper equipment such as International Harvester test components FES64-7-6 (pump), FES 64-7-1 (test body), FES64-7-5 (gage), FES64-7-4 (petcock) and FES64-7-2 (plug) is available. Relief valve should open at approximately 1500 psi. Bear in mind when testing valves in this manner that pressures will be on the low side of the pressure range due to the small volume of fluid being pumped.

system as outlined in paragraph 26B.

NOTE: If necessary, the pressure setting of the safety relief valve can be checked and/or adjusted as outlined in paragraph 29G.

FLOW DIVIDER VALVE

Series 504-2504

46B. R&R AND OVERHAUL. Removal of flow divider valve is obvious after an examination of the unit and reference to Figs. IH1314B and IH1329.

NOTE: Valve shown in Fig. IH1329B was used on early Farmall and International series tractors. Valve with lock-out assembly, shown in Fig. IH-

1329C, replaced the early type valve. Lock-out is used to shut off oil flow to power steering circuit during some stationary operations where increased oil flow (volume) is needed for the auxiliary circuit. On models with lockout, BE SURE to restore flow to power steering circuit so power steering system will be operative before driving tractor.

To overhaul early type flow divider valve, refer to Fig. IH1329B and proceed as follows: Remove plug (P) and "O" ring, then remove valve spool (V) and spring (S) from body (B). Remove adaptor (A) and "O" ring using a pair of needle-nose pliers, pull safety (relief) valve cartridge (R) and "O" ring from valve body. Inspect spool and bore in body

ENGINE AND COMPONENTS (NON-DIESEL)

R&R ENGINE AND CLUTCH

Farmall 340

47. To remove the engine and clutch as an assembly, first remove the steering gear unit and front axle as outlined in paragraph 18; then, proceed as follows: On models equipped with a front mounted hydraulic pump, remove hydraulic lines between hydraulic pump and control valve support. Disconnect fuel lines, wiring harness and controls from engine and engine accessories.

Swing engine in a hoist, unbolt engine from clutch housing and move engine forward and away from tractor.

Farmall 504

47A. To remove engine and clutch as an assembly, proceed as follows: Remove muffler, precleaner, hood and hood skirts and side plates from radiator. Disconnect steering shaft center support and separate steering shaft. Attach split stand to side rails. If equipped with power steering, disconnect power steering lines (oil cooler)

at base of steering gear support and loosen retaining clips. Remove the two nuts from studs on top side of fuel tank front support and the two cap screws from front ends of instrument panel housing, then pry heat shield and tank upward and off the two studs. Disconnect air cleaner inlet tube bracket from fuel tank rear support, then remove tube and air cleaner assembly. Remove fuel supply line and governor control rod. Disconnect tachometer cable from drive unit. Remove air cleaner cup, then disconnect heat indicator sending unit. Disconnect wiring from generator, resistor, coil, oil pressure switch and horn (if so equipped), loosen clips on tappet cover and move wiring rearward. Disconnect choke control from carburetor. Remove the bottom dust cover from clutch housing. Remove the lower engine to clutch housing cap screws, then loop a chain under rear of engine and attach to side rails to provide support for rear of engine as the tractor is split. Place a jack under clutch housing and support clutch housing. Remove remaining engine to clutch housing cap screws and the side rail cap screws and separate tractor.

Remove radiator brace and disconnect upper radiator hose. Remove fan and fan spacer. Disconnect lower radiator hose. Remove tappet cover and rocker arms assembly, then attach a lifting chain under two end cylinder bolts to provide balance. Attach a hoist to lift chain and take engine weight, then remove engine front mounting bolts and lift engine from side rails. See Fig. IH1329D.

International 330-340-504-2504

48. To remove the engine and clutch as an assembly, first drain cooling system and if engine is to be disassembled, drain oil pan. Remove the hoods; disconnect radiator hoses and head light wires. Support the tractor under the clutch housing and make suitable provisions for supporting and moving the radiator and axle assembly. Disconnect power steering lines on series 504 and 2504. Disconnect the radius rod pivot bracket from clutch housing and on series 330 and 340, both drag links from the steering (knuckle) arms. Unbolt the bolster support brackets from engine and roll bolster, wheels, axle and radiator assembly forward and away from tractor.

Disconnect the heat indicator sending unit, hydraulic lines, fuel lines, oil pressure gage line, or wire on 504 and 2504, wiring harness and controls from engine and engine accessories.

Swing engine in a hoist, unbolt engine from clutch housing and move engine forward and away from tractor. Note: The two long bolts retaining top of clutch housing to engine on series 330 and 340 are unscrewed gradually as engine is moved forward. This procedure eliminates the need of removing the steering gear unit and usually saves considerable time.

Fig. IH1329D — Engine being removed from Farmall 504.

CYLINDER HEAD
All Models

49. To remove the cylinder head, first drain cooling system and remove hood assembly. On series 340, 504 and 2504, remove air cleaner and inlet tube. Disconnect wiring clips on tappet cover and if there is not enough slack in wiring loom to clear cylinder head, disconnect wires from generator. Disconnect temperature sending unit. On series 330 and 340, disconnect thermostat housing from cylinder head and by-pass hose from water pump. On series 504 and 2504, disconnect water pump by-pass hose and radiator top hose. Remove tappet cover, rocker arms assembly and push rods. Disconnect fuel supply line and choke control from carburetor, then remove carburetor and manifold assembly from cylinder head. Remove cylinder head bolts and lift cylinder head from cylinder block.

When reinstalling cylinder head, refer to Fig. IH1329E for tightening sequence and tighten cylinder head bolts to a torque of 80-90 ft.-lbs. for the C-135 engines, or 70-75 ft.-lbs. for the C-153 engines.

VALVES AND SEATS
All Models

50. Intake and exhaust valves are not interchangeable and on all models except 504 and 2504, seat directly in the cylinder head. Series 504 and 2504 exhaust valves seat on renewable inserts which are available in

standard size as well as oversizes of 0.015 and 0.030. All valves have a seat angle of 45 degrees. Seat width is 0.070-0.080. Valves on some of the early model tractors may be equipped with stem safety retainers to prevent valve from dropping into combustion chamber should a valve stem break. Valves have a stem diameter of 0.3405-0.3415 and a clearance of 0.0015-0.0035 in the guides with a maximum allowable clearance of 0.006.

Intake and exhaust valve tappet gap should be set to 0.014 hot for C-135 engines prior to engine serial number 100501. The C-135 engines, serial number and 100501 and up, and all C-153 engines have the intake valves set at 0.016 and the exhaust valves set at 0.020 hot.

When removing valve seat inserts, use the proper puller or pry them out with the edge of a large chisel. Do not attempt to drive chisel under seat insert as counterbore will be damaged. Chill new seat insert with dry ice or liquid Freon and when new insert is properly bottomed, it should be 0.008-0.030 below edge of counterbore. After installation, peen the cylinder head material around the complete outer circumference of the valve seat insert.

VALVE GUIDES AND SPRINGS
All Models

51. Intake and exhaust valve guides are interchangeable. Valve guides are pre-sized and if not distorted during installation, will require no final siz-

Fig. IH1329E — Tighten cylinder head bolts in the sequence shown and refer to text for torque values.

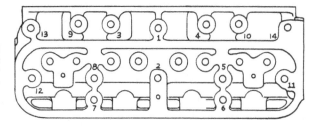

ing. Installed height of new guides is $1\frac{3}{16}$-inch above cylinder head surface. The 0.3405-0.3415 diameter valve stems should have a clearance of 0.0015-0.0035 in the installed guides. Maximum allowable stem clearance in guides is 0.006.

Intake and exhaust valve springs are interchangeable on valves without "Rotocap" valve rotators. Renew any spring which is rusted, discolored or does not meet the following specifications:

Free length, approximately

  With "Rotocap" ..... 2 1/4 inches
  Without "Rotocap" . 2 47/64 inches
Test load (lbs.) @ Length (in.)
  With "Rotocap" .. 72.4-79.4 @ 1.514
  Without "Rotocap"   49-54.6 @ 1.683

VALVE TAPPETS
(CAM FOLLOWERS)
All Models

52. The 0.560-0.561 diameter mushroom type tappets operate directly in the unbushed crankcase bores and can be removed after removing the camshaft as outlined in paragraph 58. Clearance of tappets in the crankcase bores should be 0.0005-0.0030 and should not exceed the I&T suggested limit of 0.005. Oversize tappets are not available.

VALVE LEVERS
(ROCKER ARMS)
All Models

53. The valve levers and hollow shaft assembly is pressure lubricated from the center camshaft bearing via an oiler stud in top surface of cylinder head.

Replacement valve levers are available only with installed bushings. Inside diameter of valve lever bushing should be 0.751-0.752 to provide 0.002-0.004 clearance on the 0.748-0.749 lever shaft. Maximum allowable clearance of lever on shaft is 0.006 and excessive clearance is corrected by renewing lever and/or shaft.

VALVE ROTATORS
All Models

54. Positive type exhaust valve rotators ("Rotocaps") are factory installed on some early models. Later models (C-153) are factory equipped with "Rotocoil" valve rotators on all valves.

Normal servicing of the valve rotators consists of renewing the units. It is important, however, to observe

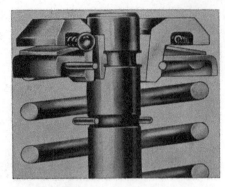

Fig. IH1330—Cut-away view showing typical installation of a "Rotocap" valve rotator.

the valve action after the engine is started. Rotator action can be considered satisfactory if the valve rotates a slight amount each time the valve opens. A cut-away view of a typical "Rotocap" installation is shown in Fig. IH1330.

VALVE TIMING
All Models

55. Valves are properly timed when single punch marked tooth on camshaft gear is meshed with the single punch marked tooth space on crankshaft gear as shown in Fig. IH1331.

TIMING GEAR COVER
All Models

56. To remove the crankcase front cover from Farmall 340 tractors, first drain cooling system and remove the complete front end assembly as outlined in paragraph 18.

On Farmall 504 tractors, remove fan and fan spacer and the lower splash pan.

Fig. IH1331 — View of the engine timing gear train showing the valve timing marks correctly aligned.

To remove the crankcase front cover from 330, 340, 504 and 2504 International tractors, first drain cooling system, remove hood and proceed as follows: Disconnect the upper and lower radiator hoses. Unbolt the fan from the hub and allow same to drop into radiator shrouds. Unbolt radiator from front support assembly and detach the radiator rear support bracket. Lift the radiator assembly from tractor.

On all models, remove governor housing. Remove crankshaft nut, attach suitable puller and remove crankshaft pulley and Woodruff key. Unbolt and remove cover from engine.

Extra care must be taken when installing the oil seal in the crankcase front cover, so as not to distort or bend the cover. Install seal with lip of same facing inward toward timing gears.

When reassembling, leave the cover retaining cap screws loose until crankshaft pulley has been installed; this will facilitate centering the seal with respect to the pulley.

NOTE: A new front oil seal has been introduced which incorporates a wear ring and this required a change in the hub section of crankshaft pulley. Also changed at the same time were the seal retainer plate, felt plug and gasket. Use caution when ordering new parts. If new oil seal assembly is used with early cover (IH part 46122-DC), enlarge hole to $2\frac{3}{32}$ inch and use sealant under wear ring.

TIMING GEARS
All Models

57. To renew the camshaft gear and/or crankshaft gear, it is necessary to use a suitable press after the respective shaft has been removed from engine.

When reassembling, mesh the single punch marked tooth on camshaft gear with the single punch marked tooth space on crankshaft gear and the double punch marked tooth on camshaft gear with the similarly marked tooth space on the governor and ignition unit drive gear.

CAMSHAFT
All Models

58. To remove the camshaft, first drain cooling system and remove the crankcase front cover as outlined in paragraph 56, and on 504 Farmalls, also remove the radiator. Remove the valve cover, valve levers and shaft assembly, push rods, oil pan and oil

pump. Push tappets up into their bores. Working through openings in camshaft gear, remove the cap screws retaining the shaft thrust plate to crankcase and carefully withdraw the camshaft and gear unit from engine. Gear can be removed from camshaft by using a suitable press.

Normal camshaft end play of 0.003-0.012 is controlled by the thrust plate located between the cam gear and crankcase. Excessive end play is corrected by renewing the plate.

The three camshaft bearing journals ride directly in the crankcase bores on early C-135 engines with a normal diametral clearance of 0.0009-0.0054. Maximum allowable clearance is 0.006.

NOTE: On later production C-135 engines and replacement engine blocks, and all C-153 engines, a bushing is used at the front camshaft journal. The intermediate and rear camshaft journals ride directly in the crankcase bores. Camshaft journal diameters and operating clearances have not changed.

Shaft journal sizes are as follows:
No. 1 (front) 1.811-1.812
No. 2 ....1.577-1.578
No. 3 ....1.499-1.500

Oil leakage at the shaft rear bearing journal is prevented by an expansion plug. Renewal of this plug requires splitting engine from clutch housing as outlined in paragraph 136, 136A or 137 and removing the flywheel.

ROD AND PISTON UNITS
All Models

59. Connecting rod and piston units are removed from above in the conventional manner after removing the cylinder head, oil pan and oil pump. Cylinder numbers are stamped on connecting rod and cap. When reassembling, make certain that the numbers are in register and face toward camshaft side of engine. Tighten the rod bolt self-locking nuts to a torque of 43-49 ft.-lbs. for the C-135 engines, and the self-locking bolts for the C-153 engine to 40-45 ft.-lbs.

PISTONS, SLEEVES AND RINGS
Series 330-340

60. Pistons are not available as individual replacement parts, but only as matched units with the wet type sleeves for standard compression ratio engines, special compression ratio engines and engines for operation at 5000 and 8000 foot altitudes. The matched units are available individually or in sets of four.

Recommended diametral clearance

of new pistons in new sleeves is 0.0011-0.0019 when measured between piston skirt and cylinder sleeve at 90 degrees to piston pin.

61. The wet type cylinder sleeves should be discarded when the out-of-round exceeds 0.006, or taper exceeds 0.005.

Special pullers are available to remove the wet type sleeves from above after the pistons have been removed. Before installing sleeves, check to make certain that the counterbore at top and sealing ring groove at the bottom are clean and free from foreign material. All sleeves should enter crankcase bores full depth and should be free to rotate by hand when tried in bores without sealing rings. After making trial installation without sealing rings, remove the sleeves and install new sealing rings dry into the grooves in crankcase. Wet the end of the sleeve with a thick soap solution or equivalent and install sleeves. If sealing ring is in place and not pinched, very little hand pressure is required to press the sleeve completely into place. Normally, the top of the sleeves will extend 0.003-0.007 above the machined top surface of the cylinder block. If sleeve stand out is excessive, check for foreign material under the sleeve flange.

Note: The cylinder head gasket forms the upper cylinder sleeve seal, and excessive sleeve stand out will result in coolant leakage. To test lower sealing rings for proper installation, fill crankcase (cylinder block) water jacket with cold water and check for leaks near bottom of sleeves.

62. All pistons are fitted with two compression rings and one oil ring. All compression rings should have an end gap of 0.010-0.020. Oil control ring should have an end gap of 0.018-0.028. Side clearance of rings in the piston grooves is 0.003-0.0045 for the top compression ring; 0.0015-0.003 for the other compression ring and the oil control ring.

Rings stamped top should be installed with word "TOP" facing toward top of engine.

Replacement ring sets are available to correct high oil consumption without renewing the pistons and sleeves. These rings, however, are not to be used with new pistons and sleeves nor if the wear exceeds the values which follow:

Sleeve out-of-round 0.006
Sleeve taper 0.005
Top ring clearance in groove...0.009

PISTONS AND RINGS
Series 504-2504

62A. The cam ground pistons operate directly in the block bores and are available in standard size as well as oversizes of 0.010, 0.020, 0.030 and 0.040. With pistons removed from engine, measure cylinder bores both parallel and at right angle to the crankshaft centerline. If taper from top of cylinder to bottom of piston travel exceeds 0.005, or if out-of-round exceeds 0.002, rebore cylinder to next larger size.

NOTE: When reboring, bore cylinder to within approximately 0.001 of desired size to allow finish honing.

To fit pistons in bores, attach a ½ x 0.0015 feeler ribbon to a spring scale, then invert piston and position feeler ribbon at 90 degrees from the piston pin hole. Insert piston and feeler ribbon into cylinder bore until piston is about three inches below top of cylinder block. Keep piston pin hole parallel with crankshaft. Now withdraw the feeler ribbon by pulling straight up on the spring scale and note reading on scale as feeler ribbon is being withdrawn. Pistons are correctly fitted to the normal 0.001-0.002 clearance when the spring scale pull reads 2-6 lbs.

Pistons are fitted with two compression rings and one oil control ring. The two compression rings should have an end gap of 0.010-0.020 and the oil control ring end gap should be 0.018-0.028. Side clearance of rings in piston grooves is 0.003-0.0045 for the top compression ring and 0.0015-0.003 for the second compression ring and the oil control ring.

Rings stamped top should be installed with word "TOP" facing top of piston.

Piston rings are available in standard size as well as oversizes of 0.010, 0.020, 0.030 and 0.040. Be sure to use the same ring size as that of the piston being used.

PISTON PINS
All Models

63. The full floating type piston pins are retained in the piston bosses by snap rings and are available in standard size as well as one oversize of 0.005. The 0.8591-0.8593 diameter pins should have a diametral clearance of 0.0002 in piston bosses and a diametral clearance of 0.0004 in the connecting rod bushing. Maximum allowable clearance of pin in piston is 0.0025 and in rod is 0.003.

CONNECTING RODS
AND BEARINGS
All Models

64. Connecting rod bearings are of the slip-in, precision type, renewable from below after removing the oil pan and connecting rod bearing caps. When installing new inserts, make certain that the projections on same engage slots in connecting rod and cap and that the cylinder identifying numbers on rod and cap are in register and face toward camshaft side of engine. Connecting rod bearings are available in standard size as well as undersizes of 0.002, 0.010, 0.020 and 0.030.

Bearing inserts should have a diametral clearance of 0.0009-0.0034 for C-135 engines prior to serial number 100501; 0.0009-0.0039 for C-135 engines after serial number 100501 and all C-153 engines, and connecting rods should have a side play of 0.005-0.014 on the crankshaft crankpins. Crankpin diameter for C-135 engines prior to serial number 100501 is 1.809-1.810 and for C-135 engines after 100501 and all C-153 engines, crankpin diameter is 2.059-2.060. Tighten the rod bolt self-locking nuts to a torque of 43-49 ft.-lbs. for C-135 engines and 40-45 ft.-lbs. for C-153 engine rod bolts.

CRANKSHAFT
AND MAIN BEARINGS
All Models

66. The crankshaft is supported in three slip-in, precision type main bearings, renewable from below after removing the oil pan, rear oil seal lower retainer plate and main bearing caps. Normal crankshaft end play of 0.004-0.010 on C-135 engines prior to serial number 100501; or 0.006-0.010 on C-135 engines after 100501 and all C-153 engines is controlled by the flanged rear main bearing inserts. Renew rear main bearing insert when end play exceeds 0.012. Renewal of the crankshaft requires R&R of engine. Main bearings are available in standard size as well as undersizes of 0.002, 0.010, 0.020 and 0.030.

Bearing inserts should have a diametral clearance of 0.009-0.0039 with a maximum allowable clearance of 0.0055 on the crankshaft main journals. Crankshaft main journal diameter for C-135 engines prior to serial number 100501 is 2.244-2.245. On C-135 engines after serial number 100501, and all C-153 engines, crankshaft main journal diameter is 2.6235-2.6245. Tighten the main bearing bolts to a torque of 75-80 ft.-lbs.

Fig. IH1333—View of the rear of the engine showing two piece crankshaft rear oil seal retainer plates installed.

CRANKSHAFT REAR OIL SEAL
All Models

67. The two piece felt rear oil seal contained in a two piece retainer as shown in Fig. IH1333 is used on C-135 engines prior to serial number 100501. Late production C-135 engines and all C-153 engines use a one-piece retainer and seal as shown in Fig. IH1333A. The procedure for renewing the seal is evident after splitting engine from clutch housing and removing the clutch and flywheel.

When installing the two piece retainer, be sure to renew all gaskets; and tighten the two vertical screws (S) before tightening those securing the retainer halves to crankcase.

FLYWHEEL
All Models

68. The flywheel can be removed after splitting engine from clutch housing as outlined in paragraph 136, 136A or 137 and removing the clutch. To install the flywheel ring gear, heat same to approximately 500 degrees F. It is not necessary to remove the flywheel to renew the ring gear.

Fig. IH1333A — View showing late type oil seal retainer and lip type oil seal.

OIL PUMP AND
RELIEF VALVE
All Models

69. The gear type oil pump is gear driven from a pinion on the camshaft and can be removed after removing the oil pan. The procedure for overhauling the pump is evident after an examination of the unit and reference to Fig. IH1334. Gaskets (3), between cover and pump body can be varied to obtain the recommended body gear end play. Check the pump parts against the values which follow:

Gear to body clearance...0.007-0.013
Body gears end play.....0.0035-0.006
Body gear backlash.......0.003-0.006
Relief valve spring free length..2.398
Lbs. test @ 1.674 inches.........24.2
Oil pressure at
  governed speed ....45-55 psi

The pump driving gear and shaft (11) are available only as a unit.

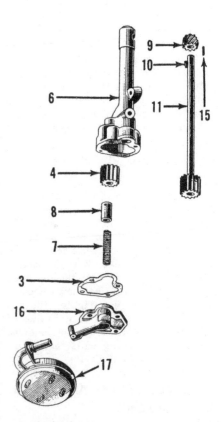

Fig. IH1334—Exploded view of oil pump. Number of gaskets (3) can be varied in order to obtain the desired body gear end play of 0.0035-0.006.

3. Gasket	10. Key
4. Idler gear	11. Driving gear and
6. Pump body	shaft
7. Relief valve spring	15. Pin
8. Relief valve	16. Cover
9. Drive pinion	17. Strainer

ENGINE AND COMPONENTS
(DIESEL)

R&R ENGINE AND CLUTCH
Farmall 340

70. To remove the engine and clutch as an assembly, first remove the steering gear unit and front axle as outlined in paragraph 18; then, proceed as follows: On models equipped with a front mounted hydraulic pump, remove hydraulic lines between hydraulic pump and control valve support. Disconnect fuel lines, wiring harness and controls from engine and engine accessories.

Swing engine in a hoist, unbolt engine from clutch housing and move engine forward and away from tractor.

Farmall 504

70A. To remove engine and clutch as an assembly, proceed as follows: Remove muffler, precleaner, hood and hood skirts and side plates from radiator. Disconnect steering shaft center support and separate steering shaft. Attach split stand to side rails. If equipped with power steering, disconnect power steering lines at base of steering gear support and loosen retaining clips. Remove the two nuts from studs on top side of fuel tank front support and the two cap screws from front end of instrument panel housing, then pry heat shield and tank upward and off the two studs. Disconnect air cleaner tube and remove tube and air cleaner. Shut off fuel and remove fuel supply line. Disconnect both ends of injection pump control rod and its bracket (guide) and remove control rod. Disconnect wire to glow plugs. Disconnect heat indicator sending unit. Disconnect wires from generator and oil pressure switch, disengage loom from clips and move wiring harness rearward out of the way. Remove bottom dust cover from clutch housing. Remove the lower engine to clutch housing cap screws, then loop a chain under rear of engine and attach to side rails to provide support for rear of engine as the tractor is split. Place a floor jack under clutch housing and support clutch housing. Remove remaining engine to clutch housing cap screws and the side rail cap screws and separate tractor.

Remove radiator brace and disconnect upper radiator hose. Remove fan and fan spacer. Disconnect lower radiator hose from radiator or water outlet elbow. Remove tappet cover and rocker arms assembly, then attach a lifting chain under two end cylinder head bolts to provide balance. Attach a hoist to lifting chain and take engine weight, then remove engine front mounting bolts and lift engine from side rails.

International 340-504-2504

71. To remove the engine and clutch as an assembly, first drain cooling system and oil pan. Remove the hoods; disconnect radiator hoses and head light wires. Support the tractor under the clutch housing and make suitable provisions for supporting and moving the radiator and axle assembly. On series 504 and 2504, disconnect power steering lines. Remove oil pan and disconnect the radius rod pivot bracket from clutch housing and on series 340, disconnect both drag links from the steering (knuckle) arms. Unbolt the bolster support brackets from engine and roll bolster, wheels, axle and radiator assembly forward and away from tractor.

Disconnect the heat indicator sending unit, hydraulic lines, fuel lines, oil pressure gage line, wiring harness and controls from engine and engine accessories.

Swing engine in a hoist, unbolt engine from clutch housing and move engine forward and away from tractor. Note: On series 340 the two long bolts retaining top of clutch housing to engine are unscrewed gradually as engine is moved forward. This procedure eliminates the need of removing the steering gear unit and usually saves considerable time.

CYLINDER HEAD
Series 340-504-2504

72. To remove the cylinder head on diesel tractors, first remove the hood and drain cooling system. Re-

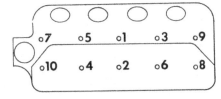

Fig. IH1335 — The diesel engine cylinder head retaining cap screws should be tightened in the sequence shown and to a torque of 110-120 Ft.-Lbs.

move high pressure fuel lines from injection pump to nozzles and cap all openings.

NOTE: All fuel connections should be capped **immediately** after they are disconnected to prevent entrance of dirt or other foreign material.

Remove hose from inlet manifold to air cleaner and both manifolds. Disconnect the coolant by-pass hose from thermostat housing and remove generator. Remove thermostat housing and temperature gage bulb from cylinder head and disconnect main glow plug wire from junction block. On series 504 and 2504, disconnect oil cooler line from cylinder head. Detach fuel filters from cylinder head. Remove rocker arm cover. On International models, unbolt and raise front of fuel tank. Remove rocker arms assembly. Remove the remaining cylinder head retaining cap screws, attach a hoist and lift off head.

Guide studs should be used to position the gasket and head. The cylinder head retaining screws should be tightened in the sequence shown in Fig. IH1335 and to a torque of 110-120 ft.-lbs. The head retaining cap screws should be retorqued after the engine has been run and is at operating temperature and again after engine has been run 100 hours. Tappet gap should be adjusted to 0.027 hot for both the inlet and exhaust.

VALVES AND SEATS
Series 340-504-2504

73. Inlet and exhaust valves are not interchangeable. Both the inlet and the exhaust valves seat directly in the cylinder head with a face and seat angle of 45 degrees. Valve seats can be narrowed if necessary by using a 15 degree stone. Both the inlet and the exhaust valves are equipped with "Roto-Cap" (early) or "Roto-Coil" (late) valve rotators. Check the valves and seats against the specifications which follow:

Inlet and Exhaust
Stem diameter 0.3715-0.3725
Stem to guide diametral
　clearance 0.0015-0.004
Seat width 3/32
Total valve run-out, less than..0.002
Total valve seat run-out,
　less than 0.002
Tappet gap (hot) 0.027

VALVE GUIDES AND SPRINGS
Series 340-504-2504

74. The inlet and the exhaust valve guides are interchangeable. Inlet and the exhaust guides should be pressed in head until top of guide is 15/16-inch above the spring recess in the head. Guides are pre-sized and if carefully installed should need no final sizing. Inside diameter should be 0.3740-0.3755 and valve stem to guide diametral clearance should be 0.0015-0.004.

Inlet and exhaust valve springs are interchangeable. The spring has 7¾ coils, a free length of 2 7/16 inches and should exert 146-156 pounds when compressed to 1.592 inches. Renew any spring which is rusted, discolored or does not meet the pressure test specifications.

VALVE TAPPETS
(CAM FOLLOWERS)
Series 340-504-2504

75. Tappets are of the barrel type and operate directly in the unbushed bores of the cylinder block. Tappet diameter is 0.9975-0.9985 and should have a normal operating clearance of 0.0014-0.003 in the 0.9999-1.0005 diameter tappet bore. Oversize tappets are not available and excessive tappet clearance is corrected by renewing tappets and/or cylinder block. Tappets can be removed from the side of the crankcase after removing valve tappet levers (rocker arms), push rods and side cover plate.

VALVE TAPPET LEVERS
(ROCKER ARMS)
Series 340-504-2504

76. **REMOVE AND REINSTALL.** Removal of the rocker arms assembly may be accomplished in some cases on series 340, without raising front of fuel tank. The rocker shaft bracket hold down screws are also the left row of cylinder head cap screws. Removal of these screws will allow the left side of the cylinder head to raise slightly and damage to the head gasket will in most cases result. Because of the probable damage to the head gasket, it is recommended that the cylinder head be removed as in paragraph 72 when removing the rocker arm assembly.

77. **OVERHAUL.** The rocker arm shaft has an outside diameter of 0.748-0.749. The bushings in the rocker arms are not renewable and the inside diameter of bushings is 0.7505-

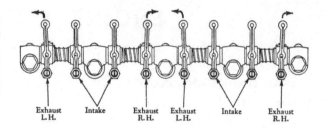

Fig. IH1335A—On diesel engines, the inlet rocker arms are interchangeable; but, the exhaust valve rocker arms should be off-set toward the nearest shaft bracket as shown.

Exhaust L.H. Intake Exhaust R.H. Exhaust L.H. Intake Exhaust R.H.

0.752 which provides a normal operating clearance of 0.0015-0.004. Renew levers and/or shaft if clearance exceeds 0.006. All inlet rocker arms are the same; but, the exhaust rocker arms are offset towards the nearest rocker shaft bracket (refer to Fig. IH1335A). Rocker shaft brackets numbers 2 and 4 are provided with dowel sleeves.

VALVE ROTATORS
Series 340-504-2504

78. Positive type valve rotators, "Roto-Caps" (early) or "Roto-Coil" (late), are installed on both inlet and exhaust valves of diesel engines.

Normal servicing of the valve rotators consists of renewing the units. It is important, however, to observe the valve action after engine is started. The valve rotator action can be considered satisfactory if the valve rotates a slight amount each time the valve opens.

VALVE TIMING
Series 340-504-2504

79. Valves and injection pump drive gear are properly timed when timing (punched) marks are properly aligned as shown in Fig. IH1335B. Timing marks can be seen after removing cover as outlined in paragraph 80.

Fig. IH1335B — The diesel engine timing gear train and timing marks.

CA. Camshaft gear
CR. Crankshaft gear
ID. Idler gear
IN. Injection pump drive gear

TIMING GEAR COVER
Series 340-504-2504

80. **REMOVE AND REINSTALL.** To remove the timing gear cover on series 340 Farmall tractors, first remove steering gear and radiator assembly as outlined in paragraph 18.

To remove the timing gear cover on series 504 Farmall tractors, it is not necessary to remove the front bolster and radiator.

To remove the timing gear cover on International tractors, first remove hood and the hood side sheets (skirts). Drain coolant and disconnect radiator hoses. Support tractor under clutch housing and sling the radiator and front end assembly in a hoist. On series 340, disconnect the drag links from the steering knuckle arms. Disconnect the radius rod pivot bracket from the clutch housing. Drain and remove engine oil pan. Remove cap screws which attach front axle support to engine and move complete radiator and axle assembly forward away from tractor.

On all models, remove fan, water pump, generator drive belts and generator. Remove the retaining nut and pull crankshaft pulley from shaft using a suitable puller.

Remove the timing gear cover retaining cap screws, then pull cover forward off dowels and remove from engine.

TIMING GEARS
Series 340-504-2504

81. **CRANKSHAFT GEAR.** The crankshaft gear is keyed and press fitted to the crankshaft. The gear can be removed using a suitable puller after first removing the timing gear cover as outlined in paragraph 80.

Before installing, heat gear in oil; then, buck-up crankshaft with a heavy bar while drifting heated gear on shaft. Make certain timing marks are aligned as shown in Fig. IH1335B.

82. **CAMSHAFT GEAR.** The camshaft gear is keyed and press fitted to camshaft. Backlash between the camshaft gear and the crankshaft gear should be 0.0032-0.0076. The camshaft

gear can be removed using a suitable puller after first removing the timing gear cover as outlined in paragraph 80 and the gear retaining nut.

Before installing, heat gear in oil; then, buck-up camshaft with heavy bar while drifting heated gear on shaft. The gear should butt up against a shoulder on the camshaft. Tighten the gear retaining nut to a torque of 110-120 Ft.-Lbs. Make certain timing marks are aligned as shown in Fig. IH1335B.

83. **IDLER GEAR.** To remove the idler gear, it is first necessary to remove timing gear cover as outlined in paragraph 80. The idler gear shaft is attached to the front of the engine block by a cap screw.

The idler gear shaft diameter should be 2.0610-2.0615 and clearance between shaft and the renewable bushing in gear should be 0.0014-0.003. Make certain that the oil passage in shaft is open and clean. Backlash between idler gear and crankshaft gear should be 0.003-0.006.

When reinstalling, make certain that the timing marks are aligned as shown in Fig. IH1335B. Make certain that the dowel in the shaft engages the hole in the engine front plate. The shaft retaining cap screw should be torqued to 85-95 ft.-lbs.

84. **PUMP DRIVE GEAR.** To remove the injection pump drive gear it is first necessary to remove the timing gear cover as outlined in paragraph 80. The drive gear and shaft can be removed by merely withdrawing the unit from injection pump. The gear can be removed from the hub after the three retaining cap screws are removed. The hub can be pressed from the shaft after the retaining nut is removed.

CAUTION: Take care not to damage the double opposed lip seals on the pump drive shaft, the thrust plunger and spring assembly in the forward end of the shaft and/or the bushing in the injection pump housing.

Backlash between pump drive gear and idler gear should be 0.003-0.009.

Before installing the pump drive shaft and hub unit, observe rear end of shaft where a small off-center drilled hole is located. During assembly this hole must register with a similar hole in the injection pump rotor or the pump will be 180 degrees out of phase. Lubricate seals with grease prior to installation. When

installing, make certain that the timing marks on all gears are exactly aligned as shown in Fig. IH1335B. Refer to paragraph 113 and retime the injection pump.

CAMSHAFT AND BEARINGS
Series 340-504-2504

85. **CAMSHAFT.** To remove the camshaft, first remove the rocker arms assembly as outlined in paragraph 76 and the timing gear cover as in paragraph 80. Remove the engine side cover and remove the cam followers (tappets). Working through openings in camshaft gear, remove the camshaft thrust plate retaining cap screws and withdraw camshaft from engine.

Recommended camshaft end play of 0.002-0.010 is controlled by the thrust plate.

Check the camshaft against the values which follow:

Journal diameter
No. 1 (front) 2.109-2.110
No. 2 ....2.089-2.090
No. 3 (rear) 2.069-2.070

When installing the camshaft, reverse the removal procedure and make certain that the valve timing marks are in register as shown in Fig. IH-1335B. The camshaft thrust plate retaining cap screws should be torqued to 35-40 ft.-lbs.

86. **CAMSHAFT BEARINGS.** To remove the camshaft bearings, first remove the engine as in paragraph 70, 70A or 71; then, remove camshaft as in the preceding paragraph 85. Remove clutch, flywheel and the engine rear end plate. Extract the plug from behind the camshaft rear bearing and remove the bearings.

Using a closely piloted arbor, install the bearings so that oil holes in bearings are in register with oil holes in crankcase. The chamfered end of the bearings should be installed towards the rear.

The camshaft bearings are pre-sized and if carefully installed should need no final sizing. The camshaft bearing journals should have a diametral clearance in the bearings of 0.0005-0.005. Maximum allowable clearance is 0.006.

When installing the soft plug at rear camshaft bearing, use Permatex or equivalent to obtain a better seal.

ROD AND PISTON UNITS
Series 340-504-2504

87. Connecting rod and piston assemblies can be removed from above after removing the cylinder head as

outlined in paragraph 72 and the engine balancer as in paragraph 91.

Cylinder numbers are stamped on the connecting rod and cap. Numbers on rod and cap should be in register and face toward the camshaft side of engine. The arrow stamped on the tops of the pistons should point toward front of tractor. Tighten the connecting rod bolts to a torque of 40-50 ft.-lbs.

PISTONS, LINERS AND RINGS
Series 340-504-2504

88. The cam ground, aluminum pistons are fitted with two compression rings and two oil control rings and are available in standard size only.

Install the compression rings with the tapered side towards the top of the piston. Install the oil rings with the scraper edges towards the bottom of the piston.

With the piston and connecting rod unit removed from block, use a suitable puller to remove the dry type cylinder sleeve (liner). Clean the mating surfaces of block and liner before new liner is installed. Top of liner should be from 0.002-0.005 for the D-166 engine, or 0.001-0.004 for the D-188 engine, above the top of the cylinder block. Sleeve stand-out adjustment shims of 0.003 and 0.005 thickness are available. Shims should be inserted in sleeve counterbore. Variation of stand-out between adjoining cylinders should be held to below 0.002. Excessive stand-out will cause leakage at head gasket. Check pistons, rings and sleeves against the values which follow:

Ring end gap
　Top compression ......0.015 -0.025
　Second compression ...0.010 -0.020
　Oil ....0.010 -0.023
Ring side clearance
　Top compression ..... tapered
　Second compression ...0.0025-0.0040
　Oil 0.0025-0.0040
Cylinder liner I.D., new..3.6875-3.6891
　Renew liner if I.D. exceeds...3.6971
　Renew liner if taper exceeds..0.008
Desired diametral clearance
　between piston (at right
　angles to piston pin at
　bottom of skirt) and cyl-
　inder liner 0.004-0.0056

PISTON PINS
Series 340-504-2504

89. The full floating type piston pins are available in standard and 0.005 oversize. Check piston pin against the values which follow:

Piston pin diameter.....1.1247-1.1249
Piston pin diametral clearance in piston (pin at
  70°F.; piston at 170°F.) 0.0000-0.0004
Piston pin diametral clear-
  ance in rod bushing ..0.0005-0.0009

ENGINE BALANCER
Series 340-504-2504

90. Diesel engines are equipped with an engine balancer which is mounted on the underside of the engine crankcase and driven by a gear on the crankshaft located just behind the center main bearing journal. Refer to Fig. IH1335C. The balancer consists of two unbalanced gear weights which rotate in opposite directions at twice crankshaft speed, thus setting up forces which tend to counteract the vibration which is inherent to four cylinder engines with a firing order of 1-3-4-2. Therefore, it is extremely important that the balancer weights be correctly timed to each other and the complete unit be correctly timed to the crankshaft. Timing marks are provided to accomplish this.

91. **R&R AND OVERHAUL.** To remove the engine balancer, drain oil and remove oil pan. Remove oil pump. Position number one piston at top dead center, then unbolt and remove balancer unit and shims from cylinder block. Before disassembling balancer, check backlash between weight gears which should be 0.003-0.009. If backlash exceeds 0.009, renew weight gears when reassembling.

To disassemble unit, first clean with a suitable solvent, then drive out the roll pins which retain weight gear shafts in housing. Press or bump shafts out roll pin side of housing and lift out weight gears.

Inspect shafts, bushings, gear teeth and all other parts for excessive wear or damage. Refer to the table of specifications which follows to determine parts renewal and operating clearances. Bushings (19—Fig. IH1335C) should be reamed after installation to an inside diameter of 0.752-0.753.
Backlash-Crankshaft gear
  to weight gear 0.004-0.015
Backlash-Between weight
  gears 0.003-0.009
Weight gear bushing
  I. D. 0.752-0.753
Shaft diameter-bearing
  surface 0.7495-0.750
Weight gear-Operating
  clearance on shaft 0.002-0.0035
Weight gear-End clearance
  in housing 0.008-0.016
Reassemble by reversing the disassembly procedure, however, lubricate

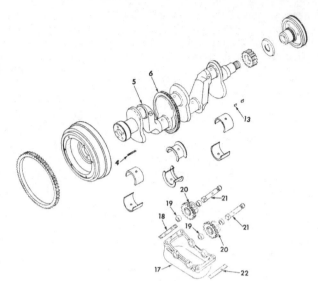

Fig. IH1335C—Exploded view of the vibration damper (balancer) assembly used on diesel engines.

4. Flywheel dowel
5. Crankshaft
6. Damper drive gear
13. Woodruff keys
17. Damper frame
18. Shims (left)
19. Bushings
20. Weight gears
21. Shafts
22. Shims (right)

parts as they are installed and BE SURE timing marks on weight gears are aligned.

When reinstalling balancer on crankcase, BE SURE that chamfered tooth of weight gear is between the chamfered teeth of crankshaft gear. Vary shims (18 and 22) to provide 0.004-0.015 backlash between crankshaft gear and weight gear.

NOTE: When installing shims be sure shims with oil hole are on side of housing which has the oil passage hole.

CONNECTING RODS AND BEARINGS
Series 340-504-2504

92. Connecting rod bearings are of the slip-in, precision type, renewable from below after removing the engine balancer and rod bearing caps. When installing new bearing shells, make certain that the rod and bearing cap numbers are in register and face toward camshaft side of engine. Bearing inserts are available in standard size as well as undersizes of 0.002, 0.010, 0.020 and 0.030. Check the crankshaft crankpins and the connecting rod bearings against the values which follow:

Crankpin diameter 2.373-2.374
Max. allowable out of round....0.0015
Max. taper 0.0015
Diametral clearance 0.0009-0.0034
Side play 0.007-0.015
Rod bolt torque, ft.-lbs. 40-50

CRANKSHAFT AND MAIN BEARINGS
Series 340-504-2504

93. The crankshaft is supported in three main bearings and the end

thrust is taken by the center bearing. Main bearings are of the shimless, non-adjustable, slip-in, precision type, renewable from below after removing the oil pan, balancer and main bearing caps. Renewal of crankshaft requires R&R of engine. Check crankshaft and main bearings against the values which follow:

Crankpin
  diameter 2.373-2.374
Main journal diameter.....2.748-2.749
Max. allowable out of round...0.0015
Max. taper 0.0015
Crankshaft end play 0.005-0.013
Main bearing:
  diametral clearance ..0.0012-0.0042
  bolt torque, ft.-lbs. 75-85
Main bearings are available in standard size as well as undersizes of 0.002, 0.010, 0.020 and 0.030.

CRANKSHAFT SEALS
Series 340-504-2504

94. **FRONT.** To renew the crankshaft front oil seal on 340 Farmall tractors, first remove the steering gear and radiator assembly as outlined in paragraph 18.

NOTE: On Farmall 504 tractor, sufficient working room is available and the front bolster need not be removed.

To renew the crankshaft front oil seal on International tractors, first remove hood and the hood side sheets (skirts). Drain coolant and disconnect radiator hoses. Support tractor under clutch housing and sling the radiator and front assembly in a hoist. On International 504 and 2504, disconnect the power steering lines. On International 340 tractors, disconnect drag links from steering knuckle arms. Disconnect radius rod pivot bracket from clutch housing. Remove oil pan. Remove the cap screws which attach front axle support to engine and move the complete radiator and axle assembly forward away from the tractor.

On all models, remove the generator drive belt. Remove the retaining nut and pull the crankshaft pulley from the shaft using a suitable puller.

Remove and renew the seal in the conventional manner. Check the condition of the crankshaft pulley seating surface and renew or recondition pulley if surface is not perfectly smooth.

95. REAR. To renew the crankshaft rear oil seal, first remove the flywheel as outlined in paragraph 96. The lip type seal can be removed after collapsing same. NOTE: Take care not to damage the sealing surface of the crankshaft as the rear seal is collapsed, removed and new seal is installed. The part number stamped on the seal should face towards rear.

FLYWHEEL

Series 340-504-2504

96. The flywheel can be removed after first splitting engine from clutch housing and removing the clutch as outlined in paragraph 136, 136A or 137.

To install the flywheel ring gear, it is necessary to first heat same to approximately 500 deg. F.

The flywheel retaining cap screws should be tightened to 45-55 Ft.-Lbs. of torque.

OIL PUMP AND RELIEF VALVE

Series 340-504-2504

97. The gear type oil pump, which is gear driven from a pinion on the camshaft, is accessible for removal after removing the engine oil pan.

Disassembly and overhaul of the pump is evident after an examination of the unit and reference to Fig. IH1335D. Gaskets between pump cover and body can be varied to obtain the recommended 0.0025-0.0055 pumping (body) gear end play. Refer to the following specifications:

Pumping gears recommended
backlash 0.003-0.006

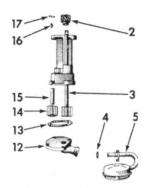

Fig. IH1335D—Exploded view of the engine oil pump. The body gear end play is controlled by the number of gaskets (13). The upper end of the drive shaft is supported by the tachometer drive shaft.

2. Drive gear	13. Gasket
3. Pump drive shaft and body gear	14. Follower gear
4. "O" ring	15. Idler shaft
5. Screen	16. Woodruff key
12. Cover	17. Pin

Fig. IH1335E — Replacement pump shaft and gear assemblies (3—Fig. IH 1335D) may need to have a ⅛-inch hole drilled as shown.

1-3/32"   1/8" hole

Pump drive gear
recommended backlash..0.000-0.008
Pumping (body) gear
end play 0.0025-0.0055
Gear teeth to body
radial clearance 0.0068-0.0108
Drive shaft clearance
in bore 0.0015-0.003
Pump mounting bolt
torque, ft.-lbs. 20-23

Service (replacement) pump shaft and gear assemblies may not be drilled to accept the pump driving gear pin. A ⅛-inch hole must be drilled through the shaft after the gear is installed on the shaft to the dimensions shown in Fig. IH1335E.

98. The spring loaded plunger type oil presure relief valve is located in the left side of the crankcase and is non-adjustable. The spring should be installed with the closed coils in the plunger. Oil pressure should be 38-46 psi at approximately 1800 engine rpm. Check the relief valve plunger and spring against the values which follow:

Valve plunger diameter....0.743-0.745
Plunger diametral clearance in bore 0.003-0.007
Spring free length
(approximately) ....3 inches
Spring test and
length ....18 lbs. @ 1¹³⁄₁₆ inches

OIL COOLER

98A. Diesel engines are fitted with an oil cooler which is mounted on left side of engine. Removal is accomplished by disconnecting coolant and oil lines and unbolting unit from engine.

No repair parts are available for the oil cooler assembly and faulty units must be renewed.

DIESEL FUEL SYSTEM

The diesel fuel system consists of three basic components; the fuel filters, injection pump and injection nozzles. When servicing any unit associated with the fuel system, the maintenance of absolute cleanliness is of utmost importance. Of equal importance is the avoidance of nicks or burrs on any of the working parts.

Probably the most important precaution that service personnel can impart to owners of diesel powered tractors is to urge them

to use an approved fuel that is absolutely clean and free from foreign material. Extra precaution should be taken to make certain that no water enters the fuel storage tanks. This last precaution is based on the fact that all diesel fuels contain some sulphur. When water is mixed with sulphur, sulphuric acid is formed and the acid will quickly erode the closely fitting parts of the injection pump and nozzles.

SYSTEM CHECKS

99. The complete diesel system should be checked as outlined in the following paragraphs whenever the diesel engine does not operate properly.

100. STATIC TIMING. To check the injection pump static timing, proceed as follows: Turn the crankshaft until number one piston is coming up on compression stroke and continue

Fig. IH1336 — The pump static timing marks can be seen when the timing hole cover is removed.

Fig. IH1337—With the screws (S) loosened, the drive shaft hub can be rotated in relation to the injection pump drive gear due to the elongated holes in the gear.

turning crankshaft until the pointer extending from timing gear cover is in register with the correct degree mark (3 degrees ATDC for series 340, or 2 degrees ATDC for series 504 and 2504) on the crankshaft pulley. Shut off fuel, remove timing window cover from side of injection pump and check to be sure that timing marks are aligned as shown in Fig. IH1336.

If pump timing marks are not aligned, loosen the two pump mounting nuts (N—Fig. IH1338) and turn pump housing either way as required to align the marks; then, retighten the nuts.

In some cases, it may be impossible to turn the injection pump enough to align the timing marks with crankshaft pulley set at the correct degree mark. If this condition is encountered, align the timing pointer with correct degree mark on crankshaft pulley and remove the gear cover plate from front of timing gear cover. Loosen the three cap screws (S—Fig. IH1337) which attach the pump drive gear to the drive shaft hub. Rotate drive shaft hub in the elongated (adjusting) holes to a point where the pump timing marks (Fig. IH 1336) can be perfectly aligned. Retighten the cap screws (S—Fig. IH1337) and reinstall cover plate.

101. **PUMP ADVANCE.** Install the special timing window (TW) as shown in Fig. IH1338 and with No. 1 piston coming up on compression stroke, set the crankshaft at 3 degrees ATDC. for series 340, or 2 degrees ATDC for series 504 and 2504. Note (or mark) the position of the pump cam timing line in relation to the marks on the timing window. Start the engine.

With the engine running at low idle no-load speed, the cam timing line should have moved down 2 marks on the window (4 crankshaft degrees advance) from its static timing position when engine was not running. Advance the throttle to high idle po-

sition and with engine running at the recommended high idle no-load speed, the cam timing line should again be 2 marks lower on the timing window (4 crankshaft degrees advance) than it was with the engine not running. If the cam line isn't 2 marks lower on the window when the engine is running at low and high idle no-load speeds, than it was with the engine stopped, loosen jam nut (JN) and turn the guide stud (GS) as required. Tighten the jam nut and recheck the advance.

NOTE: Always check the amount of advance with the jam nut (JN) tight; because, when the jam nut is loose the guide stud may raise up slightly due to the looseness in the threads and the advance reading may be incorrect.

Accelerate engine by moving the throttle control lever quickly from the low idle speed position to the high idle speed position, and notice the movement of the cam timing line. The cam timing line should move from full **advance** (2 marks or 4 crankshaft

Fig. IH1338—A special transparent timing window (TW) which has timing marks on it should be installed as shown and used to adjust the injection pump timing advance. A six cylinder pump is shown; however, four cylinder models are similar.

1. High speed adjusting screw
2. Low speed adjusting screw
3. Stop screw
4. Run screw
JN. Jam nut
GS. Governor spring guide stud
N. Nuts

Fig. IH1339 — A vacuum gage can be installed as shown to check for restrictions or air leaks in line between the tank and the pump. Although a 660 International tractor is shown installation is similar on 4 cylinder tractors.

degrees) at low idle speed, to full **retard** (static timing position) during acceleration, then return to full **advance** when the engine reaches high idle speed.

If the cam remains in the advanced position during quick acceleration, back-out the guide stud slightly until the correct adjustment is obtained.

If a dynamometer is used, the timing line on the cam should be advanced 2 crankshaft degrees (1 mark from the static position) at 40-60 per cent of rated loaded power with the hand throttle in the high speed position and should be fully retarded (static timing position) at full loaded power.

If guide stud adjustments do not correct the operation of the advance, check for low transfer pump pressure, restriction in the return line, seized advance piston or malfunctioning advance check valve.

102. **VACUUM GAGE TESTS.** To check for a restriction in the line between pump and fuel tank, make certain that fuel tank is full and check the flexible fuel line between pump and filter for being clogged or deteriorated. Install a combination pressure and vacuum gage as shown in Fig. IH1339 and run the engine at high idle rpm. If the gage registers a vacuum equal to more than 10 inches of mercury, check for clogged filter, fuel lines or fittings. A vacuum equal to 5-10 inches of mercury indicates that a restriction is beginning.

To check for air leaks between the transfer pump and the fuel tank, a combination pressure and vacuum gage should be installed as shown in Fig. IH1339. Run the engine at low

idle speed and close the shut-off valve at tank. When the engine stalls, note the vacuum gage reading, which should be equal to 18-24 inches of mercury.

If the vacuum drops off rapidly, it indicates an air leak. If the vacuum drops off rapidly but an air leak can't be found, check for a stuck pressure regulating valve plunger (78—Fig. IH-1341) or a malfunctioning fuel return check valve (located at the tank end of the fuel return line).

If the vacuum gage reading is less than 18 inches of mercury when the engine stalls but doesn't drop off rapidly; check for a stuck pressure regulating valve plunger (78), damaged "O" rings (76 and 80), weak or broken springs (75 & 79) or damaged transfer pump blades (70).

103. **TRANSFER PUMP PRESSURE.** To check the transfer pump

Fig. IH1340—A pressure gage installed in place of the Allen head plug in the bottom of the injection pump end plate (as shown) can be used to check transfer pump pressure.

pressure, remove the Allen head plug from bottom of injection pump end plate and install a pressure gage of at least 150 psi capacity in place of the plug (Fig. IH1340). Start engine, run at high idle rpm and note pressure reading on gage. Pressure read-should be 48-53 psi for series 340, or 56-60 for series 504 and 2504.

If pressure is low when tested at high idle rpm, remove the plug (81—Fig. IH1341) and install a plug with the next size higher shoulder. Plug (81) is available with shoulder heights from 0.020-0.090 in graduations of 0.010. If varying the shoulder height doesn't remedy incorrect pressure, check for a broken spring (79), a stuck plunger (78) or damaged transfer pump blades (70). Spring (79) should have a free length of 0.525-0.565 inch and color coded by one end being yellow.

104. **NOZZLES.** If the engine does not run properly and a faulty nozzle is suspected or if one cylinder is misfiring, locate the faulty nozzle by loosening the high pressure line fitting on each nozzle holder in turn, thereby allowing fuel to escape at the union rather than enter the cylinder. As in checking spark plugs in a spark ignition engine, the faulty nozzle is the one which, when its line is loosened, least affects the running of the engine. The malfunctioning nozzle should be removed as outlined in paragraph 108 and tested as outlined in paragraph 107.

105. **DELIVERY VALVE AND ROTOR.** To check the delivery valve and rotor, connect a nozzle test pump and gage to the number one high pressure injection line as shown in Fig. IH1342. Loosen the connection between the nozzle injection line and the test pump hose, actuate the test pump handle until fuel flows out of the loose connection, crank the engine several revolutions with the starting motor until fuel flows out of the loose connection, then retighten the connection. Disconnect the high pressure line from the number 2 or 3 injector. Install the clear timing window (TW) as shown in Fig. IH1338 and turn the engine until the timing line on the governor weight retainer is 2 marks higher on the timing window (4 crankshaft degrees before the end of injection) than the mark on the cam. This will align the rotor discharge port with the number 1 dis-

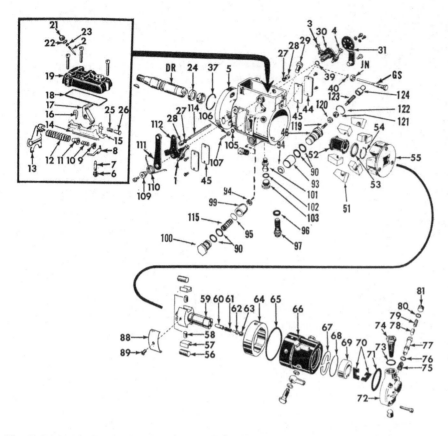

Fig. IH1341—Exploded view of a Roosa Master Injection pump. Overhaul should not be attempted by inexperienced personnel.

DR. Drive shaft
GS. Governor spring guide stud
JN. Jam nut
1. High speed adjusting screw
2. Low speed adjusting screw
3. Shut-off screw
4. Run screw
5. Housing
6. Spring
7. Fuel metering valve
8. Fuel metering valve arm
9. Governor idling spring retainer
10. Governor idling spring
11. Governor idling spring guide
12. Governor control spring
13. Governor arm
14. Governor linkage spring
15. Governor linkage hook
16. Fuel shut-off cam
17. Throttle shaft lever
18. Gasket
19. Governor control cover
21. Nut
22. Washer
23. Seal
24. Drive shaft oil seals
25. Governor linkage hook action damper sleeve spring
26. Governor linkage hook action damper sleeve
27. Seals
28. Governor arm pivot shaft retaining nuts
29. Hydraulic head locking screws

30. Fuel shut-off shaft
31. Fuel adjusting shut-off arm
37. Pilot tube oil seal
39. "O" ring
40. "O" ring
45. Gaskets
51. Governor weights (6 used)
52. Governor thrust sleeve
53. Governor thrust washer
54. Snap ring
55. Governor weight retainer
56. Roller
57. Shoe
58. Pump plunger
59. Rotor
60. Delivery valve
61. Delivery valve spring
62. Delivery valve stop
63. Delivery valve retaining screw
64. Cam ring
65. Seal
66. Hydraulic head
67. Distributor rotor retainer
68. Distributor rotor retainer ring
69. Transfer pump liner
70. Transfer pump blades
71. "O" ring
72. Injection pump end plate
73. "O" ring
74. Fuel strainer
75. Pressure regulating valve plunger retaining spring
76. "O" ring
77. Pressure regulating valve plunger sleeve

78. Pressure regulating valve plunger
79. Pressure regulating valve compression spring
80. "O" ring
81. Sleeve retaining plug
88. Plunger limiting leaf spring
89. Plunger limiting leaf spring adjusting screw
90. Seals
93. Load advance power piston
94. Load advance slide washers
95. Advance spring shims (0.0155-0.0175, 0.004-0.006 and 0.032-0.034)
96. Seal
97. Screw
99. Load advance spring piston
101. Cam advance screw
102. Seal
103. Plug
105. "O" ring
106. Nut
107. Torque screw
109. Throttle arm spring retainer
110. Throttle arm spring
111. Throttle arm
112. Throttle shaft
114. Governor arm pivot shaft
115. Load advance spring
119. Advance adjusting spring
120. Spring guide
121. Lock nut
122. "O" ring
123. Advance adjusting screw
124. Cap

charge port in the hydraulic head.

Operate the test pump, maintain a pressure of 2000-2500 psi and record the number of drops of fuel which escape from the disconnected number 2 or 3 injection line in 30 seconds. If more than 25 drops of fuel escape from the disconnected pressure line, the rotor and hydraulic head may be scored or grooved at the discharge ports.

If the amount of escaped fuel is satisfactory for the 30 second interval, operate the test pump until 2500 psi of pressure is built up. If the pressure drops below 700 psi in 10 seconds, the delivery valve is probably stuck open.

106. BLEEDING. Each time the filter element is renewed, if fuel lines are disconnected or if tractor has run out of fuel, it will be necessary to bleed air from the system.

To bleed the fuel filter, open the bleed valve on the top cover of filter. Open the fuel shut-off valve at bottom of tank. When all air has escaped and a solid flow of fuel is escaping from air bleed hole, close bleed valve.

Loosen the pump inlet line and allow fuel to flow from the connection until stream is free from air bubbles; then, tighten the connection.

Loosen connection of fuel return line at top of pump and allow fuel to flow from the connection until stream is free from air bubbles; then, tighten the connection.

Loosen the high pressure fuel line connections at the injectors and crank engine with the starting motor until fuel appears. Tighten the fuel line connections and start engine.

INJECTION NOZZLES

WARNING: Fuel leaves the injection nozzles with sufficient force to penetrate the skin. When testing, keep your person clear of the nozzle spray.

107. TESTING AND LOCATING FAULTY NOZZLE. If the engine does not run properly, and a faulty injection nozzle is suspected or if one cylinder is misfiring, locate the faulty nozzle by loosening the high pressure line fitting on each nozzle holder in turn, thereby allowing fuel to escape at the union rather than enter the cylinder. As in checking spark plugs in a spark ignition engine, the faulty nozzle is the one which, when its line is loosened, least affects the running of the engine.

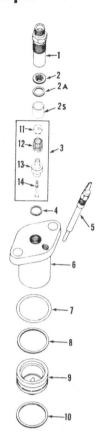

Fig. IH1342 — A nozzle test pump and gage similar to that shown can be installed to check for a stuck delivery valve or a scored rotor. Refer to text for testing procedure. A six cylinder engine is shown, but four cylinder models are similar.

Fig. IH1344 — Exploded view of the IH injection nozzle and precombustion chamber. Late type nozzles do not have screen (2) and have instead an additional gasket (2A).

1. Nozzle fitting
2. Screen
2A. Gasket
2S. Spacer
3. Nozzle valve
4. Gasket
5. Glow plug
6. Nozzle body
7. Dust seal
8. Gasket
9. Precombustion chamber
10. Gasket
11. Nozzle spring seat
12. Spring
13. Valve seat
14. Valve

Remove the suspected nozzle as in paragraph 108, place nozzle in a test stand and check the nozzle. New nozzle opening pressure should be 950-1050 psi (850 psi permissible for used nozzles).

If nozzle requires overhauling, refer to paragraph 109.

108. REMOVE AND REINSTALL. To remove any injection nozzle, first remove dirt from nozzle, injection pump and cylinder head; then, disconnect and remove the injector pipe. Cover all openings with tape or composition caps to prevent the entrance of dirt or other foreign materials. Remove the two nozzle retaining stud nuts, lift nozzle from cylinder head and remove the nozzle body dust seal. An OTC HC-689 puller or equivalent can be used for withdrawing a stuck nozzle.

When reinstalling, tighten the nozzle hold down screws evenly to a torque of 20-25 ft.-lbs.

109. OVERHAUL. Remove the nozzle fitting (1—Fig. IH1344) from the nozzle body. Remove the screen (2) (if so equipped), gasket (2A) and spacer (2S). Remove the nozzle valve unit (3) and gasket (4). Thoroughly clean and inspect all parts, and renew any which are damaged. The gaskets should be renewed each time the nozzle is subjected to complete or partial overhaul.

NOTE: Latest type injection nozzles do not include screen (2). Screen is replaced with an additional gasket (2A).

The nozzle valve assembly (3) is available as a complete unit. To disassemble and clean the nozzle valve and reset the valve opening pressure, refer to paragraph 110.

NOTE: It is recommended that the precombustion chambers be removed and cleaned whenever the nozzles are removed for service. Refer to paragraph 111.

110. To disassemble the nozzle valve assembly for cleaning and/or adjusting the opening pressure, press down on the nozzle spring seat (11—Fig. IH1344) until pressure is relieved from upper end of pintle; then, use a screw driver to push upper end of pintle sideways, thereby releasing the nozzle spring seat. Withdraw parts from valve body and clean in a suitable solvent.

The pintle (14) and seat can be lapped using Number 400 lapping compound.

Nozzle spring seats (11) with flange thicknesses of 0.101-0.102, 0.103-0.104, 0.105-0.106, 0.107-0.108, 0.109-0.110, 0.111-0.112, 0.113-0.114 and 0.115-0.116 are available to adjust the opening pressures. Opening pressure should be adjusted to 950-1050 psi and valve should not leak at 700 psi for 10 seconds.

The nozzle fitting should be tightened to 45-50 ft.-lbs. of torque.

PRECOMBUSTION CHAMBERS

111. REMOVE AND REINSTALL. Precombustion chambers can be pulled from cylinder head after removing the respective nozzle assembly. The use of a special pre-cup puller may be necessary.

When installing chamber, make certain that side stamped "TOP" is installed toward top of engine.

GLOW PLUGS

112. REMOVE AND REINSTALL. To remove the glow plugs, remove the hood right side sheet (skirt), disconnect the attached wire, unscrew and withdraw plug.

INJECTION PUMP AND DRIVE

The subsequent paragraphs will outline ONLY the injection pump service work which can be accomplished without disassembly of the injection pump. If additional service work is required, the pump should be turned over to an official diesel service station for overhaul.

113. TIMING. To check the injection pump static timing, refer to paragraph 100. To check the injection pump timing advance refer to paragraph 101.

114. PUMP UNIT—R&R. To remove the complete injection pump unit, first shut off the fuel supply and thoroughly clean dirt from pump, fuel lines and connections. Turn crankshaft clockwise (viewed from front) until number one piston is coming up on compression stroke and continue turning the crankshaft until pointer extending from timing gear cover is in register with the 3°ATDC for series 340; or 2° ATDC for series 504 and 2504 static timing degree mark on the crankshaft pulley. Disconnect the fuel lines and controls, remove the two pump mounting nuts and withdraw pump from engine.

Before reinstalling the pump, remove the timing hole cover from side of injection pump and make certain the pump timing lines are aligned as shown in Fig. IH1336. Mount pump on engine and connect fuel lines and controls.

NOTE: Be very careful when installing the pump over the drive shaft, to prevent damage to the double opposed seals on the shaft.

Recheck the pump timing as in paragraph 100, and bleed the fuel system as in paragraph 106.

115. SPEED ADJUSTMENTS. To adjust the engine governed speeds, first start engine and bring to normal operating temperature. Move the speed control hand lever to the wide open position, loosen the jam nut and turn the high speed adjusting screw (1—Fig. IH1338) either way as required to obtain an engine high idle no-load speed of 2140-2220 rpm for series 340; or 2360-2440 rpm for series 504 and 2504. Tighten the adjusting screw jam nut. With the high idle speed properly adjusted, the full load engine speed should be 2000 rpm for series 340; or 2200 rpm for series 504 and 2504. Move the speed control hand lever to the low idle speed position, loosen the jam nut and turn adjusting screw (2) either way as required to obtain an engine slow idle speed of 625-675 rpm for series 340; or 700-750 rpm for series 504 and 2504. Tighten the adjusting screw jam nut.

NOTE: It may be necessary to vary the length of the speed control rod in order to obtain full travel in either the low idle or the high idle speed position.

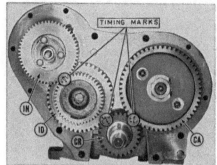

Fig. IH1349C—The diesel engine timing gear train and timing marks.

CA. Camshaft gear   ID. Idler gear
CR. Crankshaft gear   IN. Injection pump
                                drive gear

Screws (3 and 4) are provided to set the limits of shut-off arm travel and should not normally require adjustment in the field. Screw (3) adjusts for maximum travel toward the "shut-off" position; whereas, screw (4) adjusts for "run" position. Adjustment of either screw requires removal of control cover and should be done only

by experienced diesel service personnel.

116. PUMP DRIVE GEAR AND SHAFT—R&R. To remove the injection pump drive shaft, it is first necessary to remove the pump as outlined in paragraph 114. Remove gear front cover plate from the timing gear cover and withdraw the thrust plunger and spring from the pump drive shaft. Mark the relative position of the gear on the pump drive shaft hub. Remove the pump adapter plate and the three screws which attach the gear to the injection pump drive shaft hub. The shaft and hub can be withdrawn from the rear; however, to remove the drive gear it is necessary to remove the timing gear cover as outlined in paragraph 80. Refer to Fig. IH1349C for timing marks if gear is removed.

When reinstalling the drive shaft, make certain the previously scribed marks on gear and hub are aligned. Refer to paragraphs 113 and 114 for reinstalling and timing the injection pump and paragraph 106 for bleeding the fuel system.

NON-DIESEL GOVERNOR

The centrifugal flyweight type governor is mounted on front face of engine and is driven by the engine timing gear train. Before attempting any governor adjustments, check the operating linkage and remove any binding or lost motion.

All Models

116A. ADJUSTMENT. To adjust the governor proceed as follows: With engine stopped, place the speed change lever in wide open position and remove clevis pin (12—Fig. IH1350) from the governor rockshaft arm. Hold the rockshaft arm (13) and the carburetor throttle rod (10) as far toward carburetor as they will go; at which time, pin (12) should slide freely into place. If pin holes are not in alignment, adjust the length of rod (10) until proper register is obtained.

With engine running and speed change lever in wide open position, turn screw (8) on top of governor housing either way as required to obtain the recommended speeds which follow:

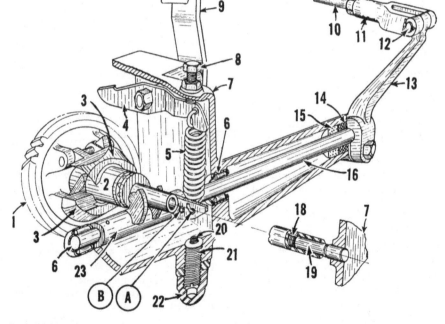

Fig. IH1350—Cut-away view of a typical governor assembly. The unit is driven from the engine timing gear train. Governor spring (5) should be hooked in hole (A) as shown.

1. Ignition-governor gear
2. Sleeve and thrust bearing
3. Governor weights
4. Spring lever
5. Governor spring
6. Needle bearings
7. Governor housing
8. Speed adjusting screw
9. Speed change lever
10. Throttle rod
11. Throttle rod clevis (yoke)
12. Clevis pin
13. Rockshaft arm
14. Felt seal
15. Oilite bushing
16. Rockshaft
18. Thrust spring
19. Thrust pin
20. Surge (bumper) spring
21. Surge spring adjusting screw
22. Spring body cap
23. Rockshaft lever
A-B. Spring position holes

COOLING SYSTEM

Series 330-340

Engine High Idle
  (No Load) rpm 2175-2225
Engine Rated (Loaded) rpm.1990-2010

Series 504-2504

Engine High idle
  (no load) rpm 2475-2525
Engine Rated (loaded) rpm 2200

Engine low idle rpm is 400-450 for all models.

Hunting or unsteady running can be eliminated by removing cap (22) and turning the bumper spring screw (21) in. Do not turn the bumper spring screw in too far as it will interfere with the low idle speed adjustment. CAUTION: Stop the engine before making the bumper spring adjustment.

116B. **R&R AND OVERHAUL.** Remove the magneto or distributor and mark location of slots in the ignition unit drive coupling on governor gear (1—Fig. IH1350) in relation to crankcase. This procedure will facilitate reinstallation of governor gear in proper timing mesh with camshaft gear if position of crankshaft is not changed while governor is removed. Remove governor speed control rod then unbolt and remove the governor assembly from timing gear cover.

When disassembling, do not bend hooks of governor spring (5). The spring lever (4) should be removed from the speed change lever shaft to remove and install the spring. Upper hook of spring should be inserted in lever shaft so that open end of hook will face the side of governor housing. Place the lower hook of spring in hole (A) of rockshaft lever.

Two needle bearings (6) and one "Oilite" bushing (15) support the rockshaft (16) and can be renewed when worn. The felt seal (14) which is assembled outside the bushing should not place any drag on the rockshaft.

Clearance between governor weights (3) and pins should be 0.001-0.004. If clearance exceeds 0.006, renew pins and/or weights.

The bushing in the crankcase which supports the governor and ignition unit drive gear hub may be renewed when governor and ignition unit are off. The I&T recommended clearance of gear hub in bushing is 0.0015 to 0.002. Governor spring part No. is 369 686 R2.

RADIATOR

All Models

117. To remove radiator, first drain cooling system, then remove hood, radiator side plates and grille. Disconnect both radiator hoses and radiator brace. On series 504 and 2504, remove grille housing (guard) and disconnect oil cooler from radiator. On 340 series diesel tractors and series 504 and 2504 International tractors, disconnect fan shroud from radiator. Remove support bolts and lift radiator from tractor.

WATER PUMP

All Models

118. To remove water pump, first drain cooling system and on all except Farmall 504 tractors, remove hood skirt and generator.

NOTE: Tractor model will determine whether water pump is removed from left or right side of tractor and this fact will also determine which hood skirt is to be removed and whether or not generator removal is required.

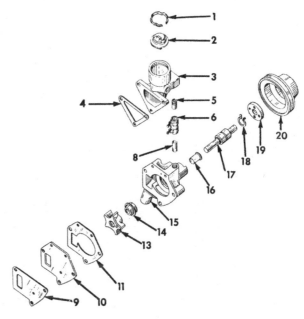

Fig. IH1351 — Exploded view of the series 340 non-diesel water pump. The pump shaft and bearing are available as an assembled unit only. Series 504 and 2504 are similar.

1. Retainer	8. Nipple	16. Slinger
2. Thermostat	9. Gasket	17. Shaft and bearing
3. Outlet elbow	10. Plate	assembly
4. Gasket	11. Gasket	18. Snap ring
5. Nipple	13. Impeller	19. Hub
6. By-pass hose	14. Seal	20. Pulley
	15. Pump body	

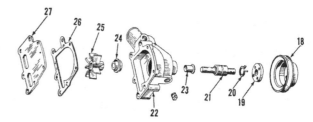

18. Pulley
19. Pulley hub
20. Snap ring
21. Shaft and bearing
22. Body
23. Slinger
24. Seal
25. Impeller
26. Gasket
27. Rear plate

Fig. IH1351A—Exploded view of the series 340 diesel engine water pump. Series 504 and 2504 are similar.

On diesel model tractors, unbolt fan blades from hub and let fan blades rest in fan shroud. Disconnect inlet hose and by-pass hoses, then unbolt water pump and remove from side of tractor.

On Farmall 504 tractors, sufficient room is available and pump can be removed after removing fan and fan spacer and disconnecting inlet and by-pass hoses.

Disassembly and overhaul procedures will be evident after an examination of the unit and reference to Figs. IH1351 and IH1351A. Clearance from face of body to face of impeller is 0.031 when unit is assembled.

LP-GAS SYSTEM

Series 504 and 2504 tractors are available with LP-Gas equipment manufactured by Ensign. Carburetor is a model XG and regulator is a model W. As with other LP-Gas systems, these systems are designed to operate with the fuel tank no more than 80% filled. It is important when starting these tractors, to open the vapor valve on the supply tank slowly; if opened too fast, the fuel supply to the regulator will be automatically shut off.

CAUTION: Before disconnecting fuel lines or removing any of the system components, close both fuel tank withdrawal valves and run engine until all fuel is exhausted from the fuel lines and components and engine stops.

ADJUST SYSTEM

Series 504-2504

119. The LP-Gas system has four adjustments; three are located on the carburetor and one on the regulator. Adjustments located on the carburetor are main fuel (load) adjustment, starting fuel adjustment and the throttle idle stop screw. The idle mixture adjusting screw is located on top side of regulator.

The adjustments are pre-set at the factory and normally do not need readjustment; however, if adjustments have been disturbed, or any of the components serviced, readjust system as follows: Loosen lock nuts on starting fuel adjustment and main fuel adjustment on carburetor. Turn both of these adjustment screws and the idle mixture adjusting screw on the regulator all the way in; then make the following initial adjustments. Starting adustment screw 1¼ turns open. Main fuel adjustment screw 1½ turns open. Idle mixture adjustment screw (on regulator) 1½ turns open. Start engine and bring to operating temperature, then place throttle control lever in low idle position. Adjust throttle stop screw to obtain an engine low idle speed of approximately 425 rpm, then turn the idle mixture needle on the regulator in or out as required to obtain the highest and

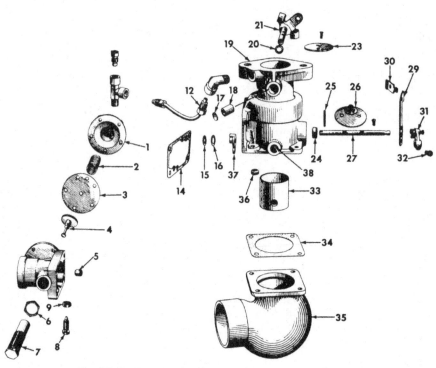

Fig. IH1351B — Exploded view of Ensign XG carburetor.

1. Economizer cover	14. Gasket	25. Lever pin
2. Economizer spring	15. Valve lever washer	26. Choke disc
3. Economizer diaphragm	16. Spring washer	27. Choke shaft
4. Diaphragm plunger	17. Expansion plug	31. Choke lever
5. Economizer bleed screw	18. Throttle shaft bushing	32. Set screw
6. Lock nut	19. Body	33. Venturi
7. Fuel adjusting screw	20. Throttle shaft seal	34. Gasket
8. Starting adjustment screw	21. Throttle shaft	35. Intake elbow
9. Lock nut	23. Throttle disc	36. Expansion plug
12. Balance line	24. Dust washer	37. Valve lever
		38. Plug

smoothest engine operation. Readjust throttle stop screw, if necessary. Place hand throttle in the high idle position and turn the main fuel (load) adjustment screw in or out as required until the smoothest engine operation is obtained.

NOTE: The load screw adjustment may have to be altered slightly after engine is loaded. If a dynamometer is available, use it when making the load screw adjustment.

Starting adjustment should be approximately correct; however, it may be varied as needed if cold starts are not satisfactory. Counter-clockwise rotation of the needle richens the mixture.

CARBURETOR

Series 504-2504

119A. **R&R AND OVERHAUL.** Removal and reinstallation procedure for the LP-Gas carburetor is obvious after an examination of the unit.

With unit removed, refer to Fig. IH1351B and completely disassemble the carburetor. Wash all parts in a suitable solvent and blow out all passages with compressed air. Be sure economizer diaphragm is in satisfactory condition with no ruptures or tears. Use new gaskets when reassembling.

After carburetor is assembled and installed, adjust same as outlined in paragraph 119.

REGULATOR

Series 504-2504

119B. **HOW IT OPERATES.** Fuel from the supply tank enters the regulating unit inlet (A—Fig. IH1351C) at tank pressure and is reduced from tank pressure to about 4 psi at the high pressure reducing valve (C) after passing through the strainer (B). Flow through high pressure reducing valve is controlled by the adjacent spring and diaphragm. When the liquid fuel enters the vaporizing chamber (D) via the valve (C) it expands rapidly and is converted from a liquid to a gas by heat from the water jacket (E) which is connected to the cooling system of the engine. The vaporized gas then passes (at a pressure slightly below atmospheric pressure) via the low-pressure reducing valve (F) into the low-pressure chamber (G) where it is drawn off to the carburetor via outlet (H). The low pressure reducing valve is controlled by the larger diaphragm (T) and small spring.

Fuel for the idling range of the engine is supplied from a separate outlet (J) which is connected by tubing to a separate idle fuel connection on the carburetor. Adjustment of the carburetor idle mixture is controlled by the idle fuel screw (K) and the calibrated orfice (L) in the regulator. The balance line (M) is connected to the air inlet horn of the carburetor so as to reduce the flow of fuel and thus prevent over-richening of the mixture which would otherwise result when the air cleaner or air inlet system becomes restricted.

TROUBLE SHOOTING

119C. SYMPTOM—Engine will not idle with Idle Mixture Adjustment Screw in any position.

CAUSE AND CORRECTION — A leaking valve or gasket is the cause of the trouble. Look for leaking low pressure valve caused by deposits on valve or seat. To correct the trouble wash the valve and seat in gasoline or other petroleum solvent.

If the foregoing remedy does not correct the trouble, check for leak at high pressure valve by connecting a low reading (0 to 20 psi) pressure gage at point (R) on the regulator. If the pressure increases **after** a warm engine is stopped, it proves a leak in the high pressure valve. Normal pressure is 3½-5 psi.

119D. SYMPTOM — Cold regulator shows moisture and frost after standing.

CAUSE AND CORRECTION—Trouble is due either to leaking valves as per paragraph 119C or the valve levers are not properly set. For information on setting of valve lever refer to paragraph 119F.

REGULATOR OVERHAUL

If an approved station is not available, the model W regulator can be overhauled as outlined in paragraphs 119E and 119F.

119E. Remove the unit from the engine and completely disassemble using Fig. IH1351D as reference. Thoroughly wash all parts and blow out all passages with compressed air. Inspect each part carefully and discard any that are worn.

119F. Before reassembling the unit, note dimension (X—Fig. IH1351C) which is measured from the face on the high pressure side of the casting to the inside of the groove in the

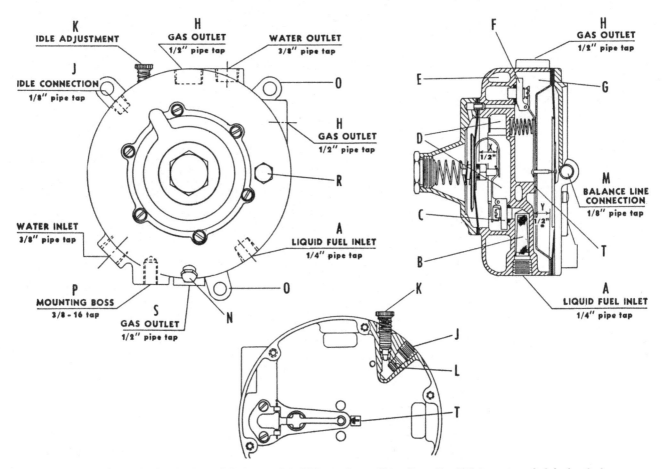

Fig. IH1351C — Sectional view of Ensign model "W" regulator. Note dimension "X" in center of right hand view.

valve lever when valve is held firmly shut as shown in Fig. IH1351E. If dimension (X) which can be measured with Ensign gage No. 8276, or with a depth rule, is more or less than ½-inch, bend the lever until this setting is obtained. A boss or post (T—Fig. IH1351F) is machined and marked with an arrow to assist in setting the lever. Be sure to center the lever on the arrow before tightening the screws holding the valve block. The top of the lever should be flush with the top of the boss or post (T).

LP-GAS FILTER
Series 504-2504

119G. Filters (Fig. IH1351G) used on these systems are designed for a working pressure of 375 psi and should be able to stand this pressure without leakage. Filter element should be renewed when indications are that element is plugged sufficiently to prevent proper fuel flow. A clogged element causes a pressure drop and results in vaporization of the fuel which may cause regulator freezing and subsequent fuel starvation of the engine. When major engine work is being performed it is advisable to remove the lower part of the filter, thoroughly clean the interior and renew the treated paper cartridge if same is not in good condition.

Fig. IH1351E — Using Ensign gage No. 8276 to measure dimension (X) on model "W" regulator. Dimension (X) is shown in Fig. IH1351C.

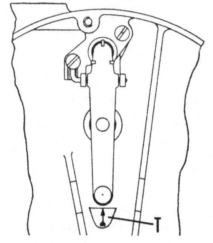

Fig. IH1351F—Top of low pressure lever should be flush with top of post and centered with arrow. Bend lever, if necessary.

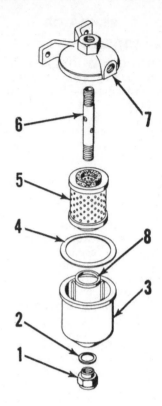

Fig. IH1351G — Exploded view of Ensign LP-Gas filter.

1. Stud nut	5. Element
2. Gasket	6. Stud
3. Bowl	7. Cover
4. Gasket	8. Magnetic ring

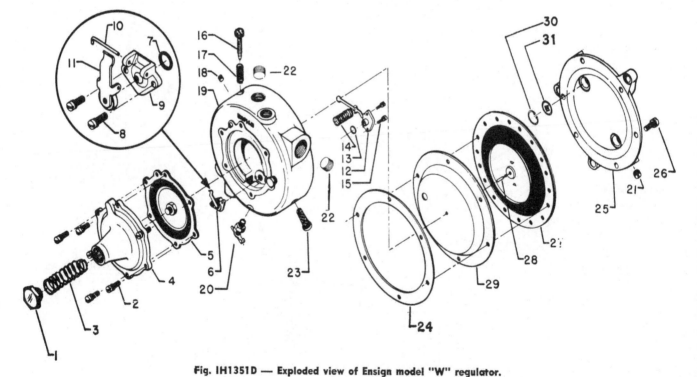

Fig. IH1351D — Exploded view of Ensign model "W" regulator.

1. Spring retainer	10. Pivot pin	17. Spring	24. Gasket
3. Pressure spring	11. Valve lever	18. Bleed screw	25. Back cover plate
4. Cover	12. Low pressure valve	19. Body	27. Diaphragm
5. Diaphragm	13. "O" ring	20. Drain cock	28. Push pin
6. High pressure valve	14. Low pressure valve	21. Plug	29. Partition plate
7. "O" ring	spring	22. Plug	30. Retainer ring
9. High pressure valve	16. Idle adjusting screw	23. Strainer	31. Compensator assembly
seat			

IGNITION AND ELECTRICAL SYSTEM

CARBURETOR
(Gasoline)

120. Zenith and Marvel-Schebler carburetors have been used. Models and specification data are as follows:

Zenith 67x7

Assembly No.	12225
Discharge jet	.........C66-114-50
Main jet	...........C52-6-24
Idle jet	...............C55-22-11
Inlet needle & seat	....C81-17-35
Gasket set	..............C181-329
Repair kit	..............K12225
Well vent jet	..........C77-18-25
Float setting*	.........1 5/32 in.

Zenith 267x9

Assembly No.	12685
Basic repair kit	..........K12685
Gasket set	..............C181-325
Inlet needle & seat	.....C81-65-55
Idle jet	..............C55-22-14
Main jet	..............C52-7-25
Discharge nozzle	.....C66-102-2-65
Well vent jet	..........C77-18-23
Float setting*	............1 5/32 in.

Zenith 267x9

Assembly No.	12758
Basic repair kit	.........K12758
Gasket set	..............C181-325
Inlet needle & seat	....C81-65-55
Idle jet	...............C55-22-14
Main jet	..............C52-7-25
Discharge nozzle	.....C66-102-2-70
Well vent jet	.........C77-18-23
Float setting*	............1 5/32 in.

* Float setting is measured from farthest face of float to gasket surface of bowl cover.

Marvel-Schebler TSX748

Basic repair kit	286-1245
Gasket set	16-613
Inlet needle & seat	233-536
Idle jet	.................49-101-L
Discharge nozzle	47-490
Power jet	49-269
Float setting*	.............¼ in.

Marvel-Schebler TSX857

Basic repair kit	286-1423
Gasket set	16-667
Inlet needle & seat	233-604
Idle jet	.................49-101-L
Discharge nozzle	.........47-A82
Power jet	49-293
Float setting*	.............¼ in.

* Float setting is measured from gasket surface to nearest surface of float.

Fig. IH1352—The crankshaft pulley is provided with timing marks as shown which indicate crankshaft degrees. Retard timing is TDC for all models.

DISTRIBUTOR

129. **INSTALLATION AND TIMING.** To install and time the battery ignition distributor, first crank the engine until the number 1 (front) piston is coming up on the compression stroke and continue cranking slowly until the TDC (0 degree) mark on the crankshaft pulley is aligned with the pointer extending from the front face of the timing gear cover as shown in Fig. IH1352.

Turn the distributor drive shaft until rotor arm is in the No. 1 firing position and mount the ignition unit on the engine, making certain that lugs on ignition unit engage slots in the drive coupling.

Note: If the driving lugs on the battery ignition unit will not engage the coupling drive slots, when rotor is in No. 1 firing position, it will be necessary to remesh the drive gears as follows: Grasp the distributor drive shaft and pull same outward to disengage the gears. Turn drive shaft until lugs will engage the drive slots, then push drive shaft inward to engage the gears.

Adjust the breaker contact gap to 0.020. Loosen the distributor mounting cap screws and retard distributor about 30 degrees by turning distributor assembly in same direction as the cam rotates. Disconnect coil secondary cable from distributor cap and hold free end of cable 1/16-1/8-inch from distributor primary terminal. Advance the distributor by turning the distributor body in opposite direction from cam rotation until a spark occurs at the gap. Tighten the distributor mounting cap screws at this point.

With engine running at high idle, the advanced timing can be checked with a timing light. Advance data is given in paragraph 130.

130. **OVERHAUL.** Defects in the battery ignition system may be approximately located by simple tests which can be performed in the field; however, complete ignition system

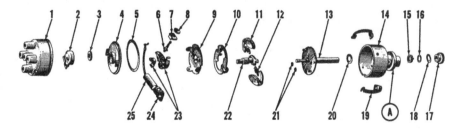

Fig. IH1352A—Exploded view of typical International Harvester battery ignition distributor. The distributor identification symbol is stamped on the outside diameter of mounting flange (A). Shaft (13) rides directly in unbushed housing (14); wear is corrected by renewal of one or both parts.

1. Distributor cap	6. Primary terminal	12. Governor spring	20. Thrust washer
2. Rotor	screw	13. Distributor shaft	21. Weight arm spacer
3. Breaker cover	7. Insulator	15. Oil seal retainer	22. Cam
felt seal	8. Insulating washer	16. Oil seal	23. Breaker contact
4. Breaker cover	9. Breaker plate	17. Distributor gear	set
5. Breaker cover	10. Governor weight	18. Thrust washer	24. Condenser clamp
gasket	guard	19. Cap retaining	25. Condenser
	11. Governor weight	spring and support	

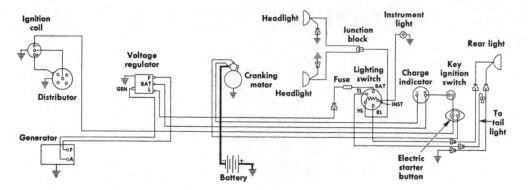

Fig. IH1352B — Wiring diagram typical of series 340 battery ignition distributor equipped tractors.

analysis and component unit tests require the use of special testing equipment. The distributor has an automatic spark advance obtained by a centrifugal governor built into the unit.

Identification and advance curve data follows. Advance data are given in distributor degrees and distributor rpm. For crankshaft degrees and rpm, double the listed values.

Distributor symbol X
Advance data (degrees @ rpm)
  Start ....0-1 @ 200
  Intermediate ....4.5-6.5 @ 400
  Intermediate ....9-11 @ 600
  Intermediate ....13-15 @ 800
  Maximum ....15-15.50 @ 900
Distributor symbol ....AE
Advance data (degrees @ rpm)
  Start ....0-0.5 @ 200
  Intermediate ....3.5-5.5 @ 400
  Intermediate ....7-9 @ 600
  Intermediate ....10-12 @ 800
  Maximum ....12-13 @ 1000
Distributor symbol ....AF
Advance data (degrees @ rpm)
  Start ....0-0.5 @ 200
  Intermediate ....2-4 @ 400
  Intermediate ....5-7 @ 600
  Intermediate ....8-10 @ 800
  Intermediate ....10-11.5 @ 1000
  Maximum ....10.5-11.5 @ 1100
Distributor symbol ....AG
Advance data (degrees @ rpm)
  Start ....0-0.5 @ 200
  Intermediate ....1-3 @ 400
  Intermediate ....3.5-5.5 @ 600
  Intermediate ....6.5-8.5 @ 800
  Intermediate ....8.5-10 @ 1000
  Maximum ....9-10 @ 1100
All distributors have a breaker arm spring pressure of 21-25 ounces and a breaker contact gap of 0.020.

MAGNETO
All Models So Equipped

131. **INSTALLATION AND TIMING.** To install and time the magneto, first crank engine until number 1 (front) piston is coming up on compression stroke and continue cranking slowly until the TDC (0 degree) mark on the

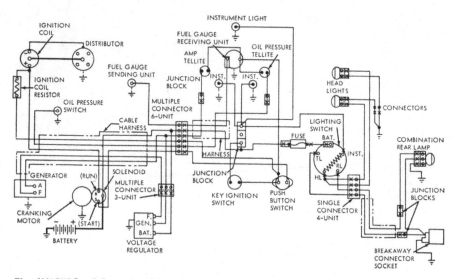

Fig. IH1352C—Schematic wiring diagram for Farmall 504 gasoline and LP-Gas tractors.

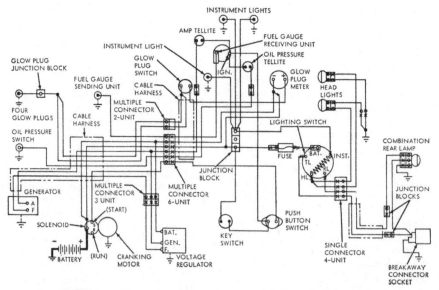

Fig. IH1352D — Schematic wiring diagram for Farmall 504 diesel tractors.

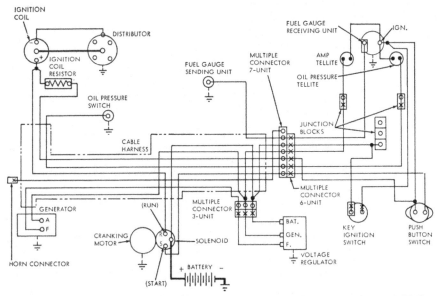

Fig. IH1352E — Schematic wiring diagram for International 504 gasoline and LP-Gas tractors. Series 2504 gasoline and LP-Gas tractors are similar.

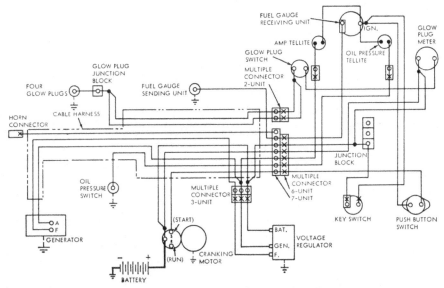

Fig. IH1352F — Schematic wiring diagram for International 504 diesel tractors. Series 2504 diesel tractors are similar.

crankshaft pulley is aligned with the pointer extending from the front face of the timing gear cover as shown in Fig. IH1352.

Turn the magneto drive lugs until rotor arm is in the No. 1 firing position and mount ignition unit on engine, making certain that lugs on ignition unit engage slots in drive coupling.

Adjust the breaker contact gap to 0.013.

Loosen the magneto mounting bolts

and retard the magneto by turning top of magneto in the normal direction of rotation. Crank engine one complete revolution and align timing marks as before. Advance magneto by turning top of magneto opposite to normal rotation until impulse coupling snaps; then, tighten the magneto mounting bolts. To check magneto timing, crank engine until No. 1 piston is again coming up on compression stroke and impulse coupling snaps. At this time, the aforementioned timing marks should be in register. Assemble the spark plug cables

to the distributor cap in the proper firing order of 1-3-4-2.

Running timing can be checked with a timing light and with engine running faster than the impulse coupling cut out speed. Impulse coupling lag angle is 35 degrees.

Refer to the following specifications when servicing magnetos:

Model H4
Breaker point gap............... 0.013
Breaker arm spring pressure . 19-23 oz.
Rotation C
Impulse trips .... TDC
Lag angle .... 35°

GENERATOR, REGULATOR AND STARTING MOTOR

All Models

132. All models are equipped with Delco-Remy electrical units and specifications data are as follows:

Generator 1100042-1100052
Brush spring tension—oz. 28
Field draw
  Volts 6.0
  Amperes 1.85-2.03
Output
  Hot or Cold .... Cold
  Amperes 35.0
  Volts 8.0
  RPM 2950

Generator 1100402-1100409
Brush spring tension—oz. 28
Field draw
  Volts 12.0
  Amperes 1.58-1.67
Output
  Hot or Cold .... Cold
  Amperes 20.0
  Volts 14.0
  RPM 2300

Generator 1100403
Brush spring tension—oz. 28
Field draw
  Volts 12.0
  Amperes 1.5-1.62
Output
  Hot or Cold .... Cold
  Amperes 25.0
  Volts 14.0
  RPM 2710

Regulator 1118999
Cut-Out Relay
  Air gap 0.020
  Point gap 0.020
  Closing voltage range .. 11.8-14.0
    Adjust to 12.8
Voltage Regulator
  Air gap 0.075
  Setting (volts) range ... 13.6-14.5
    Adjust to 14.0

Regulator 1119270E

Cut-Out Relay
Air gap 0.020
Point gap 0.020
Closing voltage range ...11.8-13.5
Adjust to 12.6

Voltage Regulator
Air gap 0.060
Setting (volts) range
@ 85° F 14.2-15.2
Adjust to 14.7

Current Regulator
Air gap 0.075
Setting (amps.) range
@ 85° F 49.0-53.5
Adjust to 51.2

Regulator 1119575

Cut-Out Relay
Air gap 0.020
Point gap 0.020
Closing voltage range ...5.9-7.0
Adjust to 6.4

Voltage Regulator
Air gap 0.075
Setting (volts) range ...6.6-7.2
Adjust to 6.9

Starting Motor 1107169
Volts 6.0
Brush spring tension—oz. 24
No Load Test
Volts 5.6
Amperes 80.0*
RPM 5500
Lock Test
Volts 3.3
Amperes 550
Torque—ft.-lbs. 11

Starting Motor 1107229
Volts 12.0
Brush spring tension—oz. 35
No Load Test
Volts 10.6
Amperes (min. & max.) ..49-76*
RPM (min. & max.) ..6200-9400
Resistance Test
Volts 4.3
Amperes (min. & max.) .270-310*

Starting Motor 1107543
Volts 12.0
Brush spring tension—oz. 35
No Load Test
Volts 10.6
Amperes (min. & max.) ..75-95*
RPM (min. & max.) ..6400-9500

Resistance Test
Volts 3.5
Amperes (min. & max.) 520-590*

Starting Motor 1108657-1108673
Volts 12.0
Brush spring tension—oz. 24
No Load Test
Volts 11.8
Amperes (min. & max.) ...40-70
RPM (min. & max.) ..6800-9200
Lock Test
Volts 5.9
Amperes 615
Torque—ft.-lbs. 29
*Includes solenoid.

ENGINE CLUTCH

ADJUSTMENT

All Models (Without "Torque-Amplifier")

133. Adjustment to compensate for lining wear is accomplished by adjusting the clutch pedal linkage, not by adjusting the position of the clutch release levers.

To adjust the linkage, loosen lock nut (C—Fig. IH1353), remove clevis pin (A) and turn clevis (B) either way as required to obtain the correct free travel of 1-1⅛-inches.

All Models (With "Torque Amplifier")

134. Adjustment to compensate for lining wear is accomplished by adjusting the clutch pedal linkage, not by adjusting the position of the clutch release levers.

The engine clutch linkage and the "Torque Amplifier" clutch linkage should be adjusted at the same time. The adjustment procedure is given in paragraph 142 and 143.

REMOVE AND REINSTALL

All Models

135. To remove the engine clutch, it is first necessary to detach (spilt) engine from clutch housing as outlined in paragraph 136, 136A or 137. The clutch can then be unbolted and removed from flywheel in the conventional manner.

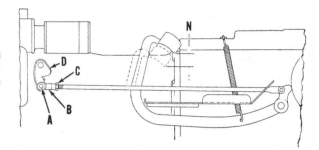

Fig. IH1353 — The clutch pedal free travel (N) should be 1-1⅛-inches as shown.

TRACTOR SPLIT

Farmall 340

136. To detach (split) the engine from the clutch housing, first drain cooling system, remove the hood and disconnect the steering shaft universal joint. On models so equipped, disconnect hydraulic lines between the forward mounted hydraulic pump and control valve support, radiator shutter control rod and tachometer. On all models, disconnect fuel lines, wiring harness and controls from engine and accessories. Support both halves of tractor and remove the clutch housing cover. Remove engine to clutch housing bolts and separate the tractor halves.

Farmall 504

136A. To split engine from clutch housing, proceed as follows: Drain cooling system. Remove precleaner, muffler and hood assembly. Disconnect steering shaft center support and

separate steering shaft. Attach split stand to side rails. If equipped with power steering, disconnect power steering lines (oil cooler) at base of steering shaft support and loosen retaining clips. Remove the two nuts from studs on top side of fuel tank front support and the two cap screws from front ends of instrument panel housing, then pry heat shield and tank upward and off the two studs.

On non-diesel models, disconnect air cleaner tube bracket from fuel tank rear support, then remove air cleaner cup and disconnect temperature indicator sending unit.

On all models, disconnect fuel lines, wiring harness and controls from engine and accessories. Loop a chain under aft end of engine and fasten to side rails to provide support for engine as tractor is separated. Support both halves of tractor, remove clutch housing and side rail bolts and separate tractor.

International 330-340-504-2504

137. To detach (split) engine from clutch housing, first drain cooling system and remove hoods. Disconnect head light wires, radius rod pivot bracket from clutch housing and on series 330 and 340, both drag links from the steering (knuckle) arms. Remove the fuel tank on series 330 and 340. Disconnect the heat indicator sending unit, fuel lines, oil pressure line or switch wire, wiring harness and controls from engine and accessories. On series 504 and 2504, disconnect oil cooler lines and power steering cylinder lines.

Attach hoist to engine half of tractor in a suitable manner and securely block rear half of tractor so it will not tip. Unbolt engine from clutch housing and separate the tractor halves.

NOTE: The two long bolts retaining clutch housing to top of engine on series 330 and 340 should be unscrewed gradually as engine is moved forward. This procedure eliminates the need of removing the steering gear unit and usually saves considerable time. In cases where piping interferes with right hand top bolt on International 504 and 2504, follow same procedure.

OVERHAUL CLUTCH

All Models

138. The procedure for disassembly, adjusting and/or overhauling the removed clutch cover assembly is conventional. Dimensions (A) and (B) in the table are shown in Fig. IH1355. Overhaul data are as follows:

OVERHAUL DATA

International Harvester Clutch

Cover setting (A)........0.851 inch
Lever height (B)...............
.... 2 13/64-2 15/64 inches
Springs
Number used 9
Lbs. test @ height
Regular ....140 @ 1.44 inches
Heavy duty ....165 @ 1.44 inches
Free length
Regular ....2.75 inches
Heavy duty ....2.61 inches

Rockford Clutch

Cover setting (A)........0.851 inch
Lever height (B)...............
.... 2 13/64-2 15/64 inches
Springs
Number used 9
Lbs. test @ height...........
.... 143 @ 1 7/8 inches
Free Length ....2.75 inches

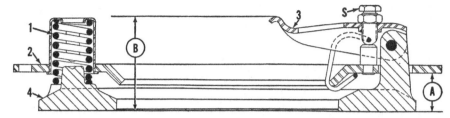

Fig. IH1355—Sectional view of a typical spring loaded clutch cover, showing the release lever adjustment dimensions. Dimension (A) is the position of the pressure plate in relation to the cover plate which must be maintained when adjusting the release lever height (B).

S. Lever adjusting screw	1. Pressure spring
	2. Cover plate
3. Release lever	4. Pressure plate

Fig. IH1356 — Bearing cage (26), independent power take-off drive shaft (25) and clutch shaft (24) removed from front of clutch housing.

CLUTCH SHAFT

All Models (Without "Torque Amplifier")

139. To remove the engine clutch shaft on models without "Torque Amplifier," detach (split) engine from clutch housing as outlined in paragraph 136, 136A or 137; then, withdraw the clutch shaft forward and out of clutch housing.

All Models (With "Torque Amplifier")

140. To remove the clutch shaft on all models with a "Torque Amplifier", first detach (split) engine from clutch housing as outlined in paragraph 136, 136A or 137 and remove the engine clutch release bearing and shaft. On models with independent power take-off, unbolt bearing cage (26—Fig. IH1356) from clutch housing and withdraw the independent power take-off drive shaft (25), bearing cage (26) and clutch shaft (24). Note: A tapped hole is provided in the end of the clutch shaft to aid in its removal. The pto drive shaft rear bearing remains in the housing. To remove this bearing, it is necessary to first remove the pto driven gear.

On some models with transmission driven power take-off, the engine clutch shaft and the power take-off drive gear are an integral unit. To remove the shaft and gear, unbolt the bearing cage from clutch housing and withdraw the bearing cage and shaft assembly. Remove shaft from bearing cage. The two shaft oil seals (one located in bearing cage and the other in the clutch housing) can be renewed at this time. The two ball type shaft carrier bearings can be pulled from shaft if renewal is required.

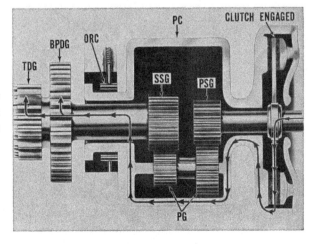

Fig. IH1358 — When the "Torque Amplifier" clutch is engaged, the system is in direct drive.

TORQUE AMPLIFIER UNIT

Torque amplification is provided by a planetary gear reduction unit located between the engine clutch and the transmission. The unit is controlled by a single plate, spring loaded clutch. When the clutch is engaged as in Fig. IH1358, engine power is delivered to both the primary sun gear (PSG) and the planet carrier (PC). This causes the primary sun gear and the planet carrier to rotate as a unit and the system is in direct drive. When the clutch is disengaged as shown in Fig. IH1359, engine power is transmitted through the primary sun gear to the larger portion of the compound planet gears (PG), giving the first gear reduction. The second gear reduction is provided by the smaller portion of the compound planet gears driving the secondary sun gear (SSG). As a result of the two gear reductions, an overall gear reduction of approximately 1.48:1 is obtained.

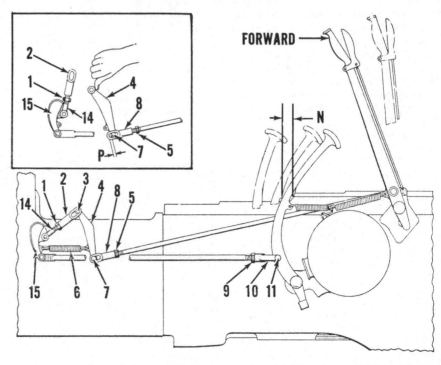

Fig. IH1360—Adjusting points for 340 Farmall engine and "Torque Amplifier" clutch linkage. Farmall 504 tractors are similar.

T A CLUTCH

141. **ADJUST.** Adjustment to compensate for lining wear is accomplished by adjusting the clutch actuating linkage, not by adjusting the position of the clutch release levers. The engine clutch linkage and the "Torque Amplifier" clutch linkage should be adjusted at the same time. Refer to paragraph 142 for Farmall 340 and 504 tractors and to paragraph 143 for International 330, 340, 504 and 2504 tractors.

142. **FARMALL.** Refer to Fig. IH1360. Remove spring (6), loosen lock nut (1) and remove clevis pin (3). Loosen lock nut (9) and remove clevis pin (11). Turn clevis (10) until clutch pedal free travel (N) is 7/8-1 1/4 inches.

Dimension (N) is measured horizontally from point of contact of clutch pedal lever and rear frame cover. After adjustment is complete, tighten lock nut (9).

After the engine clutch pedal linkage is properly adjusted, place the "Torque Amplifier" control lever in the forward position as shown and proceed to adjust the "Torque Amplifier" clutch linkage as follows:

Loosen lock nut (5), remove clevis pin (7) and turn lever (4) counterclockwise as far as possible without

forcing. See inset. This places the TA clutch release bearing against the clutch release levers. Now, adjust clevis (8) to provide a space (P) of 3/16-inch between the inserted pin (7) and the forward end of the elongated hole in clevis (8). Tighten lock nut (5) and reinstall spring (6). Adjust the length of rod (14) with clevis (2) so that rod (14) is the shortest possible length that will not change the position of levers (4 and 15) when pin (3) is inserted.

143. **INTERNATIONAL.** Refer to Fig. IH1361, remove spring (6), loosen lock nut (1) and remove clevis pin (3). Loosen nut (9), remove clevis pin (11) and turn clevis (10) either way as required to obtain a pedal free travel (N) of 7/8-1 1/4 inches as shown.

After the engine clutch linkage is properly adjusted, place the "Torque Amplifier" control lever in the forward position as shown and proceed to adjust the TA clutch linkage as follows: Loosen lock nut (5), remove pin (7) and turn lever (4) counter-clockwise as far as possible without forcing. See inset. This places the TA clutch release bearing against the clutch release levers. Now, adjust clevis (8) to provide a space (P) of 3/16-inch be-

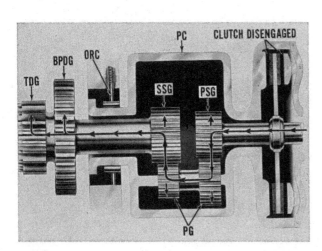

Fig. IH1359 — When the "Torque Amplifier" clutch is disengaged, an overall gear reduction of approximately 1.48:1 is obtained.

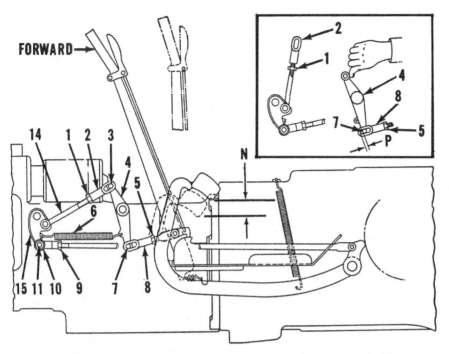

Fig. IH1361—Adjustment points on 330 and 340 International engine and "Torque Amplifier" clutch linkage. International 504 and 2504 tractors are similar.

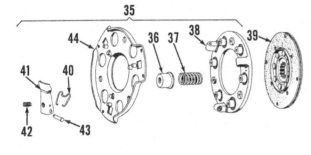

Fig. IH1362 — Exploded view of the "Torque Amplifier" clutch.

36. Spring cup
38. Pressure plate
39. Driven disc
40. Lever spring
41. Release lever
42. Adjusting screw
43. Lever pin

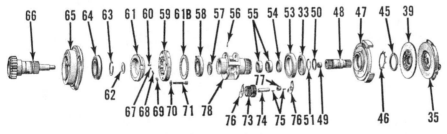

Fig. IH1363—Exploded view of the "Torque Amplifier" unit. Planetary gears (73) are available in sets only.

33. Oil seal	51. Snap ring	61B. Front thrust washer	67. Pin
35. Clutch cover assembly	53. Bearing	62. Thrust washer	68. Spring
39. Driven plate	54. Oil seal	63. Snap ring	69. Plug
45. Nut	55. Needle bearings	64. Bearing	70. Retainer ring
46. Locking washer	56. Roll pins	65. Transmission drive shaft bearing cage	71. Special screw
47. Clutch carrier	57. Spacer		73. Planetary gears
48. Primary sun gear	58. Bearing	66. Transmission drive shaft and secondary sun gear	74. Gear shaft
49. Needle bearing	59. Over-running clutch ramp		75. Needle bearing
50. Thrust washer	60. Clutch roller		76. Thrust plate
	61. Rear thrust washer		77. Bearing spacer
			78. Planet carrier

tween the inserted pin (7) and the forward end of the elongated hole in clevis (8). Tighten lock nut (5) and reinstall spring (6). Adjust the length of rod (14) with clevis (2) so that rod (14) is the shortest possible length that will not change the position of levers (15 and 4) when pin (3) is inserted.

144. R&R AND OVERHAUL. To remove the "Torque Amplifier" clutch cover assembly and lined plate, first detach (split) engine from clutch housing as outlined in paragraph 136, 136A or 137 and proceed as follows:

On 330 and 340 International tractors, remove the steering gear and fuel tank assembly as in paragraph 19 or 43. On 330 and 340 tractors so equipped, remove the hydraulic system control valves from the top of clutch housing. Remove the clutch housing top cover.

On International 504 and 2504 tractors, drain cooling system and remove hood, battery and starting motor. Disconnect heat indicator sending unit, fuel lines, oil pressure switch wire, wiring harness and controls from engine and engine accessories. Disconnect headlight and tail light wires. Disconnect hydraulic lines from control valve and flow divider valve. Disconnect the hydraulic lift pressure line. Unbolt fuel tank support and steering gear support and using a hoist, lift fuel tank, Hydrostatic hand pump assembly and instrument panel assembly from tractor. Remove clutch housing top cover.

On Farmall 340 tractors, remove fuel tank, steering shaft support and hydraulic control valves from the top of clutch housing. Remove the clutch housing top cover. Farmall 504 tractors are similar except that no control valves are involved.

On all models, remove the clutch shaft as in paragraph 140. Disconnect linkage from the TA clutch release shaft and remove lock screw from the TA clutch release fork. Remove snap ring from right end of the TA clutch release shaft; then, withdraw shaft, Woodruff keys, fork and release bearing and carrier. Use three $\frac{5}{16}$"-18 by $\frac{7}{16}$" cap screws and plain washers and screw them into the tapped holes provided in the pressure plate to keep the assembly under compression; then, unbolt the TA clutch cover assembly (35—Fig. IH1362) from carrier and withdraw the clutch cover assembly and lined plate.

If carrier (47—Fig. IH1363) is to be removed, bend tang of locking washer (46) out of notch in nut (45), remove spanner nut (Fig. IH1364) and bump carrier from splines of the primary sun gear.

Examine the driven plate for being warped, loose or worn linings, worn hub splines and/or loose hub rivets. Disassemble the clutch cover assembly and examine all parts for being excessively worn. The six pressure springs should have a free length of $2\frac{1}{32}$ inches and should require 156-166 lbs. to compress them to a height of $1\frac{1}{4}$ inches. Renew pressure plate if it is grooved or cracked. Renew back plate (44—Fig. IH1362) if it is worn around the drive lug windows.

When reassembling, adjust the release levers to the following specifications. With a back plate to pressure plate measurement (K—Fig. IH1365) of $\frac{19}{32}$-inch, the release lever height (L) from friction face of pressure plate to release bearing contacting surface of release levers is $1\frac{5}{8}$ inches.

When reassembling, observe the clutch carrier and back plate for balance marks which are indicated by an arrow and white paint. If the balance marks are found on both parts, they should be assembled with the marks as close together as possible. If no marks are found, or if only one part is marked, the clutch balance can be disregarded. Install the remaining parts by reversing the removal procedure.

PLANET GEARS, SUN GEARS AND OVER-RUNNING CLUTCH

145. R&R AND OVERHAUL. To overhaul the torque amplifier gear

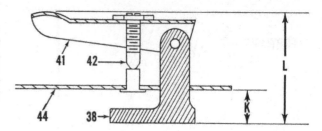

Fig. IH1365 — When adjusting the release lever height (L) of $1\frac{5}{8}$ inches, the back plate to pressure plate measurement (K) of 19/32-inch must be maintained.

set and over-running clutch, first remove the TA clutch cover assembly, lined plate and clutch carrier as outlined in paragraph 144 and proceed as follows: Remove all other parts attached to clutch housing. Remove the cover plate from bottom of clutch housing. Support rear half of tractor under rear frame and attach a chain hoist around clutch housing. Remove top bolt on each side of clutch housing and install aligning dowels. Remove remaining bolts retaining clutch housing to transmission case and separate the units. Note: Lower center bolt connecting clutch housing to main frame is accessible through the lower cover plate opening.

Unbolt the transmission drive shaft bearing cage from clutch housing and withdraw the complete TA unit. Remove the small retainer ring (70—Fig. IH1363 and IH 1366) from each of the four special over-running clutch screws (71—Fig. IH1363). Clamp the complete unit in a soft jawed vise and remove the four cap screws as shown in Fig. IH1367. Note: A cutout is provided in the planet carrier for this purpose. Separate the transmission drive shaft and bearing cage assembly from planet carrier (78—Fig. IH1363). Remove snap ring (63) from front of transmission drive shaft and press the transmission drive shaft rearward out of bearing and cage. Bearing (64) can

Fig. IH1366 — The retaining rings (70) should be installed as shown.

Fig. IH1364 — Installing the spanner nut. The clutch carrier should be blocked to prevent its turning. Remove the nut using a similar procedure.

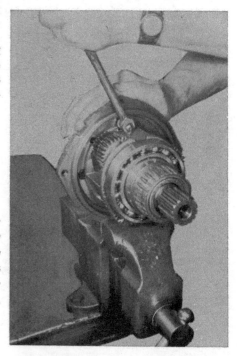

Fig. IH1367 — Removing the special over-running clutch screws.

be inspected and/or renewed at this time. Inspect ramp (59), springs (68) and rollers of over-running clutch. Renew damaged parts. Using OTC bearing puller attachment 952-A or equivalent, press front and rear bearings (53 & 58) from the planet carrier. It is important, when removing the front bearing to use a piece of pipe and press against the planet carrier and **not** against the primary sun gear shaft. Using a small punch and hammer, drive out the Esna roll pins retaining the planetary gear shafts in the planet carrier. Refer to Fig. IH-1369. Using OTC dummy shaft No. ED3259, push out the planet gear shafts and lift gears with rollers and dummy shaft out of the planet carrier. Be careful not to lose or damage the thrust plates (76—Fig. IH1363) as they are withdrawn. After the three compound planetary gears are removed, the primary sun gear and shaft can be withdrawn from the planet carrier.

Inspect splines, oil seal surface, bearing areas, pilot bearing and sun gear teeth of the primary sun gear and shaft for excessive wear or damage. If only the pilot bearing (49) is damaged, renew the bearing. If any other damage is found, renew the complete unit which includes an installed pilot bearing. Note: When installing a new pilot bearing, use OTC tool No. ED-3251 (show in Fig. IH-1370) and press the bearing in until surface (X) is even with rear edge of primary sun gear.

Inspect the planet carrier for rough oil seal surface, worn over-running clutch roller surface and elongated planet gear shaft holes.

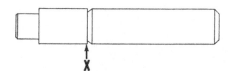

Fig. IH1370—OTC tool number ED-3251. Press pilot bearing in the primary sun gear until surface (X) is even with rear edge of the primary sun gear.

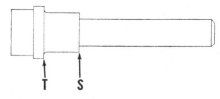

Fig. IH1371—OTC tool number ED-3250 is used to install needle bearings in the planet carrier. Refer to text.

Inspect the primary sun gear shaft needle bearings (55—Fig. IH1363) and the shaft oil seal (54). If bearings and/or seal are damaged, and planet gear carrier is O. K., drive out the faulty parts with a brass drift and install new bearings using OTC driving collar ED-3250 (shown in Fig. IH-1371). Press rear bearing in from front until surface (S) is even with front of planet carrier. Press the front needle bearing in from front until surface (T) is even with front of planet carrier. Install oil seal (54—Fig. IH1363) with lip toward rear until front of oil seal is even with front of planet carrier.

Install snap ring (51) on the primary sun gear shaft and thrust washer (50) immediately ahead of the snap ring. Using OTC oil seal protector sleeve No. ED-3245, install the primary sun gear and shaft in the planet carrier as shown in Fig. IH1372.

Inspect teeth of planet gears for wear or other damage. If any one of the three gears is damaged, renew

all three gears which are available in a matched set only. These planetary gears are manufactured in matched sets so the gears will have an equal amount of backlash when installed and no one gear will carry more than its share of the load. Note: The International Harvester Co. specifies that when the planet gears are removed, the planet gear shafts and needle rollers should always be renewed. The planet gear shafts are available in sets of three and the needle rollers are available in sets of 138.

Using chassis lubricant and OTC dummy shaft ED-3259, install twenty-three new needle bearings in one end of a planet gear. Slide dummy shaft into bearings and install bearing spacer (77—Fig. IH1363). With the aid of chassis lubricant, install twenty-three new needle bearings in the other end of the planet gear and slide the dummy shaft completely into the gear, thereby holding the needle bearings and spacer in the proper position. Assemble one thrust plate (76) to each end of the planet gear and install planet gear, dummy shaft and thrust plates assembly in the planet carrier.

Fig. IH1373—Chassis lubricant facilities installation of needle bearings in the planetary gears.

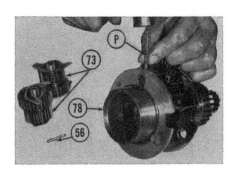

Fig. IH1369 — Using a punch and hammer to drive out the roll pins which retain the planet gear shafts in the planet carrier.

P. Punch   73. Planet gears
56. Roll pins   78. Planet carrier

Fig. IH1372 — The seal protector sleeve is used when installing the primary sun gear in the planet carrier.

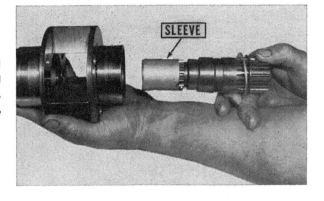

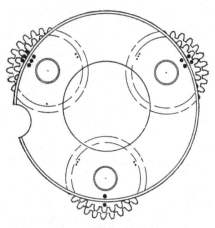

Fig. IH1374 — When planetary gears are installed properly, punch marks on gears and planet carrier will be in register.

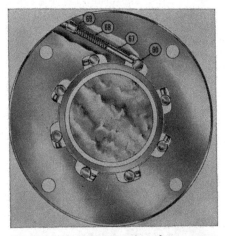

Fig. IH1375—Cut-away view shows proper installation of over-running clutch rollers (60), pins (67), springs (68) and plugs (69) in ramp.

Using one of the three new planet gear shafts, push out the dummy shaft and install roll pin securing planet gear shaft to the planet carrier. Observe rear face of the planet carrier at each planet gear location where timing marks will be found. The timing marks are punched dots (Refer to Fig. IH1374). One location has one punch mark, another location has two punch marks and the other location has three punched marks. Turn the primary sun gear shaft until the timing mark on the rear face of the installed planet gear are in register with a similar mark on the planet carrier. Now assemble the other planet gears, needle bearings, spacers, thrust plates and dummy shaft and install them so that timing marks are in register. When all three planet gears are installed properly, the single punch mark on planet carrier will be in register with single punch mark on one of the planet gears, double punch marks on carrier will register with double punch marks on one of the gears and triple punch marks on carrier will be in register with triple punch marks on one of the gears as shown in Fig. IH1374.

Install thrust washer (61B—Fig. IH-1363) on planet carrier. Assemble the pins (67—Fig. IH1375), springs (68) and rubber plugs (69) into the over-running clutch ramp and place ramp on the planet carrier. Using a small screw driver, push pins (67) back and drop rollers (60) in place as shown. Install bearing (64—Fig. IH1376) with snap ring in the rear bearing cage (65), press the transmission drive shaft (66) into position and install snap ring (63).

Place the planet carrier on the bench with rear end up and lay thrust washer (62) on the primary sun gear. Place the over-running clutch thrust washer (61) on the ramp so that polished surface of thrust washer will contact rollers. Install the assembled transmission drive shaft and bearing cage and secure in position with the four special screws (71). After the cap screws are tightened to a torque of 40 ft.-lbs., install the small retainer rings as shown in Fig. IH1366.

Inspect the large oil seal (33—Fig. IH1376) in the clutch housing and renew if damaged. Lip of seal goes toward rear of tractor.

Using OTC oil seal protector sleeve No. ED-3253 over splines of planet carrier, insert the assembled TA unit and tighten the transmission drive shaft bearing cage cap screws securely. Assemble the remaining parts by reversing the disassembly procedure.

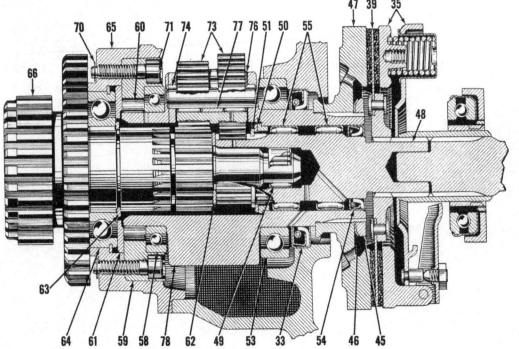

Fig. IH1376 — Typical sectional view of assembled "Torque Amplifier" unit. Bearings in the unit are non-adjustable.

33. Oil seal
35. Clutch cover assembly
39. Driven plate
45. Nut
46. Locking washer
47. Clutch carrier
48. Primary sun gear
49. Roller bearing
50. Thrust washer
51. Snap ring
53. Bearing
54. Oil seal
55. Needle bearings
58. Bearing
59. Over-running clutch ramp
60. Clutch roller
61. Thrust washer
62. Thrust washer
63. Snap ring
64. Bearing
65. Transmission drive shaft bearing cage
66. Transmission drive shaft and secondary sun gear
70. Retainer ring
71. Special screw
73. Planetary gear
74. Gear shaft
76. Thrust plate
77. Bearing spacer
78. Planet carrier

DIRECTION REVERSER

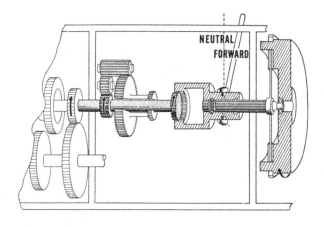

Fig. IH1380 — Schematic drawing of the direction reverser in position for forward travel.

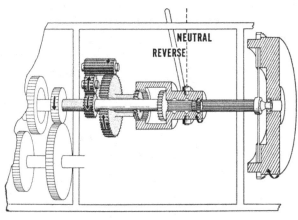

Fig. IH1381 — Schematic drawing of the direction reverser in position for reverse travel.

Series 340-504-2504

Series 340, 504 and 2504 tractors are available with a selective, sliding spur gear reversing unit located between the regular 5-speed transmission and the engine clutch. The unit is controlled by a hand lever which moves the sliding gear to the forward, neutral or reverse position. Refer to Figs. IH1380 and IH1381, which show the principles of operation.

146. **R&R AND OVERHAUL.** To remove the direction reversing mechanism, first remove the steering gear and fuel tank assembly as in paragraph 19, 43 or 144. On tractors so equipped, remove the hydraulic system control valves from the top of the clutch housing. On all models, remove the clutch housing top cover. Split the engine from the clutch housing as in paragraph 136, 136A or 137 and the clutch housing from the transmission as in paragraph 150. Remove the engine clutch release shaft, bearing and fork. On models so equipped, remove the independent PTO drive shaft (48—Fig. IH1382), bearing cage (51), driven gear and shaft. On all models, remove the direction reverser shifter fork (10) and shaft (8). Remove snap ring (39) and remove the clutch shaft (38).

1. Bracket
2. Grease fitting
3. Snap ring
4. Handle
6. Rod
7. Clevis
8. Shaft
9. "O" rings
10. Fork
11. Transmission drive shaft
12. Bearing carrier
13. Ball bearing
14. Snap ring
15. Spacer
16. Needle bearing
17. Planet carrier
18. Planet gear shafts (6 used)
19. Thrust plates (6 used)
20. Rollers (207 used)
21. Spacers (3 used)
22. Long planet gear (3 used)
23. Short planet gear (3 used)
24. Spacer (3 used)
25. Snap ring
26. Ball bearing
27. Bearing cage
28. Needle bearing
29. Reverse sun gear
30. Coupling
31. Transmission drive shaft retainer
32. Cap screw
33. Lock plate
34. Shifter coupling
35. Collar
36. Poppet balls (2 used)
37. Spring
38. Clutch shaft
39. Snap ring
40. Bearing cage
41. Ball bearing
42. Snap ring
43. Oil seal
44. Oil seal carrier
45. Ball (3/16-dia.)
46. Snap ring
47. Oil seal
48. IPTO drive shaft
49. Needle bearing
50. "O" ring
51. Bearing cage
52. Gasket
53. Ball bearing
54. Snap ring
55. Oil seal
56. Bearing retainer

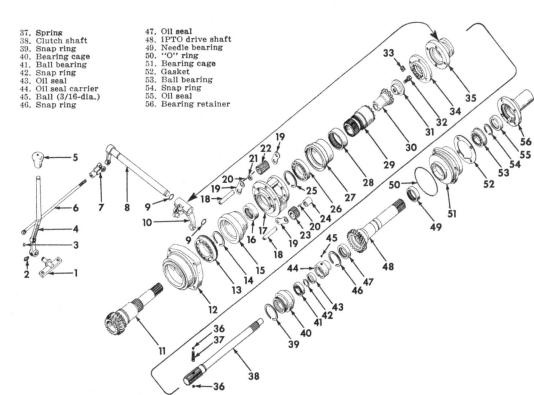

Fig. IH1382—Exploded view of the direction reverser assembly. Planet gears (22) and (23) are available in matched sets of three.

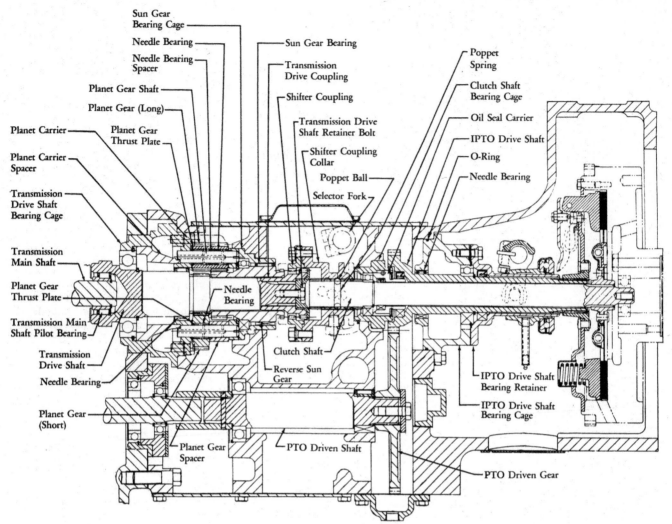

Fig. IH1383—Cross sectional drawing of 340 International clutch and direction reverser. Refer to Fig. IH1382 for exploded view. Series 504 and 2504 are similar.

Forward end of shaft is tapped to accommodate a puller. Unbolt the collar (35) from the coupling (34) and remove the collar. Remove cap screw (32), drive shaft retainer (31) and coupling (34). Remove the four retaining cap screws and withdraw the bearing cage, planet carrier and transmission drive shaft assembly out through the rear of clutch housing. The remainder of the disassembly procedure will be evident after an examination of the unit. Refer to Fig. IH1382.

Check all parts for visible damage and renew all questionable parts. Refer to the data which follows:
Poppet balls (36) diameter........$\frac{5}{16}$"
Poppet spring (37)
    Free length ....1.086"
    Pounds test @ length..10.2 @ 0.90"
Backlash
    Between planets)
      (22 & 23).........0.0010-0.0034"
    Between planets (23) and
      driven sun gear
      (11) ....0.0010-0.0034"

Bore for needle bearing in
  planets (22 & 23)
    Diameter ....0.7834-0.7839"
Planet gear shafts (18)
    Diameter ....0.5952-0.5955"
    Length ....3.210"
Combined wear of shaft (18)
  OD and gear (22 or 23) ID
  should not exceed..........0.003"

Check the clutch housing to see if there is an oil drain hole in the position shown in Fig. IH1384.

Reassemble and install in reverse of the removal and disassembly procedure. Refer to Figs. IH1382 and IH1383. Cap screw (32—Fig. IH1382) should be torqued to 52-59 Ft.-Lbs.

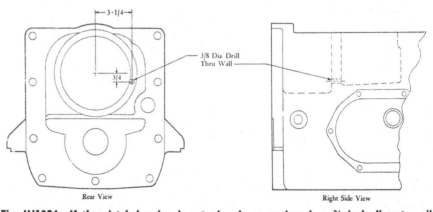

Rear View   Right Side View

Fig. IH1384—If the clutch housing is not already so equipped, a ⅜-inch diameter oil drain hole should be drilled through the housing wall between the spline shifter coupling and the planet carrier compartments as shown.

TRANSMISSION

The transmission, differential and final drive gears are all contained in the same case which is called the rear frame. A wall in the case separates the bull gear and differential compartment from the transmission gear set. Shifter rails and forks are mounted on underside of the rear frame (transmission) cover.

All bearings supporting the transmission shafts are of the non-adjustable ball type.

TOP COVER

International 330

147. To remove the transmission top cover, it is necessary to first remove the seat and hydraulic reservoir. Unbolt and lift the cover from the tractor.

NOTE: The additional work of detaching the rockshaft from the top cover may be necessary on some tractors.

Series 340

148. To remove the transmission top cover, it is necessary to first remove the seat and bracket, hydraulic control valves cover and battery. Unbolt the top cover and using a suitably attached hoist, lift same from tractor.

NOTE: The additional work of removing or disconnecting hydraulic lines, hydraulic control valve linkage and the rockshaft may be necessary in some cases depending upon the equipment or combination of equipment that the tractor is provided with.

Series 504-2504

148A. To remove the transmission top cover, along with the hydraulic lift housing, proceed as follows: Remove seat and drain lift housing. If additional working room is desired, remove both fenders. Remove both platforms. Disconnect pto control rod at pto and remove pto shifter lever and bracket. Disconnect lift links from rockshaft arms, and if so equipped, disconnect the break-away coupling bracket. Disconnect draft control rod from bellcrank, unbolt torsion bar and remove torsion bar and bellcrank. Remove torsion bar bracket. Mark the cap screw holes used and remove the quadrant. Disconnect rear junction block bracket, remove banjo bolts, then remove junction block and lines. Remove banjo bolts from front port lines and remove lines. Remove control valve through bolts and remove control

Fig. IH1394—Before the tractor halves are rejoined, a rubber band should be positioned around the rollers of the transmission main shaft front bearing to hold the rollers in position and keep them from being pushed out of the bearing.

valves and levers and transfer block. Disconnect lines from front of hydraulic lift housing. Unbolt rear frame top cover from rear frame, attach hoist and lift assembly from tractor.

NOTE: It is not required that control valves be removed but most mechanics prefer to do so in order to provide working room.

TRACTOR SPLIT

International 330

149. To detach (split) the transmission from the clutch housing, drain the transmission and "Torque Amplifier" housings. Remove the hoods and disconnect or remove all hydraulic tubes and wires that would hinder the separation of the tractor. Support the tractor halves, remove the cap screws attaching the clutch housing to the transmission case; then, separate the tractor.

Before the tractor halves are rejoined, a rubber band should be positioned around the rollers of the transmission main shaft front bearing. (Similar to Fig. IH1394.) The use of a rubber band will generally serve to hold the bearing rollers in position; however, care should be exercised when joining the tractor halves to prevent a roller (or rollers) from being pushed out of the bearing.

Series 340-504-2504

150. To detach (split) the transmission from the clutch housing, drain the transmission and "Torque Amplifier" housings. Remove the hoods and the hydraulic control valves cover on series 340. Disconnect or remove all hydraulic tubes, wires, control rods and control cables that would hinder the separation of the tractor. On 504 Farmall, disconnect front end of steps from "Torque Amplifier" housing. Support the tractor halves, remove lower clutch housing cover and the cap screws attaching the clutch housing to the transmission case; then, separate the tractor.

Always renew the sealing ring (S—Fig. IH1394) as it seals the power steering and hydraulic pump suction tube. The gasket between the transmission case and the clutch housing should not cover the oil passages (O).

Before the tractor halves are joined, a rubber band should be positioned around the rollers of the transmission main shaft front bearing as shown in Fig. IH1394. The use of a rubber band will generally serve to hold the bearing rollers in position; however, care should be exercised when joining the tractor halves to prevent a roller (or rollers) from being pushed out of the bearing. Use guide studs in top hole on each side of transmission housing.

OVERHAUL

All Models

Data on overhauling the various transmission components are outlined in the following paragraphs. In general, the following paragraphs apply to all models; however, in some cases, the overhaul procedures differ on models equipped with and without torque amplifier or direction reverser as well as those equipped with and without independent power take-off. Where these differences are encountered, they will be mentioned.

151. **SHIFTER RAILS AND FORKS.** Shifter rails and forks are retained to bottom side of the rear frame (transmission) cover and are accessible for overhaul after removing the cover as outlined in paragraph 147, 148 or 148A. The overhaul procedure is conventional and evident after an examination of the unit.

152. TRANSMISSION DRIVING SHAFT. On models with torque amplifier or direction reverser, the transmission driving gear and shaft (66—Fig. IH 1395) is integral with the secondary sun gear and is normally serviced in conjunction with overhauling the "Torque Amplifier" or direction reverser unit.

On models without torque amplifier or direction reverser, the transmission driving shaft is considered a part of the transmission. To remove the drive shaft, it is first necessary to detach (split) the transmission housing from the clutch housing as outlined in paragraph 149 or 150.

With the transmission detached from the clutch housing, the procedure for removing and overhauling the drive shaft is evident after an examination of the unit and reference to Figs. IH-1396, 1397 and 1398.

153. MAINSHAFT PILOT BEARING. To remove the mainshaft pilot bearing (9 — Fig. IH 1398), detach (split) the transmission housing from the clutch housing as outlined in paragraph 149 or 150. On models with no "Torque Amplifier" or direction reverser, remove transmission input shaft. Remove the bearing retaining snap ring (11) or the capscrew, lock plate and washer (10) from front end

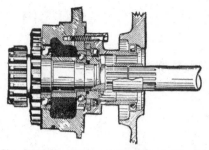

Fig. IH1396—On tractors without "Torque Amplifier" or direction reverser, the area shown is different from Fig. IH1395.

of mainshaft and using OTC bearing puller or equivalent as shown in Fig. IH1399, remove the pilot bearing.

Fig. IH1397 — On tractors without "Torque Amplifier" or direction reverser, independent power take-off, hydraulic lift and power steering; the area shown is different from Fig. IH1395.

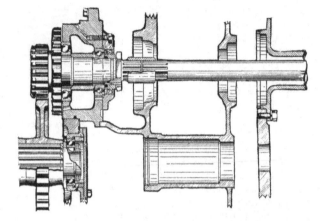

When installing the pilot bearing, the chamfered end of the inner race should be towards the forward end of the shaft.

154. MAINSHAFT (SLIDING GEAR OR BEVEL PINION SHAFT). To remove the transmission mainshaft, first detach (split) transmission from clutch housing as in paragraph 149 or 150 and remove the transmission top cover.

Remove the three cap screws retaining the mainshaft rear bearing retainer

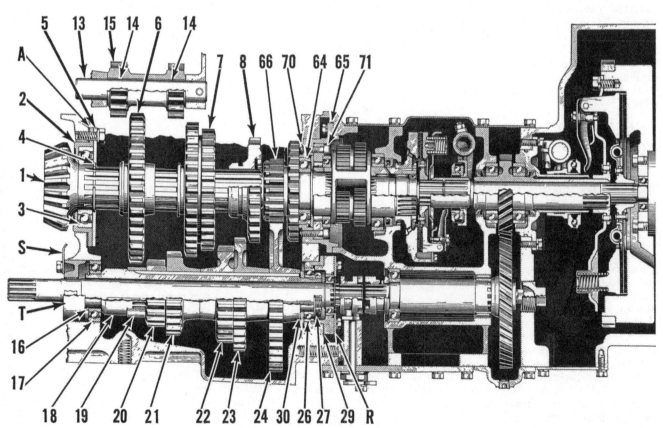

Fig. IH1395—Sectional view of a typical transmission with "Torque Amplifier" and independent power take off. Refer to Fig. IH1398 for legend.

(5 — Fig. IH 1398) to the main case dividing wall, move the mainshaft assembly forward and withdraw the unit, rear end first, as shown in Fig. IH1400.

Remove the mainshaft pilot bearing (9—Fig. IH 1398) and slide gears from shaft. Remove snap ring (4) and press or pull rear bearing and retainer from shaft. When reassembling, install bearing (3) so that ball loading grooves are toward front or away from the bevel pinion gear. Use Figs. IH1395 and IH1398 as a guide when installing the sliding gears and if the same mainshaft is installed, be sure to use the same shims (A) as were removed. If a new mainshaft is being installed, use the same shims (A) as a starting point, but be sure to check and adjust if necessary the main drive bevel gear mesh position as outlined in the main drive bevel gear section.

155. **COUNTERSHAFT.** To remove the transmission countershaft, first remove the mainshaft as outlined in paragraph 154 and proceed as follows: On models with independent power take-off, remove the four cap screws retaining the independent power take-off extension shaft front bearing cage to main frame and withdraw the extension shaft, bearing cage and retain-

er as shown in Fig. IH 1401. Working in the bull gear compartment of the main frame, remove the independent power take-off coupling shaft and the extension shaft rear bearing carrier retainer strap (S—Fig. IH1395).

On models without power take-off, remove the cap (46—Fig. IH 1398).

On all models, remove the nut from forward end of countershaft and while bucking up the countershaft gears, bump the countershaft rearward until free from front bearing. Remove the countershaft front bearing and cage.

Withdraw the countershaft from rear and remove gears from above. The rear bearing can be removed from countershaft after removing snap ring (16—Fig. IH1398). When reassembling, use Fig. IH1395 as a guide and make certain that beveled edge of constant mesh gear spacer (30) is facing toward front of tractor.

156. **REVERSE IDLER.** With the countershaft removed as outlined in paragraph 155, the procedure for removing the reverse idler is evident. Bushings (14—Fig. IH1398) are renewable and should be reamed after installation, if necessary, to provide a recommended clearance of 0.003-0.005 for the idler gear shaft.

Fig. IH1399 — Using a puller to remove pilot bearing from front of transmission mainshaft.

Fig. IH1400 — Removing the transmission mainshaft assembly.

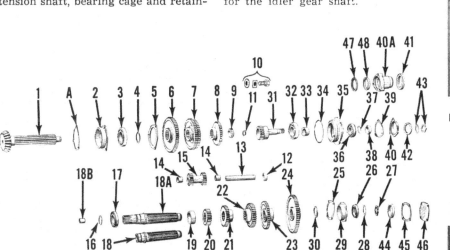

Fig. IH1398—Exploded view of transmission shafts, gears and associated parts. All bearings are of the non-adjustable ball type. Main drive bevel pinion position is controlled by shims (A). Countershaft (18A) and bushing (18B) are used on some early 330 International tractors without independent pto or "Torque Amplifier". Washer, lock and cap screw (10) is used on early tractors instead of snap ring (11).

A. Shims (0.003, 0.007, 0.015 and 0.030)	12. Cup plug 13. Reverse idler shaft 14. Reverse idler bushings	23. Fourth speed driving gear 24. Constant mesh gear
1. Mainshaft and bevel pinion	15. Reverse idler gear	26. Bearing 27. Nut
2. Bearing cage	16. Snap ring	29. Bearing cage
3. Rear bearing	17. Bearing	30. Spacer
4. Snap ring	18. Countershaft	31. Transmission
5. Bearing retainer	18A. Countershaft	drive shaft
6. First and reverse sliding gear	18B. Bushing 19. Spacer	32. Ball bearing 33. Spacer
7. Second and third sliding gear	20. First speed driving gear	34. Seal ring 35. Bearing cage
8. Fourth and fifth sliding gear	21. Second speed driving gear	36. Ball Bearing 37. Seal ring
9. Bearing	22. Third speed driving gear	38. Spacer
10. Washer, lock and cap screw		39. Gasket
11. Snap ring		

40. Retainer, (Used with no IPTO, TA or direction reverser)
40A. Retainer, (Used with IPTO & no TA)
41. Oil seal
42. Oil seal
43. Nut and lock washer
44. Spacer
45. Gasket
46. Cap
47. Spacer, (Used with IPTO & no TA or DR)
48. Seal, (Used with IPTO & no TA or DR)

Fig. IH1401 — Removing the independent power take-off extension shaft, front bearing and cage.

MAIN DRIVE BEVEL GEARS AND DIFFERENTIAL

After renewing a bevel pinion or ring gear, the gear mesh position and backlash as well as the differential carrier bearings should be adjusted as follows:

157. CARRIER BEARING ADJUSTMENT. There is no adjustment, as such, for the ball type differential carrier bearings used on series 330 and early series 340, but it is important that the shims (B—Fig. IH1410) located under the differential bearing carriers be varied to eliminate all end play from the differential unit without causing any binding tendency in the carrier bearings. To accurately check the adjustment, first install more than enough shims (B) under each carrier and proceed as follows: Bump the differential toward left side of tractor and make certain there is some backlash between the ring gear and pinion. Then check and record the amount of backlash. Now, bump the differential toward right side of tractor and again check and record the amount of backlash. Subtract the first backlash reading from the second; then, remove shims (B), equal to ½ the difference between the two backlash readings, from under each bearing carrier.

For example, suppose the second backlash reading was 0.029 and the first was 0.015; the difference between the two readings is 0.014. In this case, remove one 0.007 thick shim (B) from each side.

Proceed to paragraphs 158 and 159 and adjust the bevel gear backlash and mesh position.

Late series 340 and all 504 and 2504 series tractors are equipped with tapered roller differential carrier bearings as shown in Fig. IH1413A.

To adjust bearings (9), remove the bull gears as outlined in paragraph 163, then adjust bearing preload to a rolling torque of 10-20 inch pounds (excluding drag of bull pinion shaft and oil seals) by varying shims (items 13 through 16—Fig. IH1413A) under bearing cages (17 and 24). Shims are available in thicknesses of extra light, light, medium and heavy.

Reinstall bull gears and adjust bevel drive gear backlash and mesh position as outlined in paragraphs 158 and 159.

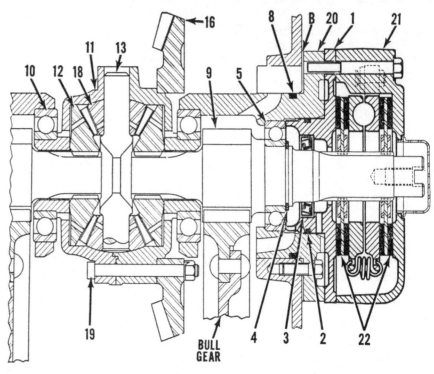

Fig. IH1410—Sectional view of differential and double disc brakes. Refer to Fig. IH1413 for legend. Late model tractors have tapered bearings on differential instead of the ball bearings shown.

NOTE: A tapered shim is used only under right hand bearing cage on series 504 and 2504. A shim should also be used on other model tractors which have been modified to incorporate the tapered roller differential bearings. Only one tapered shim is used per tractor and if additional shims are required under the right hand bearing cage, they should be positioned between the tapered shim and the final drive housing (rear frame). The tapered shim is used to increase bearing life by compensating for deflection of the differential when under load.

158. BACKLASH ADJUSTMENT. After the carrier bearings are adjusted as outlined in paragraph 157, the backlash can be adjusted as follows: Transfer shims from under one bearing carrier to the other to provide 0.006-0.015 for series 330 and 340; or 0.006-0.009 for series 504 and 2504 backlash between teeth of the main drive bevel pinion and ring gear. To increase backlash, remove shim or shims from carrier on ring gear side of housing and install same under carrier on opposite side. Only transfer shims, do not remove shims or the previously determined carrier bearing adjustment will be changed.

159. MESH POSITION. The mesh position of the bevel pinion and ring gear must be adjusted when renewing bevel pinion, bevel pinion bearings and/or ring gear. Before setting the mesh position, adjust the bearings and backlash as outlined in paragraphs 157 and 158.

The next step is to arrange shims (A —Fig. IH1395), located between the transmission mainshaft rear bearing cage and the rear frame, to provide the proper tooth contact (mesh position) of the bevel gears.

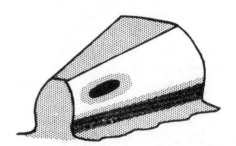

Fig. IH1412 — The correct mesh position of the bevel gears will be indicated by removal of the Prussion blue or red lead as shown by the dark spot on the drawing.

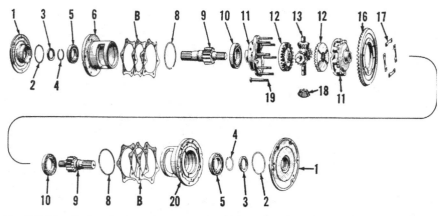

Fig. IH1413—Exploded view of early type differential, bull pinions and associated parts. Shims (B) control bearing adjustment and backlash of the main drive bevel gears. Also see Fig. IH1413A.

B. Shims
1. Bearing retainer (inner brake plate)
2. Seal ring
3. Bull pinion shaft oil seal
4. Snap ring
5. Ball bearing
6. Left bull pinion shaft bearing cage
8. Seal ring
9. Bull pinion
10. Differential carrier bearing
11. Differential case half
12. Differential side gear
13. Spider
16. Bevel ring gear
17. Lock plates
18. Pinions
19. Case bolts
20. Right bull pinion shaft bearing cage

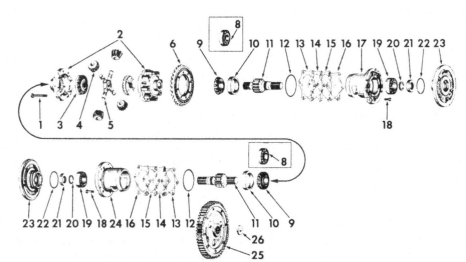

Fig. IH1413A — Exploded view of differential used in series 504 and 2504 tractors. Note tapered bearings (9) which have replaced ball bearings (8). Late models of the 340 series will be similar.

1. Case bolts
2. Case halves
3. Bevel gear
4. Pinion gear
5. Spider
6. Bevel ring gear
8. Ball bearing (not used)
9. Bearing cone
10. Bearing cup
11. Bull pinion shaft
12. "O" ring
13-16. Shims
17. Bearing cage (RH)
19. Roller bearing
20. Snap ring
21. Oil seal
22. "O" ring
23. Bearing retainer
24. Bearing retainer (LH)
25. Bull gear
26. Retainer washer

Paint the pinion teeth with Prussian blue or red lead, rotate the ring gear by hand in normal direction of rotation and observe the contact pattern on the tooth surfaces.

The area of **heaviest contact** will be indicated by the coating being **removed** from the pinion at such points.

After obtaining the tooth contact pattern shown in Fig. IH1412, recheck the backlash and if not within the desired limits of 0.006-0.015 for series 330 and 340; or 0.006-0.009 for series 504 and 2504, adjust by transferring a shim or shims from behind one differential bearing cage to the other bearing cage until desired backlash is obtained.

RENEW BEVEL GEARS

All Models

160. To renew the main drive bevel pinion, follow the procedure outlined in paragraph 154 for overhaul of the transmission main shaft. To renew the main drive bevel ring gear, follow the procedure outlined for overhaul of the differential assembly (paragraph 161).

DIFFERENTIAL AND CARRIER BEARINGS

All Models

Differential unit is of the four pinion type mounted back of a dividing wall in the rear frame (transmission case). The differential case halves are held together by bolts which also retain the bevel ring gear.

161. **R&R AND OVERHAUL.** To remove the differential and the main drive bevel ring gear assembly, first remove the final drive bull gears as outlined in paragraph 163 and proceed as follows: Remove the brake housings, brake discs and the inner brake plates (1—Figs. IH1410 and IH1413) or (23—Fig. IH1413A). Remove the bull pinion shaft bearing cages and lift the differential and bevel ring gear assembly from tractor. Save and do not mix shims. Note: The differential carrier bearings can be renewed at this time.

The procedure for overhauling the removed differential unit is evident after an examination of Fig. IH1413 or IH1413A. Note that differential case bolts on 504 and 2504 tractors have self locking nuts instead of lock plates.

When reinstalling the differential unit, assemble bull pinion shaft to left hand bearing cage and install the unit. Lower differential unit into rear frame and enter same over splines of bull pinion shaft. Install opposite (right) bearing cage and bull pinion unit.

Adjust the differential carrier bearings as outlined in paragraph 157 and check the bevel gear mesh and backlash as in paragraphs 158 and 159.

NOTE: Refer to note at end of paragraph 157 concerning the use of tapered bearing cage shims.

FINAL DRIVE

As treated in this section, the final drive will include the bull pinions and integral shafts, both bull gears and wheel axle shafts.

BULL PINION (DIFFERENTIAL) SHAFTS

All Models

162. **REMOVE AND REINSTALL.** To remove either bull pinion shaft (9 —Fig. IH1413) or (11—Fig. IH1413A), remove the respective brake unit, remove the bull pinion shaft bearing retainer (inner brake drum) and withdraw the bull pinion shaft and bearing from the bearing cage.

BULL GEAR

All Models

163. **REMOVE AND REINSTALL.** To remove either bull gear, first remove the transmission top cover as in paragraph 147, 148 or 148A. Remove the rear wheel and cap screw (M— Fig. IH1414) retaining bull gear to inner end of wheel axle shaft.

Note: On some models, bull gear may be retained to axle by a snap ring. Remove cap screws retaining rear axle housing or sleeve to transmission housing and withdraw carrier and shaft as a unit from the rear frame. Bull gear can now be removed from rear frame (transmission housing).

WHEEL AXLE SHAFT

All Models

164. **REMOVE AND REINSTALL.** The wheel axle shaft can generally be removed by removing the pto rear unit (or cover) and removing the bull gear retaining cap screw and washer (or snap ring) by working through the rear opening. Remove the cap screws which attach the rear axle housing (sleeve) to the transmission housing and separate the axle and housing from the transmission housing.

NOTE: In cases where the bull gear is stuck on the axle shaft it may be necessary to remove the transmission top cover in order to work the bull gear loose.

With the axle shaft and carrier unit off tractor, remove outer bearing retainer and remove shaft from carrier by bumping on its inner end. Refer to Fig. IH1414.

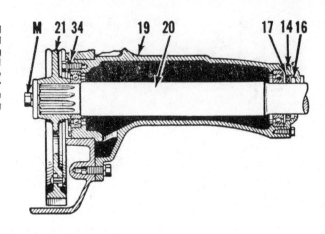

Fig. IH1414 — Sectional view showing a typical rear axle and housing used on all models. Bull gear (21) is retained to inner end of axle shaft by cap screw (M). On some models, a snap ring is used.

M. Cap screw
14. Bearing retainer
16. Felt seal
17. Oil seal
19. Axle housing
20. Axle shaft
21. Bull gear
34. Bearing retainer

BRAKES

All Models

165. Brakes are of the double disc, self energizing type, which are splined to the outer ends of the bull pinion shaft. The molded linings are bonded to the brake discs. Procedure for removing the lined discs will be evident after an examination of the unit and reference to Fig. IH1420.

166. **ADJUSTMENT.** To adjust the brakes, loosen jam nut (1—Fig. IH-1421 or IH1422) and turn adjuster (2— Fig. IH1421) or (2B—Fig. IH1422) either way as required to obtain the correct pedal free travel of 1¾-inches for 340 tractors; 1⅜-inches for 330 International tractors; 1-inch for Farmall 504 tractors or 1⅞ inches for International 504 and 2504 tractors. Measurement should be taken between pedals and rear frame cover. Brakes can be equalized by loosening the tight brake. When adjustment is completed, tighten jam nut (1).

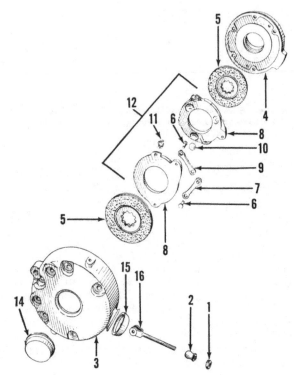

Fig. IH1420 — Exploded view of the typical brake assembly. Three balls (10) are used in each brake unit. Later brakes will include a brake positioning spring and anchor which is not shown.

1. Jam nut
2. Adjusting nut
3. Housing
4. Oil seal retainer
5. Brake discs
6. Studs
7. Yoke actuating link
8. Actuating discs
9. Plain actuating link
10. Balls
11. Extension spring
12. Actuating disc assembly
14. Plug
15. Boot
16. Operating rod

POWER TAKE-OFF

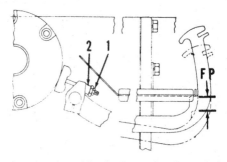

Fig. IH1421—The brake pedal free play (FP) can be obtained on series 340, 504 and 2504 by turning the adjusting nut (2). Refer to text for free play measurements.

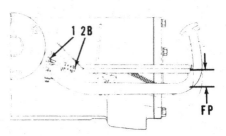

Fig. IH1422—Brake pedal free play of 1⅜ inches on International 330 tractors can be obtained by turning the adjusting bolt (2B).

BELT PULLEY

A rear mounted belt pulley unit which obtains its drive from the pto shaft, is used.

167. **OVERHAUL.** The procedure for disassembling and reassembling the removed belt pulley unit is evident after an examination of the unit and reference to Fig. IH1425. Shims (15 and 24) control the mesh position and backlash of the bevel gears. Recommended bevel gear backlash is 0.008-0.010.

ADJUSTMENTS

Planetary Type

170. **REACTOR BANDS.** To adjust the reactor bands, remove the adjusting screw cover, loosen lock nuts (N—Fig. IH1426) and back off the adjusting screws (M) approximately four turns. Hold the operating lever so that the pawl on the control lever is centered in the middle land of the quadrant in a manner similar to that shown at (X).

Turn the adjusting screws in (clockwise) until screws are reasonably tight and set the lock nuts finger tight. Move the operating lever back and forth several times; then, back-off the adjusting screws approximately one turn and tighten the lock nuts.

Fig. IH1426 — When adjusting the early type power take-off reactor bands, the pawl on the handle should be centered on the quadrant land as shown at (X).

Using a $\frac{5}{16}$-inch rod approximately 8 inches long as a lever in hole of pto shaft, check to make certain that shaft is free to turn only when the pawl is exactly centered on the middle land of the quadrant as shown at (X—Fig. IH-1426). The shaft should not turn with lever in any other position.

If shaft turns freely with the control lever in any other position the adjusting screws are not evenly adjusted and readjustment is necessary.

If shaft is not free to turn in any position of the lever, turn the adjusting screws **out** (counter-clockwise) one-half turn, lock and recheck.

Tighten lock nuts when adjustment is complete.

Clutch Type

171. **OVERCENTER CLUTCH.** To adjust the overcenter clutch on the early models of the clutch type pto, the pto rear unit must be removed. Remove the two cap screws which attach the two halves of the rear unit together and separate the halves. Place the rear housing cover and the clutch assembly in a vise. Disengage the clutch, mount a dial indicator as shown in Fig. IH1427, zero the indicator face, then engage the clutch and note the indicator reading. If the amount of vertical movement at the outside edge of the Belleville washers isn't 0.035-0.045 as the clutch is engaged, raise the lock plate (36P—Fig. IH1428) and turn it slightly so as to engage the lock plate in the thread relief (TR) as shown. Turn the spider (35) either way as re-

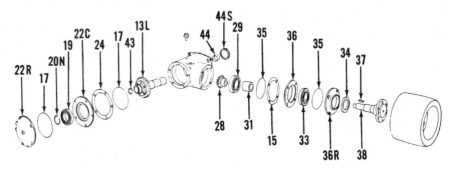

Fig. IH1425—Exploded view of typical rear mounted belt pulley unit. Drive is taken from the pto shaft.

13L. Drive gear and shaft	22C. Bearing cage	31. Spacer	37. Woodruff key
15. Shims (0.007, 0.012 and 0.030)	22R. Bearing retainer	33. Ball bearing	38. Pulley shaft
17. "O" rings	24. Shims (0.007, 0.012 and 0.030)	34. Oil seal	43. Cup plug
19. Ball bearing	28. Driven bevel gear	35. "O" ring	44. Drive shaft bushing
20N. Snap ring	29. Ball bearing	36. Bearing cage	44S. Oil seal
		36R. Bearing retainer	

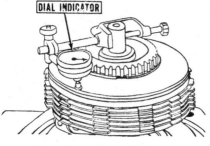

Fig. IH1427—With a dial indicator mounted as shown, measure the vertical movement at the outside edge of Belleville washers. If the vertical movement isn't 0.035-0.045 refer to text and adjust.

quired to obtain the desired movement as the clutch is engaged. Each notch of the adjusting spider changes the deflection approximately 0.010. Engage the lock pin with the spider. Reassemble and install the unit in the reverse of the removal procedure. Refer to paragraph 172 and adjust the linkage.

171A. On late model clutch type pto, the overcenter clutch can be adjusted with the unit on the tractor as follows: Remove the large plug from top side of pto housing. Place control lever in middle (neutral) position. Rotate pto output shaft by hand until flat portion of brake plate is in line with plug hole. Now insert the special adjusting tool (furnished with tractor)) into plug hole and between brake plate and housing and into slot of adjusting spider. See Fig. IH1428A. Hold front of adjusting tool downward, push tool forward and disengage lock pin from adjusting spider. Hold tool and lock pin in this position and turn pto output shaft counter-clockwise to tighten clutch. Adjust clutch one notch (locking position) at a time and recheck after each adjustment. Usually one or two notches is sufficient. Over-adjusting will prevent full engagement of clutch.

After adjusting pto clutch, a brake (anti-creep) adjustment must also be made. Refer to paragraph 172 or 172A.

172. **LINKAGE.** To adjust the pto linkage on early models of the clutch type pto, vary the length of the operating rod (R—Fig. IH1429) by turning the clevis (CL) until the pto output shaft stops turning when the control lever locking pawl engages the stop position notch in the quadrant.

172A. Adjust the linkage (brake) on the late externally adjusted clutch pto as follows: Disconnect control rod clevis from pto cross shaft lever and

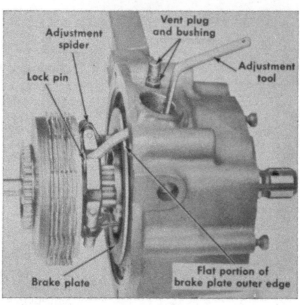
Fig. IH1428A — View showing method of adjusting the late model clutch type pto. Unit is shown removed for illustrative purposes.

adjust clevis until front (lower) edge of pawl is 1 16-inch above top edge of pawl notch quadrant when clevis pin is installed. This will provide adequate braking when pawl is moved into its notch.

OVERHAUL

Transmission Driven PTO

173. On International 330 tractors so equipped, the procedure for overhauling the transmission driven pto shaft and associated parts will be evident after an examination of the unit and reference to Fig. IH1429A.

Independent Type

The occasion for overhauling the complete independent power take-off system will be infrequent. Usually, any failed or worn part will be so positioned that localized repairs can be accomplished. The subsequent paragraphs will be outlined on the basis of local repairs.

Fig. IH1429 — The linkage should be correctly adjusted as described in text.

CL. Clevis   R. Rod

Planetary Type

174. **RENEW REACTOR BANDS.** To renew the reactor bands (34—Fig. IH1430), remove the pto shaft guard and the band adjusting screw cover. Unbolt bearing retainer (47) from the housing cover and remove the retainer. Unlock the pto shaft nut (45), place the operating lever in the forward position to prevent shaft from turning and remove nut (45). Remove the cap screws retaining the rear cover to the housing, place operating lever in the neutral position and turn the pto shaft to align threaded holes in the rear drum (38) with the unthreaded holes in the rear cover.

Using OTC puller ED-3262 or equivalent as shown in Fig. IH1431, remove the rear drum and cover. Loosen the band adjusting screws and remove bands. Inspect bands for distortion, lining wear and for looseness of strut pins. When reassembling, adjust the bands as outlined in paragraph 170.

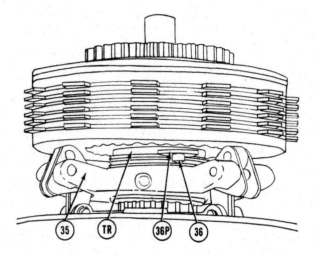
Fig. IH1428 — The lock plate (36P) is engaged with the thread relief (TR). When the lock plate is released, the lock pin (36) should engage a hole in the spider (35).

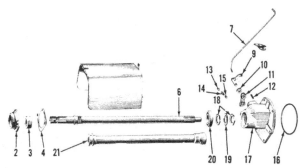

2. Oil seal retainer
3. Oil seal
4. Gasket
6. PTO shaft
7. Shift rod
9. Shifter lever and shaft
10. Oil seal
11. Shifter arm
12. Pin
13. Plug
14. Spring
15. Poppet
16. "O" ring
17. Housing
18. Nuts
19. Washer
20. Bearing
21. Tube

Fig. IH1429A—Exploded view of the transmission driven pto shaft and associated parts used on some International 330 tractors.

Fig. IH1431 — Using a puller to remove cover and rear drum from the early type independent power take-off rear unit.

175. **PLANET GEARS, SUN GEARS AND SHAFTS.** To overhaul the pto rear unit, first drain transmission, then remove the reactor bands as outlined in paragraph 174. Unbolt and remove the unit from the tractor.

With the unit on bench, remove cap screws securing bearing cage (12—Fig. IH1430 or IH1434) to housing and remove bearing cage with ring gear and shaft. Remove the front bearing retainer (16) and seal (17). Remove snap ring (14) and press the ring gear and shaft from bearing (13). The front bearing (13) can be removed from cage at this time. To remove the ring gear rear bearing (9), remove snap ring (10) and using a punch through the two holes in the ring gear hub, bump bearing from shaft. Inspect needle roller pilot bearing (7). If the bearing is damaged, it can be removed, using a suitable puller as shown in Fig. IH1432.

Withdraw the planet carrier and pto shaft from housing and mark the rear face of each planet gear so it can be installed in the same position. Using a punch as shown in Fig. IH1433, remove the roll pins which retain the planet gear shafts in the planet carrier and remove the shafts, gears, spacers and needle bearings. Remove snap ring (26—Fig. IH1430 or IH1434) from sun gear and press sun gear (36) and drum (38) from housing. Bearing (25) can be removed from housing at this time. If the rear bearing is damaged, use a punch through holes in hub and drift the bearing from the sun gear as shown in Fig. IH1435. Remove operating linkage from side of housing. To disassemble the spring retainer plug assembly, turn spring anchor block (57—Fig. IH1437) clockwise to relieve spring pressure and remove snap ring (56).

Fig. IH1432 — Removing the needle pilot bearing from ring gear.

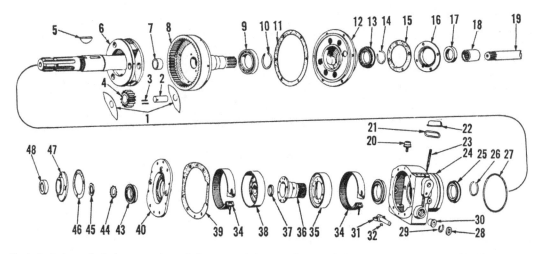

Fig. IH1430—Exploded view of the rear section of the early (planetary) type independent power take-off. Planetary gears (4) are available in sets only.

1. Thrust washers	8. Ring gear and shaft	14. Snap ring
2. Planet gear shaft	9. Drive shaft rear bearing	15. Gasket
3. Needle bearing rollers (72)	10. Snap ring	16. Bearing retainer
4. Planet gear	11. Gasket	17. Oil seal
5. Key	12. Bearing cage	18. Coupling
6. Planet carrier and pto shaft	13. Drive shaft front bearing	19. Coupling shaft
7. Needle bearing		20. Breather
		21. Gasket

22. Anchor bolt cover	30. Bushing
23. Bolt	31. Lever
24. Housing	32. Key
25. Bearing	34. Bands
26. Snap ring	35. Brake drum
27. Seal ring	36. Sun gear
28. Oil seal	37. Spacer
29. Snap ring	38. Creeper drum

39. Gasket	
40. Housing cover	
43. Bearing	
44. Lock washer	
45. Nut	
46. Gasket	
47. Bearing retainer	
48. Oil seal	

Caution: If the anchor bolt is broken or if spring tension cannot be relieved, use care when removing the snap ring.

Inspect all parts and renew any which are excessively worn. The planet gears are available only in sets as are the planet gear needle bearings and the gear shafts. Friction surface of drums should not be excessively worn.

When reassembling, reverse the disassembly procedure. OTC dummy shaft No. ED-3258-1 is used to assemble the needle bearings in the planet gears and to install the planet gears, with thrust plates to the planet carrier. With the planet gear, thrust plates and dummy shaft in position in the planet carrier, push the dummy shaft out with the new shaft. Secure the planet gear shafts to the planet carrier with the roll pins.

When installation is complete, adjust the reactor bands as outlined in paragraph 170.

175A. COUPLING SHAFT. To renew the pto coupling shaft (19—Fig. IH-1434), remove the complete pto rear unit (Fig. IH1436) and withdraw coupling shaft from rear frame.

176. EXTENSION SHAFT. To remove the pto extension shaft (Fig. IH-1438), first detach (split) clutch housing from rear frame as outlined in paragraph 149 or 150. Remove the sea-

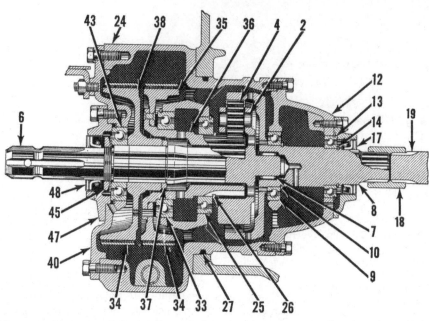

Fig. IH1434—Sectional view of the early (planetary) type power take-off rear unit. Refer to legend under Fig. IH1430.

sonal disconnect coupling from the front of shaft (if so equipped). Unbolt the extension shaft front bearing retainer and cage from transmission and withdraw the extension shaft assembly as shown in Fig. IH1401.

Remove the coupling shaft. Remove cap screw and strap which retains the extension shaft rear bushing carrier (Fig. IH1438) in the rear frame and remove the bushing carrier. Bushing can be renewed if it is worn.

Reinstall the extension shaft by reversing the removal procedure.

177. DRIVEN SHAFT AND GEAR. Refer to Fig. IH1438. To remove the pto driven shaft and gear, first detach the transmission from the clutch housing as outlined in paragraph 149 or 150 and proceed as follows:

Remove the driven gear cover and the clutch and flywheel opening cover from bottom of clutch housing and the large pipe plug which is located directly in front of the driven shaft. Remove cap screw and washer retaining driven gear to shaft and remove snap ring from behind the driven shaft rear bearing. Withdraw the driven shaft rearward and the gear from below. NOTE: In some cases it may be necessary to detach the clutch housing from the engine and use a brass drift to bump the shaft rearward. If the driven shaft front needle bearing is damaged, it can be renewed in a con-

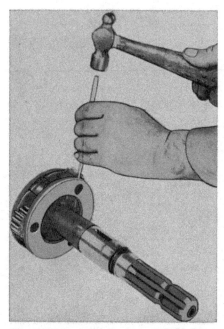

Fig. IH1433—Removing the roll pins which secure the planet gear shafts in the planet carrier.

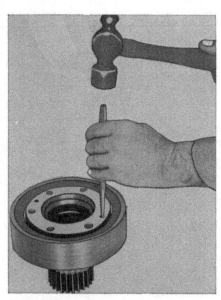

Fig. IH1435—Using a punch through holes of sun gear to remove rear bearing.

Fig. IH1436 — Rear view of tractor, showing a typical installation of the early (planetary) type power take-off rear unit.

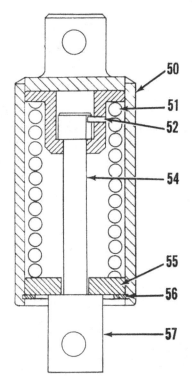

Fig. IH1437 — Sectional view of spring retainer plug assembly. Refer to caution in text before disassembling.

50. Spring sleeve
52. Roll pin
54. Anchor bolt
55. Retainer plate
56. Snap ring
57. Anchor block

ventional manner at this time. The needle bearing race on shaft can also be renewed if damaged. Remove the race with a brass drift to avoid damaging the shaft.

When reassembling, reverse the disassembly proceure and use sealing compound around pipe plug in front of driven gear.

178. **DRIVE SHAFT.** Refer to Fig. IH1438. To remove the driving shaft and integral gear, first detach (split) engine from clutch housing as outlined in paragraph 136, 136A or 137 and remove the engine clutch release bearing and shaft. Unbolt the drive shaft front bearing cage and withdraw the drive shaft and bearing cage from clutch housing. The need and procedure for further disassembly is evident.

Clutch Type

179. **REAR UNIT.** To overhaul the rear pto unit, it is necessary to first remove the unit. Remove the two cap screws which attach the two halves of the rear unit together and separate the halves. Refer to Fig. IH1440. Remove the two bearing retaining screws

(55) and withdraw the input shaft (49) and bearing (50) from the housing (53).

The remainder of disassembly and overhaul is obvious. Refer to the following table for information on the rear unit.

Pressure plate thickness...0.215-0.220
    Max. allowable warpage......0.012
Internal splined disc
    thickness 0.060-0.065
    Max. allowable wear........0.010
External splined disc
    thickness 0.110-0.115
Max. allowable wear of
    all discs 0.125
Brake plate springs
    Free length ....2.250 in.
    Test length ....1.649 in.
    Test load ....135 lbs.
Lock pin spring
    Free length ....1.250 in.
    Test length ....1.000 in.
    Test load .... 5 lbs.

Reassemble in reverse of disassembly procedure and adjust the clutch as in paragraph 171 or 171A. Reinstall in reverse of the removal procedure and adjust the linkage as in paragraph 172 or 172A.

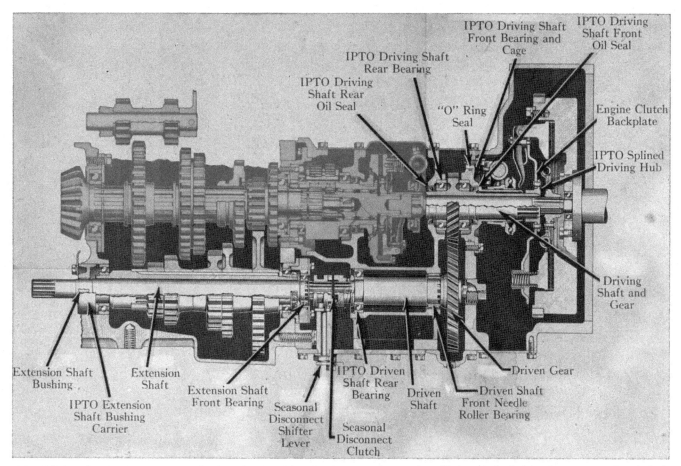

Fig. IH1438—Sectional view of the clutch and transmission housing, showing the installation of the independent power take-of shafts and gears. Later models will not have seasonal disconnect.

180. COUPLING SHAFT. To renew the pto coupling shaft (19—Fig. IH-1440), remove the complete pto rear unit (Fig. IH1441) and withdraw coupling shaft from rear frame.

181. EXTENSION SHAFT. To remove the pto extension shaft (Fig. IH-1438), first detach (split) clutch housing from rear frame as outlined in paragraph 149 or 150. Remove the seasonal disconnect coupling from front of shaft (if so equipped). Unbolt the extension shaft front bearing retainer and cage from transmission and withdraw the extension shaft assembly as shown in Fig. IH1401.

Remove the coupling shaft. Remove cap screw and strap which retains the extension shaft rear bushing carrier (Fig. IH1438) in the rear frame and remove the bushing carrier. Bushing can be renewed if worn.

Reinstall the extension shaft by reversing the removal procedure.

182. DRIVEN SHAFT AND GEAR. Refer to Fig. IH1438. To remove the pto driven shaft and gear, first detach the transmission from the clutch hous-

ing as outlined in paragraph 149 or 150 and proceed as follows:

Remove the driven gear cover and the clutch and flywheel opening cover from bottom of clutch housing and the large pipe plug which is located directly in front of the driven shaft. Remove cap screw and washer retaining driven gear to shaft and remove snap ring from behind the driven shaft rear bearing. Withdraw the driven shaft rearward and the gear from below. NOTE: In some cases it may be necessary to detach the clutch housing from the engine and use a brass drift to bump the shaft rearward. If the driven shaft front needle bearing is damaged, it can be renewed in a conventional manner at this time. The needle bearing race on shaft can also be renewed if damaged. Remove the race with a brass drift to avoid damaging the shaft.

When reassembling, reverse the disassembly procedure and use sealing compound around pipe plug in front of driven gear.

183. DRIVE SHAFT. Refer to Fig. IH1438. To remove the driving shaft

Fig. IH1441 — Rear view of the 340 tractor, showing the installation of the late (clutch) type power take-off rear unit. Other models are similar.

and integral gear, first detach (split) engine from clutch housing as outlined in paragraph 136, 136A or 137 and remove the engine clutch release bearing and shaft. Unbolt the drive shaft front bearing cage and withdraw the drive shaft and bearing cage from clutch housing. The need and procedure for further disassembly is evident.

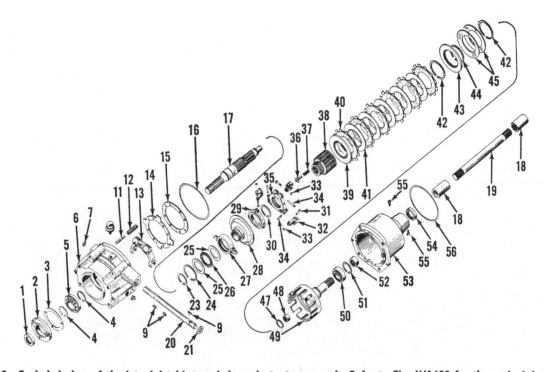

Fig. IH1440—Exploded view of the late (clutch) type independent pto rear unit. Refer to Fig. IH1430 for the early (planetary) type.

1. Oil seal	12. Brake plate	23. Snap ring	31. Link (3 used)	39. Pressure plate	48. Needle bearing
2. Bearing retainer	springs (4 used)	24. Snap ring	32. Actuating lever	40. Lined discs	49. Input shaft
3. Gasket	13. Clutch fork	25. Bearing races	(3 used)	(6 used)	50. Ball bearing
4. Snap ring	14. Brake plate	(2 used)	33. Retainer ring	41. Discs (5 used)	51. Snap ring
5. Ball bearing	15. Brake facing	26. Needle thrust	(6 used)	42. Snap rings	52. Needle bearing
6. Housing cover	16. "O" ring	bearing	34. Pins (3 used)	43. Pressure plate	53. Housing
7. Plug	17. Output shaft	27. Release bearing	35. Spider	44. Snap ring	54. Oil seal
9. Woodruff keys	18. Couplings	28. Release sleeve	36. Lock pin	45. Belleville washers	55. Bearing retaining
11. Dowel pins	19. Coupling shaft	29. Actuator	37. Spring	(2 used)	screws (2 used)
(4 used)	20. Cross shaft	30. Snap ring	38. Clutch hub	47. Snap ring	56. "O" ring
	21. Oil seal				

HYDRAULIC LIFT SYSTEM
"HYDRA-TOUCH"

Series 330-340

NOTE: The maintenance of absolute cleanliness of all parts is of utmost importance in the operation and servicing of the hydraulic system. Of equal importance is the avoidance of nicks or burrs on any of the working parts.

LUBRICATION

Series 330

184. It is recommended that only IH "Hy-Tran" fluid, or a mixture of IH Torque Amplifier additive and SAE 10W engine oil in the ratio of one quart of additive to each four gallons of engine oil, be used in the hydraulic system and the reservoir fluid level should be maintained at the "Full" mark on dip stick. Whenever the hydraulic lines have been disconnected, reconnect the lines, fill the reservoir and with the reservoir filler plug removed, cycle the system several times to bleed air from the system; then, refill reservoir to "Full" mark on dip stick and install the filler plug.

Series 340

185. The transmission and differential case is also the fluid reservoir for the hydraulic power lift and power steering systems. The fluid capacity is approximately 10 gallons. IH "Hy-Tran" fluid, or a mixture of IH Torque Amplifier additive and SAE 10W engine oil in the ratio of one quart of additive to each four gallons of engine oil, should be used in the system. The oil should be changed every 1,000 hours of operation or once a year, whichever comes first. Air can be bled from the system by cycling the system several times, then recheck fluid level in transmission housing.

FILTER

Series 340

186. The working fluid for the hydraulic power lift and power steering systems is also the lubricating oil for the "Torque Amplifier", transmission and differential; therefore, it is very important that the oil be kept clean and free from foreign material. The oil passes through a wire screen filter element (Fig. IH1445) before it enters the pump.

The filter should be cleaned in kerosene **at least** every 250 hours of operation. If the tractor remains in operation after the filter is clogged, the filter element may collapse rendering the element useless and possibly damaging the pump (or pumps).

NOTE: Early filters were of the wire screen type and had a by-pass valve (V— Fig. IH1446). Later filters were of the wire screen type but did not have the by-pass valve. The latest type (Fig. IH1447) incorporates the wire screen and two renewable filter elements.

When an early type filter (with the by-pass valve) is encountered, it should be discarded and the latest type should be installed.

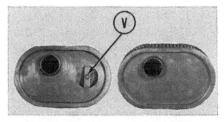

Fig. IH1446 — Very early type filter elements were equipped with a by-pass valve (V). A later filter element should be installed in place of the early type. Refer to Fig. IH1447 for an exploded view of the latest type.

TESTING

Series 330-340

187. The unit construction of the Hydra-Touch system permits removing and overhauling any component of the system without disassembling the others. However, before removing a suspected faulty unit, it is advisable to make a systematic check of the complete system to make certain which unit (or units) are at fault.

NOTE: High pressure (up to 1200-1500 psi) in the system is normal only when one or more of the control valves is in either lift or drop position. When the control valve (or valves) are returned to neutral, the pressure regulator valve is automatically opened and the system operating pressure is returned to a low, by-pass pressure. The low, by-pass pressure is not a factor in the test procedure.

Should improper adjustment, overload or a malfunctioning pressure regulator valve prevent the system from returning to low pressure, continued operation will cause a rapid temperature rise in the hydraulic fluid, damaging "O" rings and seals. Excessively high temperatures will sometimes be indicated by discoloration of the paint on the hydraulic manifold.

Before proceeding with the test, first make certain that the reservoir is filled to the correct level with the proper fluid.

188. Install a pressure gage of sufficient capacity (at least 3000 psi) and a shut-off valve in the pressure line between the pump and the control valves. NOTE: The shut-off valve should be located between the gage and the control valve (refer to Fig. IH1448). With the shut-off valve **open**, start the engine. Move a control valve lever to a raise (or lower) position and notice the pressure as indicated by the gage. At 900-1200 psi the con-

Fig. IH1445 — The power steering and hydraulic system filter should be cleaned in kerosene at least every 250 hours of operation.

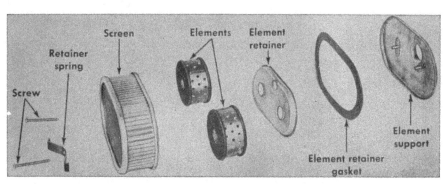

Fig. IH1447—The latest type filter uses a wire screen and two renewable elements.

trol valve handle should automatically return to the neutral position. With the control valve lever held in the raise (or lower) position, to prevent the lever from moving to neutral, the gage should register 1200-1500 psi.

189. If the pressures as checked in the preceding paragraph 188 are correct, the pump and valves are O.K. and any trouble is located in the lines between the control valves and the cylinder or in the cylinder itself.

If the gage pressure is more than specified, a valve is probably stuck in the closed position.

190. If the pressures were too low when checked as in paragraph 188, proceed as follows: Start engine and with same running, carefully shut the valve off and notice the gage reading which should be at least 1500 psi.

NOTE: The pump may be seriously damaged if the shut-off valve is left in the closed position for more than a few seconds.

If the gage reading is above 1500 psi, with the shut-off valve closed, the pump is O. K. and the trouble may be due to maladjusted and/or leaking relief valves.

If the gage reading remains low with the shut-off valve closed and the filter is known to be O. K., observe the oil. If the oil has a milky appearance, an air leak in the suction line can be suspected. On 340 tractors with a transmission driven pump, leakage at the pump intake pipe seal can be corrected by removing the pump and mounting bracket and renewing the pipe and seal (24—Fig. IH1449). If, after a reasonable length of time, the oil is still milky, the transmission should be split (detached) from the clutch housing and seal (S—Fig. IH-1394) should be renewed. On 340 tractors with crankshaft or camshaft driven pumps, air leakage may be caused by a leak in suction line or by a faulty oil seal (S—Fig. IH1394).

On 330 tractors, air leakage may be due to low oil level in the reservoir.

TROUBLE SHOOTING
Series 330-340

191. The following trouble shooting chart lists troubles which may be encountered in the operation and servicing of the hydraulic power lift system and should be used after testing as in paragraphs 187 through 190. The procedure for correcting many

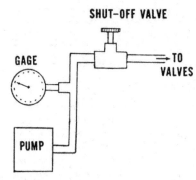

Fig. IH 1448—To check the pump and system operating pressures, connect a gage and a shut-off valve in the pressure line as shown and refer to text.

of the causes of trouble is obvious. For those remedies which are not so obvious, refer to the appropriate subsequent paragraphs.

A. System unable to lift load, gage shows pressure O. K. Could be caused by:
1. System is overloaded
2. Damaged hydraulic cylinder
3. Implement damaged in a manner to restrict free movement
4. Interference restricting movement of cylinder or implement
5. Hose couplings not completely coupled

B. System unable to lift load, gage shows little or no pressure. Could be caused by:
1. Clogged filter (340)
2. Leaking pump suction line
3. Faulty pressure regulator valve
4. Failure of safety valve to close
5. Leakage past cylinder piston seals
6. Pump failure
7. Loss of fluid

Fig. IH1449—View of Cessna transmission driven hydraulic lift and power steering pumps and mounting flange. The longer smooth section of the suction tube should be pressed in the hydraulic pump until it bottoms. Although Cessna pumps are shown, Thompson pumps are similar.

C. System lifts load slowly, gage shows low pressure. Could be caused by:
1. Clogged filter (340)
2. Leaking pump suction line
3. Faulty pressure regulator valve
4. Failure of safety valve to close
5. Pressure regulator orifice enlarged or loose in block
6. Pump failure

D. With all control valves in neutral, gage shows high pressure. Could be caused by:
1. Pressure regulator piston stuck in its bore
2. Regulator orifice plugged

E. Loss of hydraulic fluid, no external leakage. On tractors with a camshaft driven pump, could be caused by:
1. Failure of the pump drive shaft seal

F. Operating pressure exceeds 1500 psi. Could be caused by:
1. Safety valve piston stuck in its bore
2. Failure of safety valve spring

G. Control valve will not latch in either lift or drop position. Could be caused by:
1. Broken garter spring
2. Orifice in rear unlatching piston plugged
3. Unlatching valve leakage

H. Control valve cannot be readily moved from neutral. Could be caused by:
1. Control valve gang retaining bolts and nuts too tight, causing valves to bind in valve bodies
2. Control valve linkage binding
3. Orifice in rear unlatching piston plugged
4. Scored control valve and body

I. Control valve unlatches before cylinder movement is completed, gage shows pressure O. K. Could be caused by:
1. System is overloaded
2. Damaged hydraulic cylinder
3. Implement damaged in a manner to restrict free movement
4. Interference restricting movement of cylinder or implement
5. Hose couplings not completely coupled

J. Control valve unlatches before cylinder movement is completed, gage shows low pressure. Could be caused by:
1. Weak unlatching valve spring
2. Unlatching valve leakage

K. Control valve unlatches from lift
position but not from drop position. Could be caused by:

1. Valve set for single acting cylinder, wherein unlatching valve
   is inoperative in drop position
2. Channel between front and rear
   unlatching valve is plugged

L. Control valve will not center itself in neutral position. Could be
caused by:

1. Control valve gang retaining
   bolts and nuts too tight, causing
   valves to bind in valve bodies
2. Control valve linkage binding
3. Scored control valve and body
4. Centering spring weak or brok-
   en
5. Unlatching pistons restricted in
   movement

M. Control valve will not automatically unlatch from either lift or
drop position. Could be caused by:

1. Clogged filter (340)
2. Leaking pump suction line
3. Faulty pressure regulator valve
4. Plugged channels in control
   valve which lead to unlatching
   valve
5. Leakage past the unlatching
   pistons
6. Loose rear unlatching piston
   retainer
7. Pump failure
8. Loss of fluid

N. Noisy operation. Could be caused
by:

1. Clogged filter (340)
2. Leaking pump suction line
3. Damaged pump
4. Insufficient fluid in reservoir
5. Pump manifold tubes contacting some foreign part of tractor
6. Plugged intake screen

O. Cylinder will not support load.
Could be caused by:

1. External leakage from cylinder, hoses or connections
2. Leakage past cylinder piston
   rings
3. Internal leaks in control valve

PUMP

Series 330-340

International 330 tractors may be
equipped with either a camshaft gear
driven pump or a front mounted,
crankshaft driven pump. Series 340
tractors may have a camshaft gear
driven pump, crankshaft driven pump
or a transmission pump.

Fig. IH1450—View of the right side of a
340 tractor showing the installation of the
camshaft driven hydraulic pump.

Camshaft Driven

192. **REMOVE AND REINSTALL.**
To remove the gear type hydraulic
pump, which is gear driven from the
camshaft gear, drain the hydraulic
system and remove the hydraulic
lines (manifold). Remove the pump
attaching cap screws and lift pump
from tractor.

Install the hydraulic pump by reversing the removal procedure and
bleed the system as outlined in paragraph 184 or 185. Refer to Fig. IH1450.

193. **OVERHAUL.** Overhaul is limited to disassembling, cleaning and renewing gasket, seals, bearing spring
and check valve spring and ball. If
other parts are excessively worn or
damaged, it will be necessary to renew the complete pump unit.

To disassemble the pump, proceed
as follows: Mark the pump body and
cover so they can be reassembled in
the same relative position and remove
the drive gear and cover. Mark the
exposed end of the driven (idler) gear
so it can be installed in the same position and remove the drive and driv-
en gears. Identify the bearings with
respect to the pump body and cover

so they can be reinstalled in the same
position and remove the bearings. The
procedure for further disassembly is
evident after an examination of the
unit.

Check the pump parts for damage
or wear. If any of the parts are excessively worn, the pump should be
renewed.

Note: Small nicks and/or scratches
can be removed from the body cover,
shafts and gears, and bearings by
using crocus cloth or an oil stone.
When dressing the bearing flanges,
however, make certain that the flange
thickness of both bearings in either
pair are identical. Be sure that the
small drilled passages in pump body
and cover are open and clean.

Lubricate all pump parts with hydraulic fluid prior to reassembly. Use
care when installing the drive shaft
to avoid damaging the shaft seal lip.
A seal jumper can be made to facilitate installation of the shaft.

Crankshaft Driven

194. **REMOVE AND REINSTALL.**
The procedure for removing the gear
type hydraulic pump, which is driven
from the crankshaft pulley, is self-
evident after an examination of the
installation. After reinstalling, bleed
air from the system as outlined in
paragraph 184 or 185.

195. **OVERHAUL.** Overhaul is limited to disassembling, cleaning and
renewing seals (2, 4, 5 & 11—Fig.
IH1451), bearing spring (6) and check
valve spring (14) and ball (13). If
other parts are excessively worn or
damaged, it will be necessary to renew the complete pump unit.

To disassemble the pump, proceed
as follows: Mark the pump body and

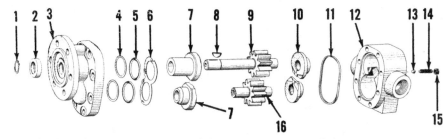

Fig. IH1451—Exploded view of typical front mounted crankshaft driven pump. Pumps are
available with 12 gpm and 16 gpm capacities. Parts indicated by an * are not available
for service.

1. Snap ring	6. Bearing spring	*10. Body bearings
2. Oil seal	*7. Cover bearings	11. Seal
*3. Cover	8. Woodruff key	*12. Body
4. Washers	*9. Driving gear	13. Check valve ball
5. "O" rings	and shaft	

14. Spring
15. Plug
*16. Follower gear
and shaft

cover so they can be reassembled in the same relative position. Remove the pump drive flange, then remove pump cover. Mark the exposed end of driving and driven gears so they can be installed in the same position and remove the driving and driven gears. Identify the bearings with respect to the pump body and cover so they can be reinstalled in the same position and remove the bearings. The procedure for further disassembly is evident after an examination of the unit.

Check the pump parts for damage or wear. If any of the parts are excessively worn, the pump should be renewed.

Note: Small nicks and/or scratches can be removed from the body cover, shafts and gears, and bearings by using crocus cloth or an oil stone. When dressing the bearing flanges, however, make certain that the flange thickness of both bearings in either pair are identical. Be sure that the small drilled passages in pump body and cover are open and clean.

Lubricate all pump parts with hydraulic fluid prior to reassembly. Use care when installing the drive shaft to avoid damaging the shaft seal lip.

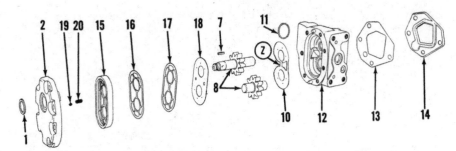

Fig. IH1454—Exploded view of a Series 340, Cessna transmission driven hydraulic pump. Pumps of 12 and 17 gpm capacities are available.

1. Seal	10. Diaphragm	14. Cover	17. Phenolic gasket
2. Cover	11. "O" ring	15. Diaphragm seal	18. Diaphragm
7. Key	12. Body	16. Back-up gasket	19. Ball
8. Pumping gears and shafts	13. Gasket		20. Spring

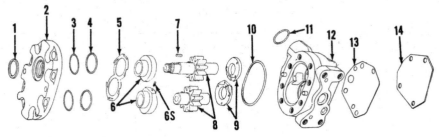

Fig. IH1455—Exploded view of a Series 340, Thompson transmission driven hydraulic pump. Pumps of 12 and 17 gpm capacities are available.

1. Seal	5. Spring	8. Pumping gears and shafts	11. "O" ring
2. Cover	6. Pressure bearing package	9. Bearings	12. Body
3. Back-up washers	7. Key	10. "O" ring	13. Gasket
4. "O" rings (2 used)			14. Cover

Transmission Driven

Both Cessna (Fig. IH1454) and Thompson (Fig. IH1455) transmission driven hydraulic pumps are used on 340 tractors. Both pumps are of the gear type, available in 12 or 17 gpm capacity and are mounted on the manifold (flange). On tractors that are equipped with power steering, the power steering pump is mounted on the rear of the hydraulic pump.

196. REMOVE AND REINSTALL. The pump mounting flange (manifold) with the pump or pumps attached can be removed from the left side of the tractor after the hydraulic lift pressure line and, if so equipped, the power steering pressure line are detached from the mounting flange and the flange retaining cap screws are removed. See Fig. IH1449.

On tractors so equipped, the power steering pump can be separated from the hydraulic lift pump after removing the four attaching cap screws. Take care to prevent damage to the sealing surfaces when separating power steering pump from the hydraulic lift pump.

On all models the hydraulic lift pump is attached to the mounting flange with four cap screws. Each time the pump is removed from the mounting flange, the sealing rings should be renewed. On tractors so equipped, the cap screws which attach the power steering pump to the hydraulic pump should be torqued to 25 ft.-lbs.

When reinstalling the pump (or pumps) and mounting flange on the tractor, make certain that the longer smooth section of the suction pipe (24 —Fig. IH1449) is pressed completely into the hydraulic lift pump. The lip of the seal which is integral with the pipe should face toward the left (pump) side of the tractor. Vary the number and thicknesses of the gaskets which are located between pump mounting flange and clutch housing until the backlash between the pump driven gear and the PTO gear is 0.002-0.022. Gaskets are available in two thicknesses (0.011-0.013 and 0.021-0.023).

197. **OVERHAUL CESSNA.** To overhaul the removed Cessna pump, first remove the power steering pump, or cover (14—Fig. IH1454), from rear of pump and nut, drive gear and key (7) from front. The remainder of the disassembly procedure will be evident.

The bushings in the cover and body are not available separately. The pumping gears, shafts and snap rings are available only as a complete set (8).

12 GPM Pump

O.D. of shafts at bushings .0.8101 min.
I.D. of bushings in body
    and cover..............0.8161 max.
Thickness of gears........0.5711 min.
I.D. of gear pockets in body 2.002 max.

17 GPM Pump

O.D. of shafts at bushings .0.8101 min.
I.D. of bushings in body
    and cover ....0.8161 max.
Thickness of gears........0.8121 min.
I.D. of gear pockets in body 2.002 max.

When reassembling, use new diaphragms (10 and 18), phenolic gasket (17), back-up gasket (16), diaphragm seal (15), ball (19), spring (20) and all oil seals. With open part of diaphragm seal (15) towards cover (2) work same into grooves of cover using a dull tool. Press the phenolic gasket (16) and the back-up gasket (17) into the relief in the diaphragm seal (15). Install the check ball (19) and spring (20) in cover; then, install the diaphragm (18) with bronze face toward gears. NOTE: The diaphragm (18) must fit inside the raised rim of the diaphragm seal (15). Dip gears and shafts assemblies (8) in oil and install them in cover. Position the diaphragm (10) in the body (12) with the bronze side toward gears and the cut-out (Z) toward inlet side of pump body. Install the pump body over the gears and shafts. Check the pump rotation. Pump should have a slight amount of drag but should rotate evenly.

198. OVERHAUL THOMPSON. To overhaul the removed Thompson pump, first remove the power steering pump, or cover (14—Fig. IH1455), from rear of pump and nut and drive gear key (7) from front. Remove cap screws retaining cover to pump body, bump pump drive shaft on a wood block to loosen cover from pump body and remove cover. Remove seal rings (4), fiber washers (3), spring (5) and ring gasket (10). Press drive shaft seal (1) out of pump cover. Tap drive shaft on a wood block to loosen bearings (6) from pump body, then remove the bearings.

Remove gears (8). Tap body on wood block to remove bearings (9). Identify the bearings so they can be installed in their original position.

Clean all metal parts in a suitable solvent and dry them with compressed air. If seal contacting surfaces on drive gear shaft are not perfectly smooth, polish them with fine crocus cloth and rewash the drive gear and shaft.

Dimensions for both pumps are: Shaft O.D., 0.8120; bearing I.D., 0.8155; gear pocket I.D., 1.7718; gear width (12 gpm), 0.600; gear width (17 gpm), 0.7765.

When reassembling, lubricate all parts with clean oil and use new gaskets and seals.

Install bearings (9) in their original position with milled slot on pressure side. Install gears and shafts (8) in the pump body. Install bearings (6), seal ring (10) and the spring (5). Install seal (1) in cover so that lip of seal faces center of pump. Install back-up washer (3) and seal ring (4) in cover. Install pump cover carefully

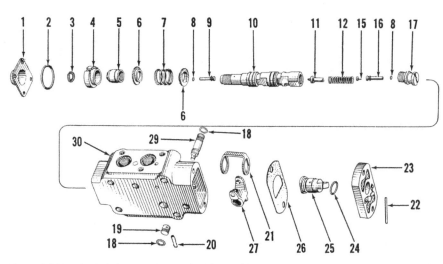

Fig. IH1457—Exploded view of a typical hydraulic system control valve. The garter spring (3) may be one piece or two pieces.

1. Valve cap	7. Control valve	15. Control valve	22. Roll pin
2. Seal ring	centering spring	orifice plug	23. Body cover
3. Garter spring	8. Seal ring	and screen	24. Seal ring
4. Garter spring	9. Unlatching piston	16. Unlatching piston	25. Indexing bushing
sleeve	10. Control valve	17. Unlatching piston	26. Gasket
5. Unlatching piston	spool	retainer	27. Yoke
retainer	11. Unlatching valve	18. Seal ring	29. Control valve
6. Control valve	12. Unlatching valve	19. Bushing	lever shaft
centering spring	spring	20. Roll pin	30. Control valve body
retainer		21. Guide	

to avoid damaging the seal. Install cover cap screws and tighten them to a torque of 25 ft.-lbs. Install the drive shaft Woodruff key and drive gear. Check the pump rotation. Pump should have a slight amount of drag but should rotate evenly.

CONTROL VALVES

Series 330-340

199. All tractors may be equipped with single, dual or triple control valves. On series 340, all valves are identical on multiple valve systems. On International 330 tractors, all valves are similar except that the control valve lever shaft and yoke assembly may be reversed depending upon position which valve is mounted. This section will cover the overhaul of only one valve.

200. R&R AND OVERHAUL. The removal procedure will vary depending upon the number of valves and the tractor model; however, procedure will be self-evident. Be sure to mark the relative position of the small lever with respect to the shaft (29—Fig. IH1457) to insure correct assembly.

Thoroughly clean the removed control valve unit in a suitable solvent, refer to Fig. IH1457 and proceed as follows:

Remove body cover (23) and gasket

(26). Drift out roll pin (22) and remove indexing bushing (25). Remove roll pin (20) and withdraw lever shaft (29), bushing (19) and yoke (27). Lift out the valve guide (21). Remove cap (1), garter spring sleeve (4) and garter spring (3). Push the control valve spool assembly from the valve body. Clamp the spool in a soft jawed vise and remove the rear retainer (17); then, withdraw the unlatching valve (11) and its spring (12). Remove the unlatching piston (16) from retainer (17). Unscrew the orifice plug (15) from piston (16). Turn the spool over in the vise and remove retainer (5), retainers (6), centering spring (7) and unlatching piston (9).

The unlatching valve spring (12) should have a free length of 1⅝ inches and should test 12 lbs. at ⅞-inch. The garter spring (3) should be renewed if outer diameter shows excessive wear. Note: Garter springs in later control valves are of two-piece construction but are interchangeable with early springs. Renew the matched spool and body units if clearance is excessive or if either part shows evidence of scoring or galling. Thoroughly clean channels and the small bores in the valve spool (10). Inspect the unlatching valve (11) and its seat in the control valve spool for damage.

Clean the orifice plug and screen (15) with compressed air and make certain that bore in rear piston (16) as well as the passages in body (30) are open and clean.

When reassembling, dip all parts in clean hydraulic fluid, then use new "O" ring seals and gaskets and reverse the disassembly procedure. Special bullet tool No. ED-3396 will facilitate installation of the garter spring (3) and spring sleeve (4). When installing guide (21), make certain that strap of guide is toward same side of body as shown in Fig. IH1457. Install yoke (27) with serrated end of yoke bore toward same side of body as bushing (19). Insert shaft (29) with pin hole up; then, install bushing (19) and roll pin (20). Install body cover (23) so that indexing bushing (25) engages guide (21). Install the small lever on outer serrations of shaft (29) in its original position. Note: On multiple valve systems, the relative position of the lever with respect to shaft (29) must be the same on all valves.

When installing the control valve, tighten the retaining bolts to a torque of 25 ft.-lbs. Over-tightening will result in distorted body and binding valve spool.

REGULATOR AND SAFETY VALVE BLOCK

Series 330-340

201. On tractors with a front mounted pump (either crankshaft or camshaft driven) and power steering, the regulator and safety valve block is fitted with a flow control valve composed of items (16 through 26—Fig. IH1458). On models without power steering and 340 tractors with a transmission driven hydraulic system pump, the unit is similar except the flow control valve is not used.

202. **R&R AND OVERHAUL.** The procedure for removing the valve block will vary depending upon the model of tractor and the number of hydraulic control valves. In all cases removal procedure will be evident after examining the installation.

Thoroughly clean the removed control valve unit in a suitable solvent, refer to Fig. IH1458 and proceed as follows:

Remove two diagonally opposite cap screws retaining the cover (1) to the body and insert in their place 2-inch long cap screws and tighten them finger tight. Remove the two remaining short cap screws; then, relieve the pressure of the safety valve spring gradually by alternately unscrewing

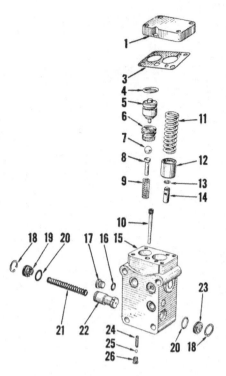

Fig. IH1458—Exploded view of a typical regulator, safety and flow control valve used on models with "Hydra-Touch" system.

1. Cover	12. Spring retainer
3. Gasket	13. Snap ring
4. Seal ring	14. Safety valve
5. Regulator valve	piston
piston	15. Valve housing
6. Regulator valve	16. "O" ring
seat	18. Snap ring
7. Steel ball	19. Spring retainer
8. Ball rider	20. "O" ring
9. Ball rider spring	21. Spring
10. Safety valve	22. Flow control valve
orifice screen	23. Valve retainer
and plug	24. Spring
11. Safety valve	25. Steel ball 3/16"
spring	26. Seat

the 2-inch long screws. Remove cover (1) and gasket (3). Withdraw the safety valve spring (11), spring retainer (12) and safety valve (14). Remove the regulator valve piston (5), unscrew the valve seat (6) and remove the ball (7), rider (8) and spring (9). Unscrew and remove the orifice plug and screen (10).

On tractors equipped with a front mounted hydraulic pump and power steering, remove snap rings (18) and withdraw the flow control valve spring and valve. Remove the auxiliary safety valve (24, 25 and 26).

Inspect ball valve (7) and valve seat (6) for damaged mating surfaces. Piston (5) and safety valve (14) must be free of nicks or burrs and must not bind in the block bores. Clean the orifice plug and screen (10) with compressed air and make certain that bores in block (15) are open and clean.

When reassembling, dip all parts in clean hydraulic fluid, use new "O" ring seals and gaskets and reverse the disassembly procedure. Be sure that all openings in block, gasket (3) and cover (1) are aligned.

CHECK VALVE

Models So Equipped

203. The procedure for removing, overhauling and/or cleaning the check valve is evident after an examination of the unit.

"TEL-A-DEPTH SYSTEM"

Series 340

A "Tel-A-Depth" hydraulic system is available on 340, tractors equipped with an implement hitch. The system provides a means whereby the implement returns to the previously determined working depth which has been selected on the control quadrant. An adjustable stop on the quadrant allows the operator to select any desired working depth and to return to the same depth by moving control lever against stop. The stop can also be by-passed should the operator desire.

When a tractor is equipped with the "Tel-A-Depth" system, the "Tel-A-Depth" valve replaces the "Hydra-Touch" control valve used for rear implements.

Refer to Fig. IH1460 or 1460A for schematic views of the "Tel-A-Depth" system.

Many of the early tractors originally equipped with cable linkage actuated "Tel-A-Depth" have since been converted to the rod type. If damage to cable linkage parts is excessive, it is recommended that the tractor be converted instead of servicing the cable type.

SYSTEM ADJUSTMENT

Early (Cable Linkage) Type

204. Check hydraulic fluid reservoir (transmission case) and fill to proper level, if necessary. Start engine and cycle system several times; then, place hand control lever in the forward position. The system should go off pressure and the distance between center of cylinder rod pin and face of

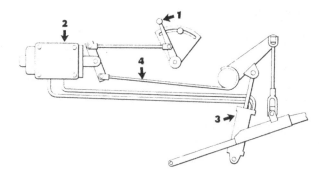

Fig. IH1460 — Schematic drawing of the early cable linkage of "Tel-A-Depth" system. Lever (1) opens valve (2) which actuates cylinder (3) to raise or lower implement. When movement of lever (1) is stopped, the follow-up cable (4) attached to rockshaft closes valve (2).

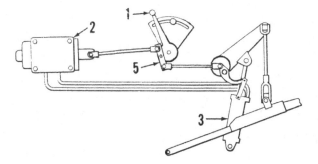

Fig. IH1460A—Schematic drawing of a "Tel-A-Depth" system typical of that used on late 340 tractors. Earlier 340 models may be changed to this later type. Movement of lever (1) opens the valve (2) and movement of the rockshaft closes it again by means of the walking beam (5).

⅜-inch dimension after lever is reinstalled. Tighten ball joint lock nut on compensating rod and, if necessary, the cap screw on compensating lever; then, reconnect ball joint to follow-up lever on valve.

Start tractor engine, cycle system several times and place hand control lever in the forward position. If system remains on pressure with hand control forward and cylinder retracted (collapsed), loosen conduit pivot clip and move the conduit back and forth until an off pressure position is found and the distance between center of cylinder pin and cylinder face is 1¾-1⅞ inches. Tighten pivot clip. Cycle system several times and recheck cylinder position with hand lever in forward position. If additional adjustment is needed to maintain the 1¾-1⅞ inches cylinder measurement, the quadrant limit stop can be adjusted until proper cylinder measurement is obtained.

NOTE: The ⅜-inch exposed cable measurement may not be maintained after making above adjustments.

With engine running, place hand control in rearward position, check to see that system is off pressure and that the distance between center of cylinder pin and face of cylinder is 9⅜-9½ inches. Adjust the quadrant stop to obtain the proper cylinder measurement.

cylinder should be 1¾-1⅞ inches as shown in Fig. IH1461. Now move the hand control lever to the rearward position. The system should again go off pressure and the distance between center of cylinder rod pin and face of cylinder should be 9⅜-9½ inches as shown in Fig. IH1462.

If above conditions are not met, adjust the system as follows: Start tractor engine and place hand control lever in its forward position on the quadrant. Disconnect the ball joint from the valve follow-up lever and fully retract (collapse) hitch cylinder by moving the follow-up lever rearward. Shut off engine and be sure cylinder remains retracted. Loosen lock nut under ball joint on compensating rod (3—Fig. IH1461); then, disconnect ball joint (2) from compensating lever (1). Move compensating lever downward to be sure that rockshaft drive unit slip clutch is engaged. If slip clutch is not engaged, a click will be heard as compensating lever is moved downward and the clutch engages. NOTE: Slip clutch should engage every 90 degrees. If clutch does not engage, on the downward movement of the compensating lever, move lever in the opposite (upward) direction until it does engage. Pull upward on compensating lever and pull follow-up cable into conduit until approximately ⅜-inch of cable is left exposed between end of conduit and back of lock nut at valve end of cable.

Hold the compensating lever in this position to maintain the ⅜-inch dimension, then, adjust length of compensating rod so that ball joint can be engaged on ball of compensating lever without binding. In some cases, it may be necessary to reposition the compensating lever on the splined shaft which extends from the rockshaft drive unit (6—Fig. IH1461). When this occurs, mark the position of lever on shaft prior to removal of lever and re-establish the previously mentioned

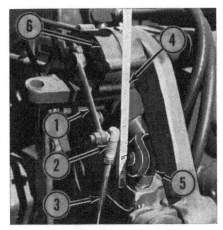

Fig. IH1461 — With hand control lever in the forward position, system should be off pressure and distance between face of cylinder and centerline of cylinder rod pin should be 1¾-1⅞ inches. A 240 tractor is shown, but measurement is same for 340.

1. Compensating lever
2. Ball joint
3. Compensating rod
4. Scale
5. Piston rod
6. Gear box

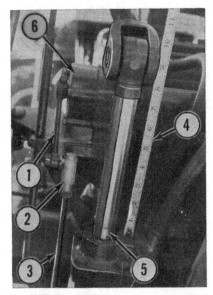

Fig. IH1462—With hand control lever in the rearward position, system should be off pressure and distance between face of cylinder and centerline of cylinder rod pin should be 9⅜-9½ inches. Refer to Fig. IH1461 for legend. A series 240 tractor is is shown, but measurement is same for 340.

Latest (Rod Linkage) Type

205. Check the hydraulic fluid in the reservoir (transmission case) and fill to the proper level, if necessary. Start engine and cycle system several times; then, place hand control lever in the forward position. The system should go off pressure and the distance between center of cylinder rod pin and face of cylinder should be 1¾-1⅞ inches. (Refer to Fig. IH1461 which shows the same measurement of the early type.) Now move the hand control lever to the rearward position. The system should again go off pressure and the distance between center of cylinder rod pin and face of cylinder should be 9⅜-9½ inches.

If the system remains on pressure when the distance between the cylinder face and the center of the cylinder rod pin is 1¾-1⅞ inches, the front actuating rod (AR—Fig. IH1463) should be **shortened**. If the system goes off pressure before the distance is 1⅞ inches, the front actuating rod should be **lengthened**.

Move the "Tel-A-Depth" control handle completely to the rear of the quadrant. If the system stays on pressure when the distance between the center of the cylinder rod pin and the face of the cylinder is between 9⅜-9½ inches or if the center of the cylinder rod pin is not 9⅜-9½ inches from the face of the cylinder, adjust the quadrant stop.

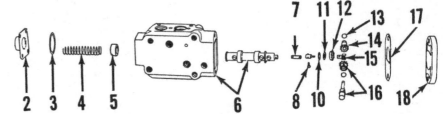

Fig. IH1464—Exploded view of a cable linkage control "Tel-A-Depth" valve used on early tractors.

2. Cap	7. Spool link
3. "O" ring	8. Pin
4. Centering spring	10. Washer.
5. Spring cap	11. Shim (light and
6. Valve body and	heavy)
spool	
12. Idler gear	15. Eyebolt shaft
13. "O" ring	16. Follow-up shaft
14. Control shaft and	and gear
gear	17. Gasket
	18. Cover

TEL-A-DEPTH CONTROL VALVE

206. **R&R AND OVERHAUL.** Disconnect control rods from control valve input and follow-up levers. Remove the nuts from the through bolts which hold ganged valves to valve support and if necessary, remove the outer "Hydra-Touch" control valve. Withdraw through bolts from right side of tractor until "Tel-A-Depth" valve is free; then, pull valve downward and remove manifold from top side of valve. Use caution during this operation not to deform manifold.

With valve removed, disassembly is as follows: Remove valve cover (18—Fig. IH1464). Valve spool can be removed for cleaning and/or inspection by removing pin which retains spool to yoke. If spool and/or valve body are damaged, renew the valve assembly as spool and body are mated parts. On valves so equipped, remove gears and shafts by driving roll pins from hubs of control lever and follow-up shaft gears. Unscrew yoke from link and separate yoke, idler gear and link. Refer to Fig. IH1465. On late 340 tractors, refer to Fig. IH-1464A and remove the levers and shafts.

NOTE: If valve is the very early type, without markings, pay close attention to the position of gear yoke, input lever and follow-up lever. Place correlation marks on shafts and levers prior to disassembly.

On later valves, the idler gear has two index marks 180 degrees apart on back face of gear and the input and follow-up gears have one index mark on the outside diameters. In addition, the input and follow-up levers and their shafts have index marks to insure correct reassembly. Refer to Fig. IH1466 or IH1466A.

On all models, centering spring (4—Fig. IH1464) and spring cup (5) can be removed after removing cap (2).

Inspect all parts for damage and/or wear. If input shaft, input gear, follow-up shaft or follow-up gear are to be renewed, it will be necessary to renew complete valve assembly as the above parts are not catalogued separately.

Fig. IH1463—View of the left side of a late 340 tractor showing the late rod linkage installed. The front actuating rod is shown at (AR).

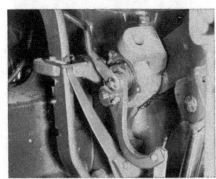

Fig. IH1463A—View of the compensating arm and rockshaft pivot plate installed on a late tractor.

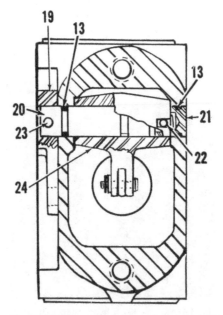

Fig. IH1464A — End view of a late 340 "Tel-A-Depth" control valve showing the parts installed which are different than the valve in Fig. IH1464.

13. "O" rings	22. Roll pin
19. Valve lever	3/32 x 7/16
20. Lever shaft	23. Roll pin
21. End bushing	5/32 x 7/8
	24. Actuating arm

Fig. IH1465 — View showing shafts and gears removed from an early "Tel-A-Depth" valve.

Reassembly is the reverse of disassembly, however, it is recommended that new "O" rings be used. Refer to Fig. IH1466 or IH1466A.

On models with drive gears, adjust yoke until idler gear is snug yet will turn freely. On the early unmarked valves be sure parts are reassembled in the same position as they were originally. Axis of yoke and gears must be parallel to cover surface. On later marked valves, mate all index marks. Refer to Fig. IH1467 for an assembled view.

On all models, it may be necessary to readjust system as outlined in paragraph 204 or 205.

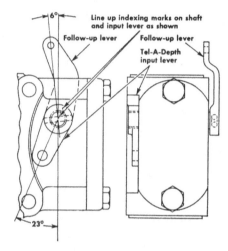

Fig. IH1466—On early 340 tractors with cable linkage, use this drawing as a guide when reassembling the "Tel-A-Depth" valve.

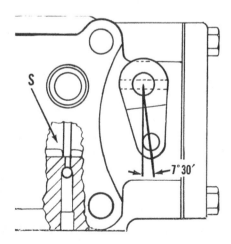

Fig. IH1466A — On late 340 tractors equipped with the rod linkage, use this drawing as a guide when reassembling the "Tel-A-Depth" valve. Spool (S) is centered.

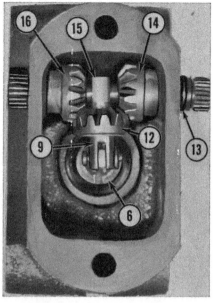

Fig. IH1467 — View of an early "Tel-A-Depth" valve with end cover removed. Note that shaft on right side is partially removed to show "O" ring (13). Refer to Fig. IH1464A for the late (rod linkage) valve.

6. Spool
9. Yoke
12. Idler gear
13. "O" ring
14. Control lever gear
15. Eyebolt (idler) shaft
16. Follow-up shaft gear

ROCKSHAFT DRIVE UNIT

Early (Cable Linkage) Type

207. **R&R AND OVERHAUL.** Remove conduit shield. Disconnect ball joint from valve follow-up lever and remove pivot clip retaining cap screw. Remove conduit from retaining clips located along tractor rear frame. Disconnect compensating rod from compensating lever and remove lever. Unbolt rockshaft drive unit from rockshaft drive side plate and pull unit rearward. Remove ball joint and rubber boot from valve end of cable assembly. Loosen conduit retaining nut from drive unit and remove cable and conduit by turning cable counterclockwise. Pull cable from conduit. Remove cover from drive unit, which

in some cases will require removal of staking. Remove parts from drive unit as shown in Fig. IH1468.

Clean all parts and inspect same for excessive wear and/or damage.

NOTE: Early units had gears made of powdered metal while later gears are steel. When renewing planet gear or input gear, order IH service package number 373 623 R91. Other parts are available as service items.

With cable and conduit cleaned, lubricate cable with Lubriplate (105-V) or equivalent, and install cable in conduit. Temporarily install ball joint on threaded end of cable and attach a low reading (1-10 lbs.) spring scale to same. Measure effort required to move cable through conduit and if effort exceeds four pounds, renew cable and conduit assembly.

Lubricate all drive unit parts with Lubriplate (105-V) or equivalent, and proceed as follows: Place drive wheel in housing and position same so cable anchor hole aligns with cable lead-in hole in housing. Hold wheel in position, install cable and turn cable clockwise until it bottoms. Rotate wheel and push cable into housing until conduit is positioned; then, tighten conduit retaining nut. Balance of reassembly is the reverse of disassembly. The planet gear carrier may be installed in any position with respect to the drive wheel.

After installing compensating lever and prior to installing the compensating rod, check operation of rockshaft slip clutch as follows: Pull up on compensating lever until ball joint butts against conduit. Attach a spring scale to lever and while keeping scale at 90 degrees to compensating lever, measure the force required to cause clutch to slip. This force should be 38 to 50 pounds. If not within these limits, refer to paragraph 208.

Connect compensating rod and lever and adjust system as outlined in paragraph 204.

Fig. IH1468—View showing the component parts of the rockshaft drive unit (gear box) and their relative position. Tractors which are equipped with the rod linkage don't use this drive unit.

1. Cover
2. Input gear
3. Coupling
4. Spring
5. Planet gear assembly
6. Cable drive wheel
7. Housing

ROCKSHAFT SLIP CLUTCH

Early (Cable Linkage) Type

208. **R&R AND OVERHAUL.** To remove the rockshaft slip clutch, disconnect the compensating rod from compensating lever, loosen the conduit retaining nut, then, unbolt the drive unit from the rockshaft side plate and swing unit away from rockshaft. Use a tool such as that shown in Fig. IH1469 and compress clutch spring and ring retainer into rockshaft. Remove the retaining snap ring, slowly release tool and remove ring retainer, spring and coupling. See Fig. IH1470.

Clean all parts including the coupling insert which is pressed into rockshaft. Inspect all parts for excessive wear and/or damage. Pay particular attention to mating surfaces of clutch coupling and insert and remove any burrs which may be present.

Apply a light coat of light oil to clutch faces and inside bore of rockshaft; then, reassemble unit by reversing the disassembly procedure. Check operation of clutch after assembly as follows: Use a torque wrench fitted with a screw driver adaptor and check the torque required to cause clutch to slip. Clutch should slip in either direction of rotation between 80-100 in.-lbs. Turn clutch coupling 180 degrees and repeat operation. If clutch does not meet the above specifications, renew assembly.

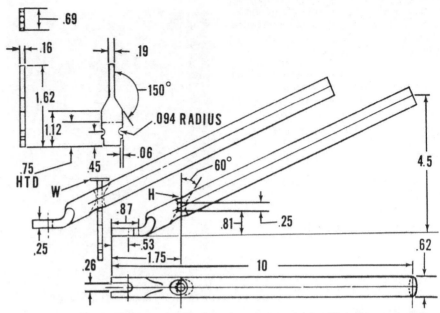

Fig. IH1469—View of special tool used to facilitate removal of rockshaft slip clutch. Tool can be made locally. "H" is hole of 0.250 diameter. "HTD" means harden this distance. "O" is washer.

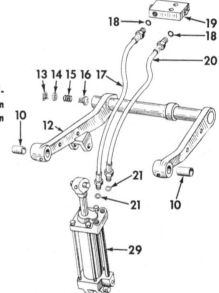

Fig. IH1470—Typical rockshaft and "Tel-A-Depth" cylinder. The slip clutch shown exploded from the rockshaft is not used on tractors with the rod linkage.

10. Bushing	16. Coupling
12. Rockshaft	18. Ring seal
13. Snap ring	19. Check valve
14. Retainer	21. Ring seal
15. Spring	29. Cylinder

HYDRAULIC LIFT SYSTEM
(Draft and Position Control)

Series 504-2504

209. The series 504 and 2504 hydraulic lift unit, which provides both draft and position control, also serves as a top cover for the differential and final drive portions of the tractor rear center frame. Contained within the hydraulic lift housing are the rockshaft, work cylinder, control valves and the linkage for operation of the components. See Fig. IH1475.

Pressurized oil for the operation of the hydraulic lift unit is supplied by a transmission driven pump (13—Fig. IH1475A). Pumps for hydraulic lift unit may be either 4½ or 7 gpm capacity, depending upon usage of tractor. In addition, tractors with power steering and auxiliary circuits are equipped with either 12 or 17 gpm pumps (10) mounted on the same pump mounting flange with the hy-

draulic lift pump. Pumps operate at a pressure of 1550-1600 psi, controlled by the system pressure relief valves. A hydraulic fluid filter is mounted in right side of clutch housing.

The hydraulic lift system draws its oil supply from the transmission and differential housing and thus, the tractor rear center frame and hydraulic lift system share a common reservoir.

LUBRICATION AND BLEEDING

210. To drain and refill the reservoir, and bleed the hydraulic system, proceed as follows: Remove drain plug from rear center frame and right rear bottom of hydraulic lift housing. Pull coil wire from center of distributor cap and crank engine briefly to clear pump and connecting lines.

Remove filler plug from transmission top cover and level plug from right side of rear center frame. Fill rear center frame to level plug opening with IH Hy-Tran Fluid, or a mixture of IH Torque Amplifier Additive and SAE 10W engine oil in the ratio of one quart of additive to each four gallons of engine oil. Start engine and with filler plug out, cycle lift system, and remote cylinders if so equipped, about ten or twelve times, then place position control lever and remote control levers in the forward position. If tractor is equipped with power steering, cycle steering system from one extreme position to the other, then place wheels in straight ahead position. Recheck fluid level and add as necessary to bring to level plug opening. Install and tighten level plug and filler plug.

FILTER

211. The working fluid for the hydraulic power lift and power steering systems is also the lubricating oil for the "Torque Amplifier", transmission and differential; therefore, it is very important that the oil be kept clean and free from foreign material. The oil passes through a filter element (Fig. IH1471) before it enters the pump.

The filter should be cleaned in kerosene and renewable elements renewed **at least** every 250 hours of operation. If the tractor remains in operation after the filter is clogged, the filter element may collapse rendering the element useless and possibly damaging the pump (or pumps).

TROUBLE SHOOTING

212. The following trouble-shooting chart lists troubles which may be encountered in the operation and servicing of the hydraulic lift system. The procedure for correcting many of the causes of trouble is obvious. For those

Fig. IH1471 — The hydraulic and power steering system filter is located in right hand side of clutch housing as shown.

remedies which are not so obvious, refer to the appropriate subsequent paragraphs.

A. Hitch will not raise. Could be caused by:
1. Unloading valve orifice plugged.
2. Unloading valve piston sticking.
3. Flow control valve spring broken, piston sticking or check ball stuck in orifice.
4. Unloading valve ball not seating.
5. Main relief valve spring broken or valve is leaking.
6. Cylinder safety (cushion) valve faulty.
7. Auxiliary valve cover "O" ring damaged.
8. Linkage disconnected from control lever or valve.

B. Hitch lifts load too slow. Could be caused by:
1. Flow control valve piston sticking or faulty valve spring.
2. Flow control valve piston stop broken.
3. Unloading valve ball seat leaking.
4. Faulty main relief valve.

5. Cylinder safety (cushion) valve leaking.
6. Scored lift cyclinder or piston "O" ring damaged.
7. Excessive load.

C. Hitch will not lower. Could be caused by:
1. Control valve spool sticking or "O" ring damaged.
2. Drop poppet valve sticking or "O" ring damaged.

D. Hitch lowers too slow. Could be caused by:
1. Drop control valve spool sticking.
2. Damaged "O" ring on drop poppet valve.
3. Linkage out of adjustment.

E. Hitch lowers too fast. Could be caused by:
1. Faulty drop control valve piston.

F. Hitch lowers too fast in slow action position. Could be caused by:
1. Drop control linkage out of adjustment.

G. Hitch raises and lowers but will not maintain position (hiccups). Could be caused by:
1. Check valve in main control valve leaking.
2. Cylinder safety (cushion) valve leaking.
3. Cylinder scored or piston "O" ring damaged.
4. Drop poppet valve sticking.
5. Check valve actuating rod adjusting screw improperly adjusted.

H. System stays on high pressure. Could be caused by:
1. Broken, disconnected or improperly adjusted linkage.
2. Control valve spools faulty.
3. Auxiliary valve not in neutral.
4. Restraining chains adjusted too short.

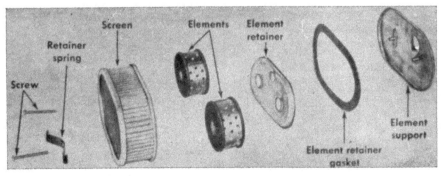

Fig. IH1472 — Exploded view of hydraulic and power steering system filter.

5. Core plug in control valve body missing or leaking badly.

I. Malfunction of position control. Could be caused by:
   1. Control valve spool or linkage binding.
   2. Control lever quadrant not mounted correctly.
   3. Incorrect valve link-to-spool adjustment.

J. Response time too slow. Could be caused by:
   1. Unloading valve piston sticking.
   2. Unloading valve orifice plugged (partially).

K. Hitch has too much depth variation (over-travels). Could be caused by:
   1. Torsion bar bearing not lubricated.
   2. Flow control valve check ball missing.
   3. Unloading valve orifice partially plugged.

L. Inadequate depth control during deep operation. Could be caused by:
   1. Foreign material between torsion bar bellbrank and its stops.
   2. Incorrect adjustment of hitch lift link.
   3. Interference between top link and rear frame.
   4. Control lever quadrant mounted in wrong position.
   5. Top link pin excessively worn.

M. Insufficient transport clearance of mounted implement. Could be caused by:
   1. Incorrect valve link-to-control valve spool adjustment.
   2. Safety shut-off improperly adjusted.
   3. Incorrect adjustment of hitch lift link.

N. Auxiliary circuit will not lift load or lifts load slowly. Could be caused by:
   1. Faulty main relief valve.
   2. Excessive leakage past valve spool.
   3. Faulty auxiliary valve check ball.
   4. Faulty cylinder relief valve (industrial valve).

O. Auxiliary valve does not automatically unlatch. Could be caused by:
   1. Incorrect unlatching adjustment.
   2. Faulty unlatching piston.
   3. Faulty detent sleeve.
   4. Spool sticking due to improperly tightened mounting bolts.

P. Auxiliary valve unlatches prematurely. Could be caused by:
   1. Incorrect unlatching adjustment.
   2. Broken unlatching spring.
   3. Faulty detent sleeve.

Q. Load drops slightly when auxiliary valve is put in lift position. Could be caused by:
   1. Valve check ball leaking or not seating.

R. Auxiliary valve will not hold load in position. Could be caused by:
   1. Excessive leakage past valve spool.
   2. Faulty cylinder or piston.
   3. Faulty cylinder relief valve (industrial valve).

S. Fluid leaking from detent breather. Could be caused by:
   1. Faulty unlatching piston "O" ring.
   2. "O" ring at detent end of valve spool leaking.

SYSTEM OPERATING PRESSURE AND RELIEF VALVE

Series 504-2504

213. The overall system operating pressure can be determined with the lift system pressure relief valve (in lift housing) and the auxiliary system relief valve (in transfer block) installed in tractor but as both relief valves are involved in the system operation it is difficult to determine which relief valve is involved when system is malfunctioning.

If a flow meter, such as the International Harvester Flo-Rater, or its equivalent, is available, an indication of which valve is malfunctioning can be determined by the amount of drop in gpm. This gpm drop will correspond to the size of pumps which the tractor is equipped with. For example, if the gpm drop is something between 4½ and 7 gpm it would indicate that the lift system relief valve is opening too soon. If the gpm drop is something between 9 and 14 gpm, it would indicate the auxiliary circuit relief valve is opening too soon.

If the overall system operating pressure test results in a low reading, it is recommended that both valves be removed and bench tested.

The overall system operating pressure can be checked as follows: Use a gage capable of registering at least 2000 psi and install gage in series with a shut-off valve in a line to which has been attached the male half of a quick coupler. Gage must be in the line between shut-off valve and male half of quick coupler. Install test as-

sembly male end in quick coupler of tractor and place open end in reservoir filler hole. Be sure to fasten open end securely in filler hole. Start engine and operate until hydraulic fluid is warmed to operating temperature; then, with engine operating at high idle speed, move auxiliary lever to the position that will direct fluid to test gage. Manually hold control valve lever in this position and close the shut-off valve only long enough to observe the pressure reading on the gage. Pressure reading should be 1550-1600 psi.

NOTE: Before removing test fixture, test auxiliary control valve unlatching pressure as follows: Operate engine at low idle speed, move auxiliary control valve lever to position that will direct fluid to test gage, then slowly close the shut-off valve and observe the pressure at which the valve control lever returns to neutral. This pressure should be not less than 1000 psi nor more than 1250 psi.

If the overall system pressure is not as specified, refer to paragraph 213A for information concerning the lift system relief valve and to paragraph 231 for information concerning the pilot relief (auxiliary circuit) valve.

For information concerning unlatching mechanism of auxiliary control valve, refer to paragraph 230.

NOTE: System relief valve, pilot relief valve, cylinder cushion (safety) valve and the flow divider relief (safety) valve can be tested after removal by using an injector tester, or a hydraulic hand pump, and the proper adapters. However, bear in mind that the pressures obtained will be toward the low side of the specified pressure ranges due to the low volume of fluid being pumped. The International Harvester test equipment components are as follows: FES64-7-1 (test block), FES64-7-2 (plug), FES64-7-4 (petcock), FES64-7-5 (gage) and FES64-7-6 (hand pump).

If flow meter equipment is available, proceed as follows: Connect meter inlet to quick disconnect that is convenient and secure the outlet hose in the rear frame filler hole. Be sure to fasten outlet hose securely in the filler hole. Start engine and run until fluid is at operating temperature, then activate auxiliary control valve to direct fluid through the flow meter and note the free flow gpm. This flow should approximate the combined output of both pumps, less the three gallons per minute taken by the flow divider valve for the power steering system. Now slowly close the restrictor valve of the test set and observe the flow meter. If both relief valves are functioning properly, the pressure gage should show the 1550-1600 psi with no appreciable decrease in the flow rate. However, if when the pressure begins to build to the higher pressures, an abrupt drop of approximately 4½ to 7 gpm occurs, it would indicate that the lift (draft and position control) system relief valve is faulty. Similarly, if a drop of approximately 9 to 14 gpm occurs, it would indicate that the auxiliary circuit relief valve in the transfer block is faulty.

For information concerning the lift system relief valve refer to paragraph 213A. For information on the auxiliary system relief valve refer to paragraph 231.

213A. The relief valve for the lift (draft and position control) system is located in the right front corner of the lift housing as shown in Fig. IH1475. Valve can be removed after removing seat and hydraulic lift housing cover and the procedure for doing so is obvious. Refer to paragraph 213 for testing procedures.

Early relief vales were factory set and staked and could not be adjusted. Later valves have component parts catalogued.

SYSTEM ADJUSTMENTS

214. Whenever hydraulic lift unit has been serviced and adjustment is required, the system should be cycled, by using the position control lever, at least ten or twelve times to insure purging any air which might be present in system. All checking and adjusting of the system should be done with a load on the hitch and the hydraulic fluid at operating temperature.

IMPORTANT: Adjustments are made with the top cover of unit removed. The system pressure relief valve located in the right front corner of the unit housing discharges upward when it relieves, therefore, it is necessary to install a shield over the relief valve when operating the system with the top cover removed.

While some early model tractors may not be so equipped, the later model tractors have a drain plug located in the left front corner of unit housing to permit lowering of the fluid level so adjustments can be made without working below the surface of fluid.

215. **DROP POPPET ADJUSTMENT.** Place the draft control lever in the full forward position, raise the hitch to its maximum position with the position control lever, then lower hitch to mid-position. System should go off high pressure with no cycling (hic-

Fig. IH1475—View of hydraulic draft and position control unit with top cover removed.

  C. Cylinder
  N. Adjusting collar
  R. Relief valve
  S. Safety (cushion) valve
  T. Timing switch
  V. Control valve

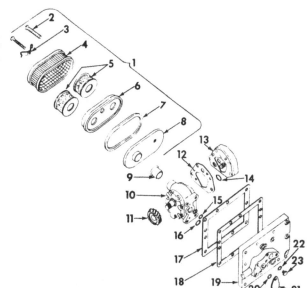

Fig. IH1475A — View showing arrangement of hydraulic pumps, pump mounting flange and filter assembly.

  1. Filter assembly
  3. Retainer spring
  4. Screen
  5. Elements
  6. Retainer
  7. Gasket
  8. Support
  9. Suction tube
  10. Pump (12 or 17 gpm)
  11. Drive gear
  12. Gasket
  13. Pump (4½ or 7 gpm)
  14. "O" ring
  15. "O" ring
  16. "O" ring
  17. Gasket
  18. Spacer
  19. Flange
  20. "O" ring
  21. Cover (no draft control)
  22. "O" ring
  23. Plug

cups). If unit cycles (hiccups), adjust the drop poppet actuating rod adjusting screw (located between legs of main control valve link) as follows: Again fully raise hitch and lower to mid-position. Turn adjusting screw in until unit begins to cycle (hiccup), then back-out screw until the point is reached where the unit stops hiccupping. See Figs. IH1476 and IH1477. Now turn the screw out an additional ¾-turn to obtain the proper clearance (0.020) between push rod and check ball.

NOTE: Move the position control handle a small amount after each turn of the adjusting screw to make sure control valve spool is in the normal centered position.

216. **POSITION CONTROL LINKAGE ADJUSTMENT.** Place draft control lever in full forward position. Loosen cap screw (B—Fig. IH1478)

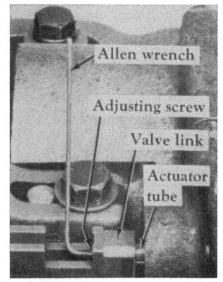

Fig. IH1476—View showing method of adjusting drop poppet actuating rod. Refer to text.

Fig. IH1477—Drop poppet actuating rod adjusting screw is located as shown. Valve link has been removed for illustrative purposes.

and slide hydraulic shut-off switch assembly (C) rearward temporarily. Place the position control lever at the offset on forward end of quadrant at which time the hitch should be fully lowered. Check hitch by by-passing the quadrant offset with the position control handle and moving it to the bottom of the slot after loosening lock nut (L) and backing out drop control screw (M). If hitch drops any further, the control valve actuator tube

must be turned into the control valve link. See Fig. IH1476. Be sure to tighten lock nut after each adjustment. This adjustment can be checked by moving the position control lever to the rearward position at which time the rockshaft lift arms should be 45-46 degrees above vertical position. Angle can be determined by measuring from center of lower link pivot pin to the center of the lift link pivot pin. This distance should be 33¾ inches with a plus or minus tolerance of ¼-inch.

Note: On some early model tractors, it is permissible to set the position control lever not more than ⅝-inch beyond offset in quadrant to put hitch in the fully lowered position. On these early models, use shims between valve link and actuator tube when adjusting, or install a new actuator tube which has the adjusting nut.

Without moving the position control lever after obtaining the 45-46 degree rockshaft arm position, move the hydraulic switch assembly (C—Fig. IH1478) forward until lever (F) is in contact with pin (G) and lever (H) is in contact with pin (E) and tighten cap screw (B). Distance between end of slot and dowel should

be approximately 11/32-inch as shown in Fig. IH1478. System should go off high pressure when the rockshaft lift arms reach the 45-46 degree above vertical position when hitch is raised with draft control lever.

With position control lever position established as outlined, set bumper stop (Fig. IH1479) on the quadrant so lever contacts it 0.020-0.030 before lever reaches end of its travel and tighten stop securely.

Cycle system several times and recheck the 45-46 degree rockshaft arm position. Rockshaft arms should have approximately 1-inch additional free travel, when lifted manually, after hitch has reached its highest point. If hitch doesn't have the additional 1-inch free travel, recheck the rockshaft lift arm angle and the actuator tube adjustment.

217. **DROP CONTROL ADJUSTMENT.** Loosen lock nut (L—Fig. IH1478) and back-out adjusting screw (M). Move position control lever to offset of quadrant, then turn adjusting screw until it just contacts the cam which operates the drop control valve spool. Tighten lock nut.

This adjustment can be checked by moving the draft control lever completely forward. The drop control

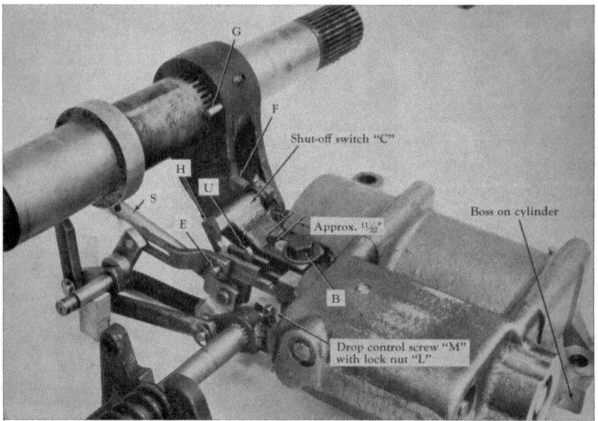

Fig. IH1478 — View of the draft and position control unit internal linkage and component parts.

valve spool should bottom when the lever is about $\frac{1}{8}$-inch from full forward.

218. **DRAFT CONTROL LINKAGE ADJUSTMENT.** Remove cap screw which retains torsion bar flange to bracket and pull torsion bar out until flange clears dowel and bellcrank is free to rotate. Set the position control lever at approximately the offset of the quadrant. Set the draft control lever at the full forward position. Rotate top of bellcrank to the rear until it contacts its stop and hold in this position with a small bar. Rockshaft should not move; however, if it does move, remove pin from draft control rod adjusting nut and turn nut out (counter-clockwise) until bellcrank, with pin installed, will bottom without causing rockshaft to move.

Continue to hold bellcrank against stop and move draft control lever to the rear edge of word "OFF" on the quadrant. The rockshaft arms should raise to the 45-46 degree angular position and the system should go off high pressure. If system does not react as stated, turn draft control rod nut in (clockwise) until operation is correct. Reinstall pin, spring washers and retaining pin. Reinstall torsion bar.

219. **CONTROL LEVERS ADJUSTMENT.** Position control lever is adjusted by the two locking rings (collars) at inside of control levers and draft control lever is adjusted by the two nuts on the outside of the levers. Procedure for adjusting is obvious. Levers should be adjusted until 4 to 6 lbs. for used friction discs; or 6 to 8 lbs. for new friction discs, is required to move levers. Measurement is taken at control lever knob.

PUMP UNIT

The draft and position control hydraulic system pump may be either Cessna or Thompson. Pump capacity may be either 4½ or 7 gpm, however, construction of the pumps remains the same except that pumping gears of the 7 gpm pump are wider. Refer to Figs. IH1479A and IH1479B for exploded views of the two pumps.

220. **REMOVE AND REINSTALL.** To remove the hydraulic pump, first drain housing, then disconnect the hydraulic lift pressure line, and if so equipped, the power steering and auxiliary system pressure line from the pump mounting flange. See Fig. IH-1479C for the location of these lines. Mounting flange can now be unbolted

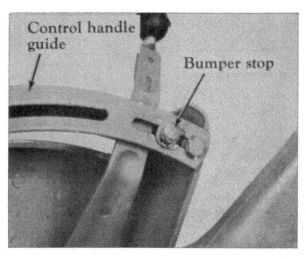

Fig. IH479 — View showing bumper stop. Refer to text for adjustment.

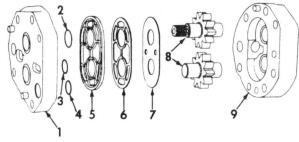

1. Pump cover
2. "O" ring
3. "O" ring
4. "O" ring
5. Diaphragm seal
6. Back-up gasket
7. Diaphragm
8. Pump gears and shafts
9. Pump body

Fig. IH1479A — Exploded view of Cessna draft and position control system pump. Pump may be either 4½ gpm or 7 gpm capacity.

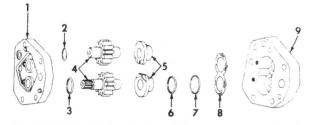

1. Pump cover
2. "O" ring
3. "O" ring
4. Pump gears and shafts
5. Bearings
6. "O" ring
7. "O" ring retainer
8. Pressure plate spring
9. Pump body

Fig. IH1479B — Exploded view of Thompson draft and position control system pump. Pump may be either 4½ gpm or 7 gpm capacity.

Fig. IH1479C — View of pump mounting flange. Small line feeds draft and position control. Large line feeds power steering and auxiliary systems via flow divider valve.

and the flange and pump, or pumps, removed. See Fig. IH1479D.

On models equipped with both draft and position control and auxiliary system, the draft and position control pump (small) can be separated from the auxiliary system pump (large) after removing the four attaching cap screws. Take care not to damage sealing surfaces when separating pumps.

On all models the auxiliary system (large) pump is attached to the mounting flange with four cap screws. Renew "O" rings each time pump is removed from mounting flange. Cap screws which secure draft and position control pump to auxiliary system pump should be tightened to a torque of 25 ft.-lbs.

NOTE: Beginning with Farmall 504 tractor serial number 5160 and International tractor serial number 1330, a 17 tooth pump drive gear is being used instead of the 15 tooth gear used prior to these serial numbers. If a new pump (or drive gear which has the 17 teeth) is being installed, it will be necessary to add a spacer to the pump mounting flange and to use a new longer pump suction tube to compensate for the increased diameter gear. It is recommended by the International Harvester Company that when noisy hydraulic pumps are encountered, the 15 tooth pump drive gear be replaced with the 17 tooth gear along with the spacer and new longer pump suction tube. New pump mounting flanges are also available to accommodate the 17 tooth gear.

When reinstalling the pump (or pumps) and mounting flange on the tractor, make certain that the longer smooth section of the suction tube (24—Fig. IH1479D) is pressed completely into the pump. The lip of the seal, which is integral with the tube, should face toward left (pump) side of tractor. Vary the number and thicknesses of the gaskets which are located between the pump mounting flange and the clutch housing to provide a backlash of 0.002-0.022 between pump drive gear and PTO gear. Gaskets are available in thicknesses of 0.011-0.019 and 0.016-0.024.

221. OVERHAUL (CESSNA). With unit removed as outlined in paragraph 220, remove retaining cap screws and separate cover (1—Fig. IH1479A) and body (9). Remove "O" rings (2, 3 and 4), diaphragm seal (5), back-up gasket (6), diaphragm (7) and the gears and shafts (8).

Bushings in body and cover are not available separately. Pump gears

Fig. IH1479D — View of Cessna pumps after removal of mounting flange. Note how pumps are bolted together. Item (24) is suction tube. Thompson pumps are similar.

and shafts are available only in pairs and when ordering the gears and shafts package, also order an "O" ring and gasket package.

Refer to the following specifications.

4½ GPM Pump
O. D. of shafts at bushings.....0.6875
I. D. of bushings in body
  and cover 0.689
Thickness (width) of gears.....0.312
I. D. of gear pockets...........1.716

7 GPM Pump
O. D. of shafts at bushings.....0.6875
I. D. of bushings in body
  and cover 0.689
Thickness (width) of gears.....0.460
I. D. of gear pockets...........1.716

When reassembling, use new diaphragm (7), back-up gasket (6) and diaphragm seal (5). Use all new "O" rings. Lubricate gear and shaft assemblies prior to installation. Bronze face of diaphragm (7) must face pump gears and must fit inside the raised rim of the diaphragm seal (5). If necessary, use a blunt tool to work diaphragm seal into position. Tighten body to cover retaining screws evenly and check rotation of pump. Pump will have a slight drag but should turn evenly with no tight spots.

Fig. IH1480—Seal tube (T) and seal (S) assembly can be removed after torsion bar bracket is off.

221A. OVERHAUL (THOMPSON). With pump removed as outlined in paragraph 220, remove retaining cap screws and separate cover (1—Fig. IH1479B) and body (9). Remove "O" rings (2 and 3) from pump cover. Remove gears and shafts (4), bearings (5), "O" rings (6), "O" ring retainers (7) and presure plate spring (8) from body.

Pump body is not available separately, and if defective, renew complete pump. Pump cover and bushings are not available separately. Pump gears and shafts (4) and bearings (5) are available in pairs only and when ordering pump gears and shafts, bearings and/or pump cover, also order an "O" ring and gasket package.

Refer to the following specifications.

4½ GPM Pump
O. D. of shafts at bushings......0.625
I. D. of bearings and
  cover bushings 0.6277
Thickness (width) of gears....0.4395
I. D. of gear pockets...........1.449

7 GPM Pump
O. D. of shafts at bushings......0.625
I. D. of bearings and
  cover bushings 0.6277
Thickness (width) of gears....0.6865
I. D. of gear pockets...........1.449

When reassembling, use all new "O" rings and gaskets. Lubricate gears and shafts prior to installation and if reusing bearings, reinstall them in their original positions. Tighten body to cover screws evenly and check rotation of pump. Pump will have a slight drag but should turn evenly with no tight spots.

HYDRAULIC LIFT UNIT

222. REMOVE AND REINSTALL. To remove the hydraulic lift unit from tractor, first remove plug from lower rear right hand corner of lift housing and drain housing. Remove seat, upper lift link and right fender. Remove retainer from right end of draft control bellcrank pin and slide pin to left until it clears draft control nut. Unbolt torsion bar flange from torsion bar bracket and remove torsion bar and bellcrank. Unbolt and remove torsion bar bracket from lift housing.

Note: At this time, bushings in torsion bar bracket, and seal and bearings in draft control rod tube (Fig. IH1480) can be renewed. Control rod tube can be removed by unscrewing after draft control rod nut is off.

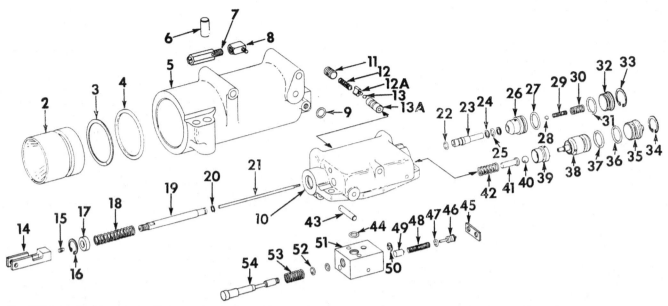

Fig. IH1481—Draft control cylinder and valve assembly showing component parts and their relative positions. Later units do not use retainer (45) and plug (46) differs slightly.

2. Piston	11. Retaining screw	18. Spring	27. "O" ring	37. "O" ring	46. Plug
3. Back-up washer	12. Spring	19. Actuator tube	28. Ball	38. Unloading valve	47. "O" ring
4. "O" ring	12A. Ball retainer	20. Retaining ring	29. Spring	piston	48. Spring
5. Cylinder	13. Check ball	21. Drop valve	30. Spring	39. Valve seat	49. Drop valve piston
6. Dowel	13A. Flow control valve	actuating rod	31. "O" ring	40. Ball	50. Retaining ring
7. Relief (safety)	14. Actuator	22. "O" ring	32. Plug	41. Ball rider	51. Drop valve body
valve	15. Drop valve	23. Pilot valve seat	33. Snap ring	42. Spring	52. "O" ring
8. Elbow	adjusting screw	24. Back-up washer	34. Snap ring	43. Pivot pin	53. Spring
9. "O" ring	16. Snap ring	25. "O" ring	35. Plug	44. "O" ring	54. Variable orifice
10. Draft control	17. Spring retainer	26. Drop poppet valve	36. "O" ring	45. Plug retainer	spool
valve body					

Disconnect hydraulic pump manifold flange at lift housing, and either disconnect manifolds and pipes from auxiliary control valves and transfer block or if desired, remove the valves and transfer block. Disconnect lift links from rockshaft arms. Remove cap screws which retain lift housing to rear center frame, attach chain and hoist and lift unit from tractor.

DRAFT CONTROL CYLINDER AND VALVE ASSEMBLY

223. REMOVE AND REINSTALL. To remove the cylinder and valve assembly, remove plug from rear lower right hand corner of lift housing and drain housing. Remove seat, seat brackets and top cover. Remove "C" ring and pin from control valve link and disconnect control linkage from control valve link, move control levers rearward, then unbolt and remove the complete cylinder and valve assembly from housing.

Note: Catch safety switch assembly as mounting cap screws are loosened. Tip forward end of cylinder and valve assembly upward when removing from housing.

NOTE: When reinstalling, it is necessary that the cylinder be seated against the boss in lift housing. Use a small bar to hold cylinder against boss while tightening retaining cap screws. In some cases it may

be necessary to reposition the draft control valve.

After unit is installed, refer to paragraph 214 for adjustment. See Figs. IH1481 and IH1482.

224. OVERHAUL. Because of the inter-relation of the cylinder and piston, draft control valve and drop control valve, they will be treated as subassemblies and each will be removed from the complete assembly, serviced and reinstalled.

225. DROP CONTROL VALVE. To remove and overhaul the drop control valve assembly, remove the complete cylinder and valve assembly as outlined in paragraph 223.

Remove the Allen screws which retain drop control valve, lift same from draft control valve and remove "O" ring. Remove "C" retainer from

end of variable orifice spool, pull spool from bore and remove the "O" rings from each end of body bore. On early valves, remove end plate (retainer) and remove the spacer, plug and guide (esna pin), spring and piston. Remove "O" ring from plug assembly. See Fig. IH1483. On later valves plate is not used and plug serves as the retainer.

Inspect spool, piston and bores for nicks, burrs, scoring and undue wear. Variable orifice spool and valve should fit bores snugly, yet slide freely. Inspect springs for fractures, distortion or signs of permanent setting. Refer to following specifications and renew parts as necessary.

Use all new "O" rings, reassemble valve by reversing disassembly procedure and install unit on draft control valve.

Fig. IH1482 — Cylinder and control valve assembly shown removed. Note that valve link is removed.

C. Cylinder
D. Drop control valve
P. Piston
V. Control valve

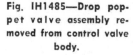

Fig. IH1485—Drop poppet valve assembly removed from control valve body.

Fig. IH1483—View of drop control valve with internal parts removed. Late valves do not use retainer plate.

Variable Orifice Spool Spring
    Free length—In. ....$1\frac{7}{8}$
    Test load lbs. at
      length—In. ....13.5-16.5 @ $\frac{7}{8}$
Piston Spring
    Free length—In. ....$1\frac{3}{32}$
    Test load lbs. at
      length—In. ....10.5-12.3 @ 61/64

Fig. IH1486 — Main control valve assembly removed from control valve body. Note drop control valve cam.

226. DRAFT CONTROL VALVE. To remove and overhaul the draft control valve, remove the complete cylinder and valve assembly as outlined in paragraph 223. Remove drop control valve from draft control valve, then remove draft control valve from work cylinder and piston assembly.

With valve removed, refer to Fig. IH1484 and remove internal snap ring, then remove plug and unloading valve assembly. Unscrew valve seat and remove ball, ball rider and ball rider spring.

Refer to Fig. IH1485 and remove internal snap ring, then remove plug, drop poppet valve spring (large), pilot valve spring (small), ball and the drop poppet valve and pilot valve seat assembly. Pull pilot valve seat from poppet valve.

Refer to Fig. IH1486 and remove link from control valve, then remove internal snap ring, and remove spring retainer, main valve spring, main valve actuator tube and valve spool. Remove the small retaining ring and pull drop valve actuating rod from actuator tube. Do not remove drop valve adjusting screw (Fig. IH1477) from outer end of actuator tube unless necessary.

Refer to Fig. IH1487 and remove plug, spring, and nylon check valve ball retainer and ball and the flow control valve.

Inspect all parts for nicks, burrs, scoring and undue wear. Refer to the following specifications and renew parts as necessary. Valve body and spool are not available separately. Use caution when renewing "O" rings to insure that proper size is installed and be sure the back-up rings used in bore of drop poppet valve are positioned on each side of the "O" ring. Be sure all check balls and their respective seats are in good condition.

Coat all parts with lubricating oil and reassemble by reversing disassembly procedure. Use new "O" ring and install valve to cylinder assembly. Install drop control valve assembly to draft control valve.

Fig. IH1484 — Unload valve assembly removed from control valve body.

Unloading Valve
  Spring free length—In. ....$1\frac{7}{16}$
  Test load lbs. at
    length—In. ....16.0-17.0 @ $1\frac{1}{32}$
Drop Poppet Valve
  Spring free length—In. 41/64
  Test load lbs. at
    length—In. ....9-11 @ $\frac{17}{32}$
  Check ball spring free
    length—In. 59/64
  Check ball spring test load
    lbs. at length—In. ...3.5-4.1 @ $\frac{3}{4}$
Control Valve
  Spring (return) free
    length—In. 2 61/64
  Test load lbs. at
    length—In. .....18.4-21.6 @ $1\frac{29}{32}$
Flow Control Valve
  Spring free length—In. ....$1\frac{1}{4}$
  Test load lbs. at
    length—In. ....4.9-5.8 @ $1\frac{1}{16}$

227. WORK CYLINDER AND PISTON. To remove and overhaul the work cylinder and piston, first remove the complete cylinder and valve assembly as outlined in paragraph 223, then remove the draft control valve and drop control valve assembly from work cylinder.

Piston can be removed from cylinder by bumping open end of cylinder against a wood block, or by carefully applying compressed air to the cylinder oil inlet port. With piston removed, the piston "O" ring and back-up washer can be removed.

Inspect piston and cylinder bore for nicks, burrs, scoring or undue wear. Small defects can be corrected by using crocus cloth. Renew parts which are unduly scored or worn. Piston outside diameter is 3.497-3.499 inches. Cylinder inside diameter is 3.500-3.502 inches.

When installing piston "O" ring and back-up washer, be sure "O" ring is toward pressure (closed) end of piston. Coat piston assembly with oil prior to installation in cylinder. Use new "O" ring between cylinder and control valve assembly.

Note: The safety relief (cushion) valve (S—Fig. IH1475) attached to left side of work cylinder can be removed after lift housing top cover has been removed and procedure for doing so is obvious. Relief valve can be bench tested by using an injector tester; however, unit is preset at the factory and is non-adjustable. Renew valve if found to be faulty. Valve is set to relieve at 1650-1900 psi.

228. CONTROL LEVERS AND SHAFTS AND CONTROL LINKAGE. Control levers, quadrant and control lever shafts can be removed as an assembly as follows: Drain hydraulic lift housing and remove seat and lift

Fig. IH1487—Flow control valve and check ball shown removed. Plug has a nylon locking insert.

housing top cover. Remove cylinder and valve assembly as outlined in paragraph 223. Remove control handle knobs, then remove cap screw from one end of control handle guide, loosen other at opposite end and allow guide to hang. Remove quadrant and note position the quadrant was mounted (i.e., which two holes used) so it can be reinstalled the same way. Pull draft control handle rearward, then drive out roll pin at inner end of shaft and remove control lever. Remove snap ring from inner end of position control lever shaft. Unbolt control lever and quadrant support from hydraulic housing and pull support and control levers from housing.

Draft control linkage can now be removed after removing adjusting nut from aft end of draft control rod. Any further disassembly of linkage will be evident after an examination of same. Removal of rockshaft bellcrank and actuating hub will require removal of rockshaft as outlined in paragraph 229.

Oil seal and bearing for control lever shaft can now be removed from lift housing, if necessary.

Any disassembly and/or overhaul required on the control levers and quadrant assembly will be obvious after an examination of the unit. Note that the inner control lever shaft is sealed to the outer control lever shaft by an "O" ring.

NOTE: The eccentric shaft can also be removed at this time. Disconnect internal linkage, if not already done, then remove outer retainer and remove eccentric shaft from inside of housing. Eccentric shaft is fitted with an "O" ring seal. Shaft bushing can also be pulled from lift housing.

When reassembling control levers and drop valve actuating lever to

shafts, be sure to align the register marks. Belleville washers on outer end of inner (draft control) shaft are installed as follows: Outer washer, dish toward inside; center washer, dish toward outside; inner washer, dish toward inside.

229. ROCKSHAFT. If rockshaft seal renewal is all that is required, seals can be renewed as follows: Remove control lever quadrant and both lift arms. Use a screwdriver, or similar tool, and pry out old seals. Use a suitable driver and drive new seals in until they bottom. Note that left seal has a smaller inside diameter than the right seal.

To remove the rockshaft, remove the right fender, in addition to the quadrant and rockshaft lift arms. Remove the cylinder and valve assembly as outlined in paragraph 223. Remove Allen screws from actuating hub and bellcrank and slide rockshaft from left to right out of housing, bellcrank and actuating hub. Remove actuating hub key as soon as it is exposed. If actuating hub sticks on rockshaft, either pry against it with a heavy screwdriver, or use a spacer between hub and housing. DO NOT drive on rockshaft without supporting actuating hub as damage to linkage will occur.

Always renew oil seals whenever rockshaft is removed; however, do not install the seals until after the rockshaft is installed. Rockshaft bushings can be removed and reinstalled using a proper sized bushing driver. Outside edge of bushings should be flush with bottom of oil seal counterbore. Inside diameter of left bushing is 2.090-2.095; inside diameter of right bushing is 2.315-2.320. Outside diameter of rockshaft at bearing surfaces is 2.085-2.087 for the left and 2.310-2.312 for the right.

Prior to reassembly, it is recommended that the following identification marks be made even though the rockshaft and rockshaft bellcrank are master splined. These marks will provide visibility and aid in reassembly. Use yellow paint and paint the "V" notch in the actuating hub key, the rockshaft master spline and the allen screw seat in the rockshaft. Also paint a line straight up from the master spline in the rockshaft bellcrank.

Start rockshaft into housing from right side of housing and start actuating hub and rockshaft bellcrank over rockshaft. Align the affixed markings (master splines) of rockshaft and bellcrank and position bellcrank until set screw seat in rockshaft is aligned with set screw hole in bellcrank, then install the allen screw. Install the actuating hub key and slide the actuating hub over key until the "V" notch is visible through set screw hole, then install the set screw. NOTE: Use a mirror during these operations. Install new oil seals. Install rockshaft lift arms and torque the retaining bolts to 170-190 ft.-lbs.

Complete reassembly by reversing the disassembly procedure.

AUXILIARY CONTROL VALVE

230. **R&R AND OVERHAUL.** To remove the auxiliary control valve, or valves, first remove right fender, then remove control lever knobs and quadrant assembly.

NOTE: Be sure to identify the cap screw holes in the quadrant before removing the cap screws. Mounting position of quadrant differs between Farmall and International model tractors.

Remove banjo bolts and disconnect manifolds from control valves. Remove through bolts and pull cover and control valves from transfer block.

When reinstalling, torque mounting bolts to 20-25 ft.-lbs. Do not over-tighten mounting bolts as valve body may be distorted and valve sticking could result.

To disassemble, use Fig. IH1488 as a guide. Remove control handle and bracket. Remove end cap (1), then unscrew the actuator (10) and remove the actuator and detent assembly. Remove sleeve (15) and pull balance of parts from body. Check ball and retainer can be removed at any time. Note the difference between early check valves (items 26 through 30) and late check valves (items 19 through 25). On late type check valves, snap ring (25) must be removed before retainer (22) can be removed.

NOTE: Some valves do not include the detent assembly. When disassembling these valves, sleeve (15) must be removed before removing actuator (10).

In addition, industrial valves have a circuit relief valve located directly below sleeve (15) and valve can be removed at any time.

Detent (3, 4, 5 and 6) can be disassembled after removing plug (2). Push unlatch piston (8) out of actuator (10) by using a long thin punch.

Inspect all parts for nicks, burrs, scoring and undue wear and renew parts as necessary. Spool (18) and body (31) are not available separately. Check detent spring (3) and centering spring (12) against the following specifications.

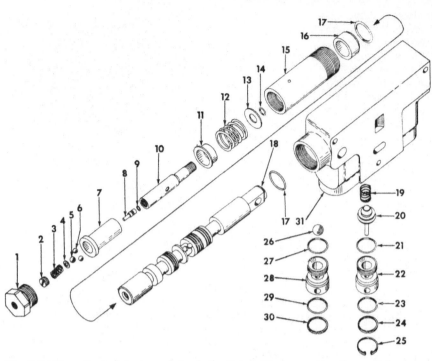

Fig. IH1488 — Exploded view of the hydraulic auxiliary control valve used for remote cylinders. Items 26 through 30 are used in early type valves. Items 19 through 25 are used in late type valves.

1. Cap	7. Position control sleeve	14. "O" ring
2. Plug	8. Unlatching piston	15. Sleeve
3. Spring	9. "O" ring	16. Retainer
4. Detent actuating ball washer	10. Actuator	17. "O" ring
5. Detent actuating ball	11. Spring retainer	18. Spool
6. Detent balls	12. Centering spring	19. Poppet spring
	13. Washer	20. Poppet
		21. "O" ring
		22. Retainer

23. "O" ring
24. Back-up washer
25. Snap ring
26. Check ball
27. "O" ring
28. Retainer
29. "O" ring
30. Back-up washer
31. Body

Detent spring

Free length—In. ....$1\frac{1}{16}$

Test load lbs. at

length—In. .....23.5-28.5 @ 45/64

Centering spring

Free length—In. ....$2\frac{5}{16}$

Test load lbs. at

length—In. ....26.5-33.5 @ 1 7/64

Use all new "O" rings and reassemble by reversing the disassembly procedure. Detent unlatching pressure is adjusted by plug (2). Unit must unlatch at not less than 1000 nor more than 1250 psi. The circuit relief valve on industrial valves is a cartridge type with the pressure setting stamped on end of body. Faulty relief valves are corrected by renewing the complete unit. Be sure filter in end cap (1) is clean (no paint) and in satisfactory condition.

TRANSFER BLOCK

231. **R&R AND OVERHAUL.** To remove the transfer block, first remove the auxiliary control valves as outlined in paragraph 230. Disconnect the supply line, then remove the two cap screws and the socket head screw and pull transfer block from lift housing.

Check ball seat can be removed by using two small screw drivers to pry it out of transfer block.

NOTE: Transfer blocks used on International 504 and 2504 tractors also have a check ball retainer and removal of same is obvious.

Pilot relief valve is removed as a unit by unscrewing plug (1—Fig. IH1489). Note the relief valve opening pressure which is stamped on valve should it be necessary to bench test the valve. Plug (1) and relief valve cartridge (2) are not available separately and if either is defective, renew complete valve.

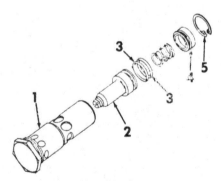

Fig. IH1489 — Exploded view of the pilot relief valve which is installed in the transfer block.

1. Plug	4. Cap
2. Relief valve	5. Retaining ring
3. "O" rings	

NOTES

NOTES

||||||||||||||||||||||||||||||||||||||||||||||||||||||||||||||||||

INTERNATIONAL HARVESTER

Models ■ 274 ■ 284

Previously contained in I&T Shop Manual No. IH-49

||||||||||||||||||||||||||||||||||||||||||||||||||||||||||||||||||

SHOP MANUAL

INTERNATIONAL HARVESTER

SERIES
274 — 284

Engine serial number is stamped on right side of engine crankcase on diesel series and on left side of engine crankcase on non-diesel series. Tractor serial number is stamped on name plate attached to right side of clutch housing of all series.

INDEX
(By Starting Paragraph)

INDEX CONT.

CONDENSED SERVICE DATA

	274 Diesel	284 Non-Diesel	284 Diesel
GENERAL			
Engine Make	Nissan	Toyo-Kogyo	Nissan
Engine Model	SD-16	M471G	SD-16
Number of Cylinders	3	4	3
Bore – mm	83	70	83
Stroke – mm	100	76	100
Displacement – Cubic centimeters	1623	1169	1623
Main Bearing, Number of	4	5	4
Cylinder Sleeves	Dry	Wet	Dry
Forward Speeds	8	8	8
Alternator and Starter Make	Hitachi	Mitsubishi	Hitachi
TUNE-UP			
Compression Pressure (2)	2944	1165	2944
Firing Order	1-3-2	1-3-4-2	1-3-2
Valve Tappet Gap (Hot) mm			
Intake	0.3-0.4	0.25	0.3-0.4
Exhaust	0.3-0.4	0.25	0.3-0.4
Valve Seat Angle (Degrees)			
Intake	45	45	45
Exhaust	45	45	45
Timing Mark Location	See Para. 93	Crankshaft Pulley	See Para. 93
Ignition Distributor Make	.....	Mitsubishi	.....
Breaker Contact Gap (mm)	.....	0.45 ± 0.05	.....
Distributor Timing	.....	See Para. 119	.....
Spark Plug Electrode Gap (mm)	.....	0.8	.....
Carburetor Make (Gasoline)	.....	Nihon Kikakki	.....
Injection Pump Make	Bosch "A"	.....	Bosch "A"
Injection Pump Timing	See Para. 93	.....	See Para. 93
Battery Terminal Grounded		Negative	
Engine Low Idle Rpm	700	650 ± 50	700
Engine High Idle Rpm, No Load	2760	2950 ± 50	2760
Engine Full Load Rpm	2600	2600	2600
(2) Approximate kPa, at sea level, at cranking speed			
SIZES – CAPACITIES – CLEARANCES			
Crankshaft Main Journal Diameter, mm	70.90-70.92	55.944-55.959	70.90-70.92
Crankpin Diameter, mm	52.91-52.93	44.940-44.955	52.91-52.93
Camshaft Journal Diameter, mm			
No. 1 (Front)	45.45	47.899	45.45
No. 2	43.91	45.899	43.91
No. 3	41.23	44.899	41.23
No. 4	.....	43.899	.....
No. 5	.....	42.899	.....
Piston Pin Diameter, mm	25.99	19.984-19.993	25.99
Valve Stem Diameter, mm			
Intake	7.97-7.99	6.95	7.97-7.99
Exhaust	7.94-7.96	6.95	7.94-7.96
Main Bearing Diametral Clearance, mm	0.15	0.025-0.050	0.15
Rod Bearing Diametral Clearance, mm	0.15	0.027-0.073	0.15
Piston Skirt Diametral Clearance, mm	0.29	0.030-0.069	0.29
Crankshaft End Play, mm	0.06-0.14	0.110-0.274	0.06-0.14
Camshaft Bearing Diametral			
Clearance, mm	0.03-0.09	0.15	0.03-0.09
Camshaft End Play, mm	0.08-0.28	0.02-0.18	0.08-0.28
Cooling System Capacity – Liters	7	5	7
Crankcase Oil – Liters	5	3	5
Transmission and Differential – Liters	13	13	13
Front Differential Housing			
(All Wheel Drive Tractor) – Liters	.....	.....	2.7
Clutch Housing Case			
(All Wheel Drive Tractor) – Liters	.....	.....	2.7
Front Final Drive Housing			
(All Wheel Drive Tractor) – Liters	.....	.....	0.9
Rear Axle Final Drive – Liters	1.4	.....	.....

FRONT AXLE SYSTEM

AXLE MAIN MEMBER

Model 274

1. The axle main member pivots on pin (23 – Fig. 1) which is retained in the mounting bracket (front sub-frame) by set bolt (24). The two pre-sized axle pivot pin bushings (25 and 26) pressed into axle main member can be renewed after removing axle main member from tractor. To remove the axle assembly turn front wheels all the way to the left, then disconnect tie rods from pitman arm (9). Raise front of tractor until front wheels are clear of ground. Support tractor under side rails with suitable jack stands. Remove front wheels. Loosen locknut on set bolt (24) then remove set bolt. Place rolling floor jack under center of axle. Drive pivot pin (23) forward with a brass drift and remove from axle. Lower axle assembly enough to clear sub-frame, then roll axle forward away from tractor.

Renew pivot bushings (25 and 26) if clearance between pivot pin (23) and bushings exceeds 0.3 mm. Reinstall by reversing the removal procedure.

Torque set bolt (24) in pivot pin mounting to 58-73 N·m. Torque tie rod slotted nuts (11) on pitman arm to 49-69 N·m. Torque wheel mounting bolts to 98-123 N·m. Adjust toe-in if necessary as outlined in paragraph 3.

Model 284

2. The axle main member pivots on pin (3 – Figs. 2 and 3) which is retained in the mounting bracket (front sub-frame) by set bolt (4). The two pre-sized axle pivot pin bushings (10 and 11) pressed into axle main member can be renewed after removing axle main member from tractor.

To remove the axle assembly, raise front of tractor until front wheels are clear of ground. Support tractor under side rails with suitable jack stands. Remove front wheels. Disconnect steering link from steering arm (15). Loosen locknut on set bolt (4) and remove bolt. On Models (S.N. 011793 and above) remove grease fittings (13 and 14 – Fig. 3). Install a slide hammer in pivot pin (3 – Figs. 2 and 3) and pull pin from axle assembly. Lower axle assembly enough to clear sub-frame, then roll axle forward away from tractor.

Renew pivot pin bushings (10 and 11 – Figs. 2 and 3) if clearance between

pivot pin and bushings exceeds 0.3 mm. Reinstall by reversing removal procedure. Torque set bolt (4) to 58-73 N·m. Torque steering link slotted nut to 49-69 N·m. Torque wheel mounting bolts to 98-122 N·m. Adjust toe-in if necessary as outlined in paragraph 3.

TIE RODS AND TOE-IN

All Models Except All-Wheel Drive

3. The procedure for removal and disassembly of tie rods on all models is

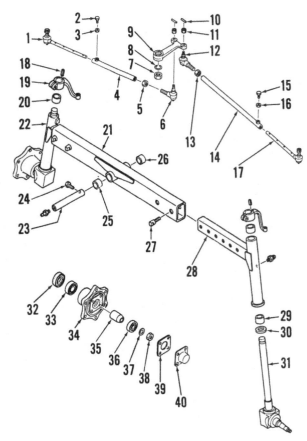

Fig. 1 – Exploded view of front axle assembly used on Model 274 Offset tractors.

1. Tie rod extension R.H.
2. Set bolt
3. Locknut
4. Tie rod tube R.H.
5. Jam nut
6. Tie rod end R.H.
7. Nut
8. Washer
9. Pitman arm
10. Pin
11. Nut
12. Tie rod end L.H.
13. Jam nut
14. Tie rod tube L.H.
15. Set bolt
16. Locknut
17. Tie rod extension L.H.
18. Key
19. Steering lever R.H.
20. Bushing (upper)
21. Axle center section
22. Axle extension R.H.
23. Pivot pin
24. Set bolt
25. Bushing
26. Bushing
27. Bolt
28. Axle extension L.H.
29. Bushing (lower)
30. Thrust bearing
31. Knuckle
32. Seal
33. Bearing (inner)
34. Hub
35. Collar
36. Bearing (outer)
37. Washer
38. Nut
39. Gasket
40. Cap

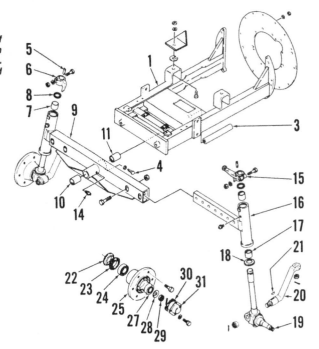

Fig. 2 – Exploded view of front axle assembly used on Model 284 tractors (S.N. 011793 and below) equipped with non-diesel engine.

1. Sub frame
3. Pivot pin
4. Set bolt
5. Key
6. Stopper
7. Bushing (upper)
8. Seal
9. Axle center section
10. Bushing
11. Bushing
14. Grease fitting
15. Steering lever
16. Axle extension (L.H.)
17. Bushing (lower)
18. Thrust bearing
19. Knuckle L.H.
20. Steering arm L.H.
21. Key
22. Seal
23. Seal
24. Bearing (inner)
25. Hub
27. Bearing (outer)
28. Washer
29. Nut
30. Gasket
31. Cover

obvious after examination of the units and reference to Figs. 1, 4 and 5. Tie rod ends are non-adjustable and faulty units will require renewal.

Adjust toe-in on Model 274 tractors to 6 mm and 4 to 10 mm for Model 284 tractors.

Adjustment on Model 274 tractors is made by separating the ball joints from pitman arm and turning ball joints in or out as necessary. When properly adjusted, reconnect ball joints to pitman arm and tighten slotted nuts to 49-69 N·m torque. Install cotter pins.

Adjustment on Model 284 tractors is made by releasing set bolt on tie rod and lengthening or shortening rod as necessary. When properly adjusted, retighten set bolt.

STEERING KNUCKLES AND HUB

Model 274

4. To remove either steering knuckle (31 – Fig. 1), raise front of tractor and remove wheel. Unbolt knuckle arm (19) and tap knuckle downward using a brass drift. Retain key (18). Remove knuckle from bottom of axle extension. Inspect upper and lower bushings (20 and 29) for excessive wear and, if necessary remove using a slide hammer and puller. New bushings should bottom on shoulder inside axle extension tube. Inspect thrust bearing (30) for wear and renew if necessary.

Remove hub and wheel bearings (items 32 through 40), inspect for wear and renew as necessary. Use new seal (32) upon assembly.

Reassemble steering knuckle and hub by reversing disassembly procedure. Lubricate steering knuckle through fitting on rear of axle extension tube with good grade lithium grease. Pack wheel bearings upon assembly. Tighten knuckle arm clamp bolt to 64-73 N·m. Tighten hub slotted nut to 79 N·m and install cotter pin.

Model 284

5. To remove either steering knuckle (19 – Figs. 2 or 3), raise front of tractor and remove wheel. On Model 284 (S.N. 011793 and below), disconnect tie rod from arm (20 – Fig. 2). Disconnect steering link from knuckle arm (15 – Figs. 2 or 3). Unbolt knuckle arm (15) and tap knuckle downward and out of axle extension tube (16) using a brass drift. Retain key (5). Inspect upper and lower bushings (7 and 17) for excessive wear and, if necessary, remove using a slide hammer and puller. New bushings should bottom on shoulder inside axle extension tube. Inspect thrust bearing (18) and oil seal (8 – Fig. 2) for wear and renew as necessary.

Remove slotted nut on hub spindle and using a suitable puller remove hub assembly (items 22 through 31 – Figs. 2

and 3). Inspect for wear and renew components as necessary. Use new oil seals upon assembly.

Reassemble steering knuckle and hub by reversing disassembly procedure. Lubricate steering knuckle through fitting on front of axle extension tube with good grade lithium grease. Pack wheel bearings upon assembly. Tighten knuckle arm clamp bolt to 64-73 N·m torque. Torque slotted nuts on tie rod and steering link to 49-69 N·m. Install cotter pins. Tighten slotted nut on hub spindle to 49-98 N·m torque. Install cotter pin and hub cap. Torque wheel mounting bolts to 98-122 N·m torque.

Fig. 4 – Exploded view of steering linkage on Model 284 tractors (S.N. 011793 and below) equipped with non-diesel engines.

1. Pitman arm
2. Drag link
3. Tie rod end
4. Adjusting tie rod
5. Set bolt
6. Steering lever
7. Steering arm
8. Key

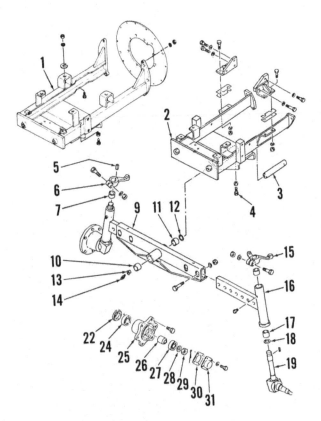

Fig. 3 – Exploded view of front axle assembly used on Model 284 tractors (S.N. 011794 and above) equipped with either non-diesel or diesel engines.

1. Sub frame (non-diesel engine)
2. Sub frame (diesel engine)
3. Pivot pin
4. Set bolt
5. Key
6. Steering lever R.H.
7. Bushing (upper)
9. Axle center section
10. Bushing
11. Bushing
12. Shim (0.2 and 0.4 mm)
13. Adapter
14. Grease fitting
15. Steering lever L.H.
16. Axle extension L.H.
17. Bushing (lower)
18. Thrust bearing
19. Knuckle
22. Seal
24. Bearing (inner)
25. Hub
26. Collar
27. Bearing (outer)
28. Washer
29. Nut
30. Gasket
31. Cap

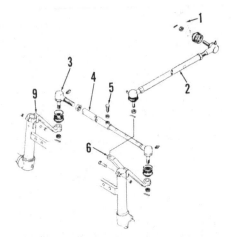

Fig. 5 – Exploded view of steering linkage on Model 284 tractors (S.N. 011794 and above) equipped with non-diesel or diesel engines.

1. Pitman arm
2. Drag link
3. Tie rod end
4. Adjusting tie rod
5. Set bolt
6. Steering lever/arm
9. Steering arm

FRONT SYSTEM ALL-WHEEL DRIVE

Model 284 tractors equipped with diesel engines are available as 4-wheel drive (All-Wheel Drive) units.

DROP HOUSING ASSEMBLY AND STEERING KNUCKLE

Models So Equipped

6. **R&R AND OVERHAUL.** Any service to either drop housing assembly may be performed without removal of complete axle assembly. If axle is mounted on tractor, raise front of tractor and support axle under side to be worked on. Remove wheel, then drain oil from bottom plug on rear of housing. If working on L.H. housing, disconnect steering link (84 – Fig. 6) from knuckle arm (56). Disconnect tie rod (85) from tie rod arm (66). Remove bolts and separate cover plates (55), packing (54) and back plates (53), noting their locations. Unbolt and remove tie rod arm (66) along with its needle bearing and rings. Unbolt and remove knuckle arm (56). Retain shims (57) noting number and positions. Attach suitable hoist and remove housing.

With drop housing assembly removed, unbolt and separate pinion drive cover (80) from drop housing (60). Pry lock tab (68) from bolts (67), then unbolt and remove bolt lock and bearing retainer (69). Press outer axle shaft (83) out of pinion drive cover (80) from the inside. Remove inner bearing (70), collar (71) and spur gear (75) from cover. Remove pinion shaft (74) and bearings (73 and 76) as an assembly. Remove snap ring (52) from pinion shaft and press off inner and outer pinion bearings (73 and 76). Press outer bearing (77) from cover. Remove and discard oil seals (72, 81 and 82).

Slide ball joint (51) from inner shaft and inspect for excessive wear or other damage. If ball joint is defective, unit must be renewed as an assembly. However if disassembly for cleaning is desired, proceed as follows: Using a brass drift, tilt bearing cage until one

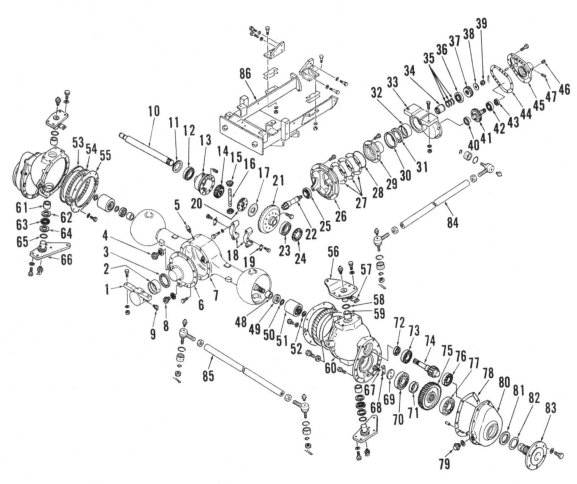

Fig. 6 – Exploded view of "ALL WHEEL DRIVE" front axle assembly available on Model 284 tractors equipped with diesel engines.

1. Front pivot housing	16. Pin	30. Collar	44. Gasket	58. "O" ring	72. Oil seal
2. Pivot bushing	17. Thrust washer	31. Oil seal	45. Cover	59. Bushing	73. Bearing
3. Dust seal	18. Bearing caps	32. Bushing	46. Plug	60. Drop housing	74. Shaft
4. Plug	19. Ring locks	33. Rear pivot housing	47. Set screw	61. Bushing	75. Gear
5. Breather	20. Side gear	34. Collar	48. Bushing	62. Bearing ring	76. Bearing
6. Cover	21. Ring gear	35. Shims (0.5 mm)	49. Oil seal	63. Needle bearing	77. Bearing
7. Axle housing	22. Pinion shaft	36. Bearing	50. Snap ring	64. Bearing ring	78. Gasket
8. Plug	23. Bearing	37. Gear	51. Ball joint	65. "O" ring	79. Plug
9. Grease fitting	24. Adjusting ring	38. Washer	52. Snap ring	66. Tie rod arm	80. Cover
10. Shaft	25. Bearing cone	39. Nut	53. Back plates	67. Bolt	81. Oil seal
11. Adjusting ring	26. Carrier	40. Bearing	54. Packing	68. Lock tab	82. Seal
12. Bearing	27. Shims	41. Gear	55. Cover plates	69. Holder	83. Shaft
13. Differential carrier	28. "O" ring	42. Bearing	56. Knuckle arm	70. Bearing	84. Drag link
14. Side gear	29. Pinion cage	43. Oil seal	57. Shim (0.2 mm)	71. Collar	85. Tie rod
15. Pinion gear					

ball can be removed as shown in Fig. 7. Repeat procedure to remove all balls. Refer to Fig. 8 and rotate bearing cage until it is 90 degrees (perpendicular) to surface of ball joint case. Roll bearing cage until its two elongated slots line up with any two opposing teeth of the ball joint case, then lift bearing cage out of case. Turn inner race at a right angle to bearing cage as shown in Fig. 9. Align any tooth of inner race with the elongated slot in bearing cage and roll inner race out of cage.

Reassemble ball joint by reversing disassembly procedure, paying attention to the following: Inner race must be inserted in the bearing cage from the narrow side of bearing cage. When installed, the flat side of inner race must face the inside of ball joint case. The narrow side of bearing cage must face inside of ball joint case. The first bearing assembly installed in ball joint case must be parallel with the end surface of ball joint case to allow installation of the remaining bearing assembly. Pack ball joint with good grade lithium grease upon reassembly.

7. Inspect swivel bushings (59 and 61 – Fig. 6) for excessive wear and renew as necessary. Check inner axle shaft (10) for excessive free play in bushing (48) and renew if necessary. If inner axle is removed for bushing renewal, discard oil seal (49) and renew upon reassembly.

Install new "O" rings (58 and 65 – Fig. 6) on knuckle and tie rod arms (56 and 66). Inspect pivoting surfaces, needle bearings (63) and bearing rings (62 and 64) for wear or damage. If renewing needle bearing assembly, install rings with chamfered sides facing away from bearing. Lubricate knuckle and tie rod arms with good grade lithium grease upon assembly.

8. Inspect all gears, bearings and shafts for excessive wear or other damage and renew as necessary. If renewing any gears, also renew bear-

ings supporting the gears and/or shafts. Check pinion drive cover and steering knuckle housing for cracks, or wear in bearing bores and renew if necessary. Lubricate all parts well before reassembling.

9. To reassemble the drop housing assembly and steering knuckle, first lightly coat the O.D. of new oil seals (72, 81 and 82 – Fig. 6) with suitable sealer and install in pinion housing cover (80) and steering knuckle (60). Place pinion housing cover over outer axle shaft (83) and press new outer bearing (77) onto shaft and into bearing bore in cover until it contacts shoulder in cover. If removed, press new bearings (72 and 76) onto pinion shaft (74). Install new snap ring (52). Install pinion shaft assembly and spur gear (75) into pinion housing cover. Install collar (71) and press inner bearing (70) onto outer axle shaft. Install bearing retainer (69), bolt lock (68) and bolts (67). Tighten bolts to 58-73 N·m and bend lock tabs against flats of bolt heads. Clean mating surfaces of pinion cover (80) and drop housing (60), install new gasket (78) then place pinion assembly into drop housing and tap into place with a soft mallet. Install mounting bolts and tighten to 30 N·m torque. Install drain plug. Slide new packing ring (54) over end of axle center housing (7). Slide ball joint (51) onto end of pinion shaft. Pull inner axle (10) out of axle housing and insert into other end of ball joint. Support drop housing assembly with suitable hoist and start inner shaft and drop housing together into axle housing. Rotate outer axle shaft (hub) to align splines of inner axle shaft (10) and differential side gear (14). Install knuckle arm (56) and new "O" ring (58) without shim pack (57) or bolts. Install

tie rod arm (66), new "O" ring (65) and bearing assembly leaving bolts loose. Install knuckle arm bolts and tighten both knuckle arm and tie rod arm bolts to 97-117 N·m torque. Place floor jack under drop housing to load needle bearings. Measure gap between knuckle arm and drop housing to determine shim pack (57) needed to achieve zero end play between drop housing and axle end. Shims are available in 0.2 mm thickness. Lubricate grease fittings in knuckle and tie rod arms with gun grease. Bolt back plates (53), packing (54) and cover plates (55) to drop housing. Connect steering link (84) to knuckle arm (L.H. side only) and tie rod (85) to tie rod arm (66) and tighten slotted nuts to 49-69 N·m torque. Install cotter pins. Fill drop housing to fill plug hole (79), (0.9 liter), with SAE 85W-140 multi-purpose gear lube. Install front wheels and tighten wheel bolts to 245-309 N·m. Check oil level in axle center section and add as necessary.

AXLE PIVOT HOUSINGS

10. **R&R AXLE ASSEMBLY.** To remove the "All-Wheel Drive" as an assembly, first disconnect the steering link from knuckle arm on L.H. drop housing. Raise front of tractor, support under sub-frame with suitable jack stands and remove front wheels. Loosen set screw (47 – Fig. 6) that secures the drive shaft cover pipe to rear pivot housing. Place a rolling floor jack under the differential housing. Remove mounting bolts securing front and rear pivot housings (1 and 33) to tractor sub-frame. Roll front axle assembly forward enough to disengage the drive shaft from rear housing.

NOTE: Do not lower axle more than is necessary to roll it forward or damage to drive shaft and related parts may occur.

When axle assembly is clear of drive shaft cover pipe. Lower axle and roll away from tractor.

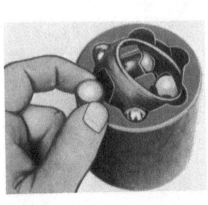

Fig. 7 – Rotate bearing cage in cylinder and remove ball as shown. Repeat rotation for removal of all balls. Refer to text.

Fig. 8 – View showing removal of ball cage from ball joint. Refer to text.

1. Elongated slot

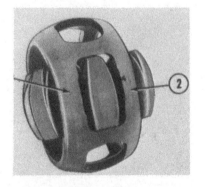

Fig. 9 – View showing positioning of inner race for removal from ball cage. Refer to text.

1. Ball cage   2. Elongated slot

Installation of axle assembly is the reverse of removal procedure. Lightly coat splines of drive shaft and coupling sleeves with Molykote lubricant. Apply Loctite #262 or equivalent to threads of axle mounting bolts. Torque mounting bolts to 244-307 N·m. Torque slotted nut on steering link to 49-69 N·m. Torque front wheel mounting bolts to 245-309 N·m.

11. R&R AND OVERHAUL PIVOT HOUSINGS. Service of the front and rear pivot housings (1 and 33–Fig. 6) requires removal of the complete axle assembly as outlined in paragraph 10. Service of the front housing is limited to replacement of the dust seal (3) and bushing (2).

NOTE: Removal of grease fitting may be necessary for removal of front pivot bushing.

When replacing front pivot bushing, line up hole in bushing with grease fitting hole. Always install new dust seal. Lubricate bushing with multi-purpose lithium grease.

To remove rear pivot housing (33–Fig. 6) first drain center housing (7) and remove bolts from housing cover (45), then using two of the bolts, turn them into jackscrew holes on cover. Turn bolts evenly to push cover from housing. Discard gasket (44). Using a brass drift, tap gear and bearings (items 40, 41 and 42) from cover (45). Inspect gear and bearings and renew if necessary. Remove and discard oil seal (43). Remove housing oil seal (31) and drive out bushing (32). Drive in new bushing and install new oil seal. Press large bearing (42) onto input shaft side of gear (41) and small bearing (40) onto housing side of gear. Tap bearing and gear assembly into bore of housing. Install new oil seal (43) and cover gasket (44), then bolt on cover (45) to housing (33). Fill center housing (5) to fill/level

plug (4), (2.7 liters), with SAE 85W-140 multi-purpose gear lube.

DIFFERENTIAL

12. R&R AND OVERHAUL. To remove the differential assembly for service, first remove axle assembly as outlined in paragraph 10, then remove drop housing assemblies as outlined in paragraph 6. Remove front and rear pivot housings as outlined in paragraph 11. Pull both axle shafts (10–Fig. 6) out of axle housing. Remove pinion cage (29) mounting bolts, then using two of the bolts, turn them into the jackscrew holes on flange of cage. Turn bolts evenly and push pinion cage (29) from carrier assembly (26). Retain shims (27) and discard "O" ring (28). Attach lifting brackets and a suitable hoist to the differential carrier assembly (26). Remove bolts and lift carrier assembly out of axle housing. Flatten tabs on ring locks (19) and remove. Unbolt and remove bearing caps (18).

NOTE: Bearing caps should be indexed for correct reassembly.

Remove bearing adjusting rings (11 and 24) and lift differential carrier assembly out of housing. Unbolt and separate ring gear (21) from carrier (13) using a soft hammer. Remove side gear (20) and thrust washer (17) from ring gear recess. Remove ring gear bearing (23) using a suitable puller. Remove pin retaining screw in differential carrier (13), then drive out pin (16) and remove side and pinion gears (14 and 15). Remove differential carrier bearing (12) using a suitable puller.

13. To remove the drive pinion assembly from the pinion cage, remove cotter pin then remove slotted nut (39–Fig. 6), washer (38) and gear (37). Press pinion shaft (22) out of small bear-

ing (36) and pinion cage (29). Press off bearing cone (25) and retain collar (34) and shim pack (35) noting thickness of shim pack after removal to aid in reassembly. Press bearing cups from pinion cage.

14. Clean and inspect all parts for excessive wear or other damage and replace as necessary. Lubricate all moving parts with petroleum jelly prior to assembly. Press large bearing cone (25) onto pinion shaft (22). Press new bearing cups, if removed, into pinion cage (29). Install collar (34) and shim pack (equal to one removed) onto pinion shaft. Insert pinion shaft with shims and collar into pinion cage and, using a suitable piece of pipe, press small bearing cone (36) onto pinion shaft. Leave pinion assembly on press and wrap a string around pinion cage, then connect a hand spring scale to string as shown in Fig. 10 to measure rolling torque. Slowly apply pressure to load bearing while pulling on scale. Bearing preload is correct when a pull of 17.6 to 30.8 N under 1.45 tons load is achieved. Add or subtract shims until correct preload is obtained. Install gear (37–Fig. 6), washer (38) and slotted nut (39) and tighten nut to 79-97 N·m torque. If necessary, tighten nut to install cotter pin.

15. PINION MOUNTING DISTANCE. The pinion mounting distance (A–Fig. 11), the distance from center line of differential bearings to the pinion back face, is a NOMINAL 103 mm. Any variation from the nominal is stamped on the face of the pinion and will be preceded by a plus (+) or minus (−). This figure must be added or subtracted from the nominal (103 mm) to determine ACTUAL mounting distance.

NOTE: There are three sets of numbers stamped on the pinion gear face. The whole number figure is the factory identification number and is also stamped on the outside edge of the ring gear. The second number is the variation number (a) of pinion mounting nominal distance

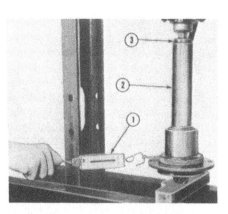

Fig. 10 – View showing procedure for checking preload of pinion bearing. Refer to text.

1. Spring scale
2. Tube
3. Plate

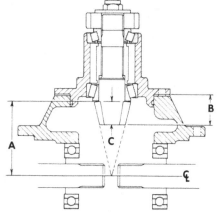

Fig. 11 – Cross sectional diagram of pinion cage and differential showing dimensions for pinion mounting distance. Refer to paragraph 15.

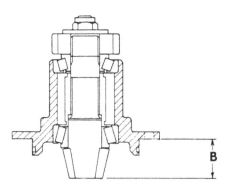

Fig. 12 – Sectional view showing dimension B representing pinion cage height. Refer to text.

(A – Fig. 11). The third number in parenthesis is the variation number (c) of the nominal pinion tooth length (C – Fig. 11).

The NOMINAL tooth length (C) of the pinion gear is 33 mm. Any variation from the nominal is the third number in parenthesis stamped on the pinion face. The number represents hundredths of a millimeter. For example: (– 1) represents – 0.01 mm. This figure must be added or subtracted from the nominal tooth length to determine the ACTUAL pinion tooth length.

To calculate mounting distance (X – Fig. 13), taken from the centerline of differential bearing bore to the pinion cage mounting face, add pinion mounting distance (A – Fig. 11) and pinion cage height (B – Fig. 12). Then, subtract the ACTUAL pinion tooth length (C + c – Fig. 11).

To calculate distance (Y – Fig. 13), install Pinion Setting Depth Tool PS 79-101 (1 – Fig. 14) or an equivalent tube that fits the differential bearing bores. Using a depth micrometer and surface plate (if needed), measure the distance (Y – Fig. 13) from the pinion mounting surface to the outside diameter of the pinion setting gage.

NOTE: If a surface plate is needed, be certain to subtract the width of the plate from the measurement.

Measure the diameter of the pinion mounting tool using a micrometer and add half the diameter of the tool to the depth measurement.

To determine shim pack thickness needed (Z – Fig. 13), subtract distance (Y) from distance (X).

Use the following example problem as a guide.

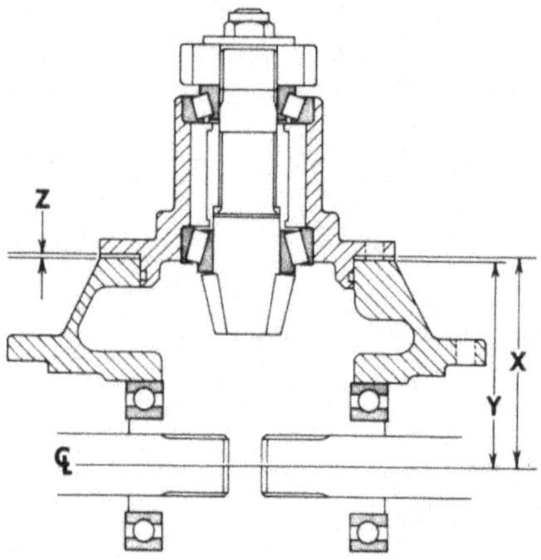

EXAMPLE:

NOMINAL distance (A) ..	103.00 mm
1st Figure on pinion face (a)	+ 0.10 mm
ACTUAL pinion mounting distance	103.10 mm
Pinion cage height (B)	42.11 mm
NOMINAL pinion tooth length (C)	33.00 mm
2nd figure on pinion face (c), (– 1) represents (– 0.01 mm) .	– 0.01 mm
ACTUAL tooth length ...	32.99 mm
ACTUAL pinion mounting distance (A + a)	103.10 mm
Pinion cage height (B)	+ 42.11 mm
	145.21 mm
ACTUAL pinion tooth length (C + c)	– 32.99 mm
Mounting distance (X)	112.22 mm
Housing face to setting tool	65.02 mm
Half tool diameter	+ 45.13 mm
Distance (Y)	110.15 mm
Distance (X)	112.22 mm
Distance (Y)	– 110.15 mm
Shim pack required (Z) ...	2.07 mm

Shims (27 – Fig. 6) are available in 1.0 mm thicknesses. Install pinion assembly with required shim pack and new "O" ring (28). Tighten pinion cage mounting bolts to 58-72 N·m.

16. Reassemble differential carrier by reversing disassembly procedure. Apply IH Thread Lock 634 017C1 or equivalent to carrier mounting bolts and tighten to 58-73 N·m torque. Place carrier assembly in carrier housing with ring gear teeth facing dowel hole in carrier flange. Install differential bearing adjusting rings and tighten hand tight against bearings. Install bearing caps and tap into place.

NOTE: Make sure not to cross thread adjusting rings when installing bearing caps.

Install bearing cap bolts and washers and hand tighten. Install a dial indicator as shown in Fig. 15.

NOTE: Make sure indicator shaft is mounted on the very end of gear tooth face.

Turn both adjusting rings equally the same direction until backlash is 0.15 to 0.20 mm. When correct backlash setting is obtained, install adjusting ring lock plates, apply IH Thread Lock 241 or equivalent and tighten lock plate bolts securely.

NOTE: Tighten adjusting rings to allow lock plate installation if necessary.

Bend tabs on lock plates over mounting bolts. Torque bearing cap bolts to 97-127 N·m.

Position differential assembly into axle housing using dowel as a guide to line up assembly. Apply IH Thread Lock 634 017 C1 or equivalent to mounting bolts. Install bolts and tighten to 53-73 N·m torque.

Balance of assembly is reversal of disassembly. Install axle assembly onto tractor as outlined in paragraph 10.

17. **TOE-IN ADJUSTMENT.** Toe-in setting is 5-10 mm. Adjustment is made by loosening jam nuts on tie rod and turning tie rod as required to achieve correct setting. After toe-in is correct

Fig. 14 – View showing depth micrometer set up to measure distance from pinion mounting surface to outside diameter of pinion setting gage. Refer to text.

1. Pinion Setting Tool PS 79-101
2. Surface plate
3. Depth micrometer

Fig. 13 – Cross sectional view of pinion cage and differential showing calculated dimensions needed to find shim pack "Z" for correct pinion mounting distance. Refer to text.

Fig. 15 – View showing installation of dial indicator to measure ring gear to pinion backlash.

tighten tie rod jam nuts to 126-165 N·m. Adjust turn limiter stop bolts so that a distance of 4 cm measured from bolt head to surface of drop housing is obtained.

DRIVE HOUSING ASSEMBLY

Models So Equipped

18. The drive assembly which transmits power to the front axle is contained in the clutch housing. For overhaul procedure of the All-Wheel Drive clutch housing assembly, refer to paragraph 126 in the Transmission Section.

MANUAL STEERING SYSTEM

ADJUSTMENTS

Model 274

20. The worm shaft/ball nut/sector shaft type steering gear is bolted to the right front sub-frame of tractor and is adjustable by turning adjusting screw (17 – Fig. 20) in or out to obtain proper backlash between the sector shaft (18) and the worm shaft "rack" (8). To adjust, disconnect tie-rods from pitman arm (20), then set up a dial indicator as shown in Fig. 21. Loosen locknut (3 – Fig. 20) and turn screw until a 0.3 to

0.4 mm free play is obtained at the pitman arm. After adjustment, tighten locknut to 19-29 N·m torque. Reconnect tie rods and torque slotted nuts to 49-69 N·m. Install cotter pins.

Model 284

21. The manual worm shaft/ball nut/ sector shaft type steering gear may be found on all Model 284 diesel and nondiesel tractors. The manual steering gear is located on top of the clutch housing of all 284 series tractors and has provision for adjustment. To adjust, loosen locknut (18 – Fig. 22) on adjustment screw (22), then turn screw in or out until rotating free play at steering wheel is 10-12 mm. When correctly adjusted, tighten locknut to 24-33 N·m torque.

R&R AND OVERHAUL

Model 274

22. To remove the steering gear, first turn steering wheel to full left position. Disconnect both tie rods from pitman arm. Unbolt and disconnect steering shaft universal joint from steering gear worm shaft. Remove radiator grill for access to two of the steering gear mounting bolts. Unbolt and remove steering gear from tractor.

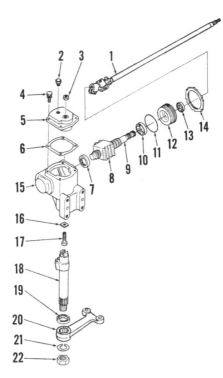

Fig. 20 – Exploded view of manual steering gear used on Model 274 Offset tractors.

1. Steering shaft	12. Side cover
2. Plug	13. Oil seal
3. Locknut	14. Lock ring
4. Cap screw	15. Gear housing
5. Top cover	16. Shim
6. Gasket	17. Adjusting screw
7. Bearing	18. Sector shaft
8. Ball nut	19. Oil seal
9. Worm shaft	20. Pitman arm
10. Bearing	21. Lockwasher
11. "O" ring	22. Nut

Fig. 22 – Exploded view of manual steering gear available on Model 284 tractors.

1. Cover
2. Nut
3. Washer
4. Steering wheel
5. Grommet
6. Steering column
7. Cap screw
8. Gasket
9. Oil seal
10. Bearing
11. Key
12. Worm shaft
13. Ball nut
14. Bearing
15. Plug
16. Gear box
17. Cap screw
18. Locknut
19. Side cover
20. Gasket
21. Shim
22. Adjusting screw
23. Sector shaft
24. Oil seal
25. Pitman arm
26. Washer
27. Nut

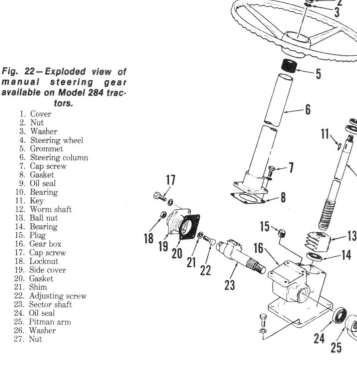

Fig. 21 – View showing procedure for adjusting steering gear box on Model 274 offset tractors.
1. Adjusting screw
2. Pitman arm
3. Dial indicator

23. With gear box removed, remove nut (22 – Fig. 20) and washer (21). Index pitman arm (20) to shaft (18) and remove pitman arm using a suitable puller. Remove breather plug (2) and drain oil from gear box. Remove locknut (3) from adjusting screw (17). Turn screw counter-clockwise one turn, then remove top cover mounting bolts (4). Holding cover (5), turn adjusting screw (17) clockwise to remove top cover. Discard cover gasket (6). Tap sector shaft (18) on splined end with a soft mallet and remove from steering box. Loosen rear cover locknut (14). Using a spanner wrench, remove rear cover (12), then remove worm shaft/ball nut assembly (8 and 9).

Clean and inspect all parts for excessive wear or other damage. Check ball nut/worm shaft assembly end play. If end play exceeds 0.10 mm, the ball nut/worm shaft assembly must be renewed. Check clearance of adjusting screw in sector shaft slot. If clearance exceeds 0.10 mm, replace shim (16 – Fig. 20) with thicker shim. Shim pack No. 973 549 C1 contains five shims of different thicknesses. Check bearings and races (7 and 10) and renew as necessary. Check clearance of sector shaft in top cover and gear box bushings. The bushings are not serviceable so if clearance is excessive, gear box and top cover must be renewed. Press new oil seals (13 and 19) into bores until flush. Renew "O" ring (11).

Lubricate all parts with IH 135 HEP gear lubricant or equivalent prior to assembly. Install ball nut/worm shaft assembly with bearing cages into gear box. Install rear cover (12) loosely into gear box. Using spanner wrench tighten rear cover until rotating torque of 0.24 to 0.54 N·m is obtained on worm shaft bearings. After correct rotating torque is obtained, tighten cover locknut (14) to 157-196 N·m. Recheck rotating torque and readjust as necessary. Place ball nut in center of worm shaft then install sector shaft lining up center tooth of sector shaft in center groove of ball nut "rack". Apply IH gasket eliminator or equivalent to both sides of top cover gasket (6). Install adjusting screw with shim in sector shaft slot. Align top cover with adjusting screw and install top cover by turning adjusting screw counter-clockwise until cover seats on gear box. Install top cover mounting bolts and torque to 17-27 N·m. Reinstall gear box on tractor. Fill gear box with 0.46 liter of IH 135 HEP gear lub or equivalent. Adjust gear box as outlined in paragraph 20.

Model 284

24. To remove manual steering gear disconnect headlight wiring, then remove muffler, radiator grill, engine hood and side panels. Shut off fuel at bottom of tank and disconnect fuel line. On diesel tractors, disconnect fuel return line. Disconnect fuel gauge wiring at tank. Remove fuel tank strap then remove tank. On non-diesel tractors, disconnect choke. Disconnect tachometer cable from engine. Disconnect governor control rod, throttle control rod and accelerator rod at fuel tank support. Unbolt governor control lever from steering gear. Using a suitable puller, remove steering wheel. Remove Woodruff key. Disconnect remaining wiring or controls necessary to allow removal of instrument panel. Unbolt and remove instrument panel and fuel tank support. Disconnect steering link from pitman arm. Unbolt and remove steering gear box from center section.

25. With steering gear box removed, register pitman arm (25 – Fig. 22) to sector shaft (23), remove nut and washer (26 and 27), then using a suitable puller remove pitman arm. Remove bolts on side cover (19). Unlock nut (18) on adjusting screw (22). Turn screw clockwise while removing side cover (19). Drain lubricant from side opening. Tap sector shaft (23) on splined end with a soft mallet and remove out of gear box. Remove bolts (7), then withdraw steering column (6) and shaft/ball nut assembly (12 and 13) from gear box. Retain shims (8) noting thickness.

Clean and inspect all parts for excessive wear or other damage. Check ball nut/worm shaft assembly end play. If end play exceeds 0.10 mm, the ball nut/worm shaft assembly must be renewed. Check clearance of adjusting screw in sector shaft slot. If clearance exceeds 0.10 mm, replace shim (21 – Fig. 22) with thicker shim. Shim pack 973 549 C1 contains five shims of different thicknesses. Check bearings and races (10 and 14) and renew as necessary. Check clearance of sector shaft in side cover and gear box bushings. The bushings are not serviceable so if clearance is excessive, gear box and side cover must be renewed. Press new oil seals (9 and 24) into bores until flush. Renew gasket (20).

Lubricate all parts with IH 135 HEP gear lube or equivalent prior to assembly. Install ball nut/worm shaft and steering column with bearing cages into gear box. Use original shim thickness (8) when assembling. Bolt steering column assembly to gear box and tighten bolts to 17-29 N·m torque. Check rotating torque of steering shaft. If torque is less than or exceeds 0.24 to 0.54 N·m, vary shim pack thickness (8 – Fig. 22) until correct rotating torque is obtained. Place ball nut in center of

worm shaft then install sector shaft lining up center tooth of sector shaft in center groove of ball nut "rack". Apply IH gasket eliminator or equivalent to both sides of side cover gasket (20). Install adjusting screw with shim in sector shaft slot. Align side cover with adjusting screw and install side cover by turning adjusting screw counter-clockwise until cover seats on gear box. Install side cover mounting bolts and torque to 17-27 N·m. Fill steering box to bottom of filler hole with IH 135 HEP or equivalent gear lube. Reinstall gear box by reversing removal procedure. Install pitman arm lining up index marks. Install washer and nut, tightening nut to 150-197 N·m. Reconnect steering link and tighten slotted nut to 49-69 N·m. Install cotter pin. Adjust gear box as outlined in paragraph 21.

POWER STEERING SYSTEM

The worm shaft/ball nut/sector shaft type steering with hydraulic power assist available on all Model 284 tractors employs an open center type hydraulic system. Pressurized oil is provided by an external gear type pump located behind the radiator grill on non-diesel models or on the right front of the engine block on diesel models. Non-diesel tractors utilize a 28.7 liters/min. (7.6 gpm) pump and diesel tractors are equipped with a 27.1 liters/min. (7.1 gpm) pump. The oil flow is directed to a priority type flow divider where approximately 4 liters/min. (1 gpm) is directed to the steering circuit. The steering gear and hitch control valve are protected by their own relief valve. An additional relief valve in the flow divider provides further protection to the steering circuit. If the power steering becomes inoperative, the tractor can be steered manually. Refer to Fig. 23 for a cross-sectional schematic view of the power steering and hydraulic circuit.

NOTE: The maintenance of absolute cleanliness of all parts is of utmost importance in the operation and servicing of the power steering system. Of equal importance is the avoidance of nicks or burrs on any of the working parts.

LUBRICATION AND BLEEDING

Model 284

26. The tractor rear frame serves as a common reservoir for all hydraulic and lubrication operations. The spin on type

filter (3–Fig. 24), IH part No. 1004 366 C1, should be renewed at 10 hours, 50 hours, 100 hours, and every 200 hours thereafter. Tractor rear frame should be drained and new fluid added every 1000 hours, or once a year, whichever occurs first.

Only IH "Hy-Tran" fluid should be used. Full capacity of rear frame should be 13.0 liters.

Whenever power steering lines have been disconnected, or fluid drained, start engine and cycle power steering system from stop to stop several times to bleed air from the system, then if necessary, check and add fluid to reservoir.

OPERATING PRESSURE AND RELIEF VALVES

Model 284

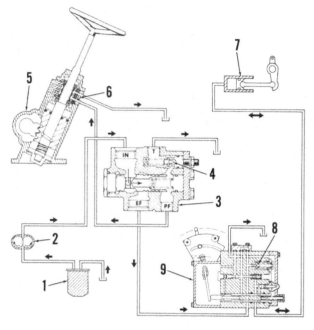

Fig. 23 — Schematic diagram of hydraulic system on Model 284 tractors equipped with power steering.

1. Filter
2. Pump
3. Flow divider
4. Steering circuit relief valve
5. Steering gear
6. Steering relief valve
7. Hitch cylinder
8. Hitch circuit relief valve
9. Control valve assembly

27. System operating pressure and relief valve operation can be checked as follows: Refer to Fig. 25 and remove the banjo bolt connecting power steering supply line (4) to flow divider. Remove reducer connection (2) from the port in the flow divider. Install a ⅜-inch NPT (male-female-female) service tee (5) into the port as shown. Use Teflon tape to seal the threads. Connect power steering supply line (4) to the tee using the removed reducer fitting and banjo bolt. Using suitable fittings, connect inlet line of a 14-51D Flo-Rater or equivalent to the service tee as shown. Place outlet line of Flo-Rater into the filler hole in hitch cover. Secure outlet hose in hole with wire. Open restrictor valve on Flo-Rater, start tractor and set engine speed to 2600 rpm. Run tractor until hydraulic fluid reaches temperature of 50°C. Hold steering wheel against one of its stops and note steering circuit flow. Flow should be 2.6 liters/min. (0.7 gpm) minimum. With steering wheel held against stop, close Flo-Rater restrictor valve and note steering circuit relief pressure of relief valve in flow divider. Allowable relief pressure for two-wheel drive tractors is 6867 to 7357 kPa (996 to 1067 psi) and for "All-Wheel Drive" tractors allowable relief pressure is 11,770 to 12,259 kPa (1707 to 1778 psi). The steering relief valve in the flow divider has provision for adjustment. If adjustment is necessary, loosen locknut on Allen head set screw at the front of the flow divider, and turn screw in or out as necessary to obtain correct relief pressure. After adjustment, tighten locknut. If relief pressure and flow can-

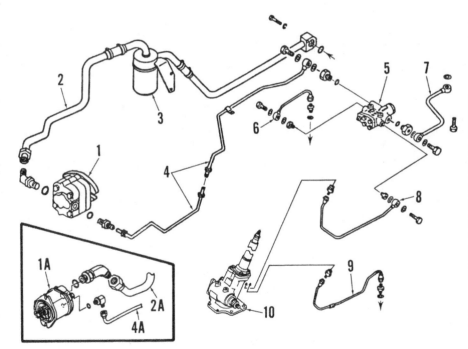

Fig. 24 — Exploded view of hydraulic and power steering systems for all Model 284 tractors so equipped.

1. Hydraulic pump (non-diesel engine)
1A. Hydraulic pump (diesel engine)
2. Low pressure line (non-diesel engine)
2A. Low pressure line (diesel engine)
3. Filter
4. High pressure line (non-diesel engine)
4A. High pressure line (diesel engine)
5. Priority valve
6. Relief by-pass return
7. Hitch control supply
8. Steering supply
9. Steering return
10. Integral steering gear

Fig. 25 — View showing priority valve with Flo-Rater installed to check system operating and relief pressure on Model 284 tractors equipped with power steering.

1. Flo-Rater inlet line
2. Reducer
3. Banjo bolt
4. Power steering supply line
5. Service tee
6. Power steering return line

not be obtained, service flow divider as outlined in paragraph 38.

28. STEERING RELIEF VALVE. To check relief pressure for relief valve in the power steering gear, fully open restrictor valve on Flo-Rater and shut off tractor. Refer to Fig. 26 and remove banjo bolt connecting the flow divider relief dump tube (2) to the flow divider. Remove the reducer fitting and install a ⅜-inch NPT plug (1) into dump port as shown. Start tractor and set engine speed at 1000 rpm. Close restrictor valve on Flo-Rater and hold the steering wheel against one of its stops. Pressure for the steering gear relief valve should be 12,749 to 13,045 kPa (1849 to 1892 psi). The steering gear relief valve has provision for adjustment. If adjustment is necessary, loosen locknut on adjustment screw located on left side of steering gear just above and ahead of outlet

Fig. 26 – View showing priority valve with Flo-Rater installed and relief dump port plugged to check relief valve in steering gear.

1. ⅜-inch NPT plug   2. Relief valve dump tube

Fig. 27 – View of Model 284 tractor upper rear frame showing rockshaft feedback linkage disconnected and secured fully rearward to check hydraulic pump free flow. Refer to text.

1. Rockshaft feedback link
2. Wire
3. Flo-Rater outlet hose

port. Turn screw in or out as necessary to obtain correct relief pressure. After adjustment, tighten locknut. If relief pressure cannot be obtained, service steering gear as outlined in paragraph 36.

29. On "All-Wheel Drive" tractors, remove plug (1–Fig. 26), reconnect relief dump line, and recheck flow divider relief pressure and readjust if necessary. After check and/or adjustment, disconnect relief dump line (2) and reinstall ⅜-inch NPT plug. Disconnect Flo-Rater inlet line from tee fitting (5–Fig. 25), remove the tee and reconnect power steering supply line (4). Make certain the seal washers are in place on each side of the banjo fitting.

30. HYDRAULIC PUMP. To check pump free flow, install a tee fitting in the system pressure line (4–Fig. 24) and on diesel engine tractors install the tee fitting between the supply line and pump outlet. Connect the inlet line of a Flo-Rater to the service tee. Open restrictor valve on Flo-Rater, start tractor and set engine speed at 1000 rpm. Move the three-point hitch position control to full raise position, then slowly close restrictor valve on Flo-Rater until hitch arms just start to raise. When hitch is fully raised, disconnect rockshaft arm feedback linkage (1–Fig. 27) from rockshaft. Pull linkage fully rearward and secure as shown in Fig. 27. This will hold the hitch circuit "on demand". With the ⅜-inch NPT plug installed in the relief dump port as outlined in paragraphs 28 and 29, hold steering wheel against one of its stops. This directs the complete hydraulic system flow through the Flo-Rater. Set engine speed to 2600 rpm and slowly close Flo-Rater restrictor valve until pressure reaches 10,294 kPa (1500 psi). For non-diesel tractors hydraulic flow must be a minimum of 21.5 liters/min. (5.7 gpm) and for diesel tractors, a minimum of 20 liters/min. (5.3

gpm). If fluid flow is not as specified, service pump as outlined in paragraphs 32 and 34.

PUMP

Model 284 Non-Diesel

31. REMOVE AND REINSTALL. To remove the power steering/hydraulic system pump, first remove front radiator grill and engine side panels. Disconnect hydraulic return and high pressure lines and plug or cap immediately to prevent dirt or foreign matter from contaminating the hydraulic system. Refer to Fig. 28 and remove nut (6) securing drive coupler (5) to pump shaft. Unbolt and remove pump. Reinstall pump by reversing the removal procedure. Bleed system as outlined in paragraph 26.

32. **OVERHAUL.** With pump removed as outlined in paragraph 31, refer to Fig. 29 and remove the six Allen head bolts (15). Remove cover (13) and discard "O" rings (12 and 14). Remove Woodruff key (7) then withdraw drive (6) and driven gears (8), bushings (5 and 9), support rings (10) and seal rings (11) from pump body (1). Discard "O" rings (4). Press out seal (2). Inspect all parts for excessive wear, scoring or other damage. Pump body (1), gears (6 and 8) and cover (13) are not serviceable and if any doubt of their condition exists, entire pump should be renewed. Support rings (10), seal rings (11), "O" rings (4, 12 and 14) and oil seal (2) are serviceable items and should be renewed. Bushings (5 and 9) should be installed with chamfered sides toward gears. "O" ring (12) should seat securely in groove of pump cover (13). Install oil seal (2) with garter spring and lip facing inside of pump body. Balance of reassembly is reverse of disassembly procedure. Install pump on tractor as outlined in

Fig. 28 – Exploded view of hydraulic pump drive shaft assembly on Model 284 tractors equipped with non-diesel engines.

1. Hydraulic pump
2. Key
3. Bracket
4. Shims
5. Coupling
6. Nut
7. Coupling
8. Coupling shaft
9. Drive coupling

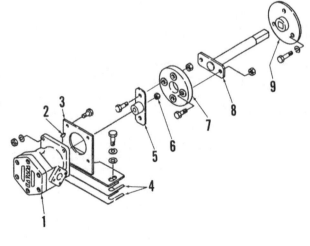

paragraph 31. Check pump flow as outlined in paragraph 30.

Model 284 Diesel

33. REMOVE AND REINSTALL. To remove pump, first disconnect high and low pressure lines from the pump. Cap or plug lines immediately to prevent dirt or other foreign matter from contaminating the hydraulic system. Unbolt and remove pump from drive housing. Remove and discard "O" ring from pump drive shaft. Install pump on tractor by reversing the removal procedure. Bleed system as outlined in paragraph 26.

34. OVERHAUL. With pump removed as outlined in paragraph 33, index front and rear covers to pump body for correct assembly. Remove the four cover bolts, then separate the covers from pump body. Discard all seals and gaskets (1, 4 and 5 – Fig. 30). Inspect gears and bushings for excessive wear, scoring or other damage. Pump body, gears and bushings are not serviceable and if any doubt of their condition exists, entire pump should be renewed. Renew all seals and gaskets on assembly. If shaft seal is renewed, install snap ring to retain seal in its bore. Bushings should be installed in body with chamfered side of bushings facing gear teeth. "O" rings should seat securely in grooves of pump covers. Reassemble pump by reversing disassembly procedure. Tighten cover bolts to 38-43 N·m torque. When installing pump on tractor, use new "O" ring between hydraulic pump and mounting surface. Install pump on tractor as outlined in paragraph 33. Check pump flow as outlined in paragraph 30.

POWER STEERING GEAR

Model 284 So Equipped

35. R&R STEERING GEAR. To remove integral steering gear assembly, first disconnect battery cables. Remove steering wheel using a suitable puller, then remove the rubber grommet between steering column and instrument panel. Disconnect steering drag link from pitman arm. Index pitman arm to sector shaft, then using a suitable puller remove pitman arm. Disconnect headlight wiring and remove front hood. Remove stay side flange bolts on rear hood. Disconnect steering gear hydraulic lines and plug immediately to prevent dirt or other foreign material from contaminating hydraulic system. Remove necessary hardware securing fuel tank mounting bracket to frame. Raise control panel/rear hood assembly enough to place two 22 mm nuts between fuel tank support and steering gear mounting plate. Unbolt and pry steering gear assembly upward enough to clear dowel pins and remove steering gear assembly from left side of tractor.

Reinstall steering gear by reversing removal procedure. Torque pitman arm locknut to 196-245 N·m and torque steering wheel nut to 23-35 N·m. Bleed system as outlined in paragraph 26.

36. OVERHAUL STEERING GEAR. With steering gear removed as outlined in paragraph 35, proceed as follows: Thoroughly clean outside of steering gear assembly. Refer to Fig. 31 and loosen steering column locknut (4), then unscrew steering column (3) and remove from steering assembly. Remove split pin (7) and roll pin (6) securing steering shaft (5) to worm shaft (34). Remove steering shaft being careful not to lose coil spring (8) between shafts. Unbolt and remove rear cover (11). Straighten the flange of the guide ring locknut. Using an adjustable wrench on the flat of the worm shaft, hold shaft and remove guide ring locknut (13). Remove upper guide rings (14 and 16), and thrust bearing (15).

Hold worm shaft horizontal and remove valve assembly (28), making certain not to lose or damage any reaction pistons (18 and 20), centering springs (19) or spool (17). Remove lower guide rings (30 and 32) and thrust bearing (31). Remove locknut (52) and packing washer (53) from sector shaft adjusting screw (41). Remove bolts from side cover (51). Turn adjusting screw clockwise and remove side cover. Tap splined end of sector shaft with soft mallet and remove sector shaft (42) from gear case (45). Remove large snap ring (69) and remove bottom cover (68) and worm shaft/ball nut assembly from gear case. Be careful not to bind the ball nut in gear case upon removal.

Remove and discard all seals, "U" packings, "O" rings, back-up rings and Teflon seals. Inspect all needle bearings for damage or wear and renew as necessary. Inspect teeth of sector shaft (42) and ball nut (59) for wear or damage. Check sliding surfaces of the ball nut for scuffing or wear. Very slight shallow scratches may be cleaned up with crocus cloth. If scratches are too deep, gear case (45) and ball nut/worm shaft assemblies must be renewed. Measure the O.D. of the sector shaft where needle bearings contact with a micrometer. If diameter is less than 38.025 mm, the sector shaft must be renewed. Measure O.D. of the ball nut and I.D. of the gear case. If clearance exceeds 0.015 mm, renew ball nut or gear case. Specified O.D. of ball nut is 70 mm. Using a dial indicator check end play of worm shaft in ball nut. If end play exceeds 0.10 mm, worm shaft/ball nut assembly must be renewed. Check rotation of worm shaft in ball nut. If roughness is present, renew worm shaft/ball nut assembly.

NOTE: Do not disassemble the ball nut and worm shaft.

Measure worm shaft at its widest diameter, if less than 28.475 mm, renew worm shaft/ball nut assembly.

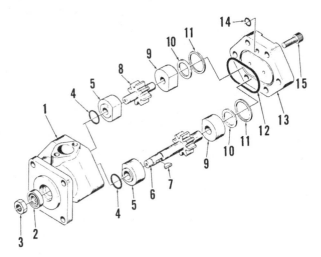

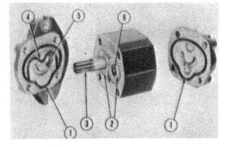

Fig. 29 – Exploded view of hydraulic pump used on Model 284 tractors equipped with non-diesel engines.

1. Pump body
2. Oil seal
3. Nut
4. "O" ring
5. Bushing
6. Drive gear
7. Key
8. Driven gear
9. Bushing
10. Support ring
11. "O" ring
12. "O" ring
13. Cover
14. "O" ring
15. Bolt

Fig. 30 – View of partially disassembled hydraulic pump used on all Model 274 and Model 284 tractors equipped with diesel engines.

1. Gasket
2. Bushings
3. Drive gear (shaft)
4. Cover seal
5. Shaft seal
6. Driven gear

Remove spool (17), reaction pistons (18 and 20) and centering springs (19) from valve body (28). Inspect edges of valve body sleeve and spool for nicks, burrs or other wear and damage. Slight nicks or burrs can be dressed with crocus cloth. Measure I.D. of valve body sleeve and O.D. of spool and if clearance exceeds 0.025 mm, valve body and spool must be renewed as an assembly. If reaction pistons do not operate smoothly in their bores, renew reaction pistons.

Check centering springs against following specifications:

Small Springs

Free length – mm 22.97
Test length – mm 20.5
Test load – .... 78N (17.6 lbs.)

Large Springs

Free length – mm 16.66
Test length – mm 15.0
Test load – .... 167N (37.5 lbs.)

Renew any springs not meeting the above specifications.

Loosen locknut (27 – Fig. 31) on steering relief valve adjusting screw (25), then remove adjusting screw and locknut as an assembly. Remove spring (24) and poppet (23) and inspect for distortion or other damage. Check poppet seat in valve body (28) for deformation. Renew parts as necessary.

Clean all parts and lubricate with IH "Hy-Tran" oil before assembling. Install new oil seal (46), "U" packing (44) and needle bearing (43) into gear case (45). Soak Teflon ball nut ring (57) in 82 to 93° C water until soft and pliable. Install back-up "O" ring (58) on ball nut, then lubricate and install Teflon ring on ball nut. Using a ring compressor, compress Teflon ring until it shrinks to its normal diameter. Install needle bearing (12), if removed, oil seal (10) and snap ring (9) in rear cover (11). Install needle bearing (66), if removed, back-up ring (65), Teflon ring (prepared as previously described) and "O" ring (67) into bottom (front) cover (68). Insert worm shaft end into Teflon ring to assure proper seating when Teflon cools. Install back-up ring (36) and Teflon ring (35) (prepared as previously described) into gear case (45). Lubricate ball nut/worm shaft assembly and install in gear case. Install bottom (front) cover (68) into case and secure with snap ring (69).

NOTE: Make certain to install worm shaft through Teflon ring (35) before it cools.

Lubricate area around "U" packings with good grade lithium grease. Install new "O" ring (54), "U" packing (38) and needle bearing (39), if removed, into side cover (51). Install sector shaft (42) with adjusting screw (41) and adjuster (40), lining up center tooth of sector shaft with center groove in ball nut. Install side cover and tighten bolts to 37-57 N·m torque.

Assemble valve body with longer chamfer of spool (17) facing downward. Install reaction pistons (18 and 20) and centering springs (19). Install relief valve (items 23 through 27). Place lower guide rings (30 and 32) and thrust bearing (31) onto worm shaft. With new "O" ring (29) seated securely in valve body, and new "O" rings (37) installed in gear housing, place valve body over worm shaft.

Fabricate a 6mm steel "set plate" of dimensions shown in Fig. 32 to aid in compressing centering springs upon assembly of valve body. Place "set plate" over reaction pistons then using 14mm nuts as spacers, tighten "set plate" against top of valve body, thus compressing centering springs. Install upper guide rings (14 and 16) and thrust bear-

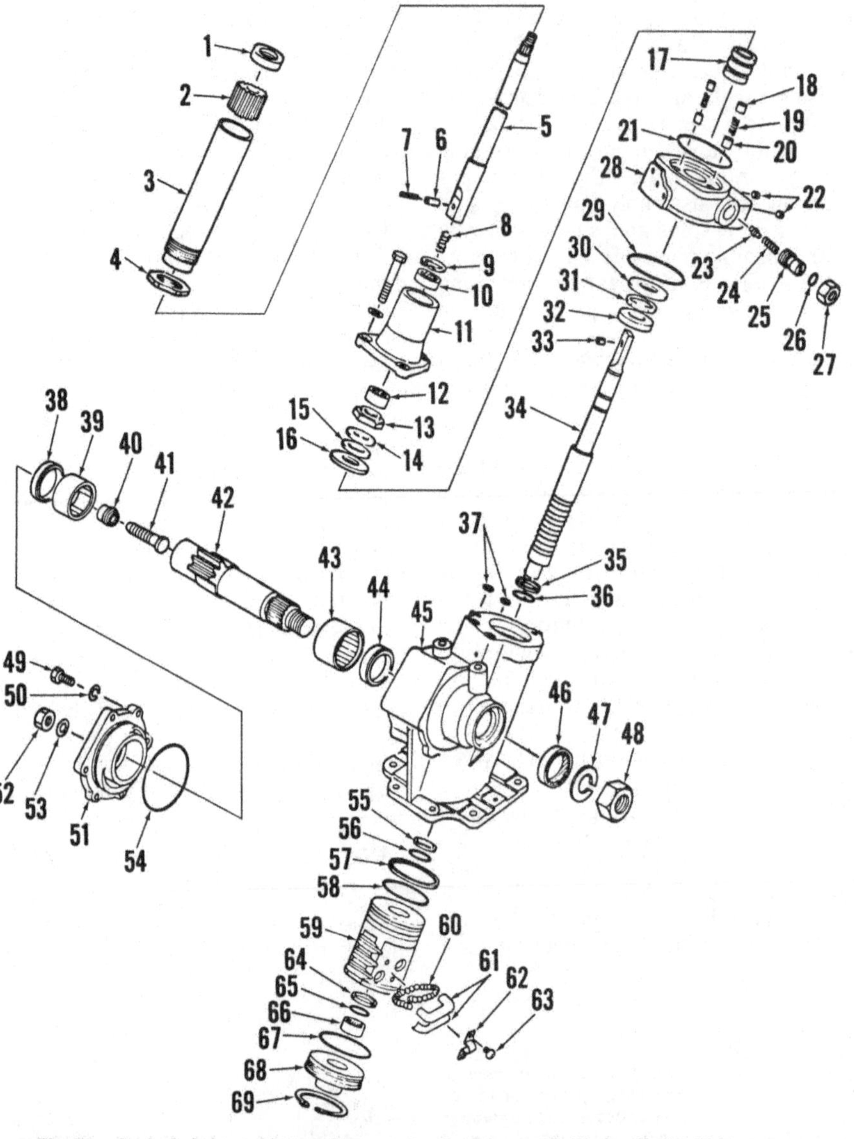

Fig. 31 – Exploded view of integral power steering gear available for Model 284 tractors.

1. Oil seal
2. Bushing
3. Column
4. Locknut
5. Steering shaft
6. Spring pin
7. Pin
8. Spring
9. Snap ring
10. Oil seal
11. Rear cover
12. Bearing
13. Nut
14. Guide ring
15. Thrust washer
16. Guide ring
17. Spool
18. Reaction piston

19. Centering spring
20. Reaction piston
21. "O" ring
22. Plug
23. Poppet
24. Spring
25. Adjusting screw
26. "O" ring
27. Locknut
28. Valve body
29. "O" ring
30. Guide ring
31. Thrust washer
32. Guide ring
33. "O" ring
34. Worm shaft
35. Teflon ring

36. Back-up ring
37. "O" rings
38. "U" packing
39. Bearing
40. Adjuster
41. Adjusting screw
42. Sector shaft
43. Bearing
44. "U" packing
45. Gear housing
46. Seal
47. Washer
48. Nut
49. Cap screw
50. Washer
51. Side cover
52. Locknut

53. Packing
54. "O" ring
55. Teflon seal
56. Back-up ring
57. Teflon seal
58. Back-up ring
59. Ball nut
60. Balls
61. Ball tube
62. Clamp
63. Screw
64. Teflon ring
65. Back-up ring
66. Bearing
67. "O" ring
68. Front cover
69. Snap ring

ing (15) over worm shaft. Turn worm shaft fully counter-clockwise and hold in this position with an adjustable wrench on the flat of the shaft, then tighten locknut (13) tight as possible by hand. Stake locknut to worm shaft. Remove "set plate", install "O" ring (21), then bolt rear cover (11) to valve body and tighten bolts to 37-57 N·m torque. Install double ribbed "O" rings (33) in worm shaft, then install steering shaft (5) with spring (8) and secure with roll pin (6) and split pin (7). Thread steering column (3) into rear cover and tighten locknut (4) to 98-127 N·m torque. Install pitman arm or sector shaft, set up a dial indicator on tip of pitman arm, and adjust sector shaft screw until pitman free play is 0.3 ot 0.4 mm at center of steering wheel travel. Install copper packing ring (53) and tighten locknut to 58-79 N·m torque. Reinstall steering gear onto tractor by reversing removal procedure as outlined in paragraph 35.

FLOW DIVIDER

Model 284 So Equipped

37. **REMOVE AND REINSTALL.** Remove shielding below gear shift lever. Clean area around flow divider and all hydraulic fittings. Disconnect fittings on flow divider and cap or plug immediately to prevent dirt or foreign substances from contaminating hydraulic system. Unbolt and remove flow divider.

Reinstall by reversing removal procedure. Use new seal washers on banjo

fittings. Bleed system as outlined in paragraph 26.

38. **OVERHAUL.** With flow divider removed as outlined in paragraph 37, refer to Fig. 33 and proceed as follows: Unbolt and remove end plate (5), then withdraw spool valve end plug (4), spring (2) and spool valve (1). Remove relief valve housing (8), poppet (9) and spring (10). Remove all plugs and fittings and discard "O" rings.

Clean and inspect all parts for scoring, nicks or burrs. Slight burrs can be dressed with crocus cloth. If valve body is scored, body and spool must be replaced as a unit. Check poppet and seat for damage. Renew all parts as necessary.

Assemble by reversing disassembly. Use new "O" rings and lubricate all parts with IH "Hy-Tran" fluid upon assembly. Reinstall flow divider as outlined in paragraph 37. Bleed power steering system as outlined in paragraph 26. Test and adjust flow divider and relief valve as outlined in paragraph 27.

ENGINE AND COMPONENTS (NON-DIESEL)

Model 284 non-diesel tractor is equipped with an engine having a bore and stroke of 70 x 76 mm and a displacement of 1169 cubic centimeters.

Engine is a four cylinder design, and is fitted with wet type cylinder sleeves. Dry type air filters are used in the air induction system.

R&R ENGINE AND CLUTCH ASSEMBLY

Model 284 Non-Diesel

40. To remove engine and clutch assembly, proceed as follows: Remove muffler, disconnect headlight wiring, then remove radiator support and engine hood with side panels. Disconnect and remove hydraulic return line from hydraulic pump and filter base. Cap or plug hydraulic lines or openings at once to prevent dirt or foreign material from contaminating hydraulic system. Disconnect all engine electrical connections. Disconnect throttle control rod, tachometer drive cable and choke control cable from carburetor. Shut off fuel at tank and disconnect fuel line at carburetor. Disconnect steering drag link from knuckle arm. Disconnect and remove high pressure hydraulic supply line from pump and flow divider. Cap or plug all openings. Wedge wooden blocks between engine side channel and front axle to prevent tipping. Place rolling floor jack under center section of tractor, support subframe with suitable jack stands.

NOTE: If tractor has front bolster weights, place jack stand under weights.

Support engine with sling and suitable hoist. Unbolt engine from center sec-

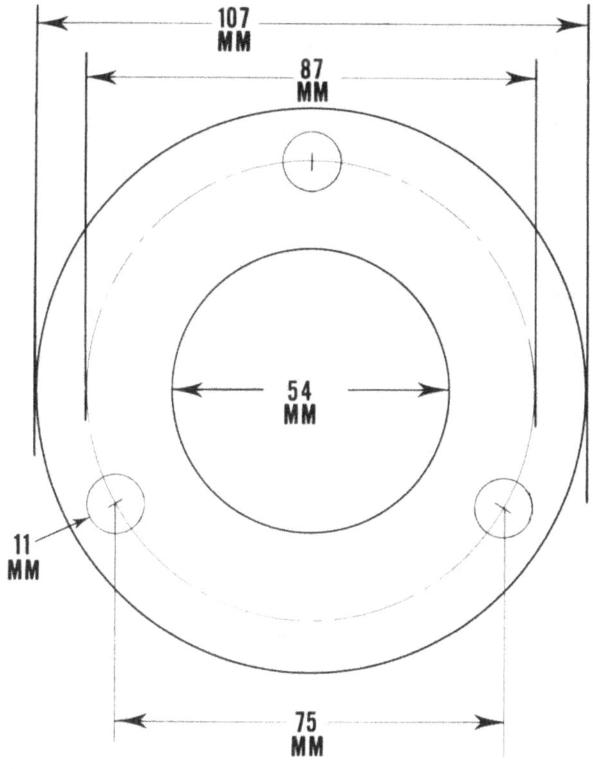

Fig. 32—Diagram showing dimensions of "set plate" needed for assembly of valve body on power steering gear.

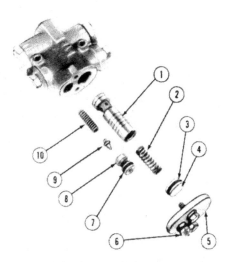

Fig. 33—Exploded view of priority valve used on Model 284 tractors equipped with power steering.

1. Spool valve	6. Relief valve adjusting screw
2. Spring	7. "O" ring
3. "O" ring	8. Relief valve housing
4. Spool valve end plug	9. Poppet
5. End plate	10. Spring

tion, then split tractor by rolling rear section rearward.

Drain cooling system, disconnect upper and lower radiator hoses, then remove radiator. Remove hydraulic pump drive from crank pulley. With engine supported, unbolt and remove engine from sub-frame. Note and retain engine and hydraulic pump mounting shims on sub-frame after engine removal.

Install engine by reversing removal procedure.

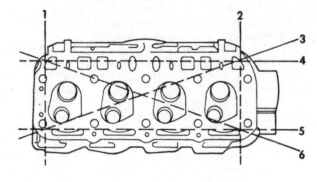

Fig. 35 — Illustration showing six areas to measure warpage of cylinder head on Model 284 non-diesel engines.

CYLINDER HEAD

Model 284 Non-Diesel

41. To remove cylinder head, first drain cooling system. Remove muffler, disconnect headlight wiring, then remove radiator grill and engine hood with side panels. Remove air cleaner and bracket assembly. Disconnect spark plug wires, unbolt coil bracket from rocker cover and lay coil and wiring harness out of the way. Disconnect upper radiator hose and bypass hose. Unbolt and remove fan shroud and fan assembly. Disconnect controls from carburetor, shut off fuel and disconnect fuel line to carburetor, then unbolt and remove inlet manifold and carburetor assembly. Unbolt and remove exhaust manifold. Remove rocker cover, rocker arms and shafts, and push rods. Remove cylinder head cap screws and lift off cylinder head.

Check cylinder head for warpage on combustion side with a straightedge and feeler gage. Warpage must not exceed 0.15 mm at six locations shown in Fig. 35. Renew cylinder head if more than 0.2 mm (maximum) surface removal is necessary to correct warpage. Standard distance from rocker arm cover surface to combustion surface of cylinder head is 76 mm.

When reinstalling cylinder head, use new head gasket and make certain that gasket sealing surfaces are clean and dry. **DO NOT** use sealant or lubricants on head gasket, cylinder head or block. When reassembling rocker arms and shafts, install longer push rods on exhaust valve side. Move exhaust side rocker arm supports so that valve stem-to-rocker arm center is offset 1 mm as shown in dimension (A – Fig. 36). Then temporarily tighten rocker support nuts. Install cylinder head hex nuts and tighten in sequence shown in Fig. 37 to 67 N·m torque. Balance of installation is reversal of removal procedure. Manifold retaining nuts should be tightened to 20 N·m torque. Adjust valve tappets as outlined in paragraph 42. Tighten rocker cover bolts to 2.2 N·m torque.

VALVES AND SEATS

Model 284 Non-Diesel

42. Inlet and exhaust valves are not interchangeable. Inlet and exhaust valves seats are not renewable. If valve seat requires refacing more than 1.5 mm from standard size for adequate reconditioning, cylinder head must be renewed. If

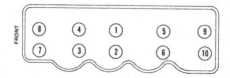

Fig. 36 — Illustration showing required rocker offset of exhaust valves on Model 284 non-diesel engines. "A" dimension is 1 mm.

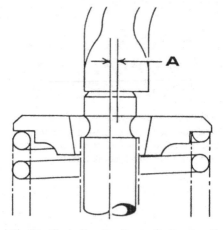

Fig. 37 — Illustration showing head bolt tightening sequence on non-diesel Model 284 tractors.

Fig. 38 — Chart shows valve tappet gap adjusting procedure used on Model 284 non-diesel engines. Refer to text.

valve or valve seat has been refaced several times or heavily for proper seating, valve position will sink below standard 40 mm depth. If valve sink exceeds 0.5 mm from standard depth, shim up valve spring with appropriate thickness washer. If valve sink exceeds 2.0 mm, renew valve or cylinder head. Valve face and seat angle for both inlet and exhaust is 45°.

Check valves and seats against the following specifications:

Inlet

Face and seat angle	45°
Stem Diameter (min)	6.95 mm
Stem to guide diametral clearance (max)	0.2 mm
Seat width	1.76 mm
Valve recession from face of cylinder head:	
Standard	40 mm
Maximum allowable	42 mm
Head diameter	33.9-34.1 mm

Exhaust

Face and seat angle	45°
Stem diameter (min)	6.95 mm
Stem to guide diametral clearance (max)	0.2 mm
Seat width	1.76 mm
Valve recession from face of cylinder head:	
Standard	40 mm
Maximum allowable	42 mm
Head diameter	28.9-29.1 mm

To adjust valve tappet gap, crank engine to position number one piston at top dead center (compression) and adjust four valves indicated on the chart shown in Fig. 38. Turn engine crank-

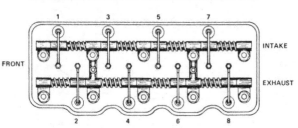

WITH	ADJUST VALVES (Engine Warm)							
No.1 Piston at T.D.C. (Compression)	1	2	3			6		
No.4 Piston at T.D.C. (Compression)				4	5		7	8

REPLACE HEAD GASKET IF ALL HEAD BOLTS ARE LOOSENED.

shaft one complete revolution to position number four piston at top dead center (compression) and adjust the remaining four valves indicated on chart. Set valve lash to 0.25 mm.

VALVE GUIDES AND SPRINGS

Model 284 Non-Diesel

43. The inlet and exhaust valve guides are not interchangeable. Inlet and exhaust guides are pressed into cylinder head until top of guide seal is 19.5 ± 0.1 mm above the spring recess of the cylinder head as shown in dimension (A – Fig. 39). Guides are pre-sized; however, since they are a press fit in cylinder head, it is necessary to ream them to remove any burrs or slight distortion caused by the pressing operation. Inside diameter should be 7.007-7.010 mm. Valve stem to guide diametral clearance should be a maximum of 0.2 mm.

Inlet and exhaust valve springs are interchangeable. Springs should have a free length of 39.3 mm and should test 207 N (46.3 lbs.) when compressed to a length of 35 mm. Renew any spring which is rusted, distorted or does not meet pressure test specifications.

NOTE: Inlet and exhaust valve keepers differ as shown in Fig. 40 and cannot be interchanged.

VALVE TAPPETS (CAM FOLLOWERS)

Model 284 Non-Diesel

44. The 21.959-21.980 mm diameter barrel type tappets operate directly in the unbushed bores of the top deck. Clearance of tappets in the bores should be 0.020-0.074 mm with a maximum allowable clearance of 0.1 mm in their bores. Tappets can be removed after removing cylinder head as outlined in

paragraph 41. Oversize tappets are not available.

VALVE ROCKER ARM COVER

Model 284 Non-Diesel

45. Removal of rocker arm cover is obvious after removal of front hood and air cleaner assemblies. Refer to cylinder head removal procedure outlined in paragraph 41.

When reinstalling rocker arm cover, use a new gasket to insure an oil tight seal. Torque rocker arm cover bolts to 2.2 N·m.

VALVE TAPPET LEVERS (ROCKER ARMS)

Model 284 Non-Diesel

Two rocker arm shafts are used to operate inlet and exhaust valves. Inlet valves operate on one shaft, exhaust valves operate on the other.

46. **REMOVE AND REINSTALL.** Removal of rocker arms and shaft assemblies is obvious after removal of rocker arm cover.

47. **OVERHAUL.** With rocker arm assemblies removed, refer to Fig. 41 and slide all parts from shafts. Keep all parts in order so they may be reassembled in their original locations. Clean and in-

spect all parts for excessive wear or damage. Measure rocker arm shafts and inside diameter of rockers. Shaft diameter should be 13.950-13.968 mm. Inside diameter of bore in rocker arms should be 14.0-14.018 mm with a maximum diametral clearance on rocker shaft of 0.1 mm. Renew parts as necessary.

Reassemble rocker arm assemblies by reversing disassembly procedure keeping the following points in mind. Be sure longer push rods are on exhaust valve side. Set exhaust side rocker arm supports so the valve stem-to-rocker arm center is offset by 1 mm as shown in Fig. 36. Set valve tappet gap as outlined in paragraph 42. Tighten rocker cover bolts to 2.2 N·m torque.

TIMING CHAIN COVER

Model 284 Non-Diesel

48. To remove timing chain cover, first disconnect battery cables, disconnect headlight wiring and remove hood and side panels. Drain cooling system, and remove radiator hoses. Remove radiator and fan shroud. Disconnect hydraulic lines from front mounted pump, then cap and plug all hydraulic openings immediately. Disconnect hydraulic pump drive and remove pump, noting number and thickness of pump mounting shims. Remove crankshaft pulley using a suitable puller. Remove

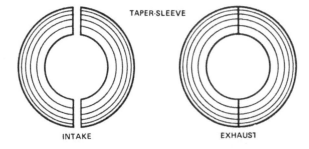

Fig. 40 – Illustration showing difference between intake and exhaust valve keepers on non-diesel Model 284 tractors.

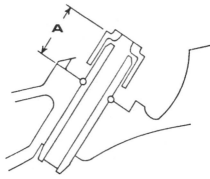

Fig. 39 – Illustration showing installation depth of valve guides in cylinder head of Model 284 non-diesel engines. Dimension "A" is 19.5 ± 0.1 mm.

Fig. 41 – Exploded view of rocker shaft and arm assembly on Model 284 non-diesel engine.

1. Shaft
2. Adjuster screw
3. Locknut
4. Gasket
5. Screw
6. Shaft support (oil gallery)
7. Support nut
8. Rocker arm (inlet)
9. Set spring
10. Shaft support
11. Washer
12. Rocker arm (exhaust)

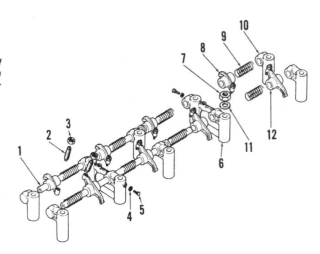

fan and water pump. Remove distributor from front cover. Remove nuts holding cover to crankcase. Remove governor supporting stud. Remove cover from engine. Reassemble by reversing removal procedure. Renew front oil seal prior to assembly as necessary. Use new gasket upon assembly. Torque front cover nuts to 17 N·m.

TIMING CHAIN AND SPROCKETS

Model 284 Non-Diesel

49. With cylinder head removed as outlined in paragraph 41 and timing chain cover removed as outlined in paragraph 48 proceed as follows: If not already done, remove cam followers. Remove oil pump drive gear and shaft assembly along with shims and thrust washer. Remove chain adjuster and vibration damper assemblies. Remove oil slinger and key from crankshaft. Unbolt and remove camshaft thrust plate. Remove camshaft and sprocket assembly, timing chain and crankshaft sprocket together.

Inspect timing chain, sprockets and related parts for excessive wear or damage and renew as necessary.

If front plate was removed, install with new gasket. Position number one piston at top dead center on compression stroke. Lubricate camshaft with clean engine oil. Align timing marks of chain and sprockets (1–Fig. 42) and position sprockets in chain. Install camshaft and sprocket, chain and crankshaft sprocket assembly rechecking alignment of timing marks. Rotate camshaft sprocket until keyway of crankshaft

Fig. 42 – View of timing marks (1) on timing chain and sprockets correctly aligned for Model 284 non-diesel engine. Refer to text for procedure.

aligns with keyway in crankshaft sprocket, then install key. Secure camshaft thrust plate. Install oil slinger on crankshaft. Secure timing chain vibration damper onto crankcase. Place new gasket on front plate and install chain adjuster. Balance of reassembly is reverse of disassembly. Use new gasket on front timing chain cover and torque nuts to 17 N·m.

CAMSHAFT AND BEARINGS

Model 284 Non-Diesel

50. The cast iron camshaft is supported in the crankcase by five non-renewable bearings that are integral with the crankcase. The tapered cam face rotates the cam followers during rotation. Remove camshaft as outlined in paragraph 49. With camshaft removed check camshaft runout. Standard runout is less than 0.01 mm with a maximum permissible runout of 0.03 mm. If runout exceeds maximum specifications, renew camshaft. Minimum allowable lobe height for both inlet and exhaust lobes is 36.389 mm. If worn beyond specifications, renew camshaft. Renew camshaft if journal diameters exceed the following minimum specifications:

Front	47.899 mm
Second	45.899 mm
Third	44.899 mm
Fourth	43.899 mm
Rear	42.899 mm

Measure camshaft journal bores in crankcase and camshaft journals to calculate running clearance. Renew camshaft or crankcase if maximum allowable running clearance exceeds 0.15 mm. With drive sprocket attached, check camshaft end play between front journal and thrust plate. Standard end play is 0.02-0.18 mm with a maximum allowable end play of 0.2 mm. Inspect distributor drive gear for excess wear or damage and renew as necessary. Inspect oil pump drive gear on camshaft for wear or damage and renew camshaft if necessary.

Installation is the reverse of removal procedure. Make sure timing marks are aligned as shown in Fig. 42.

PISTON AND ROD UNITS

Model 284 Non-Diesel

51. Connecting rod and piston assemblies can be removed from above after removing cylinder head as outlined in paragraph 41, oil pan as outlined in paragraph 60 and oil pump as outlined in paragraph 58. Remove any carbon ridge on cylinder edge before removing piston and rod assembly.

Mark piston and connecting rod

before disassembly for aid in reassembly. Piston can be heated to 50-60° C for piston pin removal. Check connecting rods for bending or distortion. Maximum allowable distortion is 0.4 mm per 100 mm of length. When renewing connecting rods, install rod of same weight code (A to K) as marked on side face of rod bearing cap. When renewing piston pin bushing, align oil hole of bushing with oil hole in connecting rod. Upon reassembly of piston and connecting rod, align "F" mark on piston with oil hole on connecting rod. When installing piston and rod assembly, "F" mark on piston must face front of engine. Do not install rings with gaps on thrust sides of piston. Tighten rod bolts to 41 N·m torque.

PISTONS, SLEEVES AND RINGS

Model 284 Non-Diesel

52. Pistons are available only in sets of four. New pistons have a diametral clearance in new sleeves of 0.030 and 0.069 mm when measured between piston skirt and sleeve at 90 degrees to piston pin.

The wet type cylinder sleeves should be renewed when out-of-round or taper exceeds 0.15 mm. Inside diameter of new sleeve is 70 mm. Maximum allowable variation between cylinder sleeves is 0.02 mm. Cylinder sleeves can usually be removed by bumping them from the bottom, with a wood block.

Before installing new sleeves, thoroughly clean counterbore at top of deck and seal ring groove at bottom. All

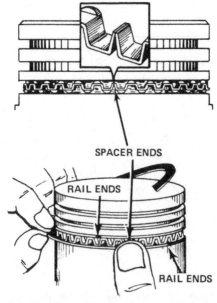

Fig. 43 – Illustration showing correct installation of oil control spacer and rail rings on non-diesel Model 284 tractors.

sleeves should enter crankcase bores full depth when tried in bores without seal rings. After making a trial installation without seal rings, remove sleeves and install new seal rings, dry, in grooves in crankcase. Wet lower end of sleeve with a soap solution or equivalent and install sleeve. Sleeve flange should extend 0.015 to 0.07 mm above deck. If sleeve standout is excessive, check for foreign material under sleeve flange.

Pistons are fitted with two compression rings and one oil control ring. Prior to installing new rings, check clearance between rings and ring grooves in piston against the following specifications:

Ring Clearance in Groove:
Top compression . . . . . 0.035-0.07 mm
Second compression . . 0.03-0.064 mm
Width of Piston Groove
for Oil Control Ring . . 4.020-4.032 mm
Width of Ring:
Top compression . . . . . . 1.97-1.99 mm
Second compression . . . 1.97-1.99 mm
Oil control rails . . . . 0.6 mm
Ring Gap:
Top and second
compression . . . . 0.2-0.4 mm
Oil control rails . . . . 0.3-0.9 mm
Maximum allowable ring
gap (in cylinder) . . . . 1.0 mm

Install oil control spacer ring with ends mating upward as shown in Fig. 43. Install oil control rail rings so that ring ends are 30 to 40 degrees to each side of spacer ends as shown in Fig. 43. Position compression rings so that end gaps are 120 degrees apart from each other and away from the thrust sides of the piston.

PISTON PINS

Model 284 Non-Diesel

53. The full floating type piston pins are retained in the piston by snap rings.

Specifications are as follows:
Piston pin diameter 19.984-19.993 mm
Piston pin length . . . . 54.95-55.05 mm
Piston pin diametral clearance
in rod bushing . . . . . 0.010-0.030 mm
Maximum clearance in rod
bushing . . . . 0.3 mm
Piston pin bushings are furnished semi-finished and must be reamed or honed for correct pin fit after they are pressed into connecting rods.

CONNECTING RODS AND BEARINGS

Model 284 Non-Diesel

54. Connecting rod bearings are of the slip-in precision type, renewable from below after removing oil pan, oil pump and connecting rod cap. Index rods and caps before removing. When installing new bearing inserts, make certain projections on same engage slots in connecting rod and that oil holes in bearing inserts align with oil holes in connecting rod and cap. Bearing caps install only one way on connecting rods, but be certain not to intermix connecting rod caps with rods. Connecting rod bearings are available in standard size and undersizes of 0.25, 0.50 and 0.75 mm. Check crankpins and connecting rod bearings against the following values:

Crankpin diameter . 44.940-44.955 mm
Crankpin journal
width . . . . 24.500-24.552 mm
Rod bearing diametral
clearance . . . . 0.027-0.073 mm
Maximum . . . . 0.1 mm
Rod side clearance . . . 0.110-0.214 mm
Maximum . . . . 0.3 mm

CRANKPIN

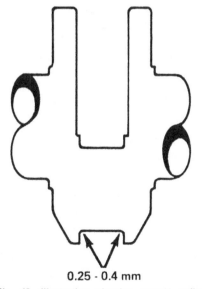

0.25 - 0.4 mm

Fig. 45 — Illustration showing correct radius finish grind for crankshaft crankpins on Model 284 non-diesel tractors.

Rod bolt torque . . . . 41 N·m
If bearing clearance is not within specifications, the crankshaft must be reground and undersize bearings installed. Refer to Figs. 44 and 45 for regrind and undersize specifications.

CRANKSHAFT AND MAIN BEARINGS

Model 284 Non-Diesel

55. Crankshaft is supported in five main bearings and crankshaft end play is controlled by thrust washers on each side of the rear main bearing. Thrust washer sets are available in standard and oversizes of 0.25, 0.50 and 0.75 mm.

NOTE: When installing thrust washers, oil grooves must face away from the bearing inserts.

Main bearings are of the non-adjustable, slip-in, precision type, renewable from below after removing oil pan, oil pump and main bearing caps. Removal of crankshaft requires removal of engine. Check crankshaft and main bearings against the values which follow:

Crankpin diameter . 44.940-44.955 mm
Main journal
diameter . . . . 55.944-55.959 mm
Crankpin width . . . 24.500-24.552 mm
Main journal width
No. 1 . . . . 25 mm
No. 2, 4 . . . . 26 mm
No. 3 . . . . 28 mm
No. 5 . . . . 29-29.052 mm
Maximum allowable out-of-round
of journals . . . . 0.041-0.056 mm
Main bearing diametral
clearance . . . . 0.025-0.050 mm
Maximum . . . . 0.08 mm
Maximum allowable crankshaft
runout . . . . 0.03 mm

REGRIND FINISH DIMENSIONS

	Crankpin Diameters
Undersize	mm
0.25mm	44.690 – 44.705
0.50mm	44.440 – 44.455
0.75mm	44.190 – 44.205

Fig. 44 — Chart showing correct regrind diameters of crankshaft crankpins for rod bearing undersizes.

REGRIND FINISH DIMENSIONS

	Crankshaft Main Journal Diameters
Undersize	mm
0.25mm	55.694 – 55.709
0.50mm	55.444 – 55.459
0.75mm	55.194 – 55.209

Fig. 46 — Chart showing correct regrind diameters of crankshaft main journals for undersize main bearings.

Crankshaft end play .0.110-0.274 mm
   Maximum . . . . .0.3 mm
   Main bearing bolt torque . . . . .61 N·m
If bearing clearance is not within specifications, the crankshaft must be reground and undersize bearings installed. Refer to Figs. 46 and 47 for regrind and undersize specifications.

CRANKSHAFT SEALS

Model 284 Non-Diesel

56. **FRONT.** To renew the crankshaft front oil seal first drain engine coolant from radiator. Disconnect headlight wiring then remove hood, radiator support and side panels. Disconnect radiator hoses and remove radiator and fan shroud. Disconnect hydraulic lines from front mounted pump, then cap and plug all hydraulic openings immediately. Disconnect hydraulic pump drive and remove pump, noting number and thickness of pump mounting shims. Remove crankshaft pulley using a suitable puller. Remove and renew oil seal in conventional manner.

Assembly is reverse of disassembly procedure. Tighten crankshaft pulley nut to 88 N·m torque.

56A. **REAR.** To renew the crankshaft rear oil seal, engine must be detached from center section as follows: Remove muffler, disconnect headlight wiring then remove radiator grill and engine hood with side panels. Disconnect and remove hydraulic return line from hydraulic pump and filter base. Cap or plug hydraulic lines or openings at once to prevent dirt or foreign material from

contaminating hydraulic system. Disconnect all engine electrical connections. Disconnect throttle control rod, tachometer drive cable and choke control cable from carburetor. Shut off fuel at tank and disconnect fuel line at carburetor. Disconnect steering drag link from knuckle arm. Disconnect and remove high pressure hydraulic supply line from pump and flow divider. Cap or plug all openings. Wedge wooden blocks between engine side channel and front axle to prevent tipping. Place rolling floor jack under center section of tractor, support sub-frame with suitable jack stands.

NOTE: If tractor has front bolster weights, place jack stand under weights.

Unbolt engine from center section, then split tractor by rolling rear section rearward.

Index clutch assembly to flywheel, unbolt and remove clutch assembly, then unbolt and remove flywheel.

Remove and renew rear crankshaft oil seal in conventional manner being careful not to mar the seal contact surface in the crankcase.

Reinstall flywheel as outlined in paragraph 57. Reinstall clutch assembly as outlined in paragraph 124. The balance of reassembly is the reverse of disassembly procedure.

FLYWHEEL

Model 284 Non-Diesel

57. To remove flywheel, first split tractor as outlined in paragraph 56. Index clutch assembly to flywheel, then

unbolt and remove clutch assembly. Remove cap screws and lift flywheel from crankshaft. Inspect flywheel for excessive wear or damage. Inspect flywheel teeth for damage and pilot bearing for smooth turning. Repair or renew as necessary.

When reinstalling flywheel on non-diesel tractors with engine S.N. 29396 and below, position flywheel on crankshaft and align reamer bolt hole on crankshaft with "O" mark on the flywheel. Install the longer reamer bolt into this hole then install other five bolts and torque them to 87 N·m.

On non-diesel tractors, engine S.N. 29397 and above, flywheel mounting bolts are all equal length and there is no "O" mark on flywheel. Torque these bolts to 87 N·m.

Reinstall clutch as outlined in paragraph 124. Balance of assembly is reverse of disassembly.

OIL PUMP

Model 284 Non-Diesel

58. The internal trochoid (gerotor) type oil pump is located in the right center of crankcase and is driven by the center camshaft gear. Lubrication is circulated to all engine parts as shown in Fig. 48. Timing chain and gears are lubricated by an oil jet at front of the crankcase. To remove the oil pump, first remove oil pan as outlined in paragraph 60. Disconnect delivery tube (10—Fig. 49) from pump body (14), then unbolt and remove pump, drive shaft (9) and pick-up tube (15) from below as an assembly. Removal of pump drive gear

MAIN JOURNAL

0.25 - 0.4 mm

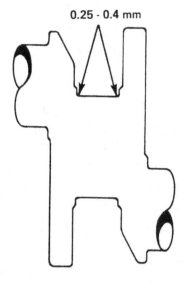

Fig. 47—*Illustration showing correct radius finish grind for crankshaft main journals on Model 284 non-diesel tractors.*

Engine Lubrication System

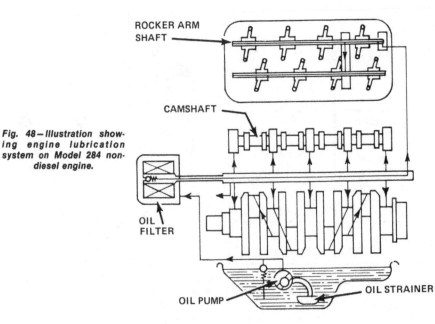

Fig. 48—*Illustration showing engine lubrication system on Model 284 non-diesel engine.*

ROCKER ARM SHAFT
CAMSHAFT
OIL FILTER
OIL PUMP
OIL STRAINER

(1 – Fig. 49) will require removal of cylinder head as outlined in paragraph 41. Clean all parts and inspect for wear as follows: Check clearance between outer rotor (13) and body (5) with a feeler gage as shown in Fig. 50. Allowable clearance is 0.3 mm. Inner (12 – Fig. 49) and outer (13) rotors are available only as a matched set. Refer to Fig. 51 and check outer rotor to pump body face clearance using a straightedge and feeler gage. Refer to Fig. 52 and check for excessive wear on pump body cover using a straightedge and feeler gage. If clearance on either exceeds 0.15 mm, lap or renew pump body.

Remove split pin (4 – Fig. 49) and remove retaining washer (8), spring (7) and plunger (6). Inspect plunger for scoring or burrs and renew as necessary. If spring free length is less than 55 mm, renew spring.

Additional pump specifications are as follows:

Shaft clearance in body . . . . . . 0.1 mm
Shaft end play
    (cover installed). . . . . . . 0.05-0.1 mm
Oil pressure
    at 3000 rpm . . . . 392 kPa (57 psi)

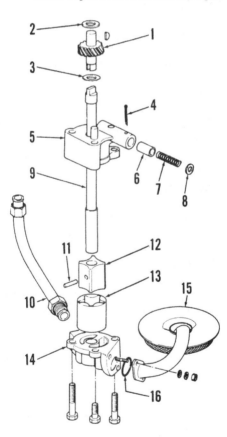

Fig. 49 – Exploded view of oil pump assembly used on non-diesel engine.

1. Driven gear	9. Shaft
2. Washer	10. Delivery tube
3. Shim	11. Pin
4. Split pin	12. Inner rotor
5. Body	13. Outer rotor
6. Plunger	14. Pump cover
7. Spring	15. Strainer
8. Seat	16. "O" ring

When reassembling pump, align pump marks of inner and outer rotors. Balance of assembly is reverse of disassembly. With pump installed in crankcase, check shaft end play between pump drive gear and pump drive cover as shown in Fig. 53. Specified clearance is 0.05 to 0.1 mm. Washers (2 – Fig. 49) are available in thicknesses of 1.5, 2.0, 2.3 and 2.5 mm. Shim (3) is available in 0.1 mm thickness. Use appropriate shims and washers to achieve proper pump shaft end play. Tighten pump drive cover nuts to 9 N·m torque. Renew "O" ring (16 – Fig. 49) upon reassembly.

OIL PRESSURE REGULATOR

Model 284 Non-Diesel

59. The oil pressure regulator valve is located in the internal oil pump. See Fig. 49 and refer to paragraph 58 for overhaul procedure and spring specifications. Filter by-pass valve is located in spin-on type oil filter. Oil pressure at 3000 rpm should be 392 kPa (57 psi).

OIL PAN

Model 284 Non-Diesel

60. Removal of oil pan is obvious after removal of tie rod and examination of

Fig. 50 – Checking clearance between outer rotor and pump body using a feeler gage. Refer to text.

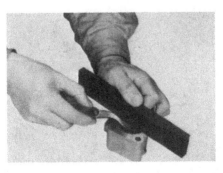

Fig. 51 – Checking clearance between rotor and face of pump body using a straight edge and feeler gage. Refer to text.

unit. Renew gasket when reinstalling and tighten oil pan nuts to 8 N·m torque.

ENGINE AND COMPONENTS (DIESEL)

Model 274 and Model 284 diesel tractors are equipped with an engine having a bore and stroke of 83 x 100 mm and a displacement of 1623 cubic centimeters.

Engines are a three cylinder design. Engines are equipped with dry sleeves. Dry type air filters are used in the air induction system.

R&R ENGINE WITH CLUTCH ASSEMBLY

Models 274 and 284 Diesel

61. To remove engine and clutch as a unit, first disconnect battery and drain cooling system. Remove muffler, disconnect headlight wiring, then remove front and rear hoods, radiator grill and grill support. Remove air cleaner assembly and support bracket. Identify and disconnect all electrical connections to engine. Shut off fuel at tank and disconnect fuel supply hose to feed pump and fuel return hose at fuel filter. Plug or cap hoses and injection lines to prevent

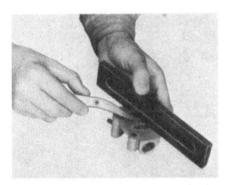

Fig. 52 – Checking wear of pump cover using a straight edge and feeler gage. Refer to text.

Fig. 53 – Checking end play clearance of pump shaft driven gear using a feeler gage. Refer to text.

contamination. Disconnect fuel shut-off cable and governor control rod at injection pump. Disconnect tachometer cable at drive assembly. Disconnect hydraulic lines at hydraulic pump and cap or plug all openings to prevent contamination. On Model 284 tractors, disconnect steering drag link at pitman arm. On Model 274 tractors, disconnect steering column at steering box. On All-Wheel Drive equipped 284 models, loosen set screw retaining drive shaft cover pipe to front axle rear support housing. Remove exhaust pipe at exhaust manifold. Support tractor center section with rolling floor jack. Support front section with sling and suitable hoist.

NOTE: If tractor is equipped with front bolster weights, support weights with suitable jack stands.

Wedge wood blocks between axle and sub-frame to prevent tipping. Remove bolts securing center section to engine and split tractor by rolling rear section rearward. Place suitable jack stands under sub-frame. Disconnect upper and lower radiator hoses, then remove radiator, air cleaner pipes and fan shroud as an assembly. Remove clutch assembly, starter motor and flywheel from engine. Remove clutch housing adaptor and rear engine plate. Attach suitable hoist to engine lift brackets, then unbolt and remove engine from sub-frame. Note and retain engine mounting shims.

When reassembling, reverse removal procedure.

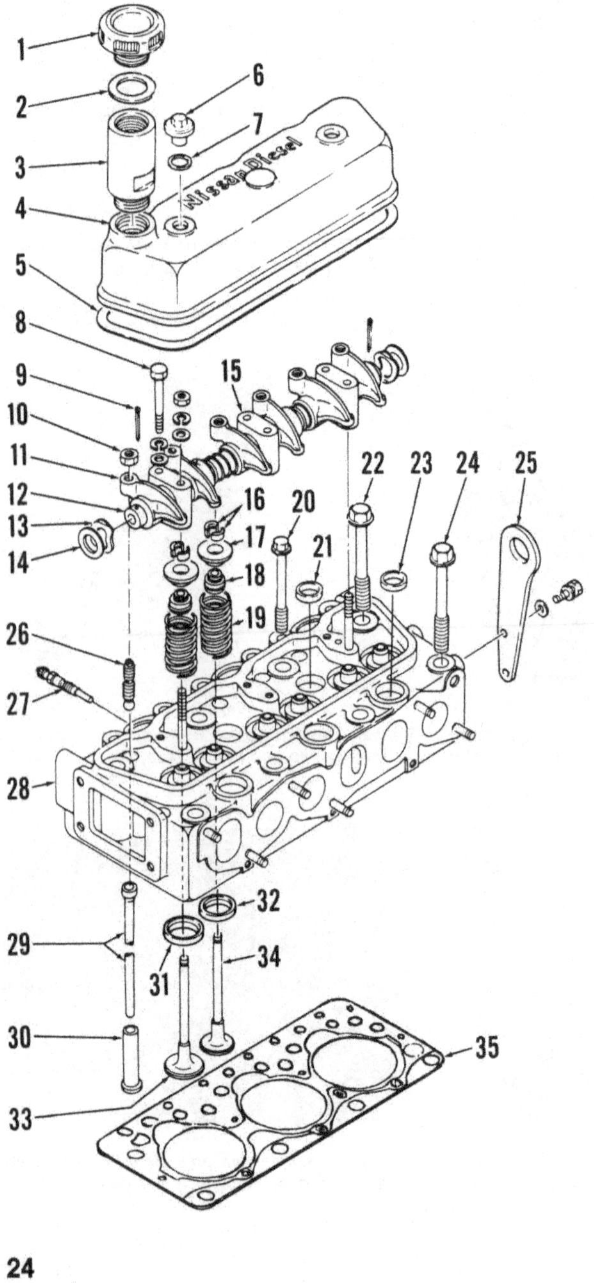

Fig. 54 — Exploded view of cylinder head and related parts on diesel engine equipped Models 274 and 284 tractors.

1. Filler cap
2. Gasket
3. Filler tube
4. Rocker cover
5. Gasket
6. Cover nut
7. "O" ring
8. Bolt
9. Split pin
10. Nut
11. Rocker arm
12. Rocker shaft
13. Outside spring
14. Washer
15. Bracket
16. Valve keepers
17. Retainer
18. Seal
19. Spring
20. Sub bolt
21. Plug
22. Head bolt
23. Plug
24. Head bolt
25. Lifting eye
26. Adjusting screw
27. Glow plug
28. Cylinder head
29. Push rod
30. Lifter
31. Inlet valve seat
32. Exhaust valve seat
33. Inlet valve
34. Exhaust valve
35. Head gasket

CYLINDER HEAD

Models 274 and 284 Diesel

62. To remove cylinder head, remove muffler, front and rear hoods. Disconnect headlight wiring and remove radiator grill, then remove radiator support and side skirt panels. Close fuel shut off valve at tank, then disconnect fuel return line at fuel filter. Disconnect and remove injection lines at injectors and injection pump. Disconnect leak-off line on injectors, then disconnect and remove leak-off return line at fuel filter. Disconnect and remove return and supply lines from injection pump and filter. Cap or plug all fuel openings and lines to prevent contamination. Remove injectors. Unbolt and remove fuel filter and mounting bracket assembly. Disconnect glow plug wiring. Disconnect water temperature unit wiring. Remove air intake hose. Remove exhaust pipe, then remove inlet and exhaust manifolds. Drain cooling system, then remove upper radiator hose and thermostat housing. Remove rocker arm cover, then remove rocker arm assembly and push rods. Remove the eight cylinder head bolts and the six sub-bolts and lift cylinder head from engine.

Remove carbon deposits and check cylinder head for warpage on combustion side with a straightedge and feeler gage. Warpage must not exceed 0.20 mm at six locations shown in Fig. 55. Renew cylinder head if more than 0.3 mm (maximum) surface removal is necessary to correct warpage. Standard cylinder head thickness is 90 mm. If cylinder head is resurfaced to correct warpage, be sure that head of valve is recessed 0.75 mm from gasket surface of cylinder head. Swirl chamber inserts should only be removed if burned, cracked or otherwise in need of renewal. Swirl chamber inserts should fit tight in head and should be retained by pin. Swirl chamber inserts can be driven out through injector nozzle hole using a 6 mm punch. Be certain not to damage seating surface. After cleaning bore, install swirl chambers by chilling with dry ice or freon, then tap in place with soft mallet, making certain to align locating

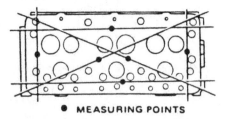

• MEASURING POINTS

Fig. 55 — Illustration showing points of measurement to check cylinder head warpage. Refer to text.

pin with slot in bore. Check for tightness after installation.

When reinstalling cylinder head, use new head gasket and make certain gasket sealing surfaces are clean and dry. Before installing cylinder head and new gasket, renew the eight rubber rings in locations shown in Fig. 56.

NOTE: Do not use sealants or lubricants on head gasket.

Install cylinder head and gasket, lubricate head bolts and tighten in sequence shown in Fig. 57 in two stages as follows:

First stage
Main bolt . . . . .58 N·m
Sub-bolt . . . . .28 N·m
Second stage
Main bolt . . . . .127 N·m
Sub-bolt . . . . .49 N·m
Balance of installation is reversal of removal procedure. Manifold retaining nuts should be tightened to 15-18 N·m. Rocker arm shaft bracket bolts and nuts should be tightened to 15-18 N·m. Adjust valve tappets as outlined in paragraph 63.

VALVES AND SEATS

Models 274 and 284 Diesel

63. Inlet and exhaust valves are not interchangeable and seat on renewable seat inserts. When renewing valve seats, chill with dry ice before installing. Press seats into cylinder head and stake at five locations shown in Fig. 58. Remove carbon from combustion side of cylinder head and measure valve recession. If valve recession exceeds 0.75 mm, renew valve and/or seat. Renew valve guide seals when new valves are installed or any valve work is done.

Check valves and seats against the following specifications:

Inlet
Face and seat angle . . . . .45°
Stem diameter . . . . .7.97-7.99 mm
Stem to guide diametral
    clearance (max) . . . . .0.15 mm

Fig. 56—View showing locations of rubber rings that should be renewed when cylinder head is removed.

Valve recession from face
    of cylinder . . . . .0.75 mm
Valve head thickness . . . . .1.3 mm
Valve contact to seat width 0.8-1.4 mm
Valve lift . . . . .9.1 mm

Exhaust
Face and seat angle . . . . .45°
Stem diameter . . . . . . .7.94-7.96 mm
Stem to guide diametral
    clearance (max) . . . . .0.20 mm
Valve recession from face
    of cylinder . . . . .0.75 mm
Valve head thickness . . . . .1.3 mm
Valve contact to seat width 0.8-1.4 mm
Valve lift . . . . .9.1 mm

To adjust valve tappet gap, crank engine to position number one piston at top dead center of compression stroke and adjust four valves indicated on chart shown in Fig. 60. Turn engine crankshaft one complete revolution to position number one piston at top dead center of exhaust stroke and adjust two remaining valves indicated on chart. Set valve lash to 0.3-0.4 mm, engine hot or cold.

VALVE GUIDES AND SPRINGS

Models 274 and 284 Diesel

64. Inlet and exhaust valve guides are

Fig. 57—Tighten cylinder head retaining bolts in sequence shown. Refer to text.

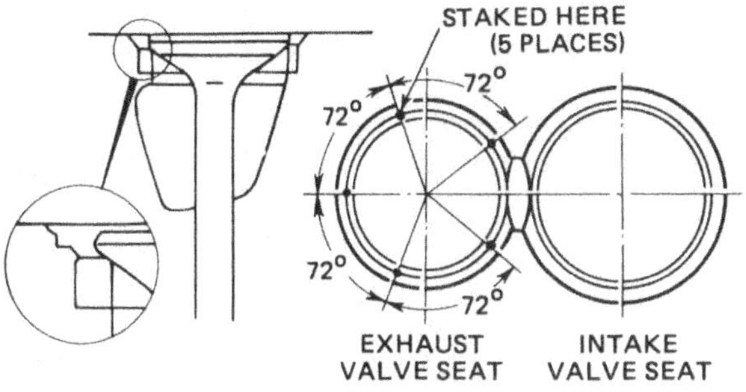

Fig. 58—When renewing valve seats, stake at points shown.

Fig. 60—Chart showing valve tappet gap adjusting procedure on diesel engine equipped Model 274 and 284 tractors. Refer to text.

not renewable. Standard bore size is 8.0 to 8.02 mm. Renew cylinder head if valve guides exceed bore dimensions. All valves are equipped with cup type valve stem seals which should be renewed whenever valves are serviced.

Valve springs are interchangeable for inlet and exhaust valves. Renew springs which are distorted, heat discolored or fail to meet test specifications which follow:

Standard free length . . . . .50.2 mm
Minimum free length . . . . . . .48.0 mm
Test pressure N. at mm . . .280-315 at 39
Minimum test load . . . . .254 N.

VALVE TAPPETS (CAM FOLLOWERS)

Models 274 and 284 Diesel

65. The 12.67-12.68 mm mushroom type tappets operate directly in the unbushed bores of crankcase. Clearance of tappets in the bores should be 0.10 mm maximum. Tappets can be removed from below after removal of camshaft as outlined in paragraph 76. Check surface of tappets that contacts camshaft and renew tappets if worn. Also check camshaft carefully if tappet shows wear and

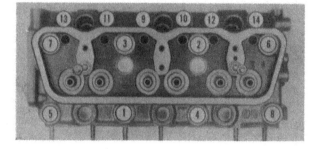

STAKED HERE
(5 PLACES)
72°   72°
72°
72°
EXHAUST VALVE SEAT   INTAKE VALVE SEAT

With	Adjust Valves			
No. 1 Piston at TDC (Compression)	1	2	4	6
No. 1 Piston at TDC (Exhaust)		3	5	

renew camshaft if chipped, broken, scored or otherwise worn. Oversize tappets are not available.

VALVE ROCKER ARM COVER

Models 274 and 284 Diesel

66. Removal of rocker arm cover is obvious after removal of front hood. When reinstalling rocker arm cover, use new gasket to insure an oil tight seal. Torque rocker arm cover nuts to 2.2 N·m.

VALVE TAPPET LEVERS (ROCKER ARMS)

Models 274 and 284 Diesel

67. Removal of rocker arms and shaft assembly is obvious after removal of rocker arm cover. With rocker arm assembly removed, refer to Fig. 61 and remove split pin (1) and slide all parts from shaft. Keep all parts in order so they may be reassembled in their original locations. Clean and inspect all parts for excessive wear or damage. Measure rocker arm shaft and inside diameter of rockers. Shaft diameter should be 19.98-20.0 mm. Inside diameter of bore in rocker arms should be 20.02-20.04 mm with a maximum diametral clearance on rocker shaft of 0.15 mm. Rocker shaft runout should not exceed 0.3 mm. Renew any parts not meeting specifications.

Reassemble by reversing disassembly procedure. Tighten rocker arm assembly retaining nuts and bolts to 19-24 N·m torque. Adjust tappet gap as outlined in paragraph 63.

VALVE TIMING

Models 274 and 284 Diesel

68. Valves are properly timed when "X" and "Y" timing marks are aligned as shown in Fig. 62. Refer to the following additional specifications:

Inlet opens–degrees . . . . 28 BTDC
Inlet closes–degrees . . . . 67 ABDC
Exhaust opens–degrees . . . . . 67 BBDC
Exhaust closes–degrees . . . . . 28 ATDC

TIMING GEAR COVER

Models 274 and 284 Diesel

69. To remove timing gear cover, first disconnect battery cables and remove muffler, disconnect headlight wiring, then remove front hood, side panels and radiator support. Drain cooling system and remove radiator hoses. Remove air inlet hose, pipe and air cleaner assembly. Remove radiator. Disconnect tachometer drive cable. Disconnect water temperature wiring. Disconnect wire to alternator, then unbolt and remove alternator and bracket. Remove thermostat housing and water pump and fan assembly. Remove crank nut and using a suitable puller, remove crank pulley. Drain engine oil and remove oil pan. Remove front pto mounting bolts and support hydraulic pump. Remove remaining mounting bolts and carefully pry timing gear cover from engine.

Crankshaft oil seal can be renewed at this time. Reassembly is reverse of disassembly procedure. Use new cover gasket and oil pan gasket upon reassembly. Tighten crank pulley nut to 294-323 N·m torque.

TIMING GEARS

Models 274 and 284 Diesel

70. With timing gear cover removed as outlined in paragraph 69 proceed as follows: Refer to exploded view in Fig. 63 and check backlash between timing gears (1, 2 and 6) and front pto driven gear (5). Specified backlash between all gears should be 0.07-0.13 mm. If backlash between any two gears exceeds 0.30 mm, renew gears.

71. **INJECTION PUMP DRIVE GEAR.** Removal of pump timing gear (6 – Fig. 63) can be accomplished without removal of complete timing gear cover (14) by means of removal of automatic timer cover (10). After removal of automatic timer cover or complete timing gear cover, remove tachometer drive nut (11) and timer mounting nut (12) from timer assembly. Secure tachometer drive nut to a slide hammer using nuts smaller than thread diameter of drive nut. Screw tachometer drive nut fully into timer assembly and pull assembly from pump shaft. Reinstall timer pump drive gear assembly by pressing on pump shaft using timer mounting nut (12 – Fig. 63) to press assembly onto pump shaft while tightening. Torque timer mounting nut to 58-68 N·m. Torque tachometer drive nut (11) to 77-88 N·m. Make certain timing marks are aligned as shown in Fig. 62. If timing assembly cover (10) was removed, use new "O" ring (13) upon reinstallation.

72. **CAMSHAFT GEARS.** With timing gear cover removed as outlined in paragraph 69, camshaft end play should be checked before removing cam gears. Specified camshaft end play is 0.08-0.28 mm with maximum allowable end play of 0.50 mm. If end play exceeds 0.50 mm, renew camshaft thrust plate (24 – Fig. 63). To remove camshaft gears (2 and 3), first remove injection pump drive gear as outlined in paragraph 71. Remove mounting bolt (4) then remove gears using a suitable puller. Unbolt gear (3) from gear (2) if necessary taking

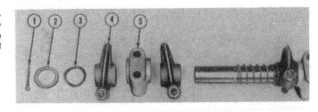

Fig. 61—View of disassembled rocker arm assembly used on diesel engine equipped Model 274 and 284 tractors.

1. Split pin
2. Washer
3. Outer spring
4. Rocker arm
5. Bracket

Fig. 62—View of timing gears with "X" and "Y" marks correctly aligned.

note of alignment pin. Press gears onto camshaft by tightening mounting bolt (4). Torque mounting bolt to 50 N·m. Reassembly is reverse of disassembly procedure being sure that timing marks align.

73. **CRANKSHAFT GEAR.** To remove crankshaft gear (1–Fig. 63) first remove pump drive gear as outlined in paragraph 71 and camshaft gears as outlined in paragraph 72. Remove crankshaft gear using a suitable puller. Heat new gear to 200° C before installing. Make certain timing marks align upon reassembly.

74. **FRONT PTO ASSEMBLY.** If not already done, remove hydraulic pump. With timing gear cover removed as outlined in paragraph 69, remove pto assembly from front engine plate (27–Fig. 63). Place front pto gear in brass jawed vice and remove nut (7), washer (8) and gear (5) from shaft (19). Remove snap ring (17) and withdraw shaft and bearings (items 18, 19 and 20) from pump adapter housing (21). Renew bearings as necessary and inspect splines in shaft (19). Reassemble by reversing disassembly procedure. Use new "O" rings (16 and 22). Tighten nut (7) to 107-127 N·m torque.

75. **ENGINE FRONT PLATE.** If necessary to remove engine front plate (27–Fig. 63), inspect for distortion. If distortion exceeds 0.20 mm renew front plate. Use new gasket (26) upon reassembly.

CAMSHAFT AND BEARINGS

Models 274 and 284 Diesel

76. To remove camshaft, first drain engine oil and remove engine as outlined in paragraph 61. Remove rocker arm assembly and push rods as outlined in paragraph 67. Remove crankshaft rear seal and flange. Remove timing gear cover. Remove pump timing gear assembly as outlined in paragraph 71. Refer to paragraph 85 and remove oil pump assembly. Unbolt thrust plate (24–Fig. 63), through access holes in camshaft gears, invert engine or prevent cam followers from dropping out of their bores, and withdraw camshaft and gear assembly.

With camshaft removed and drive gear attached, check end play between thrust plate and front journal. Specified end play is 0.08-0.28 mm. If end play exceeds 0.50 mm, renew thrust plate. Check camshaft runout. Specified runout is less than 0.03 mm. If runout exceeds 0.06 mm, renew camshaft. Specified lobe height is 37.28-37.32 mm. If worn beyond minimum allowable

height of 36.80 mm, renew camshaft. Renew camshaft if journal diameters are less than the following minimum specifications:

     Front . . . . .45.44 to 45.45 mm
     Center . . . . .43.90 to 43.91 mm
     Rear . . . . .41.22 to 41.23 mm
Remove cam plug from rear of engine.

Measure inner diameter of camshaft bushings and calculate running clearance. Renew camshaft and/or bushings if maximum allowable running clearance exceeds 0.15 mm. Standard clearance is 0.03 to 0.09 mm.

Install new bushings so that oil holes in bushings are in register with oil holes in crankcase. Camshaft bearings are pre-sized and should need no final sizing if carefully installed.

When installing cam plug at rear camshaft bearing, apply a light coat of sealing compound to edge of plug and bore.

Install camshaft by reversing the removal procedure. Tighten cam thrust plate bolts to 4-5 N·m torque. Make certain timing marks align as shown in Fig. 62.

PISTON AND ROD UNITS

Models 274 and 284 Diesel

77. Connecting rod and piston assemblies can be removed from above after removing cylinder head, oil pan and oil pump assembly as outlined in paragraphs 62, 87 and 85. Remove any carbon ridge on cylinder edge before removing piston and rod assemblies. Mark connecting rods and bearing caps as shown in Fig. 64 to aid in assembly. Mark pistons and rods for aid in reassembly. Piston can be heated to 50-60° C for piston pin removal. Check connecting rods for bending or distortion. Renew any connecting rod exceeding 0.03 mm bend or twist per 100 mm of length. When renewing any connecting rod, stamp new rod and cap as shown in Fig. 64. When installing rod and piston assemblies in engine, embossed number side of number one and three rods should face rear of engine. The embossed number side of number two rod should face front of engine. When installing new connecting rod bushing make sure oil holes align and ream new bushing to specified inside diameter of 26.03-26.04 mm. When assembling connecting rod and piston, heat piston to 50-60° C and tap piston pin in place. Install retaining rings on each side of piston pin. Make sure piston is installed on rod with double leaf pattern on piston top facing opposite the index number on rod large end as shown

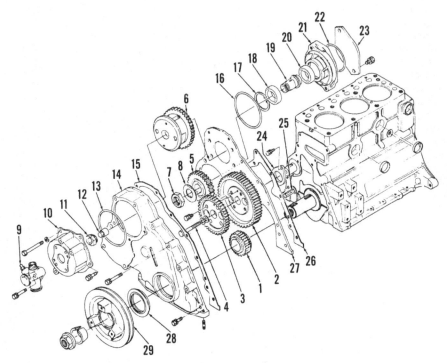

Fig. 63 – Exploded view of timing gear cover and related parts.

1. Crankshaft gear	8. Washer
2. Camshaft driven gear	9. Tachometer drive
3. Injection pump drive	housing
gear	10. Injection pump gear
4. Bolt	cover
5. Front PTO gear	11. Tachometer drive nut
6. Injection pump gear	12. Timer nut
7. Nut	13. "O" ring

14. Gear cover	22. "O" ring
15. Gasket	23. Cover
16. "O" ring	24. Thrust plate
17. Snap ring	25. Oil jet
18. Bearing	26. Gasket
19. Shaft	27. Front plate
20. Bearing	28. Seal
21. Housing	29. Pulley

in Fig. 65. When installing rod and piston assembly in cylinder block, double leaf pattern on piston top should face camshaft side of engine and marks on rod ends should face manifold side of engine.

PISTONS, SLEEVES AND RINGS

Models 274 and 284 Diesel

78. Pistons are available only in sets of three. New pistons have a diametral clearance in new sleeves of 0.09-0.20 mm when measured at a right angle to the piston pin, 50.5 mm down from top of piston. Diameter of piston should be 82.84-82.89 mm when measured at a right angle from piston pin, 50.5 mm down from piston top. If piston pin bore exceeds 26.00 mm, renew piston.

The dry type, pressed in cylinder liners should be renewed when I.D., out-of-round or taper exceeds 0.20 mm. Inside diameter of new liner is 82.98-83.04 mm. If greater than 83.20, renew liner.

Remove dry type cylinder liners using a 17-22 Hydraulic Cylinder Sleeve Puller/Installer or equivalent. With liners removed check cylinder block warp. If warpage exceeds 0.10 mm, grind cylinder block surface not to exceed 0.30 mm. Cylinder block height is 322.70 mm. If cylinder block specifications are exceeded, renew cylinder block. If cylinder liner bore in cylinder block exceeds 87.02 mm or taper and out-of-roundness exceeds 0.02 mm, renew cylinder block. Install new cylinder liners using 17-22 Hydraulic Cylinder Sleeve Puller/Installer or equivalent. Cylinder liner protrusion above cylinder block should be 0.02-0.09 mm.

Pistons are fitted with three compression rings and two oil control rings. Prior to installing new rings, check clearance between rings and ring grooves in piston against the following specifications:

Maximum Ring Clearance in Groove:
Top Compression . . . . 0.50 mm
Second and Third
Compression . . . . 0.30 mm
Oil Control (2) . . . . 0.15 mm
Check ring end gap with ring inserted in cylinder liner against the following specifications:
Ring Gap in Cylinder:
Top Compression . . . . . . 0.30-0.45 mm
Second and Third
Compression . . . . 0.20-0.35 mm
Oil Control (2) . . . . 0.15-0.30 mm
Maximum Allowable Ring Gap in Cylinder:
All Rings . . . . 1.50 mm
Install rings with gaps positioned as shown in Fig. 67.

PISTON PINS

Models 274 and 284 Diesel

79. The full floating type piston pins are retained in piston by snap rings. Piston can be heated to 50-60° C for pin removal and installation. Check piston, piston pin and rod bushings against the following specifications:
Piston pin diameter . . . . 25.99 mm
Piston pin diametral clearance
in rod bushing . . . . 0.03-0.05 mm
Maximum clearance in
rod bushing . . . . 0.10 mm
Piston pin bushing in rod
inside diameter . . . . 26.03-26.04 mm
Piston pins are furnished semi-finished and must be honed or reamed to an inside diameter of 26.03-26.04 mm after installation in connecting rods.

CONNECTING RODS AND BEARINGS

Models 274 and 284 Diesel

80. Connecting rod bearings are of the slip-in precision type, renewable from below after removing oil pan, oil pump and connecting rod caps. Mark connecting rod and cap with corresponding cylinder number on manifold side of rod and cap. When installing new bearing inserts, make certain projections on same engage slots in connecting rod. Connecting rod bearings are available in standard size and undersizes of 0.25, 0.50, 0.75 and 1.00 mm. Check crankpins and connecting rod bearings against the following specifications:
Crankpin diameter . . . 52.91-52.93 mm
Rod bearing diametral
clearance . . . . 0.04-0.10 mm
Rod side clearance . . . . . 0.10-0.20 mm
Rod bolt torque . . . . 50-55 N·m
If bearing clearance is not within specifications, the crankshaft must be reground and undersize bearings installed. Refer to Figs. 68 and 69 for regrind and undersize specifications.

CRANKSHAFT AND MAIN BEARINGS

Models 274 and 284 Diesel

81. Crankshaft is supported in four main bearings and crankshaft end play is controlled by thrust washers (8 and 23 – Fig. 66) on each side of the third main bearing (25). Thrust washer sets are available in sizes of 2.325, 2.350 and 2.375 mm.

NOTE: When installing thrust washers, oil grooves must face away from the bearing inserts.

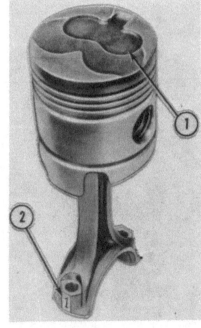

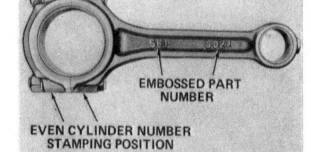

ODD CYLINDER NUMBER
STAMPING POSITION

EMBOSSED PART
NUMBER

EVEN CYLINDER NUMBER
STAMPING POSITION

Fig. 64 – View showing correct locations for indexing connecting rods on Models 274 and 284 diesel equipped tractors. Refer to text.

Fig. 65 – View showing correct position of piston to connecting rod index marks. Refer to text.

1. Double leaf pattern   2. Connecting rod large end mark

Main bearings are of the non-adjustable, slip-in, precision type, renewable from below after removing oil pan, oil pump and main bearing caps. Removal of crankshaft requires removal of engine. Check crankshaft and main bearings against the following values:

Crankpin diameter . . . 52.91-52.93 mm
Main journal diameter 70.90-70.92 mm
Main bearing diametral
   clearance . . . . 0.04-0.10 mm
   Maximum . . . . 0.15 mm
Crankshaft end play . . . 0.06-0.14 mm
   Maximum . . . . 0.40 mm

If bearing clearance is not within specifications, the crankshaft must be reground and undersize bearings installed. Refer to Figs. 68 and 69 for regrind and undersize specifications.

CRANKSHAFT SEALS

Models 274 and 284 Diesel

82. **FRONT.** To renew crankshaft front oil seal (28 – Fig. 63) first drain engine coolant. Remove muffler, disconnect headlight wiring, then remove hood, side panels and radiator support. Remove air cleaner hoses, tubes and filter assembly. Remove radiator hoses, then remove radiator and fan shroud. Remove alternator belt, then remove fan assembly. Remove crankshaft pulley nut, then using a suitable puller remove crankshaft pulley. Remove and renew oil seal in conventional manner. Coat lips of seal with clean engine oil before installing.

Reassemble by reversing disassembly procedure. Tighten crankshaft pulley nut to 294-323 N·m torque.

83. **REAR.** To renew crankshaft rear oil seal (21 – Fig. 66), engine must be detached from center section as follows: Disconnect battery, remove muffler, disconnect headlight wiring then remove front and rear hoods, radiator grill and radiator support. Identify and disconnect all electrical connections to engine. Shut off fuel at tank and disconnect fuel supply hose and return hose at tank. Disconnect fuel shut off cable and governor control rod at injection pump. Disconnect tachometer cable at drive assembly. Disconnect hydraulic lines at hydraulic pump and cap or plug all openings to prevent contamination. On Model 284 tractor, disconnect steering drag link at pitman arm. On Model 274 tractors, disconnect steering column at steering box. On All-Wheel Drive equipped 284 models, loosen set screw retaining drive shaft cover pipe to front axle rear support housing. Support tractor center section with rolling floor jack. Place suitable jack stands under subframe.

NOTE: If tractor is equipped with front bolster weights, support weights with suitable jack stands.

Wedge wood blocks between axle and sub-frame to prevent tipping. Remove bolts securing center section to engine and split tractor by rolling rear section rearward.

Remove clutch assembly, starter motor and flywheel. Remove clutch housing adapter and rear engine plate. Remove and renew rear crankshaft oil seal in conventional manner being careful not to mar the seal contact surface in seal housing. If seal housing (20 – Fig. 66) is removed, use new gasket (19) upon reinstallation. Reinstall flywheel as outlined in paragraph 84. Reinstall clutch assembly as outlined in paragraph 124. Balance of reassembly is reverse of disassembly procedure.

FLYWHEEL

Models 274 and 284 Diesel

84. To remove flywheel, first split tractor as outlined in paragraph 83. Index clutch cover to flywheel, then unbolt and remove clutch assembly. Remove mounting bolts and lift flywheel from

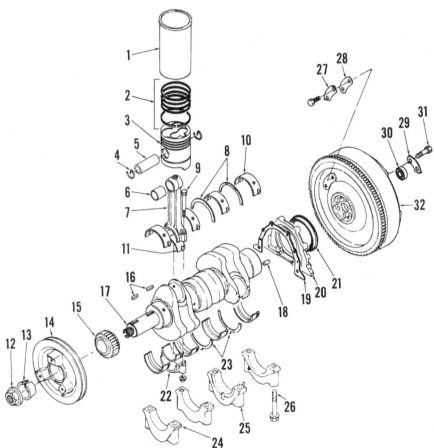

Fig. 66 – Exploded view of crankshaft and related parts.

1. Cylinder sleeve	10. Main bearing	18. Dowel	25. Thrust bearing cap
2. Piston rings	11. Rod bearing	19. Gasket	26. Bolt
3. Piston	12. Crankshaft nut	20. Rear seal cover	27. Lock plate
4. Retainer	13. Cone bushing	21. Seal	28. Counterweight
5. Piston pin	14. Pulley	22. Rod cap	29. Lock plate
6. Rod bushing	15. Crankshaft gear	23. Thrust washers	30. Bearing
7. Connecting rod	16. Keys	(lower)	31. Bolt
8. Thrust washer (upper)	17. Crankshaft	24. Main bearing cap	32. Flywheel
9. Rod bolt			

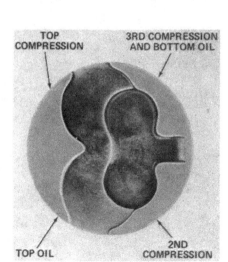

Fig. 67 – View of piston top showing correct position of ring gaps.

Crankshaft Regrind Finish Dimensions

Undersize	Main Journal Diameters	Crankpin Diameters
	mm	mm
.25 mm	70.65-70.67	52.66-52.68
.50 mm	70.40-40.42	52.41-52.43
.75 mm	70.15-70.17	52.16-52.18
1.00 mm	69.90-69.92	51.91-51.93

Fig. 68 – Chart showing correct regrind diameters for crankshaft main journals and crankpins for undersize bearings. Refer to text.

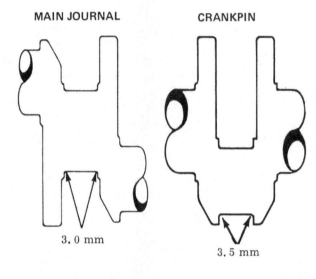

MAIN JOURNAL   CRANKPIN

3.0 mm

3.5 mm

Fig. 69 – Illustration showing correct radius finish grind for crankshaft main journal and crankpin for Models 274 and 284 diesel engine equipped tractors.

clearance exceeds 0.15 mm. Clearance between pump gear teeth and body (7) is 0.12 mm. Renew pump gears and/or pump body if clearance exceeds 0.25 mm. Pump gear backlash should be 0.25-0.40 mm. Renew gears if backlash exceeds 0.50 mm. Relief valve assembly (items 3 through 6) is non-adjustable. Inspect for scoring, burrs, nicks in plunger (6) and plunger bore in pump body. Renew relief valve and/or pump body if excessive wear or damage exists or relief pressure does not meet required specifications.

Additional pump specifications are as follows:

Pump output (litres at rpm) . 20 at 1830
Pump relief
    pressure . . . . 785 kPa (113.8 psi)

Reassembly is reverse of disassembly. Use new gasket (16 – Fig. 71) upon reassembly. Use new "O" ring (14) if cover (13) is removed. Renew oil pan gasket (29) when reinstalling.

OIL PRESSURE REGULATOR

Models 274 and 284 Diesel

86. The oil pressure regulator (items 25 through 28 – Fig. 71) and oil filter by-pass valve (items 19 through 22) are

crankshaft. Inspect flywheel for excessive wear or damage. Inspect flywheel teeth for damage and pilot bearing for smooth turning. Inspect tightness of counterweight bolts. Repair or renew as necessary.

Reinstall flywheel and torque bolts to 88-98 N·m. Bend locking plates to secure bolts. Install clutch assembly as outlined in paragraph 124. Balance of assembly is reverse of disassembly.

OIL PUMP

Models 274 and 284 Diesel

85. The gear type oil pump is located in lower right center of the crankcase and is driven by the center camshaft gear. Lubrication is circulated to all engine parts as shown in Fig. 70. Timing gears are lubricated by an oil jet (25 – Fig. 63) at engine front. To remove oil pump, first remove oil pan as outlined in paragraph 87. Refer to Fig. 71, then unbolt and withdraw pump assembly. Remove drive gear cover (13), "O" ring (14) and remove gear (15). Disassemble and clean all parts and inspect for wear as follows: Check clearance between pump gears (9 and 10) and pump cover (11) using a straightedge and feeler gage. Clearance should be 0.04-0.09 mm. Renew pump gears as a set if

LUBRICATION SYSTEM

VALVE ROCKER

ROCKER SHAFT

INJECTION PUMP

OIL MAIN GALLERY

CAMSHAFT

OIL JET

OIL PRESSURE SWITCH

CRANKSHAFT

OIL FILTER

OIL PUMP

OIL FILTER SHORTING VALVE

RELIEF VALVE

REGULATOR VALVE

OIL PAN

Fig. 70 – View of engine lubrication system used on diesel Models 274 and 284 tractors.

located in the oil filter base (24). Disassemble and clean all parts and examine for scoring, nicks or burrs. Renew pressure regulator, by-pass valve assembly and/or filter base if the following specifications are not met:

Regulator opening
  pressure....303-359 kPa (44-52 psi)
Oil filter by-pass opening
  pressure ....6.2-7.6 kPa (13-16 psi)
Oil filter (18) is a spin-on, full-flow paper type and should be renewed every 100 hours of operation. Use new gasket (23) upon reassembly.

OIL PAN

Models 274 and 284 Diesel

87. Removal of oil pan is obvious after removal of tie rod and examination of unit. Renew gasket when reinstalling and tighten pan bolts and nuts to 9-12 N·m torque.

NON-DIESEL FUEL SYSTEM

CARBURETOR

Model 284 Non-Diesel

88. Model 284 non-diesel tractors are equipped with a Nihon Kikakki 18.6 mm down draft type carburetor. Disassembly and overhaul is obvious after an examination of the unit and reference to Fig. 72.

Reassembly is reverse of disassembly while paying attention to the following: Renew any defective parts or parts in doubtful condition. Install all new gaskets. Set float height to 4 mm as shown in Fig. 73. With all connecting linkage installed and choke fully closed, throttle valve should be set at dimension (A) in Fig. 74. Adjust by bending link (B).

Parts data is as follows:

	IH PART No.
Carburetor	1 014 305 C91
Gasket set	1 014 333 C1
Main jet	1 014 250 C1
Idle needle	1 014 237 C1
Power jet	1 014 329 C1
Main air bleed jet	1 014 236 C1
Slow air bleed jet	1 014 232 C1
Check valve	1 014 212 C1
Float valve	1 014 327 C1

89. **IDLE ADJUSTMENT.** Before attempting to adjust the carburetor, first start engine and operate until completely warmed up. Shut off engine, turn idle adjusting screw (32–Fig. 72) all the way in, then unscrew it three turns. Turn throttle screw (30) in two to three turns, then start engine. Slowly unscrew throt-

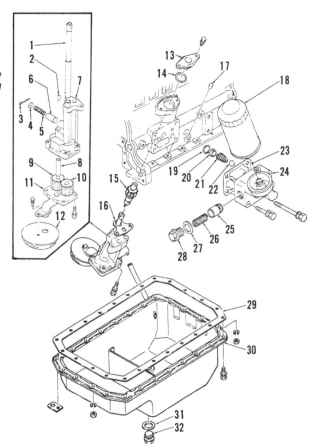

Fig. 71 — Exploded view of oil pump assembly and related parts.

1. Pump shaft
2. Key
3. Split pin
4. Spring seat
5. Spring
6. Plunger
7. Pump body
8. Shaft
9. Driven gear
10. Drive gear
11. Pump cover
12. Strainer
13. Cover
14. "O" ring
15. Driven gear
16. Gasket
17. Plug
18. Filter
19. Snap ring
20. Spring guide
21. Spring
22. Valve ball
23. Gasket
24. Filter body
25. Plunger
26. Spring
27. Washer
28. Plug
29. Gasket
30. Oil pan
31. Washer
32. Plug

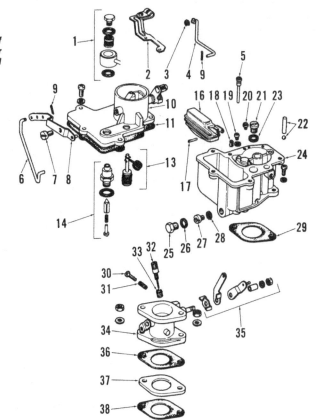

Fig. 72 — Exploded view of Nihon Kikakki carburetor used on non-diesel Model 284 tractors.

1. Strainer set
2. Choke wire holder
3. Clip
4. Rod
5. Main air bleed
6. Rod
7. Pump arm screw
8. Pump arm
9. Split pin
10. Carburetor top
11. Gasket
12. Pump plunger
14. Float valve
16. Float
17. Pin
18. Slow jet
19. Check valve
20. Slow air bleed
21. Power jet
22. Accelerator weight
23. Gasket
24. Carburetor body
25. Plug
26. Gasket
27. Main jet
28. Gasket
29. Gasket
30. Throttle adjust
31. Spring
32. Idle adjust
33. Spring
34. Flange
35. Lever assembly
36. Gasket
37. Insulator
38. Gasket

tle screw until engine speed becomes irregular. Slowly tighten idle adjusting screw until engine rpm increases, then unscrew throttle screw to lower engine speed to 650 ± 50 rpm.

FUEL PUMP

Model 284 Non-Diesel

90. Model 284 non-diesel tractors have a conventional diaphragm type fuel pump located on left rear side of crankcase. The fuel pump lever is actuated by an eccentric on engine camshaft. Procedure for removing pump is obvious after an examination of the unit. Pump discharge pressure should be 9.8-14.7 kPa (1.4-2.1 psi).

DIESEL FUEL SYSTEM

All diesel Models 274 and 284 tractors are equipped with a three plunger Bosch "A", in-line injection pump, throttle nozzles and a swirl chamber combustion chamber.

When servicing any unit of the diesel fuel system, the maintenance of absolute cleanliness is of utmost importance. Of equal importance is the avoidance of nicks or burrs on any of the working parts.

Probably the most important precaution that service personnel can impart to owners of diesel powered tractors, is to urge them to use an approved fuel that is absolutely clean and free from foreign material. Extra precaution should be taken to make certain that no water enters the fuel storage tanks.

FILTER AND BLEEDING

Models 274 and 284 Diesel

91. **OPERATION AND MAINTENANCE.** The fuel system includes a fuel filter that should be renewed at regular intervals. Filter life depends more upon careful maintenance than it does on hours or conditions of operation. Necessity for careful filling with clean fuel cannot be over-stressed. To minimize contamination of diesel fuel system, the following precautions are recommended.

Fill fuel tank after use and before storage, to eliminate presence of humid air in tank and reduce contamination due to condensation.

Check fuel strainer at bottom of fuel tank and drain off any water at least every 10 hours or once each day (more often if trouble is suspected). Allow fuel to drain until clean fuel flows. Make sure that sediment bowl fills before attempting to start engine to avoid entrance of air in injection pump. If bowl doesn't fill, open shut-off valve and loosen bowl slightly by loosening finger nut on retaining bail. Tighten finger nut when fuel has filled bowl.

Renew filter element after first 50 hours, then at least every 200 hours or immediately if water contamination is discovered. Bleed filter and lines as outlined in paragraph 92 after new filter is installed. Do not attempt to start engine until air is bled from system or injection pump may be damaged.

92. **BLEEDING.** The fuel system should be bled if fuel tank is allowed to run dry, if fuel lines, filter or other components within the system have been disconnected or removed or if engine has not operated for long period of time. If the engine fails to start or if it starts, then stops, the cause could be air in the system, which should be removed by bleeding.

To bleed, refer to Fig. 75, then remove vinyl cover (1) on priming pump, turn knob counter-clockwise until it lifts up. Loosen vent screws (2) and hand pump fuel into the filter and injection pump until fuel coming out of vent screws is free of air bubbles. Close vent screws and lock the nuts. Push priming pump down and tighten turning clockwise. Replace vinyl cover.

INJECTION PUMP

Models 274 and 284 Diesel

Because of the special equipment needed, and skill required of servicing personnel, service of injection pump is generally beyond the scope of that which should be attempted in the average shop. Therefore, this section will include only timing of pump to engine, removal and installation and the linkage adjustments which control the engine speeds.

If additional service is required, the pump should be turned over to an International Harvester facility which is equipped for diesel service, or to some other authorized diesel service station. Inexperienced personnel should NEVER attempt to service diesel injection pump.

R&R PUMP

Models 274 and 284 Diesel

93. To remove the Bosch "A" type injection pump, first thoroughly clean injection pump, fuel lines and side of engine. Shut off fuel at tank. Drain cooling system, disconnect headlight wiring, then remove front hood, side panels and radiator support. Disconnect upper and lower radiator hoses, air cleaner tube, then remove radiator, air cleaner assembly and fan shroud.

Disconnect tachometer drive cable, fuel shut-off cable, and governor control rod. Disconnect all fuel lines and hoses to injection pump and cap or plug all openings to prevent contamination of

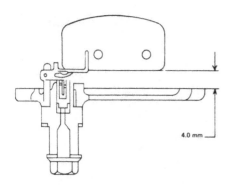

Fig. 73—Illustration showing correct float height setting. Refer to text.

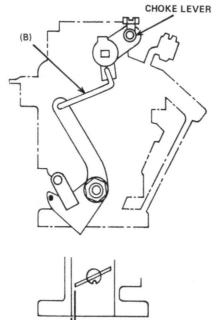

Fig. 74—Illustration showing correct throttle valve setting. Refer to text.

Fig. 75—View showing locations of injection pump bleed screws (2) and priming pump knob cover (1). Refer to text.

fuel system. Disconnect lubrication line to pump.

Remove tachometer drive housing and automatic pump timer cover. Turn crankshaft till number one cylinder is at top dead center and "Y" timing marks on pump drive and driven gears align as shown in Fig. 76. Refer to Fig. 77 and remove tachometer drive nut (4), then remove the automatic timer assembly mounting nut located behind the tachometer drive nut. Secure tachometer drive nut to a slide hammer using nuts smaller than thread diameter of drive nut. Screw tachometer drive nut fully into timer assembly (3) and pull assembly from pump shaft.

NOTE: Make certain not to turn crankshaft while removing automatic timer assembly to aid in reassembly and gear timing.

Remove injection pump mounting nuts and withdraw pump from engine front plate.

Reinstall pump by reversing removal procedure, paying attention to the following: Align mark on injection pump body with mark on engine front plate as shown in Fig. 78. When installing automatic timer assembly, align "Y" marks on drive and driven gears as shown in Fig. 76. Tighten automatic timer assembly mounting nut to 58-68 N·m torque. Tighten tachometer drive nut to 77-88 N·m torque. Bleed injection system as outlined in paragraph 92.

94. FEED PUMP PRESSURE CHECK. Install a 414 kPa (60 psi) pressure gage in the pump fuel chamber as shown in Fig. 79. Run engine at 1200 rpm and observe pressure. Operating pressure should be 177 kPa (25.6 psi). If pressure is below specifications, feed pump must be renewed.

95. ENGINE SPEED. Be sure that injection timing is correctly set as outlined in paragraph 93 and that engine is

otherwise operating correctly. Low idle speed should be 700 rpm, rated full load speed should be 2600 rpm and high idle no load speed should be 2760 rpm. Seals on speed adjusting stop screws should not be broken by unauthorized personnel.

INJECTOR NOZZLES

Models 274 and 284 Diesel

Models 274 and 284 diesel tractors use throttle type injector nozzles as shown in Fig. 80.

96. TESTING AND LOCATING FAULTY NOZZLES. If engine is missing and fuel system is suspected as being the cause of trouble, system can be checked by loosening each injector line connection in turn, while engine is run-

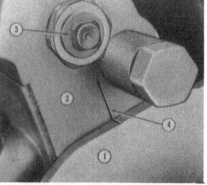

Fig. 78 — View showing fuel injection pump and engine front plate correctly aligned. Refer to text.

1. Engine front plate   3. Fuel inlet port (capped)
2. Injection pump body   4. Timing marks

Fig. 76 — View of timing gears with "X" and "Y" marks correctly aligned.

Fig. 77 — View of fuel injection pump and automatic timer assembly.

1. Feed pump
2. Priming pump knob cover
3. Automatic timer
4. Tachometer drive nut

Fig. 79 — View of 414 kPa (60 psi) gage installed to check feed pump pressure. Refer to text.

1. Gage

ning at slow idle speed. If engine operation is not affected when injector line is loosened, that cylinder is misfiring. Remove and test (or install a new or reconditioned unit) as outlined in the following paragraphs.

97. **REMOVE AND REINSTALL.** Before removing an injector or loosening injector lines, thoroughly clean injector, lines and surrounding area using compressed air and a suitable solvent. Remove high pressure line leading from pump to injector unit and disconnect the bleed line from injector. Unscrew nozzle nut (9 – Fig. 80) and withdraw nozzle from cylinder head.

Clean exterior surfaces of nozzle assembly and bore in cylinder head, being careful to keep all surfaces free of dust, lint and other small particles until nozzle is installed. Tighten nozzle holder into cylinder head to 68-88 N·m torque. Refer to paragraph 92 for bleeding information.

98. **TESTING.** A complete job of testing and adjusting the injector requires use of special test equipment. Only clean, approved testing oil should be used in tester tank. Nozzle should be tested for opening pressure, seat leakage, back leakage and spray pattern. When tested, nozzle should open

with a high-pitched buzzing sound, and cut off quickly at end of injection with a minimum of seat leakage and a controlled amount of back leakage.

Before conducting test, operate tester lever until fuel flows, then attach injector. Close valve to tester gage and pump tester lever a few quick strokes to be sure nozzle valve is not stuck, and that possibilities are good that injector can be returned to service without disassembly.

WARNING: Fuel leaves injector nozzle with sufficient force to penetrate the skin. Keep exposed portions of your body clear of nozzle spray when testing.

99. **OPENING PRESSURE.** Open valve to tester gage and operate tester lever slowly while observing gage reading. Opening pressure should be 9812 kPa (1422 psi). Opening pressure is adjusted by varying shim (4 – Fig. 80).

100. **SPRAY PATTERN.** Spray pattern should be well atomized and slightly conical, emerging in a straight axis from nozzle tip. If pattern is wet, ragged or intermittent, nozzle must be overhauled or renewed.

101. **OVERHAUL.** Hard or sharp tools, emery cloth, grinding compound or other than approved solvents or lapping compounds must never be used. An approved nozzle cleaning kit is available through a number of specialized sources.

Wipe all dirt and loose carbon from exterior of nozzle and holder assembly. Refer to Fig. 80 for exploded view and proceed as follows:

Secure nozzle in a soft jawed vise or holding fixture and remove nut (3). Place all parts in clean calibrating oil or diesel fuels as they are removed, using a compartmented pan and using extra care to keep parts from each injector together and separate from other units which are disassembled at the time.

Clean exterior surfaces with a brass wire brush, soaking in an approved carbon solvent if necessary, to loosen hard carbon deposits. Rinse parts in clean diesel fuel or calibrating oil immediately after cleaning to neutralize the solvent and prevent etching of polished surfaces.

Clean nozzle spray hole from inside using a pointed hardwood stick or wood splinter, then scrape carbon from pressure chamber using hooked scraper. Clean valve seat using brass scraper, then polish seat using wood polishing stick and mutton tallow.

Reclean all parts by rinsing thoroughly in clean diesel fuel or calibrating oil and assemble while parts are immersed in cleaning fluid. Make sure adjusting shim is intact. Tighten nozzle retaining nut (3) to a torque of 61-75 N·m. Do not overtighten, distor-

tion may cause valve to stick and no amount of overtightening can stop a leak caused by scratches or dirt. Retest assembled injector as previously outlined.

GLOW PLUGS

Models 274 and 284 Diesel

102. Glow plugs are parallel connected with each glow plug grounding through mounting threads. Start switch is provided with a **"GLOW"** position which is used to energize the glow plugs for starting. If indicator light fails to glow when start switch is held in **"GLOW"** position an appropriate length of time, check for loose connections at switch, indicator lamp, resistor connections, glow plug connections and ground. A test lamp can be used at glow plug connection to check for current to glow plug.

AIR FILTER SYSTEM

All Models

103. All models are equipped with a dry type air cleaner with a filter element (2 – Fig. 81 or Fig. 82) which should be

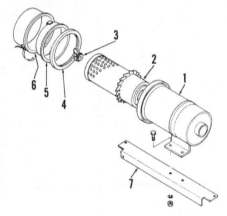

Fig. 81 – Exploded view of air cleaner assembly used on Model 274 and 284 diesel tractors.

1. Housing	
2. Element	
3. Wing nut	5. Dust cap
4. Dust seal	6. Cover
	7. Bracket

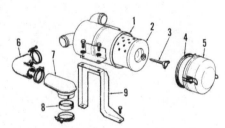

Fig. 82 – Exploded view of air cleaner assembly used on Model 284 non-diesel tractors.

1. Housing	
2. Element	6. Air hose
3. Wing bolt	7. Funnel
4. Dust seal	8. Dust seal
5. Cover	9. Bracket

Fig. 80 – Exploded view of injection nozzle.

1. Nut	
2. Banjo fitting	6. Push rod
3. Nozzle holder	7. Spacer
4. Shim	8. Nozzle assembly
5. Spring	9. Nozzle nut

renewed at least once each year.

Filter element can be cleaned by directing compressed air on the inside of the element. Air pressure must not exceed 100 psi. Renew filter element after 10 cleanings or once a year, whichever comes first.

GOVERNOR (NON-DIESEL ENGINE)

ADJUSTMENTS

Model 284 Non-Diesel

104. **SYNCHRONIZING GOVERNOR AND CARBURETOR.** If removal of carburetor, manifold, governor or governor linkage has been performed, or if difficulty is encountered in adjusting engine speeds adjust the governor to carburetor control rod as follows: Pull throttle lever down to put tension on governor spring, then disconnect control rod from governor (if necessary). Hold both carburetor and governor in the wide open (high idle) position and adjust rod clevis until pin slides freely into clevis and rockshaft lever holes. Then, remove pin and lengthen rod by unscrewing clevis one full turn. Reinstall clevis pin and tighten jam nut.

105. **HIGH IDLE SPEED.** Refer to Fig. 83 and back out stop bolt (2) about 13 mm. Start engine and operate until it reaches normal operating temperature. Move throttle lever to high idle position and check engine high idle speed, which should be approximately 2950 rpm. If engine high idle speed is not as stated, loosen jam nut and turn high idle stop bolt as required.

With engine high idle speed adjusted, move throttle control lever to low idle position, then quickly advance it to high idle position and adjust the damper screw (9 – Fig. 84) just enough to

eliminate engine surge. Repeat this operation as required until engine will advance from low idle to high idle speed without surging. Do not turn damper screw in further than necessary as engine low idle speed can be affected.

106. **LOW IDLE SPEED.** With engine running and at operating temperature, place throttle control lever in low idle position and check engine low idle speed which should be 650 rpm. If engine low idle speed is not as specified, adjust throttle stop bolt (1 – Fig. 83) on governor as required. If specified engine low idle speed cannot be obtained, recheck governor to carburetor control adjustment as outlined in paragraph 104 or carburetor idle adjustment as outlined in paragraph 89.

OVERHAUL

Models 284 Non-Diesel

107. To remove the governor unit, disconnect link from governor to carburetor. Loosen governor idler pulley to release belt tension. Unbolt and remove idler pulley bracket, governor and governor bracket as an assembly.

Disassembly of the removed governor unit will be obvious after an examination of the unit and reference to Fig. 84. Reassemble governor by reversing dis-

assembly procedure paying attention to the following: Use new "O" rings, oil seals and gaskets, coating lips of "O" rings and seals with clean engine oil. When assembling the central spring (17) on governor shaft (2), do not depress spring more than 10 mm past the need to install the seat (18), shims (19) if any and pin (1) securing the spring assembly.

NOTE: If spring is compressed more than needed to secure the assembly, the spring constant may be changed.

Install governor shaft and flyweights (2 and 15 – Fig. 84) vertically into the housing to prevent damage to the housing. Install coupling (7) on shaft and torque the hex nut to 24-30 N·m. Adjust end play on speed shaft (44) to 0.2 mm. Shims (29) are available in 0.1, 0.2, 0.3, 0.5 and 1.0 mm thicknesses. If removed, install speed arm assembly damper spring adjusting screw (33) to 45 mm as shown in Fig. 85. Using a dial indicator, check end play of governor shaft. Specified end play is 0.1-0.3 mm. Shims (19 – Fig. 84) are available in thicknesses of 0.1, 0.2, 0.3 and 0.5 mm. To install shims, housings must be separated and shims installed behind spring seat (18). Fill governor to level plug, through breather hole with engine oil.

A governor overhaul service package IH No. 1 014 449 C1 is available and a governor shaft kit IH No. 1 014 451 C1

Fig. 84 – Exploded view of governor used on non-diesel Model 284 tractors.
1. Pin
2. Shaft
3. Key
4. Bearing
5. Housing cover
6. Oil seal
7. Coupling
8. Pulley
9. Damper
10. Screw
11. Fork
12. Connector
13. Shims
14. Link
15. Flyweight
16. Shifter
17. Spring
18. Retainer
19. Shims
20. Pin
21. Fork
22. Spring
23. Pin
24. Shaft
25. Seal
26. Lever
27. Lever
28. Washer
29. Shim
30. Link
31. Lever
32. Screw
33. Adjuster
34. Low idle screw
35. Breather
36. Plug
37. Governor housing
38. High idle screw
39. Screw
40. "O" ring
41. Tachometer drive
42. "O" ring
43. Split pin
44. Control lever shaft
45. Key

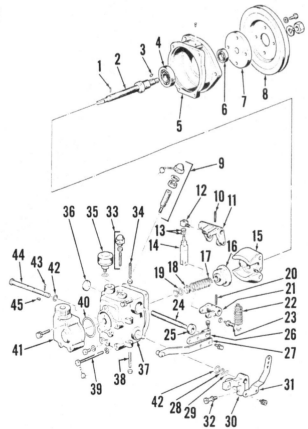

Fig. 83 – View of governor used on Model 284 non-diesel tractors showing points of adjustment for high and low idle speed. Refer to text.

1. Low idle stop bolt   2. High idle stop bolt

is also available.

After assembly and installation are complete, check and adjust the engine speed as outlined in paragraphs 104, 105 and 106.

COOLING SYSTEM

All models use a pressurized cooling system which raises coolant boiling point. An impeller type centrifugal pump is used to provide forced circulation, and a thermostat is used to stabilize operating temperature.

RADIATOR

All Models

108. Cooling system capacity is 5.0 liters for Model 284 non-diesel tractors and 7.0 liters for Model 274 and 284 diesel tractors.

To remove radiator, first drain cooling system, remove muffler then disconnect headlight wiring and remove front hood, side panels, radiator grill and radiator support. On Model 274 and 284 diesel tractors remove air cleaner assembly and air pipe. Disconnect upper and lower radiator hoses. Unbolt and remove radiator and fan shroud as an assembly.

FAN

Model 284 Non-Diesel

109. The fan is mounted on a shaft and bearing housing which is bolted to cylinder head and serves as the ther-

mostat housing cover. Two belts are used; one to drive the alternator and fan and one to drive the water pump and engine speed governor. To remove fan and housing, first drain cooling system, then remove the muffler. Disconnect headlight wiring and remove front hood, side skirts and radiator support. Loosen alternator to relieve belt tension and slide belt off fan pulley. Disconnect and remove upper and lower radiator hoses, then unbolt and remove radiator and fan shroud as an assembly. Unbolt and remove fan and housing. Disassembly is obvious after referring to Fig. 86. With fan and housing removed, thermostat (16) may be renewed at this time.

Reassemble fan assembly in reverse order, paying attention to the following: Install bearings (7 and 9) with sealed side facing away from spacer (8). Fill one third of the space between the bearings and the spacer with IH 251 HEP grease or equivalent lithium base grease. Make certain fan shaft rotates freely when assembled in the housing. If thermostat is renewed make certain the pin is at the top. When reinstalling, adjust fan belt until it deflects 10 mm with 110N exerted on longest span.

Models 274 and 284 Diesel

110. Fan is attached to the water pump and one belt drives the water pump and alternator.

To remove fan, first remove radiator as outlined in paragraph 108. Loosen

alternator to relieve belt tension and slide belt off water pump pulley. Unbolt and remove fan.

When reinstalling, tighten fan belt until it deflects 10 mm with 110 N exerted on longest span.

BELTS

All Models

111. To renew fan belt on Model 284 non-diesel tractors first loosen alternator brace bolt and tilt alternator towards engine. Slip belt off alternator pulley. Disconnect coupling of hydraulic pump drive on crankshaft pulley, slide coupling off so belt can be slipped off crankshaft pulley, then work belt over fan and remove.

To remove water pump/governor belt, loosen governor idler pulley and slacken belt tension by screwing tension bolt in. Slip belt off water pump pulley and crankshaft pulley.

NOTE: Fan/alternator belt must be removed to remove water pump/governor belt.

Install new belts in reverse order of removal. Tighten hydraulic pump drive shaft connecting bolts to 11-23 N·m torque.

Removal of fan belt on Model 274 and 284 diesel tractors is obvious after examination. Deflection of all belts should be 10 mm with 110 N applied on longest span.

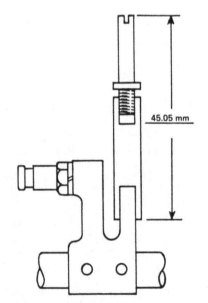

Fig. 85—Illustration showing correct setting dimension for damper spring adjusting screw. Refer to text.

45.05 mm

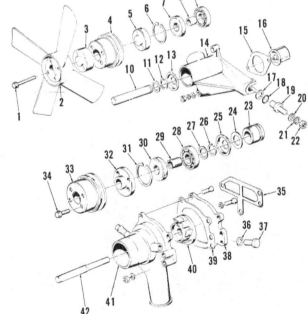

Fig. 86—Exploded view of water pump and fan housing assemblies used on Model 284 non-diesel tractors.

1. Bolt
2. Fan
3. Spacer
4. Pulley
5. Boss
6. Snap ring
7. Bearing
8. Spacer
9. Bearing
10. Shaft
11. Washer
12. Stopper ring
13. Dust seal
14. Housing
15. Thermostat gasket
16. Thermostat
17. Gasket
18. "O" ring
19. Plug
20. Washer
21. Spring washer
22. Nut
23. Water seal
24. Baffle plate
25. Dust seal
26. Retaining ring
27. Washer
28. Bearing
29. Spacer
30. Bearing
31. Snap ring
32. Boss
33. Pulley
34. Bolt
35. Gasket
36. Washer
37. Bolt
38. End plate
39. Gasket
40. Impeller
41. Pump body
42. Shaft

WATER PUMP

Model 284 Non-Diesel

112. R&R AND OVERHAUL. To remove the water pump, first drain cooling system. Remove radiator as outlined in paragraph 108. Unbolt and remove fan assembly as outlined in paragraph 109. Loosen water pump/governor belt idler pulley and remove belt and let hang out of the way. Remove by-pass hose and lower radiator hose from pump. Unbolt and remove water pump.

Refer to Fig. 86 for exploded view of water pump. The shaft and bearing assembly (items 28, 29, 30 and 42) are pressed into pump body (41) as is seal (23). The impeller (40) is also pressed onto shaft (42).

Installation is the reverse of disassembly procedure, paying attention to the following: Install bearings (28 and 30) with sealed sides facing away from bearing spacer (29). Fill approximately ⅓ of the space between bearings and spacer with IH 251 HEP grease or equivalent lithium base grease. Install new water seal (23) until bent portion of plate (24) touches the body. Apply a small amount of engine oil on sliding face of the water seal. Support the opposite end of the pump shaft, press impeller (40) on to shaft until impeller hub is flush with shaft shoulder. Be sure shaft rotates freely after installation. Support impeller end of pump shaft, press on pulley hub (32). Check for free rotation after hub installation. Install new gaskets (39 and 35). Install pump by reversing removal procedure. Adjust pump belt until it deflects 10 mm with 110 N applied at longest span.

Models 274 and 284 Diesel

113. R&R AND OVERHAUL. To remove pump, first drain cooling system. Remove radiator as outlined in paragraph 108. Loosen alternator brace bolt, tilt alternator towards engine and remove belt, then unbolt brace bar from water pump and tilt alternator out of the way. Remove fan from pump hub. Remove hoses from water pump, then unbolt and remove water pump.

Refer to Fig. 87 and disassemble pump as follows: Remove snap ring (1) and hub (2). Remove lock wire (4), cover (10) and gasket (9). Press out bearing assembly (3) from impeller side. Remove water seal (6) and floating seat (7) assembly. Renew any parts as necessary.

Reassemble by reversing disassembly procedure paying attention to the following: Renew gaskets (9 and 11), water seal and floating seat. Apply suitable bonding agent to outer edge of

water seal and floating seat assemblies and press into place. Press bearing assembly into housing until lock wire grooves on bearing assembly align with grooves on housing. Install lock wire.

Press hub onto bearing assembly shaft until distance from hub to housing end surface is 130.1 to 130.3 mm, then install snap ring. Apply thin coat of silicone grease to water seal assembly

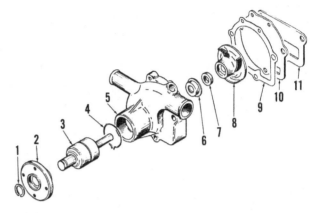

Fig. 87 — Exploded view of water pump assembly used on Models 274 and 284 diesel tractors.

1. Snap ring
2. Hub
3. Bearing assembly
4. Lock wire
5. Pump body
6. Water seal
7. Floating seat
8. Impeller
9. Gasket
10. Cover
11. Gasket

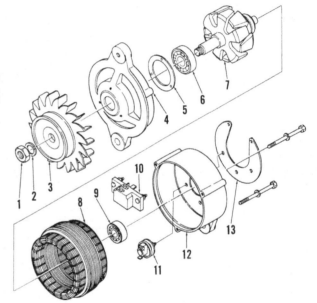

Fig. 88 — Exploded view of Mitsubishi AH 2035T2 alternator used on Model 284 non-diesel tractors.

1. Nut
2. Washer
3. Pulley
4. Front cover
5. Bearing retainer
6. Bearing
7. Rotor
8. Stator
9. Bearing
10. Brush assembly
11. Diode assembly
12. Housing
13. Cover

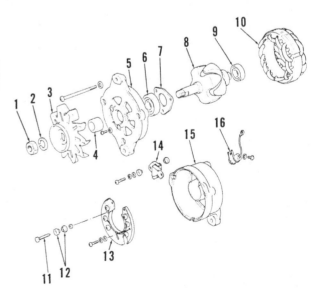

Fig. 89 — Exploded view of Hitachi LT-120-8 alternator used on Models 274 and 284 diesel tractors.

1. Nut
2. Washer
3. Pulley
4. Spacer
5. Front cover
6. Bearing
7. Bearing retainer
8. Rotor
9. Bearing
10. Stator
11. Screw
12. Insulators
13. Rectifier
14. Brush assembly
15. Housing
16. Suppressor

and floating seat assembly contact parts. Press impeller onto bearing assembly shaft until clearance between impeller and housing measures 0.4 to 0.6 mm. Install new gasket and cover.

Install pump by reversing removal procedure. Install and adjust belt tension so that it deflects 10 mm with 110 N applied midway on the longest span.

THERMOSTAT

All Models

114. The by-pass type thermostat is located behind the fan housing assembly on Model 284 non-diesel tractors and in the outlet elbow on Models 274 and 284 diesel tractors. Thermostat for Model 284 non-diesel tractors should open at 82.2-85.0° C and should be installed with the pin at the top. Thermostat for Models 274 and 284 diesel tractors should open at 71° C and should be installed with the snap ring facing downward.

ELECTRICAL SYSTEM

ALTERNATOR AND REGULATOR

All Models

115. Mitsubishi alternators (IH No. 1014 220 C91 or Mitsubishi No. AH 2035 T2) may be used on Model 284 non-diesel tractors. Hitachi alternators (IH No. 1058309C91 or Hitachi No. LT 120-8) may be used on Model 274 and 284 diesel tractors. An external regulator (IH No. 1061 462 C1) is used and has no provision for adjustment.

CAUTION: Because certain components of the alternator can be damaged by procedures that will not affect a D.C. generator, the following precautions must be observed:

a. When installing battery or connecting a booster battery, the negative post of battery must be grounded.

b. Never short across any terminal of the alternator or regulator.

c. Do not attempt to polarize the alternator.

d. Disconnect battery cables before removing or installing any electrical unit.

e. Do not operate alternator on an open circuit and be sure all leads are properly connected before starting engine.

Mitsubishi AH 2035 T2
Cold output @ specified voltage,
   Specified volts 12
   Amperes at rpm . . . . .35.0 @ 5000

Hitachi LT 120-8
Cold output @ specified voltage,
   Specified voltage 12
   Amperes at rpm . . . . .20 @ 5000

All Models

117. Mitsubishi No. 028040-28 start-

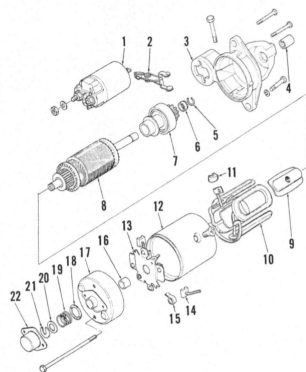

Fig. 90—Exploded view of Mitsubishi 028040-28 starting motor used on Model 284 non-diesel tractors.
1. Solenoid
2. Drive lever
3. Drive housing
4. Bushing
5. Snap ring
6. Stop collar
7. Overrunning clutch
8. Armature
9. Pole core
10. Field coil
11. Bush
12. Yoke housing
13. Brush holder
14. Brush
15. Spring
16. Bearing
17. Cover
18. Rubber ring
19. Spring
20. Washer
21. Lock plate
22. Cap

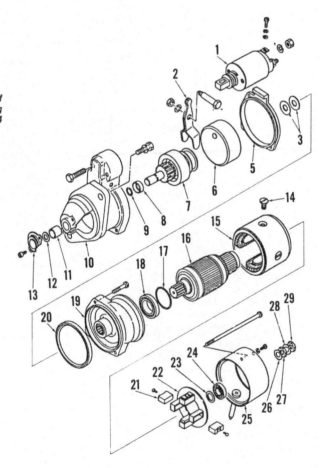

Fig. 91—Exploded view of Hitachi S13-46 starting motor used on Models 274 and 284 diesel tractors.
1. Solenoid
2. Shift lever
3. Thrust washers
4. Dust cover
6. Cover
7. Pinion assembly
8. Pinion stop
9. "O" ring
10. Case
11. Bearing
12. Thrust washer
13. Cover
14. Screw
15. Yoke
16. Armature
17. "O" ring
18. Bearing
19. Housing
20. Oil seal
21. Brush
22. Brush holder
23. Thrust washer
24. Bearing
25. Rear cover
26. Lockwasher
27. Nut
28. Lockwasher
29. Nut

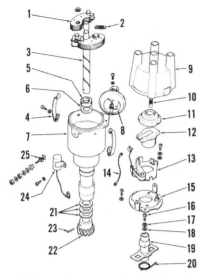

Fig. 92 – Exploded view of distributor used on non-diesel Model 284 tractors.

1. Governor weight
2. Spring
3. Shaft
4. Cap clamp
5. Oil seal
6. Washer
7. Housing
8. Vacuum control
9. Cap
10. Contact
11. Cover
12. Rotor
13. Breaker points
14. Lead wire
15. Breaker base
16. Screw
17. Spring washer
18. Washer
19. Cam
20. Hair pin spring
21. Shims
22. Gear
23. Pin
24. Condenser
25. Terminal

ing motors are used on Model 284 non-diesel tractors and Hitachi S13-46 starting motors are used on Model 274 and 284 diesel tractors. Specification data for these units is as follows:

Mitsubishi 028040-28

Volts 12
No-load test,
   Amperes (max) 50
   Rpm (max)...............5000

Hitachi S13-46

Volts 12
No-load test,
   Amperes (max) 150
   Rpm (min) 3500

STANDARD IGNITION

Model 284 Non-Diesel

118. The M4-71C non-diesel engine is equipped with standard ignition distributor and the firing order is 1-3-4-2. Overhaul procedure for the distributor is obvious after an examination of the unit and reference to Fig. 92. Breaker contact gap is 0.40-0.50 mm. Dwell angle is 55°-61°.

119. **DISTRIBUTOR INSTALLATION AND TIMING.** Position number one piston at top dead center of compression stroke so that TDC mark on crankshaft pulley (left notch facing pulley) aligns with timing pointer. Align punch mark on the distributor drive gear with groove on housing, then install distributor in timing chain cover.

NOTE: When distributor gear meshes with the camshaft drive gear, punch mark on the gear will move approximately 49 degrees counter-clockwise due to the helical angle of the gears.

Rotate distributor housing until points just break. Use a tachometer and timing light, with engine running at 550-650 rpm, timing should be set at 13 degrees BTDC, (right notch facing pulley).

CIRCUIT DESCRIPTION

120. Refer to Figs. 93 and 94 for wiring diagram typical of all models. Negative terminal of battery is grounded on all models.

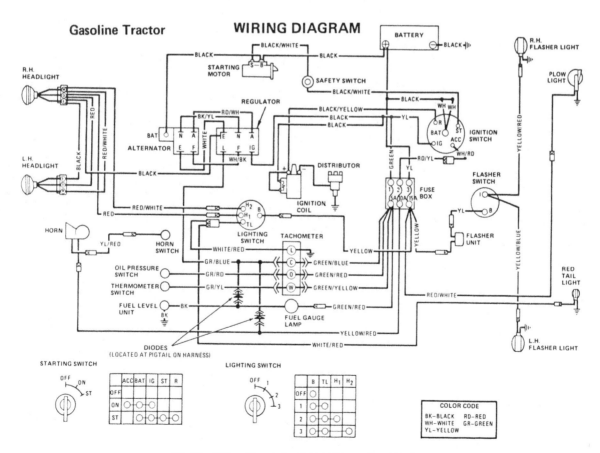

Fig. 93 – Wiring diagram for Model 284 non-diesel tractors.

Diesel Tractor

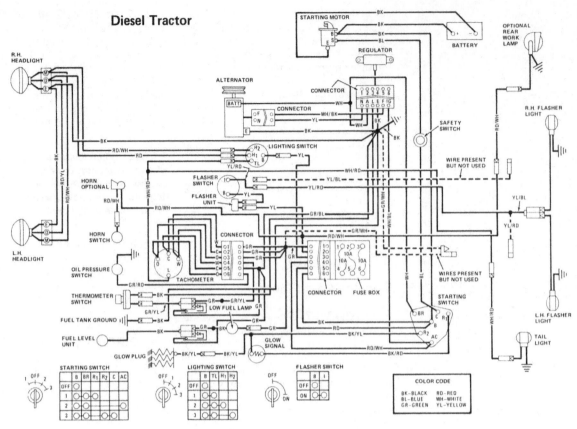

Fig. 94 — Wiring diagram for Models 274 and 284 diesel tractors.

CLUTCH

All models use a single plate, dry-disc, spring loaded clutch. Non-diesel models use a 203 mm clutch, and diesel models use a 216 mm clutch.

ADJUSTMENT

All Models

121. Clutch pedal free play should be 30-40 mm, measured at the front edge of the pedal. Pedal free play will decrease as clutch lining wears. Adjustment should be made before free play reaches 16 mm. To adjust free play, refer to Fig. 95, and turn yoke (6) in or out as necessary to obtain correct pedal free play.

STARTING SAFETY SWITCH

All Models

122. Switch is "ON" when clutch pedal is fully depressed. Adjust switch by loosening mounting bracket bolts. Switch should be on when extension tab travel is 5 mm, and have a free play of 3 mm.

REMOVE AND REINSTALL

All Models

123. CLUTCH SPLIT. To detach (split) tractor between engine and clutch housing, first remove muffler, disconnect headlight wiring then remove radiator support with engine hood and side panels. Disconnect all engine electrical connections. Disconnect and remove hydraulic return line from hydraulic pump and filter base. Disconnect and remove high pressure hydraulic supply line from pump and control valve or flow divider if so equipped. Cap or plug hydraulic lines or openings at once to prevent dirt or foreign material from contaminating hydraulic system.

On Model 284 non-diesel tractors, disconnect throttle control rod, tachometer drive cable and choke control cable from carburetor. Shut off fuel at tank and disconnect fuel line at carburetor.

On Model 274 and 284 diesel tractors, disconnect fuel shut-off cable and governor control rod at injection pump. Shut off fuel at tank and disconnect fuel supply hose to feed pump. Plug or cap hoses to prevent fuel contamination. Disconnect tachometer drive cable at drive

assembly.

On Model 284 tractors, disconnect steering drag link at pitman arm. On Model 274 tractors, disconnect steering column at steering box. On "All-Wheel Drive" equipped 284 models, loosen set screw retaining drive shaft cover pipe to center section housing. Wedge wooden blocks between engine side channel and front axle to prevent tipping. Place rolling floor jack under center section of tractor and support subframe with suitable jack stands.

NOTE: If tractor is equipped with front bolster weights, support weights with suitable jack stands.

Remove bolts securing center section to engine and split tractor by rolling rear section rearward.

Reassemble tractor by reversing removal procedure.

124. Before removing clutch assembly, index the clutch cover to the flywheel for proper assembly. Unbolt and remove clutch assembly from flywheel.

NOTE: Diesel tractors have two mounting bolts longer than the others. Note and mark holes where these bolts are removed, because these holes have recessed threads.

Inspect clutch pressure plate for cracks and deformation. Renew clutch assembly if damaged or wear is excessive. Check clutch disc for excessive wear or damage. Check transmission input shaft splines for damage and renew all parts as necessary.

When installing, long side of clutch disc should be toward rear. Use removed input shaft or suitable pilot shaft to align clutch disc while attaching pressure plate to flywheel. On diesel models, install the two longer pressure plate bolts in holes with recessed threads first. Tighten mounting bolts to 23-27 N·m torque. Refer to paragraph 125 for clutch shaft and release bearing service and to paragraph 121 for clutch pedal free play adjustment. Refer to Fig. 95 for exploded view of clutch housing, clutch assembly and release bearing assemblies.

CLUTCH SHAFT AND RELEASE BEARING

All Models

125. Separate clutch housing from engine as outlined in paragraph 123. The clutch release (throw-out) bearing (14 – Fig. 95) and housing (13) can be removed after removing wire (9) and set bolt (8). Bearing is packed at time of manufacture with heat resistant grease. **DO NOT** wash bearing in solvent or water. Old bearing can be pressed from housing and new bearing can be pressed on if renewal is required. Chamfered side of release bearing must be toward clutch fingers.

The clutch shaft (12) is the transmission input shaft and will be covered in the transmission section.

TRANSMISSION "ALL WHEEL DRIVE" CLUTCH HOUSING GEAR TRAIN

Model 284 diesel tractors equipped with "ALL-WHEEL DRIVE" are equipped with a clutch housing containing the drive gear train for the front wheel drive system. Power is taken from the pinion shaft in the main transmission and transferred through clutch housing gear train to the front drive shaft.

Models So Equipped

126. **R&R AND OVERHAUL.** To re-

move gears and shafts of "All-Wheel Drive" clutch housing gear train, tractor must be separated (split) between clutch housing and middle case as follows: Disconnect battery, remove wiring harness cover, then disconnect electrical connectors to rear of tractor and bring wires and battery cables forward. Disconnect clutch rod from clutch fork shaft. Remove four wheel drive shift lever and middle case cover as an assembly. Disconnect hydraulic high pressure pipe from the control valve. On tractors equipped with power steering, disconnect power steering supply and return lines from flow divider. Cap or plug all hydraulic lines and ports. Loosen hose clamps and slide low pressure hydraulic rear hose downward until it clears the oil filter cap. Disconnect accelerator pedal rod from the bell crank assembly. Wedge wooden blocks between subframe and front axle to prevent tipping when splitting. Support tractor under subframe with suitable jack stands. Place rolling floor jack under rear frame. Place speed range and gear shift levers in neutral position. Unbolt clutch housing from middle case and separate tractor, rolling rear section rearward.

Disconnect throttle linkage bracket from clutch housing cover and move bracket out of the way. Disconnect steering drag link from pitman arm on steering gear. Models equipped with power steering, remove pressure and return lines from steering gear, then plug ports on steering gear. Remove steering gear case mounting bolts. Lift gear case and steering column assembly and move gear case to left and clear of clutch housing cover. Support steering gear on jack stand or equivalent suitable stand. Refer to Fig. 96 and remove

clutch housing cover (63) and discard cover gasket (64). Drain lubricant from clutch housing.

127. INPUT SHAFT. To remove input shaft (22), drive out roll pin securing shifting fork (10) to shift rod (6). Remove input shaft mounting flange bolts, then using a split puller and slide hammer, pull input shaft assembly rearward until flange (24) clears its seat. Position input shaft assembly so that input shaft gear (28) is freed from shifting fork (10). Slide shifting fork forward and remove input shaft assembly. Remove snap ring (29), input shaft gear (28) and bearing retaining snap ring (27). Press mounting flange (24) and bearing (26) from input shaft (22). Press bearing (26) from flange (24), if necessary. Remove front input shaft bearing (30) from its bore in clutch housing, if necessary, using a suitable puller. Renew bearings (30 and 26), oil seal (23) and "O" ring (25) upon reassembly.

128. OUTPUT SHAFT. To remove output shaft (47 – Fig. 96), first remove flange (44) mounting bolts. Support front of tractor with suitable jack stands, loosen set bolt on front of drive shaft cover pipe (55). Loosen front axle assembly and slide forward. Unbolt rear of drive shaft cover pipe and remove cover pipe (55) and drive shaft (56). Retain couplers (54 and 57). Using a soft hammer, tap output shaft rearward until mounting flange (44) clears its bore, then remove output shaft assembly. Tap flange with soft hammer and remove output shaft and bearing assembly from flange. If necessary, remove bearings (46 and 48) from shaft (47). Renew bear-

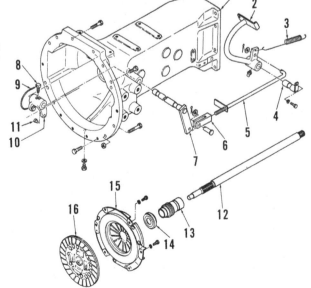

Fig. 95 – Exploded view of typical clutch housing and clutch assembly for all two-wheel drive models.

1. Clutch housing
2. Pedal
3. Spring
4. Pedal pin
5. Rod
6. Yoke
7. Release shaft
8. Set bolt
9. Wire
10. Release fork
11. Pin
12. Input shaft
13. Shifter block
14. Bearing
15. Pressure plate
16. Clutch disc

ings (46 and 48), oil seal (49) and "O" ring (45) upon reassembly.

129. IDLER SHAFT. To remove idler shaft (38 – Fig. 96), first remove idler shaft plate (36). Remove snap ring (43) and pull shaft assembly from housing while holding gear (40), spacers (39 and 42) and roller bearing (41). Remove items (39 through 42) out of top of clutch housing. Renew "O" ring (37) upon reassembly.

Inspect all gears, shafts and bearings for excessive wear or damage and renew as necessary. Inspect input and output shaft mounting flanges for cracks and damage. Check all snap rings for distortion and renew as necessary. Check shift fork and shifter rod for smooth operation. Check detent ball and spring (7 and 8 – Fig. 96) for wear and insufficient tension and renew as necessary.

Reassemble gear train by reversing removal procedure. Renew clutch housing cover gasket (64 – Fig. 96) upon reassembly. Fill clutch housing with IH "Hy-Tran" fluid to level of upper side plug (51). Join clutch housing to middle case by reversing split procedure. Adjust clutch free play as outlined in paragraph 121. If so equipped, bleed power steering system as outlined in paragraph 26.

SPEED AND RANGE TRANSMISSION

LUBRICATION

All Models

Transmission, differential and hydraulic system share a common sump. The drain plug (85 – Fig. 99) is located on the left front of the rear housing. Be sure to allow sufficient time for oil to drain from interconnected compartments. The transmission fluid filter is the common filter for the hydraulic system and service is outlined in paragraph 26.

Fill transmission through filler opening on rear of top cover with I.H. "Hy-Tran" fluid until it flows out of fill level plug hole (83). Capacity is 13 liters for all models. Be sure to allow sufficient time for oil to pass between compartments and be sure that tractor is level before checking at level plug.

Check for leaks after oil has warmed up around filter and drain plug.

REMOVE AND REINSTALL

All Models

130. To detach (split) tractor between clutch housing and front of transmission, proceed as follows: Disconnect bat-

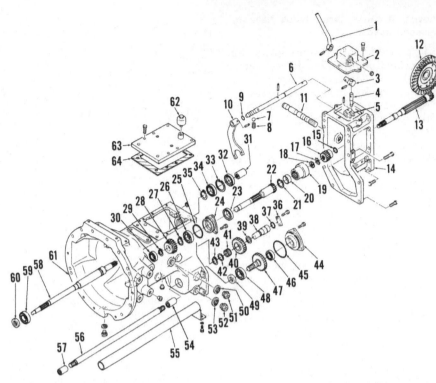

Fig. 96 – Exploded view of "All Wheel Drive" clutch housing gear train for Model 284 tractors so equipped.

1. Shift lever	18. Nut	33. Snap ring	49. Oil seal
2. Cover	19. Coupling	34. Bearing	50. Seal washer
3. Shaft	20. Collar	35. Snap ring	51. Plug
4. Rod	21. Snap ring	36. Plate	52. Plug
5. Boss	22. Shaft	37. "O" ring	53. Seal washer
6. Shaft	23. Seal	38. Shaft	54. Coupling
7. Ball	24. Flange	39. Spacer	55. Tube
8. Spring	25. "O" ring	40. Gear (idler)	56. Shaft
9. "O" ring	26. Bearing	41. Needle bearing	57. Coupling
10. Fork	27. Bearing	42. Spacer	58. Input shaft
11. Brake pedal shaft	28. Gear (shifting)	43. Snap ring	59. Bearing
12. Ring gear	29. Snap ring	44. Cover	60. Oil seal
13. Pinion shaft	30. Bearing	45. "O" ring	61. Clutch housing
14. Middle case	31. Coupling	46. Bearing	62. Breather
15. "O" ring	32. Seal	47. Gear (output)	63. Cover
16. Coupling		48. Bearing	64. Gasket
17. Washer			

Fig. 97 – Exploded view of speed and range shift linkage on Model 284 tractors (S.N. 011793 and below) equipped with non-diesel engines.

1. Shift lever
2. Pin
3. Pivot pin
4. Retainer
5. Gasket
6. Cover
7. Gasket
8. Spring
9. Ball
11. Shaft
12. Shaft
13. Shaft
14. Fork (1st and 2nd)
15. Hub (1st and 2nd)
16. Pin
17. Plug
18. Fork (3rd & 4th)
19. Hub (reverse)
20. Fork cover
21. Shaft
22. Seal washer
23. Set bolt
24. Fork (reverse)
25. Shift lever (Hi-Lo)
26. Pin
27. "O" ring
28. Link
29. Plug
30. Ball
31. Spring
32. Plug
33. Fork (Hi-Lo)
34. Set bolt
35. Seal washer
36. Shaft

tery, then disconnect rear electrical connections. Remove seat and fenders. Drain oil from rear frame. Remove hydraulic pressure line cover, then disconnect pressure line to control valve. On tractors equipped with power steering, disconnect all lines going to and from flow divider and power steering gear. Cap all openings to prevent contamination of hydraulic system. Remove right platform step and disconnect brake linkage from pedals. Disconnect brake supply lines from reservoir. Remove hydraulic return line. Disconnect clutch pedal spring. Disconnect clutch yoke from clutch shaft, then remove clutch pedal pin. Remove clutch pedal and rod from tractor. Disconnect spring to parking brake lever, then disconnect parking brake rod at the rear. Remove left platform step. Disconnect necessary electrical connections from rear frame and move them forward. Wedge wooden blocks between subframe and axle to prevent tipping. Support tractor under clutch housing with suitable jackstands. Remove hitch leveling link assemblies. Place rolling floor jack under rear frame, then raise tractor enough to remove rear wheels. Remove battery. On Model 274 tractors equipped with differential lock, unbolt pedal bracket from right axle housing, then remove pedal and bracket from operating rod. Support axle carriers with suitable hoist, then unbolt and remove. Remove inner brake rings. Unbolt and remove hydraulic hitch cover from rear frame. Support rear frame with a suitable hoist, then unbolt and split rear frame from clutch housing.

Reinstall transmission and assemble tractor by reversing split procedure. Adjust clutch free play as outlined in paragraph 121, bleed brakes as outlined in paragraph 154, fill transmission and if so equipped, bleed power steering system as outlined in paragraph 26.

OVERHAUL

All Models

131. With transmission removed as outlined in paragraph 130, proceed as follows: Remove speed transmission cover (6–Figs. 97 or 98). Remove springs and detent balls (8 and 9). Remove pins, washers, springs and differential lock fork and shaft assembly, if so equipped. Remove pins securing shift rods (11, 12 and 13–Figs. 97 or 98) and drive rods through shifting forks and out the front of transmission frame. Note and retain any spacers present on shift rods. Unbolt pto input shaft cover (5–Fig. 100) then remove using two jackscrews in threaded holes of cover. Remove pto input shaft (2). If tractor is

equipped with overrunning pto clutch, (see insert Fig. 100), release snap ring (32) from its seat, then slide washer (33), spring (34) and front clutch half (35) forward on input shaft (31). Remove snap ring (36) and slide rear half of clutch (37) off input shaft (40). Remove pto input shaft (40). Unbolt transmission input shaft cover (14–Fig. 99), remove snap ring (20) and slide transmission input gear (19) and spacer (18) rearward on shaft. Using a brass drift, drive transmission input shaft out front of frame. Bearing (23) is pressed into bore of housing.

Remove brake pistons by carefully prying out of side bores. Support differential assembly with a suitable sling hoist. Remove bolts in differential bearing retainers (2 and 23–Fig. 109), then using jack screws in threaded holes, remove bearing retainers. Note and retain shims (4 and 21) with appropriate bearing retainers. Lift out differential assembly.

132. **MAINSHAFT.** Release lock tab and remove nut (25–Fig. 99). Remove parking brake drum (27) from mainshaft. Unbolt mainshaft bearing retainer (32) and using jackscrews in threaded holes, force retainer out of frame bore. Remove bearing retainer from mainshaft using a suitable puller. Note and retain shims (33) and spacer (29). Wedge a wooden block between the 2nd and 3rd gears. install mainshaft nut and drive mainshaft out rear of speed transmission housing while withdrawing gears and associated parts out top of

opening. Remove snap ring (47) and remove mainshaft inner bearing (48).

133. **COUNTERSHAFT.** On early Model 284 non-diesel tractors (S.N. 001793 and below) remove set bolts (23–Fig. 97) and (11–Fig. 99) securing reverse fork rod and reverse idler shaft in place.

NOTE: Late Model 284 non-diesel tractors (S.N. 001794 and above) and all Model 274 and 284 diesel tractors have a longer one-piece shift fork. On these tractors, remove set bolt securing reverse idler shaft in place.

Remove reverse shift fork (24–Fig. 97 or 20–Fig. 98) and reverse fork rod (21–Fig. 97). Release snap rings (2 and 9–Fig. 99) on reverse idler shaft (7) and remove the reverse idler gear (5), thrust washers (3 and 8) and needle bearings (4 and 6).

Remove hi-lo shift rod set bolt (34–Figs. 97 or 98), then remove hi-lo shift rod (36) and shift fork (33) being careful not to lose detent ball (30) and spring (31) under plug (29).

Remove countershaft cover (52–Fig. 99) and shims (54). Using a brass hammer, tap counter shaft out towards front until outer bearing (55) clears the frame. Remove countershaft front bearing using a suitable puller. Remove countershaft assembly out of top of frame, being careful not to drop any gears, bearings or related parts off shaft. On early Model 284 non-diesel tractors (S.N. 012228 and below) remove retainer bolt, lockwasher, pin, flat washer and shims

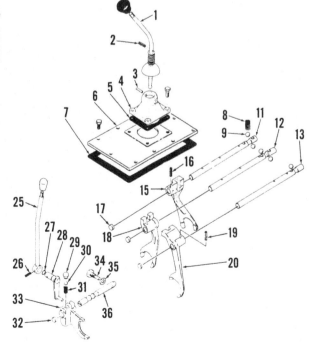

Fig. 98–Exploded view of speed and range shift linkage on Model 284 non-diesel tractors (S.N. 011794 and above) and all Model 274 and 284 diesel tractors.

1. Shift lever
2. Pin
3. Pivot pin
4. Retainer
5. Gasket
6. Cover
7. Gasket
8. Spring
9. Ball
11. Shaft (1st and 2nd)
12. Shaft (3rd and 4th)
15. Fork (1st and 2nd)
16. Pin
17. Plug
18. Fork (3rd and 4th)
19. Pin
20. Fork (reverse)
25. Shift lever (Hi-Lo)
26. Pin
27. "O" ring
28. Link
29. Plug
30. Ball
31. Spring
32. Plug
33. Fork (Hi-Lo)
34. Set bolt
35. Seal washer
36. Shaft

(insert Y – Fig. 99), then remove bearing, gears, spacers and split rings.

REASSEMBLY

134. COUNTERSHAFT. On Model 284 non-diesel tractors (S.N. 012228 and below) proceed as follows: Install split rings (60 – Fig. 99), gears and spacers (items 61 through 67).

NOTE: If split rings need renewal on tractors (S.N. J011052 and below), countershaft (68), fourth gear (61) and snap ring must be renewed due to product modification.

Press inner bearing (70) onto countershaft. Measure depth from bearing to end of shaft, (dimension "A" – Fig. 103). Select shims (1 – Fig. 103) so that thickness is same as dimension "A" minus 0.05 mm. On tractors (S.N. J011053 and above) shims are not required. Install pin (74 – Fig. 99), flat washer (75), lockwasher (76) and bolt (77) and torque bolt to 15-20 N·m.

135. On non-diesel tractors (S.N. 012229 and above) and all diesel tractors assemble countershaft as follows: Install split rings (60 – Fig. 99), gears and spacers (items 61 through 67) and sleeve (69) (diesels only) onto countershaft (68). Press inner bearing (70) onto shaft.

On all models, position countershaft assembly into rear frame. Using a brass hammer drive countershaft rearward until inner bearing bottoms against shoulder in frame. Drive outer bearing (55) into frame bore until bearing is started onto countershaft. Using a suitable size pipe, drive bearing onto countershaft until it seats.

NOTE: Make sure pipe drives against inner race of bearing only.

Refer to Fig. 102 and measure dimension "A" on countershaft cover and dimension "B" on housing. Subtract dimension "A" from "B" to obtain required shim pack "C". Install shim pack (54 – Fig. 99), install cover (52) with new "O" ring (53) and tighten mounting bolts to 31-46 N·m torque.

Install reverse idler with rear snap ring into shaft bore, then install thrust washers (3 and 8 – Fig. 99), needle bearings (4 and 6) and reverse gear (5) with tapered teeth to the rear, then secure with front snap ring (2). On non-diesel tractors (S.N. 011793 and below) install reverse fork (24 – Fig. 97) and reverse fork rod (21).

Insert hi-lo shift rod (36 – Figs. 97 or 98) into place with shift fork (33). Make sure detent ball and spring are in place under plug. Secure hi-lo shift rod in place with set bolt (34) and seal washer (35). Tighten set bolt to 25-34 N·m torque.

136. MAINSHAFT BEARING PRELOAD. If mainshaft front bearing retainer (32 – Fig. 99) or front bearings (31 and 38) are renewed, bearing preload must be checked as follows: Press bearing cups into retainer. Place outer bearing cone into retainer with a 24 mm x 100 mm bolt through bearing. Using original spacer (37) and shims (35 and 36), place inner bearing cone into retainer. Install washers and nut on bolt and tighten to 41-54 N·m torque. Support retainer in a vise, and using a string and spring scale, measure rolling torque as shown in Fig. 104. A force of 15 to 20 N should be required to rotate bearing retainer. If correct rolling torque is not obtained, select proper shim and spacer combination to achieve correct rolling torque. Spacers are available in thicknesses of 8.66, 8.68, 8.70, 8.72 and 8.74 mm. Shims are available in thicknesses of 0.30 and 0.10 mm.

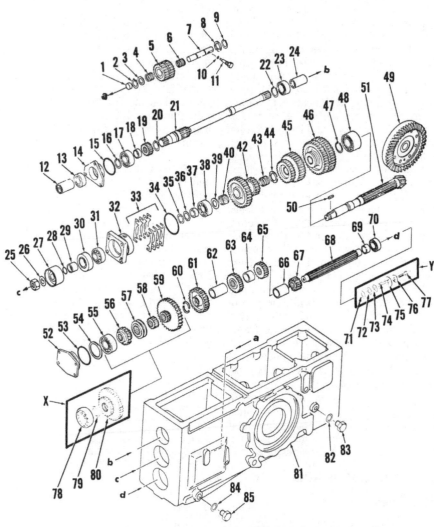

Fig. 99 — Exploded view of typical speed and range transmission for all models. Inserts X and Y are for non-diesel Model 284 tractors.

1. Plug	22. Snap ring	44. Thrust washer	65. Gear (2nd)
2. Snap ring	23. Bearing	45. Gear (3rd and 4th)	66. Spacer
3. Thrust washer	24. Coupler	46. Gear (1st and 2nd)	67. Gear (1st)
4. Needle bearing	25. Nut	47. Snap ring	68. Countershaft
5. Reverse gear	26. Washer	48. Bearing	69. Spacer
6. Needle bearing	27. Brake drum	49. Ring gear	70. Bearing
7. Shaft	28. "O" ring	50. Key	71. Shim
8. Thrust washer	29. Spacer	51. Pinion shaft	72. Shim
9. Snap ring	30. Oil seal	52. Retainer	73. Shim
10. Seal washer	31. Bearing	53. "O" ring	74. Roll pin
11. Set bolt	32. Retainer	54. Shim	75. Washer
12. Coupler	33. Shims	55. Bearing	76. Washers
13. Oil seal	34. "O" ring	56. Hub	77. Bolt
14. Flange	35. Shim	57. Sleeve	78. Bearing
15. "O" ring	36. Shim	58. Gear	79. Washer
16. Snap ring	37. Spacer	59. Gear	80. Hi-Lo gear
17. Bearing	38. Bearing	60. Snap rings	81. Transmission case
18. Spacer	39. Thrust washer	61. Gear (4th)	82. Seal washer
19. Input gear	40. Needle bearing	62. Spacer	83. Plug
20. Snap ring	42. Hi-Lo gear	63. Gear (3rd)	84. Seal washer
21. Input shaft	43. Needle bearing	64. Spacer	85. Plug

Fig. 100—Exploded view of power take-off gears, shafts and shift mechanism for all models. Insert shows overrunning clutch assembly on tractors so equipped.

1. Bearing
2. Input shaft (one-speed)
3. Bearing
4. Shims
5. Retainer
6. Bearing
7. Thrust washer
8. Slide gear
9. Snap ring
10. Washer
11. Bearing
12. Output shaft
13. Plug
14. Ball
15. Spring
16. Bushing
17. Fork
18. Link
19. "O" ring
20. Shift lever
21. Shaft
22. "O" ring
23. Pin
24. Retainer
25. Shims
26. Oil seal
27. Seal washer
28. Set bolt
29. Stop bracket
30. Rear frame
31. Input shaft
32. Snap ring
33. Spring retainer
34. Spring
35. Front clutch
36. Snap ring
37. Rear clutch

38. Washer
39. Bearing

40. Input shaft (two-speed)
41. Bearing

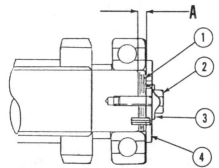

ner tapered roller bearing (38–Fig. 99) onto shaft with taper facing forward, using a suitable size pipe. Install bearing spacer (37), shims (35 or 36), if present and bearing retainer (32) with new oil seal (30) and "O" ring (34). Drive outer tapered roller bearing (31) into housing until it seats using a suitable size pipe. Install spacer (29) with "O" ring (28) on mainshaft. Install lockwasher (26) and nut (25) and torque nut to 175 N·m.

To set mainshaft pinion gear depth, proceed as follows: Refer to Fig. 106 and place a 89-143-2 bevel pinion gage bar or equivalent as shown. Brace bar across from pinion gear with bolts. Using a telescoping gage, measure distance between bar and face of pinion gear. Subtract 0.38 mm from reading to account for gap between bar and housing bore. This is actual mounting distance. The specified distance from edge of differential retainer bore to pinion gear is 26.37 mm. Three sets of numbers will be found etched on pinion gear face. Subtract the third set in parenthesis from the second set. This is the **required mounting distance.** Subtract **actual mounting distance** from the **required mounting distance** to obtain required shim pack thickness ± 0.05 mm.

137. MAINSHAFT. Reassemble mainshaft as follows: Install inner bearing (48–Fig. 99) onto mainshaft (51) and secure with snap ring (47). Install shaft through rear bearing bore and install gears, thrust washers and needle bearings onto shaft as it is installed. Using a brass hammer, drive mainshaft bearing in until it is flush in housing bore. Support mainshaft with wood block and nut and bolt as shown in Fig. 105. Drive in-

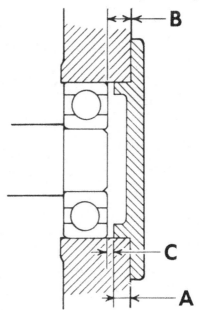

Fig. 103—Illustration showing dimensions to measure when installing shims on countershaft cover for non-diesel Model 284 tractors (S.N. 012229 and above). Refer to text.

Fig. 102—Illustration showing dimension "A" measured to determine shim pack (1) when assembling countershaft on non-diesel Model 284 tractors (S.N. 012228 and below). Refer to text.

1. Shim pack
2. Bolt

3. Lock tab
4. Washer

Fig. 104—View showing use of string and spring scale to measure mainshaft bearing preload. Refer to text.

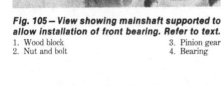

Fig. 105—View showing mainshaft supported to allow installation of front bearing. Refer to text.

1. Wood block
2. Nut and bolt

3. Pinion gear
4. Bearing

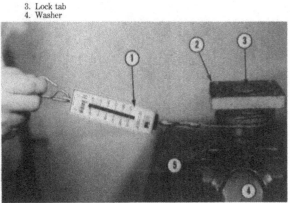

Example
Measured distance . . . . .26.05 mm
                    – .38 mm

Actual distance . . . . .25.67 mm
Stamped numbers   978 + 0.12 (−1)
NOTE: (−1) represents hundredths
        of a millimeter.
Subtract . . . . .0.12 mm
                    – .01 mm
                    ‾‾‾‾‾‾‾‾‾‾
                    0.11 mm
Add specified distance . . . . .26.97 mm
                    .11 mm
                    ‾‾‾‾‾‾‾‾‾‾
Required mounting distance 27.08 mm

Subtract "Actual" from   27.08 mm
    "Required"   – 25.67 mm
                           ‾‾‾‾‾‾‾‾‾‾
Required shim pack . . . . .1.41 mm

Shims are available in thicknesses of 0.05, 0.2, 0.3 and 0.5 mm. Install required shim pack, then install mainshaft bearing retainer mounting bolts and tighten to 54-73 N·m torque and secure with locking straps.

Reinstall differential assembly in rear frame and check differential bearing preload as outlined in paragraph 138, and backlash as outlined in paragraph 139.

Complete reassembly of tractor by reversing the disassembly procedure. Noting the following torque specifications. Torque differential bearing retainer bolts to 27-33 N·m. Torque transmission input shaft cover bolts to 16-23 N·m. Torque pto input shaft cover bolts to 27-30 N·m. Torque bolts securing rear frame to center section to 98-119 N·m. Torque hydraulic hitch cover mounting bolts to 58-69 N·m. Torque axle carrier mounting bolts to 94-122 N·m. Fill rear frame with 13 liters of IH "Hy-Tran" fluid. Bleed brakes as outlined in paragraph 154.

MAIN DRIVE BEVEL GEARS AND DIFFERENTIAL

The transmission rear frame contains the differential assembly with brake and final drive assemblies attached.

On "All Wheel Drive" equipped Model 284 diesel tractors the pinion shaft transfers power to the clutch housing gear train.

The differential lock is integral with the differential assembly.

ADJUSTMENT

All Models

138. CARRIER BEARING PRE-LOAD. To adjust preload on differential carrier bearings, the hydraulic lift assembly should be removed as outlined in paragraphs 170 or 171 and axle housings removed as outlined in paragraphs 143 or 148.

NOTE: If Model 274 tractors with drop housing axles are being worked on, the ring gear will be located on the left hand side and procedure for setting bearing preload will be the same except to substitute the word "RIGHT" for the word "LEFT" and vice versa.

Using a suitable hoist and sling, support differential in position in rear frame. Install right and left differential bearing retainers (2 and 23 – Fig. 109) without "O" rings (3 and 22) or shims (4 and 21), then secure retainers with four cap screws. Make certain lubrication holes are positioned at the bottom. Slide differential assembly to the left to obtain zero backlash. Using a feeler gage as shown in Fig. 107, measure gap between bearing and bearing retainer. Measure gap on both left and right bearing retainers. Calculate desired shim pack by **subtracting** 0.3 mm from reading of right side. Install required shims into right bearing retainer, reinstall retainer and torque bolts to 27-33 N·m. For the left retainer, calculate desired shim pack by **adding** 0.3 mm to the reading of the left side. Install required shims into left bearing retainer, install retainer and torque bolts to 27-33 N·m.

139. BACKLASH ADJUSTMENT. With differential carrier bearing preload determined as outlined in paragraph 138, backlash between ring gear and drive pinion should be checked and adjusted as follows: Set up a dial indicator as shown in Fig. 108. Be sure to block mainshaft from turning. Keep dial indicator in line with ring gear and at very tip of gear tooth. Backlash should be 0.12-0.20 mm. Take readings at two points 180 degrees apart. Adjust backlash by moving bearing retainer shims from one side to the other, as required to increase or decrease backlash. **DO NOT** add or remove shims, since this would change the bearing preload. After

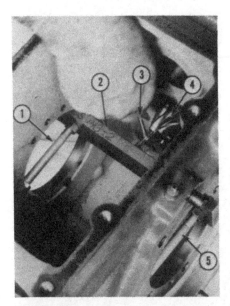

Fig. 106 — View showing installation of pinion depth measuring tool 89-143-2. Refer to text.

1. Nut and bolt
2. 89-142-2 Bevel pinion gage bar
3. Telescoping gage
4. Pinion gear
5. Nut and bolt

Fig. 107 — View showing use of feeler gage to determine correct shim pack for carrier bearing preload. Refer to text.

Fig. 108 — View showing dial indicator installed to check pinion and ring gear backlash. Refer to text.

1. Ring gear
2. Dial indicator
3. Tip of gear tooth
4. Bearing retainer

correct backlash is obtained, remove bearing retainers and install "O" rings (1, 3, 22 and 26 – Fig. 109), lubricating them with petroleum jelly. Reinstall bearing retainers and torque bolts to 27-33 N·m then secure with locking straps.

Continue reassembly by reversing disassembly procedure.

R&R BEVEL GEARS

All Models

140. The main drive bevel pinion is also the transmission mainshaft. The procedure for removing, reinstalling and adjusting pinion setting is outlined in the transmission section (paragraphs 132, 136 a 137).

To remove the bevel ring gear, follow the procedure outlined in paragraph 141 for R&R of differential. The ring gear is secured by the differential case bolts which should be tightened to a torque of 58-69 N·m.

DIFFERENTIAL

All Models

141. **R&R AND OVERHAUL.** To remove differential, tractor must be detached (split) between transmission rear frame and clutch housing as outlined in paragraph 130, then remove transmission input shaft and differential as outlined in paragraph 131.

To disassemble differential, refer to Fig. 109 and proceed as follows: Unlock locking plate (7) and remove ring gear cap screws (6) and locking plates. Pull bearing (5) from ring gear using a suitable puller. Force ring gear (8) off housing (16) using two jackscrews in threaded holes of ring gear. Remove set screw (14 – Fig. 109), then remove pinion shaft (18). Remove pinion gears (11), thrust washers (10) and inner bevel gear (12) from differential housing (16). Remove bearing (20) using a suitable puller then remove spacer (19) and differential lock (17).

Inspect gears, shaft, thrust washers and bearings for excessive wear or damage and renew as necessary. Reassemble differential by reversing disassembly procedure. Tighten case bolts to 58-69 N·m torque. Refer to paragraph 138 and 139 for bearing preload and backlash adjustments.

DIFFERENTIAL LOCK

Models So Equipped

142. The optional differential lock helps prevent wheel slip when one wheel is operating under poor traction conditions. Continual engagement is not harmful to the differential and related parts however, the differential lock must be disengaged before attempting turns.

Repair procedure for differential lock is covered in the differential overhaul procedure outlined in paragraph 141.

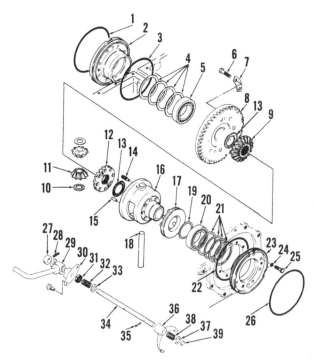

Fig. 109 – Exploded view of differential and differential lock linkage, if so equipped, typical for all models.

1. "O" ring
2. Bearing retainer (R.H.)
3. "O" ring
4. Shims
5. Bearing
6. Bolt
7. Lock plate
8. Ring gear
9. Side gear (R.H.)
10. Thrust washer
11. Pinion gear
12. Side gear (L.H.)
13. Thrust washer
14. Set screw
15. Dowel
16. Case
17. Differential lock clutch
18. Pinion pin
19. Washer
20. Bearing
21. Shims
22. "O" ring
23. Bearing retainer (L.H.)
24. Lockwasher
25. Bolt
26. "O" ring
27. Retaining ring
28. Roll pin
29. Differential lock pedal
30. Bracket
31. "O" ring
32. Spring
33. Washer
34. Shaft
35. Pin
36. Fork
37. Washer
38. Spring
39. Plug

Fig. 110 – View of ring gear and differential case showing lubrication holes.

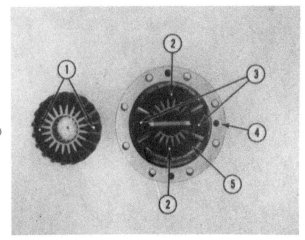

Fig. 111 – View of differential case, pinion and side gears showing lubrication holes.

1. Lube holes
2. Lube holes (inner bevel gear)
3. Lube holes (pinion gears)
4. Set screw
5. Inner bevel gear

FINAL DRIVE

AXLE, HOUSING AND FINAL REDUCTION GEARS

Model 284

143. **R&R AND OVERHAUL.** To remove either final drive assembly, first drain oil from rear frame. Disconnect necessary electrical wiring and remove fender. Remove hitch leveling link assembly. Support rear frame with floor jack and remove wheel. Support axle carrier with suitable sling and hoist. Remove axle carrier mounting bolts, then separate axle carrier from rear frame. Remove brake disc and sun shaft. Remove outer brake ring from axle housing.

Refer to Fig. 112 and remove axle retaining ring (6) for tractors (S.N. 012936 and below) or bolt (27), washer (26) and shims (25) for tractors (S.N. 012937 and above). Remove planetary assembly (items 5, 7, 8, 9, 10 and 11) from axle carrier (18). Remove ring gear (12) from axle carrier using a suitable puller. On tractors (S.N. 012936 and below) remove axle cap mounting bolts (28).

Press axle assembly out of carrier. Remove snap ring (16), then remove axle cap (tractors S.N. 012936 and below) and bearing (15) using a suitable puller. Drive out inner bearing (3), spacer (2) (tractors S.N. 012936 and below) and oil seal (1).

Inspect all parts for excessive wear or damage and renew as necessary. Use new oil seals and gaskets upon reassembly.

144. To reassemble final drive on tractors (S.N. 012936 and below), first install axle cap (29) with new oil seal (14) onto axle (24). Press on outer bearing (15) until it seats on shoulder of axle. Install snap ring (16). Place new gasket (30) on axle cap, then press axle into axle carrier. Install axle cap bolts (28) and tighten to 27-33 N·m torque. Install oil seal (1) using a suitable driver, then install spacer (2) and drive in bearing (3) until it is flush with beveled edge of the carrier.

Align retaining pin slots in ring gear (12) with pin slots in axle housing, then tap ring gear in flush with housing using a wood block. Install retaining pins (19), then using a brass drift, seat ring gear fully into housing. Install planetary car-

rier and secure with snap ring (6). Install outer brake ring into axle carrier.

145. To reassemble final drive on tractors (S.N. 012937 and above), first install new outer oil seal (14) onto axle (24), then press outer bearing (15) on axle until it seats.

NOTE: Make sure outer oil seal contacts inner race of outer axle bearing.

Press axle into axle carrier. Press inner oil seal (1) into axle carrier until approximately 6 mm past bearing seat in carrier, then press inner bearing (3) into carrier until it seats. Align retaining pin slots in ring gear (12) with pin slots in carrier. Tap ring gear flush into carrier using a wooden block. Install retaining pins (19), then using a brass drift, seat ring gear fully into housing.

To adjust axle shaft end play, proceed as follows: Refer to Fig. 113 and using a suitable depth micrometer measure distance axle extends from inner bearing; dimension "A". Measure width of planetary carrier splined portion; dimen-

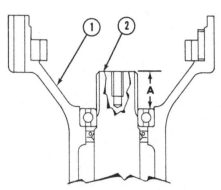

Fig. 113—Illustration showing dimension "A" measured to determine axle shaft end play for Model 284 non-diesel tractors (S.N. 012937 and above) and all Model 274 and 284 diesel tractors. Refer to text.

1. Axle carrier   2. Axle shaft

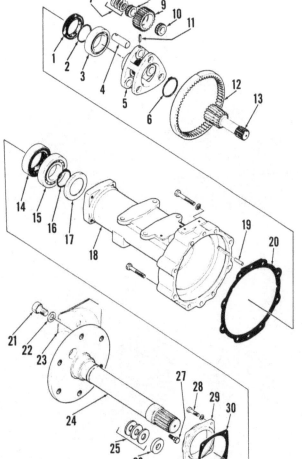

Fig. 112—Exploded view of final drive and axle housing for Model 284 tractors.

1. Oil seal
2. Spacer
3. Bearing
4. Pin
5. Planetary carrier
6. Snap ring
7. Shims
8. Bearing
9. Planet gear
10. Bearing
11. Pin
12. Ring gear
13. Sun shaft
14. Oil seal
15. Bearing
16. Snap ring
17. Retainer
18. Axle housing
19. Dowel
20. Gasket
21. Lug bolt
22. Washer
23. Wheel
24. Axle shaft
25. Shims
26. Washer
27. Bolt
28. Bolt
29. Retainer
30. Gasket

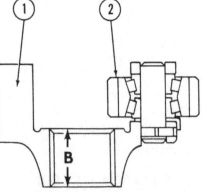

Fig. 114—Illustration showing dimension "B" measured to determine axle shaft end play on Model 284 non-diesel tractors (S.N. 012937 and above) and all Model 274 and 284 diesel tractors. Refer to text.

1. Planetary carrier   2. Planet gear

sion ("B" – Fig. 114). Subtract dimension "A" from dimension "B". Add 0.10 to 0.15 mm to remainder to calculate required shim pack (25 – Fig. 112). Install planetary carrier and secure with bolt (27), washer (26) and required shim pack (25). Shims are available in thicknesses of 0.05, 0.2, 0.3 and 0.8 mm. Torque bolt to 98-127 N·m. Use Loctite 262 or equivalent on bolt (27) prior to assembly. Install outer brake ring in housing.

146. Balance of final drive assembly and installation is reverse of disassembly procedure. Tighten axle carrier mounting bolts to 94-122 N·m torque. Fill rear frame with 13 liters of IH "Hy-Tran" fluid.

147. PLANETARY GEAR ASSEMBLY. With planetary carrier assembly removed as outlined in paragraph 143. Remove roll pins (11 – Fig. 112) by tapping into pin (4) from outside. Drive planet shafts (4) out of carrier. Remove gears, bearings and shims (items 7 through 10). Inspect all parts for excessive wear or damage and renew as necessary.

Reassemble by reversing disassembly procedure. Use original shims, then check for smooth rotation and play of gears. Add or subtract shims as necessary.

Reinstall planetary gear assembly as outlined in paragraphs 144 or 145.

Model 274

The following procedure first covers the right drop housing assembly, and then the rear axle housing. Removal of left drop housing assembly is the same as right assembly, except the brake ring is mounted in the left drop housing itself. Before removing left drop housing, oil from the transmission frame must be drained.

148. **R&R AND OVERHAUL.** If equipped with a fixed drawbar, remove bolts securing drawbar to the right drop housing. Wedge wooden blocks between front axle and engine subframe to prevent tipping. Raise tractor enough to remove right wheel, then support rear of tractor under transmission rear frame using a suitable jackstand. Disconnect necessary electrical wiring and remove right fender. Drain oil from drop housing. Using a suitable sling and hoist, unbolt and remove right drop housing assembly. Remove fixed drawbar (if so equipped), three point hitch lift links, lower links and stabilizer bar, then remove three-point hitch lower link bracket. Remove tractor seat and tool box. Disconnect and remove battery and hold down bracket. Disconnect brake reservoir mounting bracket, then move reservoir and hoses forward out of the

way. Remove bolts securing rockshaft support platform to rear axle housing. Remove bolts securing rockshaft support block to rockshaft support platform, then remove rockshaft support platform from tractor, noting number of shims under rockshaft support platform. Drain oil from rear frame. Disconnect differential lock lever from axle housing if so equipped. Using a suitable sling and hoist, unbolt and remove right axle housing.

149. OVERHAUL DROP HOUSING. With drop housing removed as outlined in paragraph 148, refer to Fig. 115 and proceed as follows: Remove drop housing lower cover (11). Unbolt hub shaft retainer (2) and loosen with soft hammer. Unlock castellated nut (38), then remove nut, tab washer (37) and seal protector plate (39). Drive hub shaft (1) out of housing using a brass drift, and supporting spur gear (8) when removing hub shaft. Remove spur gear (8), then

remove collar (35) and inner oil seal (34). Remove and discard hub shaft inner bearing (9). Remove spacer (7) from hub shaft and remove and discard shaft "O" ring (36). Using a suitable puller remove axle shaft outer bearing (6) and discard, then remove spacer (5), seal (4) and shaft retainer (2). Discard retainer gasket (3) and oil seal (4).

Remove pinion shaft cover (12) and gasket (13) and discard gasket. If present, remove coupler (right drop housing only) (26) from pinion shaft inner end. Using a brass drift, drive pinion shaft (16) out of housing from inner side. Remove snap ring (14) from outer end of pinion shaft and press outer bearing (15) off shaft. Discard bearing. Remove and discard pinion shaft inner oil seal (25) and bearing (17). On left drop housing only, remove outer brake ring.

150. REAR AXLE HOUSING. With right axle housing removed as outlined in paragraph 148, refer to Fig. 115 and

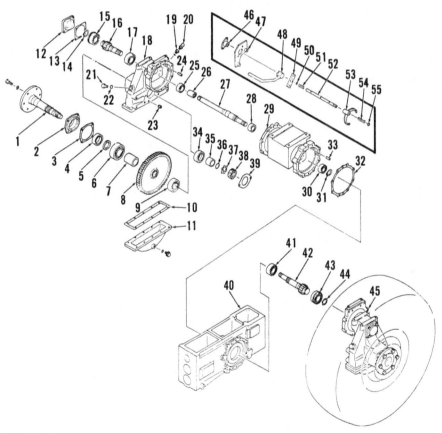

Fig. 115 – Exploded view of axle carrier and drop housing assemblies for Model 274 tractors. Insert shows differential lock shift linkage, if so equipped.

1. Axle shaft	15. Bearing	29. Axle housing (R.H.)	43. Bearing
2. Axle retainer	16. Final drive gear	30. Bearing	44. Snap ring
3. Gasket	17. Bearing	31. Snap ring	45. Drop housing assembly (L.H.)
4. Oil seal	18. Drop housing (R.H.)	32. Gasket	46. Bracket
5. Spacer	19. Plug	33. Dowel	47. Bracket
6. Bearing	20. Breather	34. Oil seal	48. Pedal
7. Spacer	21. Plug	35. Collar	49. Seal retainer
8. Final driven gear	22. "O" ring	36. "O" ring	50. "O" ring
9. Bearing	23. Plug	37. Tab washer	51. Spring
10. Gasket	24. Pin	38. Bearing locknut	52. Rod
11. Cover	25. Oil seal	39. Seal protector	53. Fork
12. Cover	26. Coupling	40. Transmission frame	54. Spring
13. Gasket	27. Shaft	41. Bearing	55. Plug
14. Snap ring	28. Oil seal	42. Final drive gear	

proceed as follows: Using a brass drift, tap axle shaft (27), toward differential end, out of axle housing. Remove snap ring (31) and remove bearing (30), if necessary. Remove outer brake ring from axle housing, if necessary.

151. Clean and inspect all parts for excessive wear or damage and renew as necessary. Use new bearings, oil seals, "O" rings and gaskets upon reassembly.

152. Reassemble by reversing the disassembly procedure, paying attention to the following: Install outer pinion bearing (15) with bearings split ring toward outer end of pinion. Use Loctite 262 or equivalent on pinion shaft cover bolts and tighten bolts to 24-34 N·m torque. Use a suitable sealer on O.D. of all oil seals. Coat lips of oil seals and all "O" rings with petroleum jelly. Position large spacer (7) of hub shaft with oil holes facing toward spur gear (8). Install spur gear (8) with "**longer**" side facing toward outside of drop housing. Tighten hub shaft retainer (2) mounting bolts to 54-73 N·m torque. Press hub shaft inner oil seal (34) into drop housing until seal is recessed 3 mm beyond chamfered edge of drop housing. Tighten castellated nut (38) to 314 N·m torque and tighten further if necessary to align tab washer (37) with nut grooves. Torque lower cover (11) mounting bolts to 24-34 N·m. Fill drop housings to level plug (23) with approximately 2.8 liters of IH "Hy-Tran" fluid.

Complete reassembly of tractor. Tighten bolt securing axle housing (right side only) to rear frame to 98-122 N·m torque. Torque drop housing mounting bolts to 98-127 N·m. Fill rear frame to correct levels with IH "Hy-Tran" fluid.

BRAKES

Brakes on all models are actuated hydraulically and are a wet-type single disc. Brake pedals can be used individually to aid in turning or latched together to provide simultaneous braking to both wheels for stopping.

Foot brakes are not intended for use in parking, or other stationary jobs since normal fluid seepage would result in brakes loosening and severe damage to equipment or injury to personnel could result. USE PARKING BRAKE when parking tractor.

BRAKE ADJUSTMENT

All Models

153. The only external adjustments that can be made on brakes are brake pedal height and free play. To adjust brake height, unlatch brake pedals and compare left and right pedal heights. If left pedal is higher or lower, loosen locknut on height adjusting screw of left pedal located on right platform of tractor. Turn bolt in or out until both pedals align, then lock the nut.

Free play of pedals should be 10 mm. To adjust free play, refer to Fig. 116 and remove pin (19), loosen locknut (17) and turn yoke (18) in or out until correct free play is obtained. Tighten locknut and replace pin. Insert cotter pin. Repeat operation for other pedal.

BLEED BRAKES

All Models

154. To bleed brakes, attach one end of a transparent hose to one of the bleed screws and place the other end in a clean glass jar. Press pedal slowly several times to bring air to bleed screw.

NOTE: Do not depress pedal more than 76 mm or damage to seal in master cylinder may occur.

Depress pedal and hold, then open bleed screw to allow air to escape. Close bleed screw and release pedal. Repeat operation until no further air bubbles appear from bleed screw. Repeat procedure for other cylinder.

Fill brake reservoir with IH "Hy-Tran" fluid.

BRAKE ASSEMBLIES

All Models

155. **R&R AND OVERHAUL.** Brake piston and disc assemblies can be withdrawn after removing final drive and axle housing units as outlined in paragraph 143 or 148. Refer to Fig. 116 and remove inner brake ring (3) (Model 284 tractors S.N. 011793 and below only). Disconnect brake pressure line. Carefully and evenly pry out inner brake piston (4). Remove brake disc (2) and outer piston (1) from axle housing. Inspect

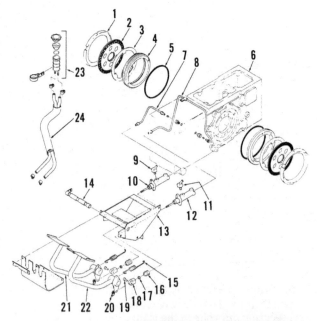

Fig. 116—Exploded view of typical brake system used on all models.

1. Outer brake ring
2. Brake disc
3. Inner brake ring (non-diesel tractors S.N. 011793 and below)
4. Brake piston
5. "O" ring
6. Transmission frame
7. Pressure line (R.H.)
8. Pressure line (L.H.)
9. Elbow (R.H.)
10. Brake cylinder (R.H.)
11. Elbow (L.H.)
12. Brake cylinder (L.H.)
13. Bracket
14. Shaft
15. Spring
16. Boot
17. Locknut
18. Yoke
19. Pin
20. Grease fitting
21. Brake pedal (R.H.)
22. Brake pedal (L.H.)

Fig. 117—View showing test kit (FES 114 installed with pressure gage to test brake operation pressure. Refer to text.

1. FES 114 hydraulic test kit
2. Axle carrier
3. Pressure gage 0-4140 kPa (0-600 psi)

brake disc, brake piston, inner brake ring (if present) and outer brake ring for cracks, uneven wear and foreign material and renew as necessary. Renew all "O" rings when reassembling.

Reassembly is reverse of disassembly procedure. Bleed brakes as outlined in paragraph 154.

PRESSURE TEST

All Models

156. Install a FES 114 Hydraulic Test Kit with a FES 544-3 adapter or equivalent testing unit as shown in Fig. 117. Fill pump with IH "Hy-Tran" fluid. Pump until housing bore is full of IH "Hy-Tran" fluid, then bleed air from system.

For Model 284 tractors, obtain 2070 kPa (300 psi) or 4116 kPa (597 psi) for Model 274 tractors, on gage and hold for 5 minutes. If pressure loss occurs, inspect piston and "O" ring for damage and renew as necessary as outlined in paragraph 155.

BRAKE CYLINDERS

All Models

157. **R&R AND OVERHAUL.** Disconnect brake fluid supply hose (24 – Fig. 116) from elbow (9 or 11) on brake cylinder (10 or 12). Quickly plug line to prevent fluid loss. Disconnect brake linkage from brake pedal. Remove necessary mounting bolts and remove brake cylinder.

Brake cylinder internal components are not available separately. Defective cylinders are to be renewed as a complete unit. However, disassembly of the unit is covered as follows: Remove dust cover (8 – Fig. 118). Place cylinder in a brass jawed vice and remove snap ring (7). Carefully remove push rod (6) out of cylinder, then remove piston (5), cup (4) and spring (3).

Inspect piston cup and piston for damage. Inspect cylinder bore for scoring. Check inlet orifices for blockage.

Reassemble by reversing disassembly procedure. Lubricate all parts with IH

"Hy-Tran" fluid prior to assembly. Reinstall cylinder on tractor and adjust pedal free play as outlined in paragraph 153 and bleed brakes as outlined in paragraph 154.

PARKING BRAKE

All Models

158. **ADJUSTMENT.** To adjust the parking brake on Model 284 non-diesel tractors, loosen or tighten the adjusting nuts on the parking brake rod. Specified pull at end of lever should be 193 N using a spring scale.

To adjust the parking brake on Model 274 and 284 diesel tractors, release brake, disconnect brake rod clevis from brake pivot lever. Turn clevis in or out as necessary. Parking brake is fully applied when lever has traveled three to five clicks up the ratchet.

159. **R&R AND OVERHAUL.** Repair of parking brake assembly will be obvious after detaching (splitting) tractor between rear frame and clutch housing as outlined in paragraph 130, and referring to Fig. 120.

POWER TAKE-OFF

The power take-off on Models 274 and 284 tractors is powered by the engine clutch. The power take-off on Model 284 tractor is a two speed design with an output of 540 or 1000 rpm. Model 274 tractors have a one speed design with an output of 540 rpm. To change output speed on Model 284 tractors, place pto lever in the neutral position, remove the two mounting bolts securing the locking plate, then turn the plate upside down and secure with mounting bolts.

Some tractors may be equipped with an overrunning clutch. See insert in Fig. 121.

R&R AND OVERHAUL

All Models

160. To remove the pto gears, shafts and shift linkage, proceed as follows: Drain hydraulic fluid from transmission rear frame. Remove seat. Disconnect and remove battery. Disconnect hydraulic pressure line from control valve or flow divider if so equipped. On Model 274 tractors, disconnect both the pressure and return lines from the auxiliary valve. Disconnect all necessary electrical wiring and move it out of the way. On Model 284 tractors, disconnect bracket securing brake reservoir to the tool box. Disconnect lift links from the three-point hitch lift yoke arms. Unbolt and remove hitch cover and pto shield.

161. **INPUT SHAFT.** On tractors not equipped with overrunning clutch, refer to Fig. 121 and proceed as follows: Remove pto input shaft retaining cover (5) by inserting two jack screws into threaded holes of cover. Retain shims (4), then withdraw input shaft (2 or 40) and bearing (3). Tap shaft and bearing out of cover. Remove bearing (3) from shaft and bearing (1) from transmission frame.

On tractors equipped with overrunning clutch, refer to insert in Fig. 121 and remove input shaft as follows: Remove pto input shaft cover (5) by inserting two jackscrews into threaded holes of cover. Retain shims (4). Dislodge snap ring (32) holding clutch spring (34) against front clutch (35). Slide snap ring (32), washer (33), clutch spring (34) and front clutch (35) forward on transmission input shaft (31).

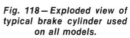

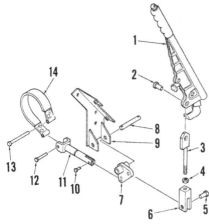

Fig. 120 – Exploded view of typical parking brake assembly used on all models.

1. Lever	8. Pin
2. Pin	9. Bracket
3. Rod	10. Pin
4. Locknut	11. Rod
5. Pin	12. Pin
6. Clevis	13. Pin
7. Bellcrank	14. Brake band

Fig. 118 – Exploded view of typical brake cylinder used on all models.

1. Inlet orifices
2. Cylinder body
3. Spring
4. Piston cup
5. Piston assembly
6. Push rod
7. Snap ring
8. Dust cover
9. Elbow

Remove snap ring (36) holding rear clutch (37) to pto input shaft (40). Pry between transmission input shaft and pto input shaft and remove rear clutch, washer and input shaft from transmission case. Tap shaft and bearing out of cover. Remove bearing (41) from shaft and bearing (39) from transmission case.

162. **SHIFT LINKAGE.** Remove retaining bolt (28 – Fig. 121) holding shift rod (21) into rear frame. Pull shift rod out of shift fork (17) being careful not to lose ball (14) and spring (15) under cap (13), then remove shift fork. Drive out pin (23) and remove shift levers (18 and 20).

163. **OUTPUT SHAFT.** Remove output shaft cover (24). Retain shims (25). Dislodge snap ring (9) which holds spacer (10) to outer bearing (11) and slide snap ring towards front on pto output shaft (12). Insert a suitable steel rod through hole in pto output shaft and using a suitable puller, pull output shaft from inner bearing (6) as shown in Fig. 122. Remove output shaft and withdraw, gear (8 – Fig. 121) and spacers from top of rear frame. Remove bearing (11) from shaft and bearing (6) from transmission case.

164. Inspect all shafts, gears and related parts for excessive wear or damage and renew as necessary. If bearings are removed, install new bearings. If new shafts, gear or bearings are used in reassembly, the output shaft end play must be checked and adjusted as follows: Refer to Fig. 123 and measure dimension "A" on output shaft cover. Using a depth micrometer, measure dimension "B" (from outer edge of rear frame to the outer bearing surface). Required shim pack is equal to "A" minus "B" plus 0.1-0.3 mm. Shims are available in thicknesses of 0.25 and 0.1 mm. Install shim pack, output shaft cover with new oil seal and torque cap screws to 27-33 N·m. Complete reassembly by reversing the disassembly procedure, paying attention to the following points:

Fig. 122 – View showing installation of puller on pto output shaft to aid in removal.

1. OTC 951 puller   3. Steel rod
2. Bolts and nuts   4. Pto output shaft

On tractors equipped with overrunning clutch, compress spring (34 – Fig. 121) fully in a vise and wire spring in compressed state. After overrunning clutch is assembled, cut and remove wire to release spring.

To set input shaft end play refer to Fig. 124 and measure dimensions "A" and "B" on input shaft cover. Using a depth micrometer, measure dimension "C" (from the outer edge of the rear frame to the outer bearing surface). The required shim pack is equal to "A" minus "B" minus "C" plus 0.1-0.3 mm. Shims are available in thicknesses of 0.25 and 0.1 mm. Install shims and input shaft cover and torque cap screws to 27-33 N·m.

Complete reassembly of tractor and fill rear frame with 13 liters of IH "Hy-Tran" fluid.

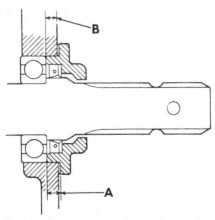

Fig. 123 – Illustration showing dimensions "A" and "B" measured on pto output shaft cover to determine shim pack when setting shaft end play. Refer to text.

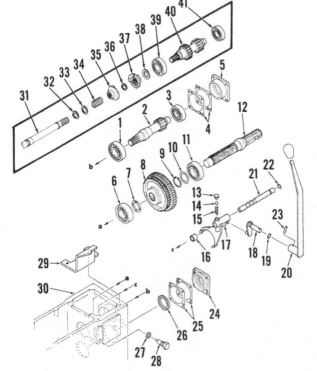

Fig. 121 – Exploded view of power take-off gears, shafts and shift mechanism for all models. Insert shows overrunning clutch assembly on tractors so equipped.

1. Bearing
2. Input shaft (one-speed)
3. Bearing
4. Shims
5. Retainer
6. Bearing
7. Thrust washer
8. Slide gear
9. Snap ring
10. Washer
11. Bearing
12. Output shaft
13. Plug
14. Ball
15. Spring
16. Bushing
17. Fork
18. Link
19. "O" ring
20. Shift lever
21. Shaft
22. "O" ring
23. Pin
24. Retainer
25. Shims
26. Oil seal
27. Seal washer
28. Set bolt
29. Stop bracket
30. Rear frame
31. Input shaft
32. Snap ring
33. Spring retainer
34. Spring
35. Front clutch
36. Snap ring
37. Rear clutch
38. Washer
39. Bearing
40. Input shaft (two-speed)
41. Bearing

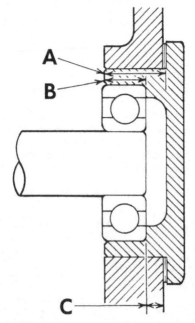

Fig. 124 – Illustration showing dimensions "A" and "B" measured on pto input shaft cover to determine shim pack when setting shaft end play. Refer to text.

HYDRAULIC LIFT SYSTEM

The open-center system utilizes a fixed displacement external gear type pump to supply flow. Flow is directed to the flow divider on models equipped with power steering or directly to the control valve on models not equipped with power steering. Specified flow from the pump is approximately 4 liters/min.

Model 284 tractors may be equipped with an auxiliary valve and Model 274 tractors have an auxiliary valve and cultivator lift cylinder as standard equipment.

FLUID AND FILTER

All Models

165. The tractor rear frame serves as a common reservoir for all hydraulic and lubricating operations. The spin-on type filter, IH part No. 1 004 366 C1, should be renewed at 10 hours, 50 hours, 100 hours and every 200 hours thereafter. Tractor rear frame should be drained and new fluid added every 1000 hours, or once a year, whichever occurs first.

Only IH "Hy-Tran" fluid should be used. Full capacity of rear frame should be 13.0 liters. When filling, allow sufficient time for oil to pass between compartments and be sure tractor is level before checking at level plug.

TEST HYDRAULIC SYSTEM

Specified hydraulic pump output at 2600 engine rpm, with 10294 kPa (1500 psi) system pressure and fluid at 50° C is 22.2 liters/min. for all tractors without power steering. Specified system relief valve pressure is 12756 kPa (1850 psi).

All Models Without Power Steering And With Auxiliary Valve

166. **RELIEF VALVE.** To test hydraulic system relief pressure, connect a 14-51D Flo-Rater or equivalent inlet line to the pressure line from the auxiliary valve and place outlet line from Flo-Rater into transmission filler hole in hitch cover and secure in place. With restrictor valve of Flo-Rater open, start engine and set at rated speed of 2600 rpm. Place auxiliary valve on demand to direct fluid through Flo-Rater. Slightly close Flo-Rater restrictor valve to warm fluid to 50° C. With auxiliary valve on demand, close restrictor valve on Flo-Rater and note system relief pressure. Specified system relief pressure should be 12,756 kPa (1850 psi). Open restrictor valve on Flo-Rater fully, then slowly close restrictor valve until a reading of 10,294 kPa (1500 psi) is obtained. A minimum of 16.6 liters/min. (4.4 gpm) must be obtained on flow gage.

NOTE: If system relief pressure is below 12,756 kPa (1850 psi), relief valve pressure must be corrected before pump output flow can be accurately measured.

All Models Without Power Steering And Auxiliary Valve

167. **RELIEF VALVE.** To test hydraulic system relief pressure, install a 9 410 202 union in the low pressure port of a 15-534 valve adapter as shown in Fig. 125. Install a 14-98-26 union adapter in high pressure port. Remove control valve cover and install the 15-534 valve adapter, making certain "O" rings are installed in each valve adapter port as shown in Fig. 126. Connect a 14-51D Flo-Rater or equivalent inlet line to high pressure port on valve adapter, and outlet line of Flo-Rater to the low pressure port.

With Flo-Rater restrictor valve open, start engine and set at rated speed of 2600 rpm. Slightly close Flo-Rater restrictor valve to warm fluid to 50° C. Close Flo-Rater restrictor valve and note system relief pressure. Specified system relief pressure should be 12,756 kPa (1850 psi). Open restrictor valve of Flo-Rater fully, then slowly close restrictor valve until a reading of 10,294 kPa (1500 psi) is obtained. A minimum of 16.6 liters/min. must be obtained on flow gage.

NOTE: If system relief pressure is below 12,756 kPa (1850 psi), relief valve pressure must be corrected before pump output flow can be accurately measured.

Model 284 With Power Steering

168. **RELIEF VALVE.** To test hydraulic system relief pressure on Model

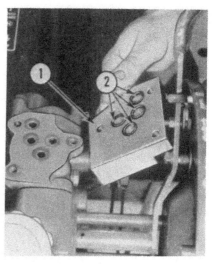

Fig. 126 — Make certain "O" rings are installed in each valve adapter port as shown.
1. 15-534 Valve adapter   2. "O" rings

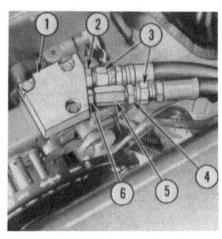

Fig. 127 — View showing Valve adapter installed on control valve and 14-51D Flo-Rater connected. Refer to text.
1. 15-534 Valve adapter   4. Union
2. Union   5. Connector
3. Pressure lines   6. Union

Fig. 125 — View of 15-534 Valve adapter used when testing hydraulic system on tractors not equipped with auxiliary valve or power steering. Refer to text.
1. 9 410 202 Union
2. 15-534 Valve adapter
3. 14-98-26 Union

284 tractors equipped with power steering, refer to procedures outlined in paragraphs 27 through 30 of Power Steering System.

CONTROL VALVE

All Models

169. **R&R AND OVERHAUL.** To remove hydraulic hitch control valve, first relieve system pressure by lowering the hitch fully. Thoroughly clean lines and area around the control valve and auxiliary valve (if so equipped). If equipped with auxiliary valve, disconnect coupler pressure lines, then unbolt and remove auxiliary valve. Disconnect position control rod (20–Fig. 130) from control valve. Disconnect high pressure line from control valve. Remove position control panel (6). Remove necessary mounting bolts and remove the control valve.

With control valve removed, remove front cover (41), then remove relief valve (items 53 through 61). If relief valve is faulty, it must be renewed as a complete unit. Remove spool and sleeve assembly (items 49 through 52). Remove check valve assembly (items 43 through 48). Ball seat (47) can be removed using a hooked wire. Back out set screw (35), then remove adjusting shaft assembly (items 31 through 34). Drive out roll pins securing control lever (10) and link (22), then remove lever and link. Remove rear cover (26), then remove unloading valve assembly (items 37 through 40).

Clean and inspect all parts for excessive wear, damage, scoring, nicks or burrs. If bores of valve body are scored, renew complete assembly. Use all new "O" rings and gaskets upon reassembly. Dip all parts in IH "Hy-Tran" fluid and reassemble by reversing disassembly procedures. If position control lever and link have been properly installed, moving the control lever back to raise position should bring the control link forward. If lever and link do not operate properly, alternately reinstall the lever or link, rotating them 180° on the shafts. Recheck for proper operation.

Install control valve on tractor by reversing removal procedure.

Bleed system by raising and lowering hitch several times.

HYDRAULIC LIFT HOUSING

Model 274

170. **R&R AND OVERHAUL.** To remove the hydraulic lift housing, first relieve system pressure by lowering hitch fully. Remove seat. Remove lift links from rockshaft arms. Disconnect hydraulic pressure line from bottom of control valve. Disconnect pressure and return lines from auxiliary control valve. Cap or plug all hydraulic openings to prevent contamination. Refer to Fig. 131 and unbolt shaft support (30) from the seat support. Note and retain shims (24) to aid in reassembly. Remove mounting bolts securing hitch cover to rear frame, attach a suitable sling and hoist and remove hydraulic lift housing.

With lift housing removed disconnect position control link from collar (27) and position control link from control lever to control valve. Disconnect control lever from auxiliary valve, if so equipped. Unbolt and remove control lever assembly. Unbolt and remove control valve assembly. Remove snap rings securing rockshaft arms (4) to rockshaft. Note timing marks and remove rockshaft arms. Remove set screw (11) in shaft support (30), then tap support off right end of rockshaft and remove bushing (29) from rockshaft. Drive out pin (26) and remove collar (27). Remove

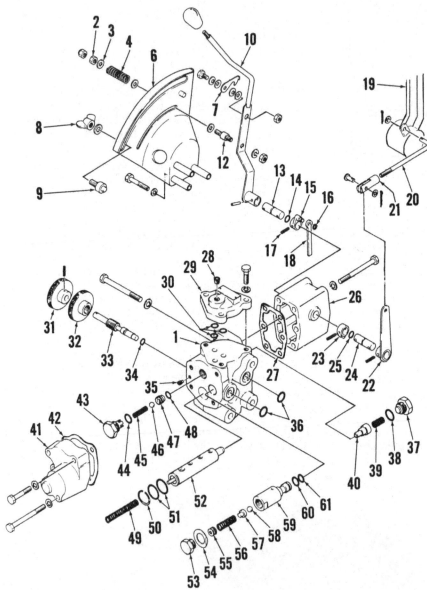

Fig. 130—Exploded view of hydraulic hitch control valve and control linkage typical of all models.

1. Valve body	18. Link	32. Lock knob	46. Ball
2. Nut	19. Rockshaft arm	33. Adjuster shaft	47. Seat
3. Washer	20. Rod	34. "O" ring	48. "O" ring
4. Spring	21. Yoke	35. Screw	50. Ring
6. Lever guide	22. Feed control lever	36. "O" rings	51. "O" rings
7. Lever lock	23. Arm	37. Plug	52. Sleeve
8. Wing nut	24. Shaft	38. "O" ring	53. Plug
9. Pin	25. "O" ring	39. Spring	54. Lock washer
10. Control lever	26. Rear cap	40. Unloading valve	55. Shims
12. Spacer	27. Collar	41. Front cap	56. Spring
13. Shaft	28. Plug	42. Gasket	57. Spring support
14. "O" ring	29. Cover	43. Plug	58. Ball
15. Arm	30. "O" rings	44. "O" ring	59. Relief valve
16. Snap ring	31. Control knob	45. Spring	60. "O" ring
17. Pin			61. "O" ring

set screw (11) from right side of hitch housing and drive out rockshaft (7) and bushing (6). Note alignment of timing spline of rockshaft to inner arm (18). Remove set screw (11) from left side of housing and drive out other rockshaft bushing. Remove inner arm (18) with piston rod (22). Remove front cover (13) and remove piston (16).

Inspect all parts for excessive wear or damage and renew as necessary. If piston bore in housing is scored, complete housing and piston must be renewed. Use all new "O" rings and dip all parts in IH "Hy-Tran" fluid prior to reassembly.

Reassemble hydraulic hitch assembly by reversing disassembly procedure paying attention to the following: When installing rockshaft into housing, align timing spline of rockshaft to center of connecting pin (19) on inner arm. When installing rockshaft arms to rockshaft, align timing marks on rockshaft end to marks on rockshaft arms.

Complete reassembly and tighten hitch cover mounting bolts to 58-69 N·m torque. Bleed hydraulic hitch system by raising and lowering hitch several times.

Model 284

171. **R&R AND OVERHAUL.** To remove hydraulic lift housing, first release system pressure by lowering hitch fully. Remove seat. Remove lift links from rockshaft arms. Disconnect hydraulic pressure line from bottom of control valve. If so equipped, disconnect and remove pressure line from the auxiliary valve. Disconnect brake fluid supply line from brake reservoir. Using a suitable sling and hoist, unbolt and remove the hitch cover from rear frame.

With hydraulic hitch assembly removed, disconnect position control link from rockshaft arm. Unbolt and remove control valve assembly from hitch cover. Refer to Fig. 132 and remove cap screws (1) and end plates (2) securing rockshaft arms (4) to rockshaft (7). Remove rockshaft arms, noting timing marks on rockshaft ends and rockshaft arms. Remove set screw (11) from left side of housing, then tap rockshaft (7) and bushing (6) out of housing using a brass hammer. Note alignment of timing spline of rockshaft to inner arm (18). Remove set screw (11) from right side of housing, and drive out other rockshaft bushing. Remove inner arm (18) with piston rod (22). Remove front cover (13) and remove piston (16).

Inspect all parts for excessive wear or damage and renew as necessary. If piston bore in housing is scored, complete housing and piston must be renewed. Use all new "O" rings and dip all parts in IH "Hy-Tran" fluid prior to assembly.

Reassemble hydraulic hitch assembly by reversing disassembly procedure paying attention to the following: When installing rockshaft into housing, align timing spline of rockshaft to center of connecting pin (19) on inner arm. When installing rockshaft arms to rockshaft, align timing marks on rockshaft end to marks on rockshaft arms.

Complete reassembly and tighten hitch cover mounting bolts to 58-69 N·m torque. Bleed hydraulic hitch system by raising and lowering hitch several times.

HYDRAULIC PUMP

All Models

172. To remove and service the hydraulic system pump for Model 284 nondiesel, follow procedure outlined in paragraphs 30 and 31. Model 274 hydraulic system pump is the same as Model 284 diesel. To remove and service Model 274 and 284 diesel hydraulic system pump, follow procedure outlined in paragraphs 33 and 34.

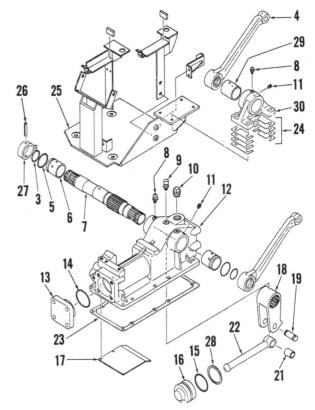

Fig. 131 — Exploded view of hydraulic hitch assembly for Model 274 tractors.

   3. "O" ring
   4. Rockshaft arm
   5. "O" ring
   6. Bushing
   7. Rockshaft
   8. Grease fitting
   9. Breather
  10. Fill plug
  11. Set screw
  12. Hitch cover
  13. Front cover
  14. "O" ring
  15. "O" ring
  16. Piston
  17. Plate
  18. Inner arm
  19. Connecting pin
  21. Bushing
  22. Piston rod
  23. Gasket
  24. Shims
  25. Bracket
  26. Pin
  27. Collar
  29. Bushing
  30. Shaft support

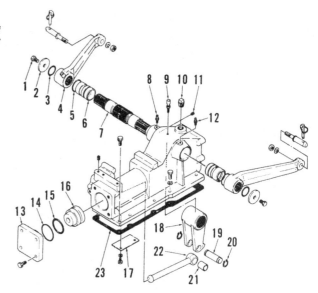

Fig. 132 — Exploded view of hydraulic hitch assembly for Model 284 tractors.

   1. Cap screw
   2. End cap
   3. "O" ring
   4. Rockshaft arm
   5. "O" ring
   6. Bushing
   7. Rockshaft
   8. Grease fitting
   9. Breather
  10. Fill plug
  11. Set screw
  12. Grease fitting
  13. Front cover
  14. "O" ring
  15. "O" ring
  16. Piston
  17. Plate
  18. Inner arm
  19. Connecting pin
  20. Snap ring
  21. Bushing
  22. Piston rod
  23. Gasket

AUXILIARY VALVE

All Models (So Equipped)

173. R&R AND OVERHAUL. To remove auxiliary valve, first relieve system pressure by lowering hydraulic hitch fully. Remove seat. On Model 284 tractors, disconnect the pressure line from auxiliary valve. On Model 274 tractors remove pressure and return lines from auxiliary valve. Cap or plug all openings to prevent contamination. On Model 274 tractors, disconnect control lever link. Unbolt and remove auxiliary valve.

Remove Allen screws (24–Fig. 133), then remove seal plate (23), wiper (21) and "O" ring (20). Remove Allen screws (1), remove rear cap (2), then withdraw spool assembly. Remove screw (4) and disassemble centering spring assembly. (items 4 through 9). Remove plug (12), then remove spring (14) and relief valve (15).

Inspect all parts for excessive wear, nicks, burrs or scoring and renew as necessary. Use new wiper rings, "O" rings and gaskets upon reassembly.

Reassemble by reversing disassembly procedure.

CULTIVATOR LIFT CYLINDER

Model 274

174. R&R AND OVERHAUL. To remove the lift cylinder, first clean hydraulic cylinder and surrounding area. Move auxiliary valve control lever back and forth to relieve system pressure. Disconnect the pressure and return lines. Cap or plug hydraulic openings to prevent contamination. Remove pins securing each end of the hydraulic cylinder to the tractor, and remove cylinder.

Refer to Fig. 134 and remove frame bolts (35). Tap rear head (10) lightly with a soft mallet to separate from cylinder (34) and return tube (23). Remove cylinder (34) and return tube (23) from front head (30). Drive out roll pin (8), remove nut (5), remove piston assembly (items 1 through 4) from piston rod (26), then slide piston rod out of front head. Remove plate (12) and remove limiter valve assembly (items 13 through 20).

Inspect cylinder tube and piston for scoring, nicks, burrs or out-of-roundness. Small scratches can be corrected using emery cloth and oil. Deeper scoring will require renewal of cylinder

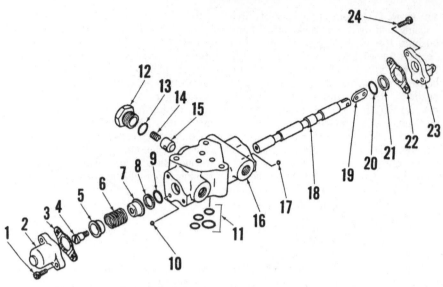

Fig. 133—Exploded view of auxiliary control valve on all models so equipped.

1. Allen screw	7. Spring cup	13. "O" ring	19. Link
2. End cap	8. Wiper ring	14. Spring	20. "O" ring
3. Gasket	9. "O" ring	15. Poppet	21. Wiper ring
4. Screw	10. Ball	16. Valve body	22. Gasket
5. Spring cup	11. "O" rings	17. Ball	23. Seal plate
6. Spring	12. Plug	18. Spool	24. Allen screw

Fig. 134—Exploded view of hydraulic cultivator lift cylinder for Model 274 tractors.

1. Piston
2. Back-up ring
3. "O" ring
4. Back-up ring
5. Nut
6. "O" ring
7. Back-up ring
8. Pin
9. Plug
10. Rear head
11. Nut
12. Seal plate
13. Limit valve
14. Wiper ring
15. "O" ring
16. "O" ring
17. Back-up ring
18. Wiper ring
19. Piston
20. Seal
21. "O" ring
22. Back-up ring
23. Tube
24. Back-up ring
25. "O" ring
26. Piston rod
27. Control collar
28. Seal
29. Oil seal
30. Front head
31. Plug
32. Back-up ring
33. "O" ring
34. Cylinder
35. Bolt

and/or piston. Use all new "O" rings and back-up rings upon assembly. Coat all parts with IH "Hy-Tran" fluid prior to assembly. Reassemble by reversing disassembly procedure. Torque frame bolt nuts to 98-122 N·m. Bleed cylinder by operating cylinder back and forth fully several times.